# Interpolation of Linear Operators

TRANSLATIONS OF MATHEMATICAL MONOGRAPHS

VOLUME 54

S. G. Kreĭn
Ju. I. Petunin
E. M. Semenov

# Interpolation of Linear Operators

American Mathematical Society

# ИНТЕРПОЛЯЦИЯ ЛИНЕЙНЫХ ОПЕРАТОРОВ

## С. Г. КРЕЙН, Ю. И. ПЕТУНИН И Е. М. СЕМЕНОВ

ИЗДАТЕЛЬСТВО «НАУКА»
ГЛАВНАЯ РЕДАКЦИЯ
ФИЗИКО-МАТЕМАТИЧЕСКОЙ ЛИТЕРАТУРЫ
МОСКВА 1978

Translated from the Russian by J. Szűcs
Translation edited by Lev J. Leifman

2000 *Mathematics Subject Classification*. Primary 46-XX.

ABSTRACT. The book is devoted to an important direction in functional analysis: interpolation theory for linear operators. The main methods for constructing interpolation spaces are expounded and their properties are studied. These methods allow one to look at a number of theorems and inequalities of classical analysis from a new standpoint. Interpolation theory for operators has numerous applications in Fourier series, approximation theory, partial differential equations, etc. Some of them are developed in the book.

**Library of Congress Cataloging in Publication Data**

Kreĭn, S. G. (Selim Grigor′evich), 1917–
Interpolation of linear operators.
(Translations of mathematical monographs; 54)
Translation of: Interpoli͡at͡sii͡a lineĭnykh operatorov.
Bibliography: p.
Includes indexes.
1. Linear operators. I. Petunin, I͡Uriĭ Ivanovich. II. Semenov, E. M. III. Title. IV. Series.
QA329.2.K7313 515.7′246 81-20637
ISBN 0-8218-4504-7 (Hard cover) AACR2
ISBN 0-8218-3176-3 (Soft cover)
ISSN 0065-9282



10 9 8 7 6 5 4 3 2 07 06 05 04 03 02

## TABLE OF CONTENTS

## FOREWORD TO THE AMERICAN EDITION

This edition includes Chapter V, "Interpolation in spaces of smooth functions", written at the same time as the preceding chapters but not included in the Soviet edition for technical reasons. This chapter expounds the abstract scheme for constructing interpolation spaces by means of an unbounded operator in a Banach space, and the corresponding approximation process. Starting points of this theory were papers by J. L. Lions, Lions and J. Peetre, and P. Grisvard, which then were extended by other authors. As an application we consider only Sobolev and Besov spaces.

The reader can get acquainted with other families of spaces in the books referred to in the foreword to the Soviet edition.

The authors express their sincere gratitude to the editor of the translation, Dr. L. J. Leifman, for his penetrating remarks that helped in eliminating a number of shortcomings.

# FOREWORD

The present book is devoted to the systematic exposition of a chapter in functional analysis that has appeared and developed in the past two decades and has found applications in various fields.

The basic objects of classical functional analysis were operators acting from one Banach space (or later from a topological linear space) into another. The spaces themselves were considered as given in advance. The change of this ideology was facilitated to a significant extent by the imbedding theorems of S. L. Sobolev, in which a number of fundamental theorems and inequalities of analysis were interpreted as assertions concerning the imbedding of one Banach space into another. Imbedding theorems arose in connection with problems of the theory of partial differential equations, in which for the study of smoothness of solutions a series of spaces is introduced; for the study of the behavior near the boundary of the domain or near some singular points other types of spaces are introduced, the study of values of solutions on manifolds of smaller dimension is performed in still other spaces, etc. The abundance of various spaces required a detailed study of the interrelations between these spaces. Thus a new level of abstraction appeared, on which the Banach spaces themselves are considered as elements of some category. The interpolation theory for linear operators expounded in the book is to a great extent connected with such an approach.

The first interpolation theorem in operator theory was obtained by M. Riesz in 1926 in the form of an inequality for bilinear forms. A sharpening and operator formulation of it were given by G. O. Thorin. An essential further step was the interpolation theorem of J. Marcinkiewicz (1939), whose proof was published by A. Zygmund in 1956. In the fifties important generalizations of the Riesz-Thorin and Marcinkiewicz theorems were obtained by E. M. Stein and G. Weiss. However, all these and other communications were concerned with $L_p$ spaces or spaces similar to them. The development of general interpolation theorems for families of abstract Hilbert and Banach spaces began in 1958 independently in several countries. The first publications are due to J. L. Lions (1958-1960), E. Gagliardo (1959-1960), A. P. Calderón (1960), and S. G. Kreĭn (1960). The work of J. A. Peetre played an essential role in the sequel. Several methods have been created for obtaining interpolation theorems, which have deep interrelations. Moreover, it became clear fairly soon that the interpolation properties of spaces intermediate between two Banach spaces are consequences of the functoriality of the methods of construction. Therefore, the main emphasis has been shifted to the study of properties of intermediate interpolation spaces obtained by various methods, and to their realization. Along with this, in the work of W. Orlicz, A. P. Calderón, G. G. Lorentz, E. M. Semenov, and others deep results have been obtained concerning the interpolation of linear operators in spaces of measurable functions.

It is impossible to expound all results of interpolation theory for linear operators in one book. We have tried to illuminate only some of the main directions in its development: the real and complex methods of constructing interpolation spaces, the method of scales of Banach spaces, and interpolation in spaces of measurable functions. Supplementary information is contained in remarks and references.

In the development of interpolation theory for operators many new general notions of functional analysis have emerged. These notions and their interrelations are studied in the first chapter of the book. The exposition is based essentially on the work of N. Aronszajn and E. Gagliardo. To read this chapter one needs to know only the basic principles of functional analysis.

The second chapter, devoted to interpolation in spaces of measurable functions, makes up a significant portion of the book. It can be read independently of the first chapter, from which only the simplest definitions are needed. The chapter contains a theorem describing all interpolation spaces between $L_1$ and $L_\infty$, and a theorem which is a further extension of the Marcinkiewicz theorem. The exposition is pursued as far as concrete applications, for example, the theory of orthogonal series: convergence properties of Fourier series and the basis property of a function system are studied. Moreover, the chapter contains much auxiliary material from the theory of functions which is discussed little in the literature. Decreasing rearrangements of measurable functions are studied in detail, function spaces symmetric in the sense of E. M. Semenov, and in particular, Lorentz and Marcinkiewicz spaces, are discussed (in the foreign literature similar spaces are called invariant with respect to permutations). Sharpenings of classical inequalities of analysis (the Hardy-Littlewood, Hilbert, and other inequaities) are given.

In the third chapter the theory of scales of Banach spaces, developed mainly in the publications of S. G. Kreĭn and Ju. I. Petunin, is expounded. The prerequisite material for this is contained in the first chapter. Important properties of the scales, in particular their "almost" interpolation properties are also expounded in the fourth chapter. In the last section of the third chapter properties of the classical scale of Hölder spaces important in applications are studied in detail.

In the fourth chapter two methods of constructing interpolation spaces enjoying the largest number of applications are described in detail: the method of complex interpolation proposed independently by A. P. Calderón and J. L. Lions and extensively developed by Calderón, and the method of constants and averages due to J. L. Lions and J. Peetre. The latter method is expounded in the more general form which it acquired in the work of V. I. Dmitriev (who took the most active part in writing the corresponding section). The fourth chapter can be read independently from the second and third chapters.

The book does not include interpolation theory in spaces of smooth functions and its applications.* This theory developed under the influence of the work on imbedding theorems by S. L. Sobolev, S. M. Nikol'skiĭ and their students and followers. The abstract theory did not rise immediately and easily to the level of concrete imbedding theorems obtained by special means. However, now such a theory has been created. Its exposition apparently needs another book. One can get acquainted with it partly in the book [7] by P. L. Butzer and H. Berens. It is expounded more completely in Hans Triebel's very recent book *Interpolation theory, function spaces, differential operators* (published by VEB Deutscher Verlag Wiss., Berlin, 1977, and by North-Holland, 1978).([1]) One can get acquainted with the applications of this theory to the study of boundary value problems for partial differential equations in the book of J. L. Lions and E. Magenes [27] and in Triebel's book mentioned above. At the end of the book there is a bibliography covering, in addition, the indicated part of interpolation theory.

As we noted above, some parts of the book were written by V. I. Dmitriev. I. Ja. Šneĭberg provided us with invaluable help. He participated in writing §1 of Chapter IV and read a significant portion of the book. His critical remarks

---

* *Editor's note.* For this translation a new Chapter V was added by the authors to cover this subject.

([1])The authors are grateful to Professor Triebel for making the manuscript of this book available.

enabled us to remove a number of inaccuracies and improve some proofs. The authors express their gratitude to both of them.

Finally, we thank all participants of the Voronezh seminar on interpolation theory for linear operators, and, in particular, M. Š. Braverman, A. A. Dmitriev, E. A. Pavlov, P. A. Kučment, and A. A. Sedaev, for their constant help in the preparation of the book.

*The authors*

# CHAPTER I

# IMBEDDED, INTERMEDIATE, AND INTERPOLATION BANACH SPACES

## §1. Imbedding of Banach spaces

**1. *Imbedded Banach spaces.***

DEFINITION 1.1. We shall say that *a Banach space $E_1$ is imbedded in a Banach space $E_0$* if the following conditions are satisfied:

1°. $x \in E_1$ implies that $x \in E_0$.

2°. The space $E_0$ induces a vector space structure on $E_1$ coinciding with the structure of $E_1$.

3°. There exists a constant $C_{01}$ such that

$$\|x\|_{E_0} \leqslant C_{01}\|x\|_{E_1} \tag{1.1}$$

for all $x \in E_1$.

The smallest possible value of the constant $C_{01}$ in (1.1) is called the *imbedding constant of $E_1$ in $E_0$*.

Sometimes the term "imbedding" is used in a wider sense. Instead of conditions 1° and 2°, it is required that there exist an injective linear mapping $j$ (the imbedding operator) mapping $E_1$ into $E_0$, and then condition (1.1) is written in the form $\|jx\|_{E_0} \leqslant C_{01}\|x\|_{E_1}$.

In such a situation we shall always identify $E_1$ with its image $jE_1$.

Condition 3° can be formulated in the following equivalent form: if $x_n \to x$ in $E_1$, then $x_n \to x$ in $E_0$. In this form the definition of imbedding can be carried over to topological linear spaces, and then we say that $E_1$ is algebraically and topologically imbedded in $E_0$.

DEFINITION 1.2. The space $E_1$ is *densely imbedded in $E_0$* if conditions 1°–3° hold, and also

4°. The set $E_1$ is dense in $E_0$.

The space $E_1$ is *compactly imbedded in* $E_0$ if conditions 1°–3° hold, and also

5°. Every set bounded in the norm of $E_1$ is relatively compact in $E_0$.

In what follows we shall denote the imbedding of $E_1$ in $E_0$ by the symbol $E_1 \subset E_0$, assuming that the symbol $\subset$ means not only set-theoretic inclusion, but imbedding having the properties 2° and 3°.

If the space $E_1$ is imbedded in $E_0$, then on $E_1$ we can introduce a new norm:

$$\|x\|^*_{E_1} = C_{01}\|x\|_{E_1}.$$

Then the space $E_1^*$ equipped with this norm is isomorphic with $E_1$, and $\|x\|_{E_0} \leqslant \|x\|^*_{E_1}$.

In connection with this we introduce the following definition.

DEFINITION 1.3. We shall say that $E_1$ is *normally imbedded in* $E_0$ if $E_1$ is dense in $E_0$ and the imbedding constant $C_{01}$ does not exceed one, i.e. $\|x\|_{E_0} \leqslant \|x\|_{E_1}$.

We consider some examples of normalized imbedding of Banach spaces.

Let $E_0 = C(0, 1)$ be the space of continuous functions and $E_1 = C^{(1)}(0, 1)$ the space of continuously differentiable functions. Then $C^{(1)}(0, 1)$ is normally imbedded in $C(0, 1)$, so that $C^{(1)}(0, 1) \subset C(0, 1)$ and

$$\begin{aligned}\|x\|_{C(0,1)} &= \max_{t\in[0,1]} |x(t)| \leqslant \|x\|_{C^{(1)}(0,1)} \\ &= \max_{t\in[0,1]} |x(t)| + \max_{t\in[0,1]} |x'(t)|.\end{aligned}$$

Besides, by the Weierstrass theorem the set $M$ of algebraic polynomials is dense in $C(0, 1)$, and consequently $C^{(1)}(0, 1) \supset M$ is also dense in $C(0, 1)$. Finally, by Arzelà's theorem, $C^{(1)}(0, 1)$ is compactly imbedded in $C(0, 1)$.

We may show analogously that the space $C^{(n)}(0, 1)$ of $n$ times continuously differentiable functions is normally and compactly imbedded in $C^{(m)}(0, 1)$ if $n > m$.

Another example of normally imbedded spaces is furnished by the spaces $L_p$.

Let $G$ be a bounded domain in $n$-space. Consider the space $L_p$ consisting of the real-valued or complex-valued functions that are $p$th power summable in $G$ ($1 \leqslant p < \infty$). We denote by $L^\alpha$ the space $L_p$ with $p = 2/(1 - \alpha)$ in which a norm is introduced by the formula

$$\|x\|_{L^\alpha} = (\text{mes } G)^{(\alpha-1)/2}\|x\|_{L_p}.$$

The space $L^\beta$ is normally imbedded in $L^\alpha$ if $\beta > \alpha$. Indeed, $L^\alpha \supset L^\beta$ for $\alpha < \beta$, and by the Hölder inequality we have

$$\begin{aligned}\|x\|_{L^\alpha} &= (\text{mes } G)^{(\alpha-1)/2}\left(\int_G |x(t)|^{2/(1-\alpha)}\, dt\right)^{(1-\alpha)/2}\\ &\leqslant (\text{mes } G)^{(\alpha-1)/2}(\text{mes } G)^{(\beta-\alpha)/2}\left(\int_G |x(t)|^{2/(1-\beta)}\, dt\right)^{(1-\beta)/2}\\ &= \|x\|_{L^\beta} \qquad (x \in L^\beta).\end{aligned}$$

The set of simple measurable functions (linear combinations of characteristic functions) is dense in any space $L^\alpha$, $-1 \leqslant \alpha < 1$, and therefore $L^\beta$ is dense in $L^\alpha$.

It is not difficult to verify that the imbedding of $L^\beta$ in $L^\alpha$ does not have the property of compactness.

It can be shown analogously that the sequence space $l_p$ is normally imbedded in $l_q$ if $p < q$.

The following circumstance should be kept in mind: the space $E_1$ can be densely imbedded in the space $E_0$, but the closure of the ball of $E_1$ in $E_0$ may contain no interior points in the sense of the norm of $E_0$. Moreover, the following assertion is well known:

LEMMA 1.1. *If a Banach space $E_1$ is imbedded in a Banach space $E_0$ and does not coincide with it, then the closure in $E_0$ of any ball of $E_1$ is nowhere dense in $E_0$.*

PROOF. We assume the contrary. Let the closure $\bar{S}_1^0$ of the unit ball $S_1$ of $E_1$ contain the ball $\sigma_{2r}(x_0)$ with center at the point $x_0$ and radius $2r$ (in the sense of the norm of $E_0$). Then the point $(y - x_0)/2 \in \bar{S}_1^0$, where $y \in \sigma_{2r}(x_0)$, runs over the ball $\sigma_r$ with center at zero. Consequently, the closure of any ball $\bar{S}_k^0$ contains the ball $\sigma_{kr}$. Let $z$ be any element of $\sigma_r$. There exists an $x_1 \in S_1$ such that $\|z - x_1\|_{E_0} \leqslant r/2$. Hence there is an $x_2 \in S_{1/2}$ such that $\|z - x_1 - x_2\|_{E_0} \leqslant r/4$. Continuing this process, we construct a sequence of elements $x_n \in S_{2^{1-n}}$ such that

$$\|z - x_1 - \cdots - x_n\|_{E_0} \leqslant r2^{-n}.$$

Then $z = \sum_1^\infty x_k$, and this series converges in the norm of $E_1$. Therefore $z \in E_1$. We have arrived at a contradiction to the assumption that $E_1 \neq E_0$.

2. ***Relative completion.*** Let $E_0$ and $E_1$ be a couple of imbedded Banach spaces ($E_1 \subset E_0$). Denote by $E_{01}$ the collection of all elements of $E_0$ which are limits in $E_0$ of sequences of elements from $E_1$ bounded in the norm of $E_1$:

$$x \in E_{01}: x = \lim_{n\to\infty} x_n \quad (\text{in } E_0) \quad \text{and} \quad \|x_n\|_{E_1} \leqslant R. \tag{1.2}$$

Obviously $E_{01}$ is a linear manifold in $E_0$. We can introduce a norm in it by putting

$$\|x\|_{01} = \inf R,$$

where the infimum is taken over those $R$ for which there exist sequences $x_n$ with property (1.2). We verify that $\|x\|_{01}$ has the properties of a norm. For this we note that (1.1) implies that $\|x_n\|_{E_0} \leqslant C_{01}\|x_n\|_{E_1} \leqslant C_{01}R$, and since $x_n \to x$ in $E_0$, we also have $\|x\|_{E_0} \leqslant C_{01}R$. This implies that $\|x\|_{E_0} \leqslant C_{01}\|x\|_{01}$, and, in particular, that $\|x\|_{01} = 0$ only if $x = 0$. It is obvious that $\|\lambda x\|_{01} = |\lambda|\ \|x\|_{01}$. Moreover, if the sequences $\{x_n\}$ and $\{y_n\}$ have properties (1.2) relative to the elements $x$ and $y$, respectively, with constants $R$ and $R_1$, then $x_n + y_n \to x + y$ in $E_0$, and $\|x_n + y_n\|_{E_1} \leqslant R + R_1$.

We obtain from this that $\|x + y\|_{01} \leqslant R + R_1$, and then, taking the infimum on the right side, that $\|x + y\|_{01} \leqslant \|x\|_{01} + \|y\|_{01}$.

We note that for $x \in E_1$ we may choose $x_n \equiv x$, and from the definition of the norm in $E_{01}$ it will follow that $\|x\|_{01} \leqslant \|x\|_{E_1}$.

The construction of $E_{01}$ may be ascribed the following geometrical meaning. Let $x \in E_{01}$ and $\|x\|_{01} = r$. By definition, the element $x$ belongs to the closure in $E_0$ of any ball of $E_1$ with radius $R > r$. Taking a sequence $R_k \to r$ and for every $R_k$, choosing an appropriate sequence of elements in $E_1$ with property (1.2), we can construct a sequence $\{x_k'\} \subset E_1$ such that $x_k' \to x$ in $E_0$ and $\|x_k'\|_{E_1} \to r$ as $k \to \infty$. Then $\bar{x}_k = r x_k'(\|x_k'\|_{E_1})^{-1} \to x$ in $E_0$, and $\|\bar{x}_k\|_{E_1} = r$. Thus, $x$ belongs to the closure in $E_0$ of the ball (and even of the sphere) of radius $r$ of $E_1$ and does not belong to the closure of balls of smaller radius.

Hence, the ball of $E_{01}$ is the closure in $E_0$ of the ball of $E_1$ with the same radius.

We prove that the normed space $E_{01}$ is complete.

Let $\{x^{(k)}\}$ be a Cauchy sequence in $E_{01}$. By the inequality $\|x\|_{E_0} \leqslant C_{01}\|x\|_{01}$, it is Cauchy in $E_0$. Let $x^{(k)} \to x$ in $E_0$. For any $\varepsilon > 0$ and sufficiently large $m$ and $l$ we have $\|x^{(m)} - x^{(l)}\|_{01} \leqslant \varepsilon$. This means that $x^{(m)} - x^{(l)}$ belongs to the closure in $E_0$ of the ball of radius $\varepsilon$ of the space $E_1$. However, $x^{(m)} - x^{(l)} \to x^{(m)} - x$ in $E_0$ as $l \to \infty$. Therefore, the element $x^{(m)} - x$ belongs to the same closure, i.e. $\|x^{(m)} - x\|_{01} \leqslant \varepsilon$. Thus, $x^{(k)} \to x$ in $E_{01}$ as $k \to \infty$, and $E_{01}$ is complete.

Summing up, we may say that we have constructed a Banach space $E_{01}$ such that $E_1 \subset E_{01} \subset E_0$; moreover, $\|x\|_{01} \leqslant \|x\|_{E_1}$ ($x \in E_1$) and

$$\|x\|_{E_0} \leqslant C_{01}\|x\|_{01} \qquad (x \in E_{01}). \tag{1.3}$$

From (1.3) it follows that if $E_1$ is normally imbedded in $E_0$, so is $E_{01}$.

The Banach space $E_{01}$ is called the *relative completion of* $E_1$ *with respect to* $E_0$.

The following facts are important.

LEMMA 1.2. *If $E_1$ does not coincide with $E_0$, then its relative completion $E_{01}$ does not coincide with $E_0$ either.*

PROOF. By Lemma 1.1 the closure in $E_0$ of any ball $S_r$ of $E_1$ is nowhere dense in $E_0$. Therefore $E_{01} = \bigcup_{r>0} S_r$ is a set of the first category in $E_0$, and consequently does not coincide with $E_0$.

LEMMA 1.3. *The completion $\hat{E}_{01}$ of $E_{01}$ with respect to $E_0$ coincides with $E_{01}$ itself.*

PROOF. If $y \in \hat{E}_{01}$, then there is a sequence of elements $y_k \in E_{01}$ such that $\|y_k\|_{01} = \|y\|_{\hat{E}_{01}}$ and $y_k \to y$ in $E_0$. For every $y_k$ there exists a sequence $x_n^{(k)} \to y_k$ in $E_0$ as $n \to \infty$ and such that $\|x_n^{(k)}\|_{E_1} = \|y_k\|_{E_{01}} = \|y\|_{\hat{E}_{01}}$. But then we can construct a sequence $x_{n_k}^{(k)}$ such that $x_{n_k}^{(k)} \to y$ in $E_0$ and $\|x_{n_k}^{(k)}\|_{E_1} = \|y\|_{\hat{E}_{01}}$, and consequently $y \in E_{01}$. Moreover, from the above it follows that $\|y\|_{01} \leqslant \|y\|_{\hat{E}_{01}}$, and since the reverse inequality is always true (see (1.3)), we have $\|y\|_{01} = \|y\|_{\hat{E}_{01}}$. The lemma is proved.

A simple example of relative completion can be obtained by setting $E_0 = L_1(0, 1)$ and $E_1 = C(0, 1)$. Then it is easy to see that $E_{01} = L_\infty(0, 1)$. Similarly, if $E_0 = C(0, 1)$ and $E_1 = C_1(0, 1)$, then $E_{01} = H_1(0, 1)$ is the space of functions satisfying a Lipschitz condition.

The following assertions concerning the relation between the spaces $E_1$ and $E_0$ are consequences of the definition of relative completion.

LEMMA 1.4. *In order that $E_1$ be isometrically imbedded in $E_{01}$ it is necessary and sufficient that the ball of $E_1$ be closed (in $E_1$) in the topology induced by the norm of $E_0$.*

PROOF. If $\|x\|_{01} \leqslant \|x\|_{E_1}$ for some $x \in E_1$, then there exists a sequence $x_n \to x$ in $E_0$ such that $\|x_n\|_{E_1} = \|x\|_{01} = a$. This means that the ball of radius $a$ in $E_1$ is not closed in the norm of $E_0$. Its limit point $x$ has norm greater than $a$.

Conversely, if the ball of radius $a$ is not closed, then there exists a point $x$ with $\|x\|_{E_1} > a$ and a sequence $x_n \to x$ in $E_0$ such that $\|x_n\|_{E_1} \leqslant a$. But then $\|x\|_{01} \leqslant a < \|x\|_{E_1}$.

LEMMA 1.5. *In order that $E_1$ be a closed subset of $E_{01}$ it is necessary and sufficient that the closure of a ball of $E_1$ in the topology induced in $E_1$ by the norm of $E_0$ be a bounded set in $E_1$.*

By the Banach theorem, $E_1$ is closed in $E_{01}$ if and only if the norms $\|x\|_{E_1}$ and $\|x\|_{01}$ are equivalent on $E_1$. After this remark the proof of Lemma 1.5 can be carried out similarly to that of Lemma 1.4.

DEFINITION 1.4. The space $E_1$ is said to be *complete with respect to* $E_0$ if $E_{01}$ coincides with $E_1$ isometrically.

By Lemma 1.3, $E_{01}$ is complete with respect to $E_0$. In particular, $L_\infty(0, 1)$ is complete with respect to $L_1(0, 1)$.

LEMMA 1.6. *If $E_1 \subset F \subset E_0$, then the completion $E_{F,1}$ of $E_1$ with respect to $F$ is imbedded in $E_{01}$ with imbedding constant not exceeding one. The completion of $E_{F,1}$ with respect to $E_0$ coincides with $E_{01}$ isometrically.*

PROOF. If $x \in E_{F,1}$, then there exists a sequence $x_n \in E_1$ with $\|x_n\|_{E_1} = \|x\|_{E_{F,1}}$ such that $x_n \to x$ in $F$. In view of the imbedding $F \subset E_0$ we then have $x_n \to x$ in $E_0$, and consequently $x \in E_{01}$ and $\|x\|_{E_{01}} \leqslant \|x\|_{E_{F,1}}$.

Moreover, the unit ball of the completion of $E_{F,1}$ with respect to $E_0$ is the closure in $E_0$ of the unit ball of $E_{F,1}$; the unit ball of $E_1$ is dense in the latter ball in the norm of $F$, and consequently in the norm of $E_0$ as well. Thus, the unit ball of the completion of $E_{F,1}$ with respect to $E_0$ coincides with the closure in $E_0$ of the unit ball of $E_1$, i.e. with the unit ball of $E_{01}$.

COROLLARY 1. *If $E_1$ is complete with respect to $E_0$, then it is complete with respect to $F$.*

For example, $L_\infty(0, 1)$ is complete with respect to all spaces $L_p(0, 1)$, $1 \leqslant p < \infty$.

COROLLARY 2. *If $E_{F,1}$ is complete with respect to $E_0$, then the completion of $E_{F,1}$ coincides with that of $E_{01}$.*

## §2. Dual spaces of imbedded Banach spaces

**1.** ***Dual spaces and relative completion.*** If the space $E_1$ is imbedded in the space $E_0$, then the restriction to $E_1$ of every continuous linear functional $f(x)$ defined on $E_0$ induces a functional on $E_1$ in a natural manner. This functional is continuous in the norm of $E_1$. Indeed,

$$|f(x)| \leqslant \|f\|_{E_0'} \|x\|_{E_0} \leqslant C_{01}\|f\|_{E_0'} \|x\|_{E_1}.$$

Thus, a linear mapping of the space $E_0'$ into $E_1'$ is obtained. If $E_1$ is not dense in $E_0$, then there exists a nonzero functional in $E_0'$ which identically vanishes on $E_1$, and consequently is mapped into zero. In this case the mapping is not injective. If, on the other hand, $E_1$ is dense in $E_0$, then the mapping is injective, and $E_0'$ can be considered imbedded in $E_1'$. Then, the imbedding constant does not exceed $C_{01}$.

The quantity

$$\sup_{x \in E_1} \frac{|f(x)|}{\|x\|_{E_1}} = \|f\|_{E_1'} \qquad (f \in E_0')$$

will be a seminorm on $E_0'$ in the first case and a norm on $E_1'$ in the second case.

Now we give another characterization of the relative completion $E_{01}$.

LEMMA 2.1. *The completion $E_{01}$ of $E_1$ with respect to $E_0$ consists of all elements of $E_0$ inducing functionals on $E_0'$, bounded in the seminorm $\|f\|_{E_1'}$, according to the formula $x(f) = f(x)$.*

PROOF. If $x \in E_{01}$, then there exists a sequence $x_n$ such that $x_n \to x$ in $E_0$ and $\|x_n\|_{E_1} = \|x\|_{01}$. Let $f \in E_0'$. We have $f(x_n) \to f(x)$ and $|f(x_n)| \leqslant \|f\|_{E_1'} \|x_n\|_{E_1} = \|f\|_{E_1'} \|x\|_{01}$. This implies that $|f(x)| \leqslant \|f\|_{E_1'} \|x\|_{01}$ also holds. Thus, the functional $x(f)$ is bounded on $E_0'$ in the seminorm $\|f\|_{E_1'}$, and its norm does not exceed $\|x\|_{01}$.

Now let $x \in E_0$ and $|x(f)| = |f(x)| \leqslant C\|f\|_{E_1'}$ $(f \in E_0')$. We show that $x$ belongs to the closure in $E_0$ of the ball $S_C$ of radius $C$ of $E_1$. Otherwise there would exist a functional $f_0 \in E_0'$ such that $\sup_{y \in S_C} |f_0(y)| < f_0(x)$ or $C\|f_0\|_{E_1'} < f_0(x)$, which contradicts the initial assumption. Thus, $x \in E_{01}$ and $\|x\|_{01} \leqslant C$.

From the proof of the lemma we obtain the following corollary.

COROLLARY 1. *The norm of the functional $x(f)$ with respect to the seminorm $\|f\|_{E_1'}$ is equal to $\|x\|_{01}$:*

$$\|x\|_{01} = \sup_{f \in E_0', \|f\|_{E_1'} \leqslant 1} |f(x)|. \tag{2.1}$$

DEFINITION 2.1. A linear manifold $M'$ of continuous linear functionals on a Banach space $E$ is said to be *normative* if

$$\|x\|_E = \sup_{f \in M', \|f\|_{E'} \leqslant 1} |f(x)| \qquad (x \in E).$$

Formula (2.1) and Lemma 1.4 imply the following theorem.

THEOREM 2.1. *In order that the restrictions to $E_1$ of all functionals from $E_0'$ form a normative set for $E_1$ it is necessary and sufficient that $E_1$ be imbedded isometrically in $E_{01}$, or, what is the same, a ball of $E_1$ be closed in $E_1$ in the topology induced by the norm of $E_0$.*

**2. *Dual spaces of densely imbedded spaces.*** If $E_1$ is densely imbedded in $E_0$, then, as we have seen above, $E_0'$ is imbedded in $E_1'$; however, this imbedding may not be dense. For example, $l_1 = E_1$ is normally imbedded in $c_0 = E_0$.

The dual spaces $E_0' = l_1$ and $E_1' = l_\infty$ are not densely imbedded. Indeed, the element $(1, 1, \ldots) \in l_\infty$ is at distance one from the linear manifold $l_1$.

We recall that a linear manifold $M' \subset E_1'$ is said to be *total* if the condition $f(x_0) = 0$ $(x_0 \in E_1)$ for all $f \in M'$ implies that $x_0 = \theta$.(1) If $E_1$ is dense in $E_0$, then $E_0'$ is a total linear manifold in $E_1'$. Indeed, if $x_0 \in E_1$ and $x_0 \neq 0$, then $x_0 \in E_0$, and so there is a functional $f_0 \in E_0'$ such that $f_0(x_0) = \|x_0\|_{E_0} \neq 0$.

If $E_1$ is reflexive, then every total manifold is dense in $E_1'$, and so in this case $E_0'$ is densely imbedded in $E_1'$.

THEOREM 2.2. *If $E_1$ is densely imbedded in $E_0$, then $E_0'$ is complete with respect to $E_1'$.*

PROOF. If $f$ belongs to the completion $(E_0')_1$ of $E_0'$ with respect to $E_1'$, then there exists a sequence $f_n \in E_0'$ such that $f_n \to f$ in $E_1'$ and $\|f_n\|_{E_0'} = \|f\|_{(E_0')_1}$. Then for any $x \in E_1$ we have $f_n(x) \to f(x)$. The sequence of the linear functionals $f_n \in E_0'$ is uniformly bounded and converges to $f$ on the set $E_1$ dense in $E_0$. By the Banach-Steinhaus theorem this implies that $f \in E_0'$ and the sequence $f_n$ weakly converges to $f$ on $E_0$. Moreover, $\|f\|_{E_0'} \leqslant \underline{\lim}\|f_n\|_{E_0'} = \|f\|_{(E_0')_1}$. Since in the case of a completion the reverse inequality always holds, we have $\|f\|_{E_0'} = \|f\|_{(E_0')_1}$.

THEOREM 2.3. *A reflexive space $E_1$ is complete with respect to any other space $E_0$ in which it is imbedded.*

PROOF. Without loss of generality we may assume that $E_1$ is dense in $E_0$, since otherwise $E_0$ could be replaced by the closure of $E_1$ in $E_0$ without changing the completion $E_{01}$. Under this assumption, the reflexivity of $E_1$ implies that $E_0'$ is densely imbedded in $E_1'$. Then by Theorem 2.2 the space $E_1'' = E_1$ is complete with respect to $E_0''$, and by Corollary 1 of Lemma 1.6 the space $E_1$ is complete with respect to $E_0$. The theorem is proved.

If $E_0'$ is densely imbedded in $E_1'$, then it is a normative set on $E_1$, and therefore Lemma 2.1 and formula (2.1) imply the following assertion.

LEMMA 2.2. *If $E_1$ is densely imbedded in $E_0$ and $E_0'$ is densely imbedded in $E_1'$, then $E_1$ is isometrically imbedded in $E_{01}$. The space $E_{01}$ can be isometrically imbedded in $E_1''$ in a natural manner, and in this natural identification we may assume that $E_{01} = E_0 \cap E_1''$.*

Now if $E_0'$ is not densely imbedded in $E_1'$, we can consider its closure $\mathring{E}_1'$ in $E_1'$. This is a subspace of $E_1'$. In this case we may reformulate Lemma 2.1 and Corollary 1 as follows.

---

(1) $\theta$ is the origin of the space.

LEMMA 2.3. *If $E_1$ is dense in $E_0$, then $E_{01}$ consists of all elements $x$ in $E_0$ inducing continuous linear functionals on $E_1'$ by the formula $x(f) = f(x)$. The mapping $x = x(f)$ is an isometric imbedding of $E_{01}$ in the dual space $(\mathring{E}_1')'$.*

DEFINITION 2.2. If $E_1$ is dense in $E_0$, and in the natural imbedding the space $E_1$ coincides with the whole space $(\mathring{E}_1')'$, then the couple of spaces $E_1$, $E_0$ is said to be *reflexive*.

It is clear from Lemma 2.3 that for a reflexive couple $E_0$, $E_1$ the space $E_1$ is complete with respect to $E_0$.

## §3. Intermediate Banach spaces

**1. *Sum and intersection of spaces in a Banach couple.***

DEFINITION 3.1. A *Banach couple** is two Banach spaces $A$ and $B$ algebraically and topologically imbedded in a separated topological linear space $\mathfrak{A}$.

This means that $A$ and $B$ are linear manifolds in $\mathfrak{A}$ and the topology induced on $A$ and $B$ by the topology of $\mathfrak{A}$ is weaker than the original topologies of the normed spaces $A$ and $B$.

It is obvious that a couple of imbedded Banach spaces $E_0$ and $E_1$, where $E_1 \subset E_{01}$, can be considered as a special case of a Banach couple, where the role of the topological space $\mathfrak{A}$ is played by the Banach space $E_0$.

With any Banach pair we may canonically associate a couple of imbedded Banach spaces in the following way:

1. The space $A \cap B$ consists of the elements common to $A$ and $B$; the norm is introduced by

$$\|x\|_{A\cap B} = \max(\|x\|_A, \|x\|_B) \qquad (x \in A \cap B). \tag{3.1}$$

2. The space $A + B$ consists of elements of the form $x = u + v$, where $u \in A$, $v \in B$ and is equipped with the norm

$$\|x\|_{A+B} = \inf\{\|u\|_A + \|v\|_B\}, \tag{3.2}$$

where the infimum is taken over all elements $u \in A$ and $v \in B$ whose sum is equal to $x$.

The first of these spaces is called the *intersection* of the spaces of the Banach couple, and the second is called the *sum* of the spaces of the Banach couple.

We show that these are Banach spaces. Let $x_1, x_2, \ldots$ be a Cauchy sequence in $A \cap B$. By virtue of the definition (3.1) of the norm, the sequence

---

**Editor's note.* In the Western literature it is usually called a *compatible couple* of Banach spaces.

$\{x_n\}$ is Cauchy in both $A$ and $B$. The completeness of $A$ and $B$ implies that $x_n \to x_A$ (in $A$) and $x_n \to x_B$ (in $B$). Since $A$ is topologically imbedded in $\mathfrak{A}$, $x_n$ converges to $x_A$ in the topology of $\mathfrak{A}$, and similarly $x_n \to x_B$ in the topology of $\mathfrak{A}$. This implies that $x_A = x_B = x$, because the topology of $\mathfrak{A}$ is separated. Consequently, we obtain that $x_n \to x \in A \cap B$ in the norms $\|\cdot\|_A$ and $\|\cdot\|_B$, and therefore in the norm of $A \cap B$.

To prove the completeness of $A + B$ we construct a Banach space isometric to it. To this end, in the product $A \times B$ of $A$ and $B$ with norm $\|(u, v)\|_{A\times B} = \|u\|_A + \|v\|_B$ we consider the subspace $L$ consisting of the elements of the form $w = (z, -z)$, $z \in A \cap B$, and the quotient space $(A \times B)/L$. If two elements $(u, v)$ and $(u_1, v_1)$ belong to the same coset modulo $L$, then $u + v = u_1 + v_1$, and consequently to this whole coset there corresponds a single element $x = u + v \in A + B$. Conversely, for any element $x \in A + B$ the pairs $(u, v)$ corresponding to all possible representations $x = u + v$ form a coset of the space $A \times B$ modulo $L$. Moreover, it is clear from the definition of the norm (3.2) that the norm of an element $x \in A + B$ is equal to the norm of the corresponding coset in the quotient space $(A \times B)/L$. Thus, $A + B$ is isometric to $(A \times B)/L$. As is well known (see [6], Fascicule de résultats, §5.5), the quotient space of a Banach space modulo a (closed) subspace of it is a Banach space, and so $A + B$ is also a Banach space. From what has been said it is clear that if $A \cap B = \theta$, then $A + B$ is isometric with the direct product of $A$ and $B$.

We note that each of the spaces $A$ and $B$ is contained in $A + B$, and moreover, by virtue of the inequalities

$$\|x\|_{A+B} \leqslant \|x\|_A + \|\theta\|_B = \|x\|_A \quad \text{for } x \in A, \tag{3.3}$$

and

$$\|x\|_{A+B} \leqslant \|\theta\|_A + \|x\|_B = \|x\|_B \quad \text{for } x \in B, \tag{3.4}$$

the spaces $A$ and $B$ are imbedded in $A + B$, and the imbedding constants do not exceed one. It follows from the preceding inequalities that $\|x\|_{A+B} \leqslant \|x\|_{A\cap B}$ holds for $x \in A \cap B$, which shows that $A \cap B$ is imbedded in $A + B$.

In the case of a couple of imbedded Banach spaces $E_1 \subset E_0$ the space $E_0 \cap E_1$ coincides with $E_1$ as a set, and the space $E_0 + E_1$ with $E_0$. Moreover, these spaces are respectively isomorphic ($E_0 \cap E_1$ with $E_1$ and $E_0 + E_1$ with $E_0$). Indeed, if $\|x\|_{E_0} \leqslant C_{01}\|x\|_{E_1}$, then for $x \in E_1$

$$\|x\|_{E_1} \leqslant \max(\|x\|_{E_0}, \|x\|_{E_1}) = \|x\|_{E_0\cap E_1} \leqslant \max(1, C_{01})\|x\|_{E_1},$$

and for $x \in E_0$

$$\min\{1, 1/C_{01}\}\|x\|_{E_0} \leqslant \min\{1, 1/C_{01}\} \inf_{x=u+v}(\|u\|_{E_0} + \|v\|_{E_0})$$

$$\leqslant \inf_{x=u+v}(\|u\|_{E_0} + \|v\|_{E_1}) = \|x\|_{E_0+E_1} \leqslant \|x\|_{E_0}$$

(the latter by virtue of (3.3) and (3.4)). Thus, $E_0 \cap E_1$ and $E_0 + E_1$ form a couple of imbedded spaces isomorphic to $E_1$ and $E_0$, respectively.

We consider the geometric structure of the unit balls $S_1(A \cap B)$ and $S_1(A + B)$ of $A \cap B$ and $A + B$. It is easy to see that $S_1(A \cap B)$ is the intersection of the unit balls $S_1(A)$ and $S_1(B)$. Indeed, if $x \in S_1(A \cap B)$, then $\|x\|_A \leqslant 1$ and $\|x\|_B \leqslant 1$, and therefore $x \in S_1(A)$ and $x \in S_1(B)$. Conversely, if $x \in S_1(A) \cap S_1(B)$, then $\|x\|_A \leqslant 1$ and $\|x\|_B \leqslant 1$, and so $\|x\|_{A\cap B} = \max\{\|x\|_A, \|x\|_B\} \leqslant 1$.

The open ball $S_1^0(A + B)$ coincides with the convex hull of the open balls $S_1^0(A)$ and $S_1^0(B)$ of $A$ and $B$:

$$S_1^0(A + B) = \operatorname{conv}(S_1^0(A), S_1^0(B)). \tag{3.5}$$

Indeed, let $\|x\|_{A+B} < 1$. Then there exist $u$ and $v$ such that $x = u + v$ and $\|u\|_A + \|v\|_B < 1$. The elements $u_1 = (\|u\|_A + \|v\|_B)u/\|u\|_A$ and $v_1 = (\|u\|_A + \|v\|_B)v/\|v\|_B$ belong to the balls $S_1^0(A)$ and $S_1^0(B)$, respectively. Moreover,

$$x = u + v = \mu u_1 + (1 - \mu)v_1, \quad \text{where } \mu = \|u\|_A/(\|u\|_A + \|v\|_B)$$

and consequently, $x \in \operatorname{conv}(S_1^0(A), S_1^0(B))$. Conversely, from the relation $x \in \operatorname{conv}(S_1^0(A), S_1^0(B))$ it follows that $x = \mu u + (1 - \mu)v$ ($u \in S_1^0(A)$, $v \in S_1^0(B)$, $0 \leqslant \mu \leqslant 1$), and hence

$$\|x\|_{A+B} \leqslant \|\mu u\|_A + \|(1 - \mu)v\|_B < 1,$$

i.e. $x \in S_1^0(A + B)$.

We note some more properties of the sum and intersection of spaces in a Banach couple.

LEMMA 3.1. *For any $x \in A$*

$$d_A(x, A \cap B) = d_{A+B}(x, B), \tag{3.6}$$

*where $d_E(x, L)$ denotes the distance in $E$ from the element $x$ to the linear manifold $L$.*

PROOF. From (3.3) and the inclusion $B \supset A \cap B$ it follows immediately that $d_A(x, A \cap B) \geqslant d_{A+B}(x, B)$. We establish the reverse inequality. If $y \in B$, then $x - y \in A + B$ and

$$\|x - y\|_{A+B} = \inf_{x-y=u+v}(\|u\|_A + \|v\|_B)$$

$$\geqslant \inf_{x-y=u+v}\|u\|_A = \inf_{x-y=u+v}\|x - y - v\|_A. \tag{3.7}$$

Since $x \in A$ and $x - y - v \in A$, we have $y + v \in A$. On the other hand, $y \in B$ and $v \in B$, and so $y + v \in B$. Then (3.7) implies that

$$\|x - y\|_{A+B} \geqslant \inf_{z \in A \cap B} \|x - z\|_A = d_A(x, A \cap B).$$

Taking the infimum of the left side over $y \in B$, we obtain (3.6).

COROLLARY 1. *$A \cap B$ is dense in $A$ if and only if $B$ is dense in $A + B$.*

Indeed, if $A \cap B$ is dense in $A$, then $d_A(x, A \cap B) = 0$, and hence also $d_{A+B}(x, B) = 0$ for any $x \in A$. But then $d_{A+B}(x, B) = 0$ for any $x \in A + B$ as well, i.e. $B$ is dense in $A + B$. The converse is obvious.

COROLLARY 2. *If $A \cap B$ is dense in both $A$ and $B$, then it is dense in $A + B$.*

Indeed, $A \cap B$ is dense in $B$, and the latter is dense in $A + B$ by virtue of what has been said above. Since $\|\cdot\|_{A+B} \leqslant \|\cdot\|_B$, the space $A \cap B$ is dense in $A + B$.

Now we assume that $A \cap B$ is not dense in $A$. Then for any $\varepsilon > 0$ there exists a nonzero element $x \in A$ such that $d_A(x, A \cap B) \geqslant (1 - \varepsilon)\|x\|_A$. Then we obtain from (3.6) that

$$(1 - \varepsilon)\|x\|_A \leqslant d_{A+B}(x, B) \leqslant \|x\|_{A+B}.$$

Comparing this inequality with (3.3), we conclude that the imbedding constant of $A$ in $A + B$ is equal to one in this case. The following deeper assertion is true.

LEMMA 3.2. *If the imbedding constant of the space $A$ in the space $A + B$ is less than one, then $A$ is imbedded in $B$.*

PROOF. Let the imbedding constant be equal to $q < 1$. For $x \in A$ with $\|x\|_A < 1$ we have $\|x\|_{A+B} < q$. By the definition of the norm in $A + B$ there exists a representation $x = u_1 + v_1$, where $\|u_1\|_A + \|v_1\|_B < q$, and consequently $\|u_1\|_A < q$ and $\|v_1\|_B < q$. We apply the same arguments to the element $u_1$ and obtain the representation $u_1 = u_2 + v_2$, where $\|u_2\|_A < q^2$ and $\|v_2\|_B < q^2$. Continuing this process, we arrive at the equality $x = u_n + \sum_1^n v_k$, where $\|u_n\|_A < q^n$ and $\|v_k\|_B < q^k$. Then

$$\left\| x - \sum_{k=1}^{n} v_k \right\|_{A+B} = \|u_n\|_{A+B} \leqslant \|u_n\|_A < q^n \to 0.$$

On the other hand, $\sum_1^\infty v_k$ converges in $B$. Since $B$ is imbedded in $A + B$, it follows that $x = \sum_1^\infty v_k \in B$. Hence $A$ is contained in $B$.

From the above it follows that

$$\|x\|_B \leqslant \frac{q}{1-q}\|x\|_A,$$

i.e., $A$ is imbedded in $B$.

The last part of the proof of the lemma could be omitted if we took account of the following lemma.

LEMMA 3.3. *If the space $A$ in a Banach couple $A$, $B$ is a linear manifold in $B$, then $A$ is imbedded in $B$.*

PROOF. If $A$ is contained in $B$, then the linear spaces $B$ and $A + B$ coincide. By (3.4) we have $\|x\|_{A+B} \leqslant \|x\|_B$. Since $B$ and $A + B$ are Banach spaces, by the Banach theorem on the inverse operator the inequality $\|x\|_B \leqslant C\|x\|_{A+B}$ holds. If now $x \in A$, then $\|x\|_B \leqslant C\|x\|_{A+B} \leqslant C\|x\|_A$.

COROLLARY. *If the spaces $A$ and $B$ in the Banach couple $A$, $B$ coincide as linear spaces, then $A$ and $B$ are isomorphic.*

In the sequel we need the following result.

LEMMA 3.4. *Given a Banach couple $A$, $B$, if there exists a subset $M \subset A \cap B$ dense in $A$ on which the inequality $\|x\|_B \leqslant C\|x\|_A$ is satisfied, then $A$ is imbedded in $B$ with imbedding constant not exceeding $C$.*

PROOF. Let $x_0 \in A$. We consider the open balls $\|x - x_0\|_A < \frac{1}{2}\|x_0\|_A$ and $\|x\|_A < \frac{1}{2}(1 + \varepsilon)\|x_0\|_A$. They intersect, and so there exists at least one element $x_1 \in M$ in their intersection. Then $\|x_0 - x_1\|_A < \frac{1}{2}\|x_0\|_A$ and $\|x_1\|_A < \frac{1}{2}(1 + \varepsilon)\|x_0\|_A$. By repeating this reasoning for $x_0 - x_1$, we find an element $x_2$ such that

$$\|x_0 - x_1 - x_2\|_A < 2^{-2}\|x_0\|_A \quad \text{and} \quad \|x_2\|_A < 2^{-2}(1 + \varepsilon)\|x_0\|_A.$$

Continuing this process, we obtain a sequence of elements $x_1, x_2, \ldots$ from $M$ such that $\|x_0 - \sum_1^m x_k\|_A < 2^{-m}\|x_0\|_A$ and $\|x_m\|_A < 2^{-m}(1 + \varepsilon)\|x_0\|_A$. Then $x_0 = \sum_1^\infty x_k$ and $\sum_1^\infty\|x_k\|_A < (1 + \varepsilon)\|x_0\|_A$. By the hypothesis of the theorem we have

$$\sum_1^\infty \|x_k\|_B \leqslant C\sum_1^\infty \|x_k\|_A < C(1 + \varepsilon)\|x_0\|_A,$$

consequently $\sum_1^\infty x_k$ converges in $B$, and by virtue of the imbedding of $A$ and $B$ in $\mathfrak{A}$ its sum is also equal to $x_0$. Consequenty $x_0 \in B$ and

$$\|x_0\|_B \leqslant \sum_1^\infty \|x_k\|_B < C(1 + \varepsilon)\|x_0\|_A.$$

Since $\varepsilon$ was arbitrary, the lemma is proved.

DEFINITION 3.2. Two Banach couples $A$, $B$ and $C$, $D$ are said to be *isomorphic* if there exists an isomorphism of $A + B$ and $C + D$ whose restrictions to $A$ and $B$ are isomorphisms of $A$ and $C$ and $B$ and $D$, respectively.

**2. *The dual spaces of a sum and an intersection.*** In order to illuminate the structure of the dual spaces of a sum and an intersection of spaces of a Banach couple we recall that the dual of the product $A \times B$ of Banach spaces with norm

$$\|(u, v)\|_{A\times B} = \|u\|_A + \|v\|_B, \tag{3.8}$$

is the space $A' \times B'$ with norm

$$\|(f, g)\|_{A'\times B'} = \max(\|f\|_{A'}, \|g\|_{B'}). \tag{3.9}$$

Similarly, if we introduce the norm

$$\|(u, v)\|_{A\times B} = \max(\|u\|_A, \|v\|_B), \tag{3.10}$$

in $A \times B$, then the norm

$$\|(f, g)\|_{A'\times B'} = \|f\|_{A'} + \|g\|_{B'} \tag{3.11}$$

arises in the dual space $A' \times B'$.

We note that in both cases the duality between $A \times B$ and $A' \times B'$ can be given by the relation

$$\langle (u, v), (f, g)\rangle = f(u) - g(v). \tag{3.12}$$

In subsection 1 we have already spoken of the fact that for a Banach couple $A, B$ the space $A + B$ is isometric to the quotient space of $A \times B$ with norm (3.8) modulo the subspace $L$ consisting of the elements of the form $(z, -z)$, where $z \in A \cap B$. According to general theorems, the dual of a quotient space is isometric to the orthogonal complement of the subspace. We explain of which functionals from $A' \times B'$ the orthogonal complement $L^\perp$ of $L$ consists. By (3.12), if $(f, g) \in L^\perp$, then $f(z) = -g(z)$ for $z \in A \cap B$.

Consequently, the space $(A + B)'$ dual to the sum $A + B$ is isometric to the subspace $L^\perp$ off $A' \times B'$ with norm (3.9), consisting of all functionals $(f, g)$ with $f(z) = -g(z)$ for $z \in A \cap B$.

The value of the functional $(f, g) \in L^\perp$ at $x \in A + B$ can be calculated by (3.12), where $x = u + v$ ($u \in A$, $v \in B$).

Similarly, if we introduce the norm (3.10) in $A \times B$, then $L$ becomes isometric to $A \cap B$. The dual space of a subspace of a Banach space is isometric to the quotient space of the dual modulo the orthogonal complement of the subspace. Hence, the space $(A \cap B)'$ dual to $A \cap B$ is isometric to the quotient space of $A' \times B'$ with norm (3.11) modulo $L^\perp$.

The picture becomes significantly simpler if we assume that $A \cap B$ is dense in both $A$ and $B$. As shown in §2.1, in this case the spaces $A'$ and $B'$ are naturally imbedded in $(A \cap B)'$, and consequently form a Banach couple. If now $(f, g) \in L^\perp$, then $g$ coincides with $-f$ as an element of $(A \cap B)'$, and since $f \in A'$ and $f \in B'$, we have $f \in A' \cap B'$. Conversely, every functional

$f \in A' \cap B'$ induces a pair $(f, -f) \in L^{\perp}$. The norm of $(f, g)$ in $A' \times B'$ with norm (3.9) coincides with the norm of $f$ in $A' \cap B'$.

Similarly, the quotient space $A' \times B'/L^{\perp}$ of $A' \times B'$ with norm (3.11) modulo $L^{\perp}$, which, as we have shown, consists of the pairs $(f, -f), f \in A' \cap B'$, is isometric to the sum $A' + B'$. Thus, we have arrived at the following assertion.

THEOREM 3.1. *If $A \cap B$ is dense in the spaces of the Banach couple $A$, $B$, then $A'$ and $B'$ form a Banach couple; the dual space $(A + B)'$ of $A + B$ is isometric to $A' \cap B'$, and the dual $(A \cap B)'$ of $A \cap B$ is isometric to $A' + B'$.*

We also elucidate how the duality between $A + B$ and $A' \cap B'$ and between $A \cap B$ and $A' + B'$ can be given. If $x \in A + B$ and $f \in A' \cap B'$, then $x = u + v$ $(u \in A, v \in B)$, and according to (3.12) we have

$$\langle x, f\rangle = \langle (u, v), (f, -f)\rangle = f(u) + f(v),$$

i.e. the functional $\langle x, f\rangle$ coincides with the linear extension of $f$ to $A + B$. If now $z \in A \cap B$ and $f \in A' + B'$, then $f = g + h$ $(g \in A', h \in B')$, and according to (3.12) we have

$$\langle z, f\rangle = \langle (z, -z), (g, h)\rangle = g(z) - h(-z) = f(z),$$

i.e., $\langle z, f\rangle$ coincides with the restriction of $f$ to $A \cap B$.

**3. *Intermediate spaces.***

DEFINITION 3.3. The Banach space $E$ is said to be *intermediate* for the spaces of the Banach couple $A$, $B$ if the imbeddings

$$A \cap B \subset E \subset A + B \tag{3.13}$$

are obtained. We recall that $\subset$ means algebraic and continuous imbedding here; however, by Lemma 3.3 in order to verify that the Banach space $E$ is intermediate it is sufficient to know that it is algebraically and continuously imbedded in $\mathfrak{A}$, contains $A \cap B$, and is contained in $A + B$.

To illustrate the above definitions, we consider a few simple examples of intermediate Banach spaces.

It is easy to see that any space $L_p(0, 1)$ is an intermediate space between $L_{p_0}(0, 1)$ and $L_{p_1}(0, 1)$ $(p_0 < p_1)$ if $p_0 \leqslant p \leqslant p_1$.

Indeed, in §1.3 we showed that $L_{p_1}(0, 1) \subset L_p(0, 1) \subset L_{p_0}(0, 1)$ and, moreover,

$$\|x\|_{L_{p_0}(0,1)} \leqslant \|x\|_{L_p(0,1)} \quad (x \in L_p(0, 1)),$$

$$\|x\|_{L_p(0,1)} \leqslant \|x\|_{L_{p_1}(0,1)} \quad (x \in L_p(0, 1)).$$

For the Hölder spaces $H_{\alpha_0}$ and $H_{\alpha_1}$ $(\alpha_0 < \alpha_1)$, for example, the Hölder spaces $H_\alpha$ are intermediate if $\alpha_0 \leqslant \alpha \leqslant \alpha_1$ (see Chapter III, §3).

Let $L_{p_0}(0, \infty)$ and $L_{p_1}(0, \infty)$ ($p_0 < p_1$) be the spaces of measurable functions whose modulus is $p_i$th power summable ($i = 0, 1$) on the semiaxis. Neither space $L_{p_i}(0, \infty)$ is imbedded in the other: it is obvious that $L_{p_0}(0, \infty) \cap L_{p_1}(0, \infty) \neq L_{p_i}(0, \infty)$. Nevertheless, $L_p(0, \infty)$ is intermediate between $L_{p_0}(0, \infty)$ and $L_{p_1}(0, \infty)$ if $p_0 \leqslant p \leqslant p_1$. Indeed, consider the separated topological vector space $S(0, \infty)$ consisting of all measurable functions defined on the semiaxis with the topology of convergence in measure (see [11], Chapter III, §2). Then each of the spaces $L_{p_i}(0, \infty)$ ($i = 0, 1$) is algebraically and topologically imbedded in the topological vector space $S(0, \infty)$, and so these spaces form a Banach couple.

We show that

$$L_{p_0}(0, \infty) \cap L_{p_1}(0, \infty) \subset L_p(0, \infty) \subset L_{p_0}(0, \infty) + L_{p_1}(0, \infty). \tag{3.14}$$

Indeed, let $x \in L_{p_0}(0, \infty) \cap L_{p_1}(0, \infty)$. We write

$$x_1(t) = \begin{cases} x(t) & \text{if } |x(t)| \geqslant 1, \\ 0 & \text{if } |x(t)| < 1, \end{cases}$$

and $x_2(t) = x(t) - x_1(t)$. Then $x_1, x_2 \in L_{p_0}(0, \infty) \cap L_{p_1}(0, \infty)$. Since $p \leqslant p_1$, we have $x_1 \in L_p(0, \infty)$. From the condition $p_0 \leqslant p$ it also follows that $x_2 \in L_p(0, \infty)$. Thus $x \in L_p(0, \infty)$, i.e. $L_{p_0}(0, \infty) \cap L_{p_1}(0, \infty)$ is contained in $L_p(0, \infty)$, and therefore, as we pointed out above, is imbedded in it. The second imbedding in (3.14) can be verified similarly.

In the next chapter we shall study the intermediate spaces for the Banach couple $L_1(0, \infty)$, $L_\infty(0, \infty)$ in detail.

**4.** ***The sum and intersection of a family of Banach spaces.*** In this subsection we shall denote by $E \overset{1}{\subset} F$ the imbedding of the Banach space $E$ in the Banach space $F$ satisfying the condition $\|x\|_F \leqslant \|x\|_E$ ($x \in E$).

Let $\mathfrak{A}$ be a separated topological linear space, and let $E_i$ ($i \in I$) be an arbitrary family of Banach spaces algebraically and topologically imbedded in $\mathfrak{A}$.

The *intersection* of the family $E_i$ is, by definition, the Banach space $\Cap E_i$ satisfying the following conditions:

1) $\Cap E_i \overset{1}{\subset} E_j$ for any $j \in I$.

2) If $F$ is a Banach space such that $F \overset{1}{\subset} E_j$ for all $j \in I$, then $F \overset{1}{\subset} \Cap E_i$.

The *sum* of the family $E_i$ is, by definition, the Banach space $\Cup E_i$ algebraically and topologically imbedded in $\mathfrak{A}$ and such that the following conditions are satisfied:

1) $E_j \overset{1}{\subset} \Cup E_i$ for any $j \in I$.

2) If the Banach space $G$ algebraically and topologically imbedded in $\mathfrak{A}$ is such that $E_j \overset{1}{\subset} G$ for all $j \in I$, then $\Cup E_i \overset{1}{\subset} G$.

We show that $\Cap E_i$ always exists. We consider the linear subset $\Cap E_i$ consisting of all elements $x$ belonging to all spaces $E_i$ for which

$$\|x\|_{\Cap E_i} = \sup_{i \in I} \|x\|_{E_i} < \infty. \tag{3.15}$$

The number on the left has the properties of a norm. We show the completeness of the space $\Cap E_i$ in this norm. Let $x_n$ be a Cauchy sequence of elements of $\Cap E_i$ in the norm (3.15). This sequence is Cauchy in every space $E_i$, and so it has a limit in $E_i$. Since all spaces $E_i$ are imbedded in $\mathfrak{A}$, these limits coincide with some element $x \in \bigcap_i E_i$. Moreover,

$$\|x - x_n\|_{\Cap E_i} = \sup_i \|x - x_n\|_{E_i} = \sup_i \lim_{m \to \infty} \|x_m - x_n\|_{E_i}$$
$$\leqslant \sup_{m > n} \sup_i \|x_m - x_n\|_{E_i} = \sup_{m > n} \|x_m - x_n\|_{\Cap E_i} \to 0 \quad \text{as } n \to \infty.$$

Thus $\Cap E_i$ is a Banach space. Now if $F \overset{1}{\subset} E_j$ for all $j \in I$ and $x \in F$, then

$$\|x\|_{\Cap E_i} = \sup_i \|x\|_{E_i} \leqslant \|x\|_F < \infty,$$

and therefore, $x \in \Cap E_i$, i.e. $F \overset{1}{\subset} \Cap E_i$.

For the construction of the sum $\Cup E_i$ we make the additional assumption that there exists a Banach space $W$ imbedded in $\mathfrak{A}$ and such that $E_i \overset{1}{\subset} W$ for all $i \in I$. As the linear system $\Cup E_i$ we take the set of all elements $x$ in $\mathfrak{A}$ representable in the form

$$x = \sum_{i \in I} u_i \qquad (u \in E_i), \quad \text{where } \sum_{i \in I} \|u_i\|_{E_i} < \infty. \tag{3.16}$$

It follows from the last condition that there are only countably many summands in $\Sigma_{i \in I} u_i$ different from zero. Moreover, since

$$\sum_{i \in I} \|u_i\|_W \leqslant \sum_{i \in I} \|u_i\|_{E_i} < \infty, \tag{3.17}$$

the series $\Sigma_{i \in I} u_i$ converges absolutely in $W$, and hence converges also in $W$. We now introduce the norm

$$\|x\|_{\Cup E_i} = \inf \sum_{i \in I} \|u_i\|_{E_i}, \tag{3.18}$$

where the infimum is taken over all possible representations of $x$ in the form (3.16). We leave the verification of the properties of a norm to the reader.

Inequality (3.17) implies that

$$\|x\|_W \leqslant \|x\|_{\Cup E_i},$$

and therefore the space $\uplus E_i$ is imbedded in $W$, and consequently also in $\mathfrak{A}$. We prove the completeness of this space. Let $x_0, x_1, \ldots$ be a Cauchy sequence in the norm (3.18). Without loss of generality we assume that this sequence is such that

$$x_0 = \theta, \qquad \sum_{k=1}^{\infty} \|x_k - x_{k-1}\|_{\uplus E_i} < \infty.$$

By the definition of the norm (3.18) there exist representations $x_k - x_{k-1} = \sum_i u_{ik}$ such that

$$\sum_i \|u_{ik}\|_{E_i} \leqslant \|x_k - x_{k-1}\|_{\uplus n E_i} + \frac{1}{2^k}.$$

Then we have

$$\sum_{i \in I} \sum_{k=1}^{\infty} \|u_{ik}\|_{E_i} < \infty. \tag{3.19}$$

This inequality implies in particular that the series $v_i = \sum_k u_{ik}$ converge absolutely in the spaces $E_i$, and $v_i \in E_i$. From (3.19) we obtain that

$$\sum_{i \in I} \|v_i\|_{E_i} < \infty,$$

and so the element $x = \sum_{i \in I} v_i \in \uplus E_i$ is defined. Next,

$$x - x_k = x - \sum_{l=1}^{k} (x_l - x_{l-1}) = \sum_{i \in I} \left\{ v_i - \sum_{l=1}^{k} u_{il} \right\},$$

and therefore

$$\begin{aligned} \|x - x_k\|_{\uplus E_i} &\leqslant \sum_{i \in I} \left\| v_i - \sum_{l=1}^{k} u_{il} \right\|_{E_i} = \sum_{i \in I} \left\| \sum_{l=k+1}^{\infty} u_{il} \right\|_{E_i} \\ &\leqslant \sum_{l=k+1}^{\infty} \sum_{i \in I} \|u_{il}\|_{E_i} \to 0 \quad \text{as } k \to \infty \end{aligned}$$

by virtue of the convergence of (3.19). Thus, the space $\uplus E_i$ is complete.

If now the Banach space $G$ is such that $E_i \overset{1}{\subset} G$, then as $W$ we can take $G$, and, as above, show that $\uplus E_i \overset{1}{\subset} G$.

## §4. Interpolation spaces and interpolation triples

**1.** ***Interpolation triples of Banach spaces.*** Let $A$, $B$ and $C$, $D$ be two Banach couples.

DEFINITION 4.1. A linear mapping $T$ acting from the space $A + B$ to $C + D$ is called a *bounded operator from the couple $A$, $B$ to the couple $C$, $D$* if the restriction of $T$ to the space $A$ ($B$) is a bounded operator from $A$ ($B$) to $C$ ($D$).

We denote by $L(AB, CD)$ the linear space of all bounded operators from the couple $A$, $B$ to the couple $C$, $D$. This is a Banach space in the norm

$$\|T\|_{L(AB,CD)} = \max(\|T\|_{A\to C}, \|T\|_{B\to D}). \tag{4.1}$$

Indeed, if the operators $T_n$ form a Cauchy sequence in $L(AB, CD)$, then their restrictions to $A$ and $C$ converge in $L(A, C)$ and $L(B, D)$ to operators $T'$ and $T''$, respectively, which obviously coincide on $A \cap B$. Then the sequence $T_n$ converges in $L(AB, CD)$ to a uniquely defined operator $T$ acting from $A + B$ to $C + D$ according to the formula $Tx = T'u + T''v$ ($x = u + v$, $u \in A$, $v \in B$).

LEMMA 4.1. *A bounded operator $T$ from the couple $A$, $B$ to the couple $C$, $D$ generates a bounded operator from $A + B$ to $C + D$ and from $A \cap B$ to $C \cap D$, and*

$$\|T\|_{A+B\to C+D} \leqslant \|T\|_{L(AB,CD)}, \tag{4.2}$$

$$\|T\|_{A\cap B\to C\cap D} \leqslant \|T\|_{L(AB,CD)}. \tag{4.3}$$

PROOF. For $x \in A + B$, by the definition of the norm in the sum of spaces of a Banach couple we have

$$\|Tx\|_{C+D} \leqslant \inf_{x=u+v}(\|Tu\|_C + \|Tv\|_D)$$

$$\leqslant \inf_{x=u+v}(\|T\|_{A\to C}\, \|u\|_A + \|T\|_{B\to D}\, \|v\|_B) \leqslant \|T\|_{L(AB,CD)}\, \|x\|_{A+B},$$

which implies (4.2). If $x \in A \cap B$, then

$$\|Tx\|_{C\cap D} = \max(\|Tx\|_C, \|Tx\|_D)$$

$$\leqslant \max(\|T\|_{A\to C}\, \|x\|_A, \|T\|_{B\to D}\, \|x\|_B) \leqslant \|T\|_{L(AB,CD)}\, \|x\|_{A\cap B}.$$

We establish an auxiliary assertion.

LEMMA 4.2. *Let $G$ and $H$ be Banach spaces, and $E$ and $F$ Banach spaces imbedded in $G$ and $H$, respectively. If a bounded linear operator $T$ from $G$ to $H$ maps $E$ into $F$, then the restriction of $T$ to $E$ is a bounded operator from $E$ to $F$.*

PROOF. It is sufficient to show that the restriction of $T$ to $E$ is a closed operator. Let $x_n \to x$ in $E$ and $Tx_n \to y$ in $F$. Then $x_n \to x$ in $G$ and $Tx_n \to y$ in $H$. Consequently $y = Tx$; the operator $T$ is closed, and thus bounded.

COROLLARY. *If $A$, $B$ and $C$, $D$ are two Banach couples, and a bounded linear operator from $A + B$ to $C + D$ maps $A$ and $B$ to $C$ and $D$ respectively, then this operator is bounded from the couple $A$, $B$ to the couple $C$, $D$.*

DEFINITION 4.2. Let $A$, $B$ and $C$, $D$ be two Banach couples, and $E$ and $F$ spaces intermediate between $A$ and $B$ and $C$ and $D$, respectively. The triple $(A, B, E)$ is called an *interpolation triple* relative to $(C, D, F)$ if every bounded operator from $A$, $B$ to $C$, $D$ maps $E$ to $F$.

If $A$ coincides with $C$, $B$ with $D$, and $E$ with $F$, then $E$ is called an *interpolation space between the spaces of the Banach couple $A$, $B$.*

REMARK. From Lemmas 4.1 and 4.2 it follows immediately that in the preceding conditions every operator $T \in L(AB, CD)$ is a bounded operator from $E$ to $F$.

LEMMA 4.3. *If a triple $(A, B, E)$ is an interpolation triple relative to $(C, D, F)$, then there exists a constant $c > 0$ (interpolation constant) such that*

$$\|T\|_{E\to F} \leqslant c\|T\|_{L(AB,CD)} \tag{4.4}$$

*for any $T \in L(AB, CD)$.*

PROOF. To every operator $T \in L(AB, CD)$ we assign its restriction to the space $E$: $\Phi(T) = T|_E$. As we proved above, $\Phi(T)$ is a bounded operator from $E$ to $F$. Consequently a linear mapping $\Phi$ of the Banach space $L(AB, CD)$ to the space $L(E, F)$ is defined. This mapping is closed. Indeed, let $T_n \to T$ in $L(AB, CD)$ and $\Phi(T_n) \to S$ in $L(E, F)$. From the first relation and Lemma 4.1 it follows that $T_n x \to Tx$ in $C + D$ for any $x \in A + B$, and in particular for $x \in E$. From the second we obtain that $T_n x \to Sx$ in $F$ for any $x \in E$. Consequently, in view of the imbedding $F \subset C + D$ we have $Tx = Sx$ for $x \in E$, i.e. $\Phi(T) = T|_E = S$. Since $\Phi$ is closed, it is bounded, which is equivalent to (4.4). The lemma is proved.

If $E$ is an interpolation space between $A$ and $B$, then (4.4) becomes the following inequality: $\|Tx\|_E \leqslant c\|T\|_{L(AB,AB)}\,\|x\|_E$. In this case $E$ can be equipped with a new, equivalent, norm

$$\|x\|_E^* = \frac{1}{c} \sup_{T\in L(AB,AB)} \frac{\|Tx\|_E}{\|T\|_{L(AB,AB)}}.$$

For this norm we have the inequality $\|T\|^*_{E\to E} \leqslant \|T\|_{L(AB,AB)}$, i.e. the interpolation constant is equal to 1.

We give an example of an intermediate space which is not an interpolation space. Let $A = l_2$ be the Hilbert space of sequences $x = (\xi_1, \xi_2, \dots)$ with the norm $\|x\|_{l_2} = (\Sigma_1^\infty|\xi_i|^2)^{1/2}$, and $B = l_2^\rho$ the Hilbert space consisting of all sequences $x$ for which

$$\|x\|_{l_2^\rho} = \left\{\sum_1^\infty i\big(|\xi_{2i-1}|^2 + |\xi_{2i}|^2\big)\right\}^{1/2} < \infty.$$

We consider the intermediate Hilbert space $E$ whose elements are all sequences satisfying the condition

$$\|x\|_E = \left\{ \sum_1^\infty \left( i|\xi_{2i-1}|^2 + |\xi_{2i}|^2 \right) \right\}^{1/2} < \infty.$$

It is obvious that $B \overset{1}{\subset} E \overset{1}{\subset} A$.

Now we define the operators $T_n$ by the equality

$$T_n x = (\xi_1, \ldots, \xi_{2n-2}, \xi_{2n}, \xi_{2n-1}, \xi_{2n+2}, \ldots).$$

It is obvious that these operators act in the spaces $A$ and $B$, and have norm 1 there, i.e. $\|T_n\|_{L(AB,AB)} = 1$. Calculating the effect of $T_n$ on $x_n$ for which $\xi_i = 0$ for $i \neq 2n$ and $\xi_{2n} = 1$, we see that $\|T_n\|_{E\to E} \geqslant \sqrt{n}$. If $E$ were an interpolation space, this would contradict Lemma 4.3.

It is useful to have in mind the following circumstance.

LEMMA 4.4. *Let a triple $(A, B, E)$ be an interpolation triple relative to $(C, D, F)$. If $\tilde{E}$ and $\tilde{F}$ denote the completions of $E$ and $F$ relative to $A + B$ and $C + D$, respectively, then the triple $(A, B, \tilde{E})$ is an interpolation triple relative to $(C, D, \tilde{F})$.*

PROOF. If $x \in \tilde{E}$, then there exists a sequence $x_n \to x$ in $A + B$ such that $\|x_n\|_E = \|x\|_{\tilde{E}}$. Then for any operator $T \in L(AB, CD)$ we have $Tx_n \to Tx$ in $C + D$, and $\|Tx_n\|_F \leqslant \|T\|_{E\to F}\|x_n\|_E = \|T\|_{E\to F}\|x\|_{\tilde{E}}$. From this it follows that $Tx \in \tilde{F}$ and $\|Tx\|_{\tilde{F}} \leqslant \|T\|_{E\to F}\|x\|_{\tilde{E}}$.

We have proved the lemma and the inequality $\|T\|_{\tilde{E}\to\tilde{F}} \leqslant \|T\|_{E\to F}$.

**2. *Good interpolation triples.* The type of an interpolation triple.***

DEFINITION 4.3. Let $(A, B, E)$ be an interpolation triple relative to $(C, D, F)$. We shall say that the triple $(A, B, E)$ is a *good interpolation triple relative to $(C, D, F)$ as applied to $A$ and $C$* if for all operators $T \in L(AB, CD)$

$$\|T\|_{E\to F} \leqslant \varphi(\|T\|_{A\to C}, \|T\|_{B\to D})$$

and $\varphi(\lambda, \mu) \to 0$ as $\lambda \to 0$ uniformly in $\mu$ on every finite interval.

THEOREM 4.1. *Let $(A, B, E)$ be a good interpolation triple relative to the triple $(C, D, F)$ as applied to $A$ and $C$, and let there exist a sequence of finite-dimensional operators $P_n$ acting in the spaces $C$ and $D$, strongly converging to $I$ in $C$ and uniformly bounded in $D$. If the operator $T \in L(AB, CD)$ is bounded as an operator from $B$ to $D$ and compact as an operator from $A$ to $C$, then it is compact as an operator from $E$ to $F$.*

---

* *Editor's note.* The original Russian term is вполне интерполяционные тройки.

PROOF. Let $S_1$ be the unit ball of $A$. The closure of the set $TS_1$ is compact in $C$. The operators $P_n$ uniformly converge to the identity operator on compact sets. Therefore

$$\|(P_n - I)T\|_{A\to C} \to 0 \quad \text{as } n \to \infty.$$

The norms $\|(P_n - I)T\|_{B\to D}$ are uniformly bounded, by the hypotheses of the theorem. Then

$$\|(P_n - I)T\|_{E\to F} \leqslant \varphi(\|(P_n - I)T\|_{A\to C}, \|(P_n - I)T\|_{B\to D}) \to 0$$

as $n \to \infty$. Thus, $T$ acting from $E$ to $F$ is the uniform limit of finite-dimensional operators, and hence is compact.

DEFINITION 4.4. The triple $(A, B, E)$ is said to be *of interpolation type* $\alpha$ $(0 \leqslant \alpha \leqslant 1)$ relative to the triple $(C, D, F)$ if it is an interpolation triple and the inequality

$$\|T\|_{E\to F} \leqslant c\|T\|_{A\to C}^{1-\alpha} \|T\|_{B\to D}^{\alpha} \tag{4.5}$$

holds.

In the case where $A$ coincides with $C$, $B$ with $D$, and $E$ with $F$, $E$ is said to be *an interpolation space of type* $\alpha$ *between* $A$ *and* $B$.

If the constant $c$ in (4.5) is equal to one, then $(A, B, E)$ is said to be a *normalized interpolation triple of type* $\alpha$ relative to $(C, D, F)$.

It can happen that an interpolation space between $A$ and $B$ is not of type $\alpha$ for any $\alpha \in [0, 1]$.

If $(A, B, E)$ is an interpolation triple of type $\alpha$, with $\alpha < 1$, relative to $(C, D, F)$, then it is a good interpolation triple as applied to $A$ and $C$, and Theorem 4.1 holds.

Theorems which establish that one triple of Banach spaces is an interpolation triple relative to another are called *interpolation theorems*. Historically, the first interpolation theorem was obtained by M. Riesz and G. O. Thorin, and the whole theory of interpolation for linear operators began to develop in the direction of generalizations of this theorem. Here we give a general formulation of this first interpolation theorem.

THE RIESZ-THORIN THEOREM. *Let* $(\Omega_1, \Sigma_1, \mu_1)$, *and* $(\Omega_2, \Sigma_2, \mu_2)$ *be two measure spaces, and let* $L_p(\Omega_i)$ $(i = 1, 2;\ p \geqslant 1)$ *be Banach spaces of complex-valued functions, pth power summable with respect to* $\mu_i$. *Then* $(L_{p_0}(\Omega_1), L_{p_1}(\Omega_1), L_p(\Omega_1))$ *is a normalized interpolation triple of Banach spaces of type* $\alpha$ *relative to the triple* $(L_{q_0}(\Omega_2), L_{q_1}(\Omega_2), L_q(\Omega_2))$, *if*

$$1/p = (1-\alpha)/p_0 + \alpha/p_1, \quad 1/q = (1-\alpha)/q_0 + \alpha/q_1 \quad (0 \leqslant \alpha \leqslant 1). \tag{4.6}$$

The proof of this theorem is based on the well-known three lines theorem in the theory of analytic functions and can be found in [50], Vol. II, Chapter XII, §1.

We recall that, for the triple, to be a normalized interpolation triple of type $\alpha$ means that for every linear operator $A$ acting boundedly from $L_{p_i}(\Omega_1)$ to $L_{q_i}(\Omega_2)$ and having norms $M_i$, $i = 0, 1$, the inequality

$$\|Tx\|_{L_q(\Omega_2)} \leqslant M_0^{1-\alpha} M_1^{\alpha} \|x\|_{L_p(\Omega_1)} \tag{4.7}$$

holds under conditions (4.6).

3. ***Optimal interpolation spaces.*** If $E$ is an intermediate space for the Banach couple $A$, $B$, and $C$, $D$ is another Banach couple, then there arises the question of describing the spaces $F$ intermediate between $C$ and $D$ such that $(A, B, E)$ is an interpolation triple relative to the triple $(C, D, F)$. It is clear that if $F$ has this property, then every space $F_1$ intermediate between $C$ and $D$ in which $F$ is imbedded has the same property. In this connection, we may wish to find a space $F$ which is minimal in a certain sense. It is clear that every space $F$ must contain all elements of $C + D$ of the form $Tx$, where $x$ runs over $E$ and $T$ over the set $L(AB, CD)$. Besides, $F$ must contain the linear span of elements of the form $Tx$. If we could introduce a Banach space structure in this span so that it became intermediate between $C$ and $D$, then this Banach space would solve our problem. Below we shall accomplish, in a more complicated way, the program outlined here.

The unit ball in $L(AB, CD)$ will be denoted by $\pi_1(AB, CD)$. Let $T$ belong to $\pi_1(AB, CD)$. Consider the kernel $N(T)$ of $T$ and the range $T(E)$. Since $T$ is continuous as an operator from $E$ to $C + D$, the kernel $N(T)$ is closed in $E$. The operator $T$ induces a canonical one-to-one correspondence between the quotient space $E/N(T)$ and the range $T(E)$. This allows us to carry the Banach space structure of $E/N(T)$ over to the linear space $T(E)$. We denote by $F_T(E)$ the resulting Banach space. According to the above, the formula for the norm of $F_T(E)$ has the form

$$\|y\|_{F_T(E)} = \inf\|x\|_E, \tag{4.8}$$

where the infimum is taken over all $x \in E$ for which $y = Tx$.

The Banach space $F_T(E)$ is imbedded in $C + D$. Indeed, according to (4.2) we have

$$\|y\|_{C+D} = \|Tx\|_{C+D} \leqslant \|T\|_{L(AB,CD)} \|x\|_{A+B} \leqslant \|x\|_{A+B} \leqslant (1/m(E))\|x\|_E,$$

where $1/m(E)$ denotes the imbedding constant of $E$ in $A + B$.

By (4.8) this implies that

$$m(E)\|y\|_{C+D} \leqslant \|y\|_{F_T(E)}.$$

In what follows, we denote by $kG$, where $G$ is a Banach space, the Banach space with norm $\|x\|_{kG} = k\|x\|_G$ ($k > 0$, $x \in G$).

With this notation, the preceding inequality can be written as

$$F_T(E) \overset{1}{\subset} m(E)(C + D). \tag{4.9}$$

We have obtained a family of Banach spaces $F_T(E)$ imbedded in the Banach space $m(E)(C + D)$. According to §3.4, we may construct the sum of spaces

$$F(E) = \biguplus_{T \in \pi_1(AB,CD)} F_T(E).$$

It follows from (4.9) that $F(E) \overset{1}{\subset} m(E)(C + D)$.

We now show that $C \cap D$ is imbedded in $F(E)$. For this, from any element $y \in C \cap D$ we construct a one-dimensional operator $T_0 x = \frac{1}{\alpha}\langle x, f\rangle y$, where $f$ is a functional from the space $(A + B)'$. According to what was said in §3.2, the restrictions of $f$ to $A$ and $B$ are linear functionals on these spaces, and $\|f\|_{(A+B)'} = \max\{\|f\|_{A'}, \|f\|_{B'}\}$. Then we obtain

$$\|T_0\|_{A \to C} = \sup \frac{\|T_0 x\|_C}{\|x\|_A} = \frac{1}{\alpha}\|y\|_C \sup \frac{|f(x)|}{\|x\|_A} = \frac{1}{\alpha}\|f\|_{A'}\|y\|_C,$$

and, similarly, $\|T_0\|_{B \to D} = \frac{1}{\alpha}\|f\|_{B'}\|y\|_D$. Therefore

$$\|T_0\|_{L(AB,CD)} = \frac{1}{\alpha}\max\{\|f\|_{A'}\|y\|_C, \|f\|_{B'}\|y\|_D\}.$$

We choose $\alpha = \max\{\|f\|_{A'}\|y\|_C, \|f\|_{B'}\|y\|_D\}$. Then $\|T_0\|_{L(AB,CD)} = 1$ and $T_0 \in \pi_1(AB, CD)$. Hence $\frac{1}{\alpha}\langle x, f\rangle y \in F_{T_0}(E)$ for any $x \in A + B$, and therefore $y \in F_{T_0}(E) \subset F(E)$. Thus, $C \cap D$ is contained in $F(E)$. To obtain the imbedding inequality for any $\varepsilon$ we choose an element $x_\varepsilon$ so that

$$\|x_\varepsilon\|_{A+B} \geqslant (1/(m(E) + \varepsilon))\|x_\varepsilon\|_E$$

(such a choice is possible since $1/m(E)$ is the imbedding constant of $E$ in $A + B$) and a functional $f_\varepsilon(x)$ so that $\|f_\varepsilon\|_{(A+B)'} = 1$ and $\langle x_\varepsilon, f_\varepsilon\rangle = \|x_\varepsilon\|_{A+B}$. Then $T_0(\alpha x_\varepsilon/\|x_\varepsilon\|_{A+B}) = y$, and therefore

$$\|y\|_{F_{T_0}(E)} \leqslant \|\alpha x_\varepsilon/\|x_\varepsilon\|_{A+B}\|_E \leqslant \alpha(m(E) + \varepsilon). \tag{4.10}$$

Next,

$$\alpha \leqslant \max(\|y\|_C, \|y\|_D)\max(\|f\|_{A'}, \|f\|_{B'}).$$

According to what was said in §3.2, the last factor coincides with $\|f\|_{(A+B)'}$, and is equal to one. Therefore $\alpha \leqslant \|y\|_{C \cap D}$, and (4.10) gives

$$\|y\|_{F(E)} \leqslant \|y\|_{F_{T_0}(E)} \leqslant (m(E) + \varepsilon)\|y\|_{C \cap D}.$$

By the arbitrariness of $\varepsilon$ this implies the following imbedding:

$$m(E)(C \cap D) \overset{1}{\subset} F(E).$$

We have arrived at the following assertion.

THEOREM 4.2. *Let $E$ be an intermediate space for the Banach couple $A$, $B$, and let $C$, $D$ be another Banach couple. There exists a space $F(E)$ intermediate for the couple $C$, $D$ and having the property that $(A, B, E)$ is an interpolation triple relative to $(C, D, F)$ if and only if $F(E) \subset F$.*

PROOF. The space $F(E)$ contains all spaces $F_T(E)$, and consequently all elements of the form $Tx$ $(x \in E, T \in \pi_1(AB, CD))$. Hence $(A, B, E)$ is an interpolation triple relative to $(C, D, F(E))$. From this it follows immediately that $(A, B, E)$ is an interpolation triple relative to $(C, D, F)$ if $F \supset F(E)$.

By construction, $F(E)$ consists of elements of the form

$$z = \Sigma y_i, \qquad \sum \|y_i\|_{F_{T_i}(E)} < \infty. \tag{4.11}$$

By the definition of the norm in $F_{T_i}(E)$ there exist elements $x_i$ such that $y_i = T_i x_i$ and $\|x_i\|_E \leqslant \|y_i\|_{F_{T_i}(E)} + 1/2^i$. Then $\Sigma\|x_i\|_E < \infty$. If now $(A, B, E)$ is an interpolation triple relative to $(C, D, F)$, then by (4.11) we have

$$\sum \|y_i\|_F = \sum \|T_i x_i\|_F \leqslant \sum \|T_i\|_{E \to F} \|x_i\|_E \leqslant c \sum \|x_i\|_E < \infty,$$

and so $z \in F$. Thus, $F(E) \subset F$.

REMARK. *$F(E)$ is an interpolation space relative to the couple $C$, $D$.*

Indeed, let $S \in \pi_1(CD, CD)$ and $T \in \pi_1(AB, CD)$. We have $\|ST\|_{L(AB,CD)} \leqslant 1$. If now $z \in F(E)$, then (4.11) holds, and since $\Sigma y_i$ converges in $C + D$, we have $Sz = \Sigma Sy_i$.

We calculate

$$\|Sy_i\|_{F_{ST_i}(E)} = \inf_{Sy_i = ST_i x} \|x\|_E \leqslant \inf_{y_i = T_i x} \|x\|_E = \|y_i\|_{F_{T_i}(E)}. \tag{4.12}$$

It then follows from (4.11) that $\Sigma\|Sy_i\|_{F_{ST_i}(E)}$ converges, and consequently $Sz \in F(E)$.

Thus, every operator from $L(CD, CD)$ maps $F(E)$ into $F(E)$, and our assertion is proved.

We note that if the space $E_1$ intermediate between $A$ and $B$ is contained in $E$, then $(A, B, E_1)$ is an interpolation triple relative to $(C, D, F(E))$, and therefore by Theorem 4.2 we have $F(E_1) \subset F(E)$.

Now let $A$, $B$ be a Banach couple, and $F$ an intermediate space in the Banach couple $C$, $D$. We can attempt to construct a largest intermediate space between $A$ and $B$ that together with $A$ and $B$ forms an interpolation triple relative to $(C, D, F)$. Let $T \in \pi_1(AB, CD)$. Consider the preimage $T^{-1}(F)$ of $F$ under the mapping of $A + B$ into $C + D$ by means of the operator $T$. On this preimage we introduce the norm

$$\|x\|_{E_{T^{-1}}} = \max\{m'(F)\|x\|_{A+B}, \|Tx\|_F\},$$

where $m'(F)$ is the imbedding constant of $C \cap D$ in $F$. With this norm $E_{T^{-1}}$ is a Banach space. Indeed, if $\{x_n\}$ is a Cauchy sequence in $E_{T^{-1}}$, then it is also a Cauchy sequence in $A + B$, and consequently it converges in $A + B$ to an element $x$, and $\{Tx_n\}$ is Cauchy in $F$ and hence converges in $F$ and in $C + D$ to an element $y$. Since $\|T\|_{A+B \to C+D} \leqslant 1$, we have $y = Tx$, and $\{x_n\}$ converges to $x$ in $E_{T^{-1}}$.

The space $E_{T^{-1}}$ is intermediate between $A$ and $B$. It is obvious that $E_{T^{-1}} \overset{1}{\subset} m'(F)(A + B)$. Let $x \in A \cap B$. If $\|x\|_{E_{T^{-1}}} = m'(F)\|x\|_{A+B}\|$, then $\|x\|_{E_{T^{-1}}} \leqslant m'(F)\|x\|_{A \cap B}$. If, on the other hand, $\|x\|_{E_{T^{-1}}} = \|Tx\|_F$, then by (4.3) we have

$$\|x\|_{E_{T^{-1}}} \leqslant m'(F)\|Tx\|_{C \cap D} \leqslant m'(F)\|x\|_{A \cap B}. \tag{4.13}$$

Thus, $m'(F)A \cap B \overset{1}{\subset} E_{T^{-1}} \overset{1}{\subset} m'(F)(A + B)$. Now we denote by $E(F)$ the intersection of the family of the spaces $E_{T^{-1}}$ for all $T \in \pi_1(AB, CD)$. We obviously have the imbeddings

$$m'(F)A \cap B \overset{1}{\subset} E(F) \overset{1}{\subset} m'(F)(A + B).$$

We recall that the norm in $E(F)$ is defined by

$$\|x\|_{E(F)} = \sup_{T \in \pi_1(AB,CD)} \max\{m'(F)\|x\|_{A+B}, \|Tx\|_F\}. \tag{4.14}$$

This formula can be simplified. Let $x_0 \in E(F)$. We construct a linear functional $f \in (A + B)'$ such that $\|f\|_{(A+B)'} = 1$ and $\langle x_0, f\rangle = \|x_0\|_{A+B}$. Using the fact that $m'(F)$ is the imbedding constant of $C \cap D$ in $F$, we choose $y \in C \cap D$ so that $\|y\|_F > (m'(F) - \varepsilon)\|y\|_{C \cap D}$. We consider the operator $T_0x = \frac{1}{\alpha}\langle x_0, f\rangle y$. If we choose $\alpha = \max\{\|f\|_{A'}\|y\|_C, \|f\|_{B'}\|y\|_D\}$, then $T_0 \in \pi_1(AB, CD)$ (see p. 24). Here we have $\alpha \leqslant \|y\|_{C \cap D}$, and

$$\begin{aligned}\|T_0x_0\|_F &= \tfrac{1}{\alpha}\|x_0\|_{A+B}\|y\|_F \\ &\geqslant \tfrac{1}{\alpha}\|x_0\|_{A+B}(m'(F) - \varepsilon)\|y\|_{C \cap D} \geqslant (m'(F) - \varepsilon)\|x_0\|_{A+B}.\end{aligned}$$

From this it follows that

$$\sup_{T \in \pi_1(AB,CD)} \|Tx_0\|_F \geqslant m'(F)\|x_0\|_{A+B},$$

and therefore (4.14) can be written in the form

$$\|x\|_{E(F)} = \sup_{T \in \pi_1(AB,CD)} \|Tx\|_F. \tag{4.15}$$

In passing we have proved that $E(F)$ consists of all elements $x \in A + B$ for which

$$\sup_{T \in \pi_1(AB,CD)} \|Tx\|_F < \infty.$$

THEOREM 4.3. *Let $A$, $B$ be a Banach couple, and $F$ an intermediate space for the Banach couple $C$, $D$. There exists a space $E(F)$ intermediate between $A$ and $B$ having the property that $(A, B, E)$ is an interpolation triple relative to $(C, D, F)$ if and only if $E \subset E(F)$.*

PROOF. It is obvious that $(A, B, E(F))$ is itself an interpolation triple relative to $(C, D, F)$, since for $x \in E(F)$ we have $Tx \in F$ for any $T \in L(AB, CD)$ by the definition of $E(F)$. This implies that for $E \subset E(F)$ the triple $(A, B, E)$ is an interpolation triple relative to $(C, D, F)$. Conversely, if $(A, B, E)$ is an interpolation triple relative to $(C, D, F)$, then by (4.15) for $x \in E$ we have

$$\|x\|_{E(F)} = \sup_{T \in \pi_1(AB,CD)} \|Tx\|_F$$
$$\leqslant \sup_{T \in \pi_1(AB,CD)} \{\|T\|_{E \to F} \|x\|_E\} \leqslant c\|x\|_E,$$

where $c$ is the interpolation constant for $E$ and $F$, i.e., $E \subset E(F)$.

REMARK. *$E(F)$ is an interpolation space between $A$ and $B$.*

Indeed, if $S \in L(AB, AB)$, then for $T \in \pi_1(AB, CD)$ we have $TS \in L(AB, CD)$, and $\|TS\|_{L(AB,CD)} \leqslant \|S\|_{L(AB,AB)}$. Thus for $x \in E(F)$ we have $TSx \in F$ and

$$\|Sx\|_{E(F)} = \sup_{T \in \pi_1(AB,CD)} \|TSx\|_F$$
$$= \|S\|_{L(AB,AB)} \sup_{T \in \pi_1(AB,CD)} \left\| \frac{1}{\|S\|_{L(AB,AB)}} TSx \right\|_F$$
$$\leqslant \|S\|_{L(AB,AB)} \sup_{T \in \pi_1(AB,CD)} \|Tx\|_F = \|S\|_{L(AB,AB)} \|x\|_{E(F)},$$

which proves our assertion.

We note that if the space $F_1$ intermediate between $C$ and $D$ contains $F$, then $(A, B, E(F))$ is an interpolation triple relative to $(C, D, F_1)$, and so by Theorem 4.3 we have

$$E(F) \subset E(F_1). \tag{4.16}$$

DEFINITION 4.5. Let $(A, B, E)$ be an interpolation triple of spaces relative to $(C, D, F)$. It is said to be an *optimal interpolation triple* if the conditions that $(A, B, \tilde{E})$ is an interpolation triple relative to $(C, D, \tilde{F})$ and $\tilde{E} \supset E$ and $\tilde{F} \subset F$ imply that $\tilde{E} = E$ and $\tilde{F} = F$.

In other words, $(A, B, E)$ is an optimal interpolation triple relative to $(C, D, F)$ if $E$ cannot be enlarged and $F$ cannot be decreased with preservation of the interpolation property.

THEOREM 4.4. *In order that $(A, B, E)$ be an optimal interpolation triple relative to $(C, D, F)$ it is necessary and sufficient that $F = F(E)$ and $E = E(F)$.*

PROOF. If $(A, B, E)$ is an optimal interpolation triple relative to $(C, D, F)$, then by Theorem 4.2 we have $F(E) \subset F$. Since $(A, B, E)$ is an interpolation triple relative to $(C, D, F(E))$, from the definition of the optimal interpolation property it follows that $F(E) = F$. Similarly, Theorem 4.3 implies that $E = E(F)$.

Now let $F = F(E)$ and $E = E(F)$, and let the triples $(A, B, \tilde{E})$ and $(C, D, \tilde{F})$ be as in Definition 4.5. Then $(A, B, E)$ is an interpolation triple relative to $(C, D, \tilde{F})$. It follows then from Theorem 4.2 that $\tilde{F} \supset F(E) = F$, and so $\tilde{F} = F$. We may show similarly that $\tilde{E} = E$.

COROLLARY. *If $(A, B, E)$ is an optimal interpolation triple relative to $(C, D, F)$, then $E$ and $F$ are interpolation spaces between $A$, $B$ and $C$, $D$, respectively.*

This follows immediately from Theorem 4.4 and the remarks after Theorems 4.2 and 4.3.

THEOREM 4.5. *Let $E$ and $F$ be intermediate between the spaces of the Banach couples $A$, $B$ and $C$, $D$, respectively. Then the following assertions are true*:

1. *$(A, B, E(F(E)))$ is the optimal interpolation triple relative to $(C, D, F(E))$.*
2. *$(A, B, E(F))$ is the optimal interpolation triple relative to $(C, D, F(E(F)))$.*
3. *If $(A, B, \bar{E})$ is the optimal interpolation triple relative to $(C, D, \bar{F})$, $\bar{E} \supset E$ and $\bar{F} \supset F$, then*

$$F(E) \subset \bar{F} \subset F(E(F)), \qquad E(F(E)) \subset \bar{E} \subset E(F).$$

PROOF. If $(A, B, \tilde{E})$ is an interpolation triple relative to $(C, D, \tilde{F})$, and if $\tilde{E} \supset E(F(E))$ and $\tilde{F} \subset F(E)$, then $(A, B, \tilde{E})$ is an interpolation triple relative to $(C, D, F(E))$, and by Theorem 4.3 we have $\tilde{E} \subset E(F(E))$, and hence $\tilde{E} = E(F(E))$. Since $(A, B, E)$ is an interpolation triple relative to $(C, D, F(E))$, by the same theorem we have $E \subset E(F(E)) = \tilde{E}$, and therefore $(A, B, E)$ is an interpolation triple relative to $(C, D, \tilde{F})$, from which by Theorem 4.2 we obtain that $\tilde{F} \supset F(E)$, and hence $\tilde{F} = F(E)$. We have proved the optimal interpolation property of $(A, B, E(F(E)))$ relative to

$(C, D, F(E))$. The analogous verification works for the triples $(A, B, E(F))$ and $(C, D, F(E(F)))$.

Now let the triples $(A, B, \bar{E})$ and $(C, D, \bar{F})$ have the properties indicated in assertion of of the theorem. Then $(A, B, E)$ and $(A, B, \bar{E})$ are interpolation triples relative to $(C, D, \bar{F})$ and $(C, D, F)$, respectively, and therefore by Theorems 4.2 and 4.3 the imbeddings $F(E) \subset \bar{F}$ and $\bar{E} \subset E(F)$ are true. By (4.16) from the first of these imbeddings we obtain that $E(F(E)) \subset E(\bar{F})$, but by Theorem 4.4 from the condition of optimality we have $E(\bar{F}) = \bar{E}$, and therefore $E(F(E)) \subset \bar{E}$. Similarly, $\bar{E} \subset E(F)$ implies that $\bar{F} = F(\bar{E}) \subset F(E(F))$.

THEOREM 4.6. *If $(A, B, E)$ is the optimal interpolation triple relative to $(C, D, F)$, then $(C, D, F)$ is an interpolation triple for $(A, B, E)$.*

PROOF. We know from the remark to Theorem 4.2 that $F$ is an interpolation space relative to $C$ and $D$ with some constant $c_F$. By Theorem 4.4 we have $E = E(F)$. If $x \in F$ and $T_1 \in L(CD, AB)$, then by (4.15) we have

$$\begin{aligned} \|T_1x\|_E = \|T_1x\|_{E(F)} &= \sup_{T \in \pi_1(AB,CD)} \|TT_1x\|_F \\ &\leqslant \sup_{T \in \pi_1(AB,CD)} c_F \|TT_1\|_{L(CD,CD)} \|x\|_F \\ &\leqslant c_F \|T_1\|_{L(CD,AB)} \|x\|_F. \end{aligned}$$

**4.** ***Complemented subcouple of a Banach couple.*** Let $A$ and $B$ be a Banach couple of spaces imbedded in a separated topological linear space $\mathfrak{A}$, and $M$ and $N$ subspaces of $A$ and $B$, respectively. Then the spaces $M$ and $N$ are imbedded in $\mathfrak{A}$, and consequently form a Banach couple, which is called a *subcouple* of $A$, $B$.

DEFINITION 4.6. The subcouple $M$, $N$ is said to be *complemented* if there exists a bounded projection operator $P$ in $A + B$ projecting $A$ onto $M$ and $B$ onto $N$.

The operator $P$ is of course bounded in $A$ and $B$. The space $A \cap B$ is projected by it onto $M \cap N$.

Let $E$ be an intermediate space between $A$ and $B$. The image $PE$ of $E$ contains $M \cap N$ and is contained in $M + N$. Reasoning in the same way as in subsection 3, in this image we can introduce a Banach space structure by making it isometric to the quotient space of $E$ modulo the kernel of $P$. Using the notation introduced in subsection 3, we denote the resulting Banach space by $F_P(E)$. Then $\|y\|_{F_P(E)} = \inf_{y=Px}\|x\|_E$. The space $F_P(E)$ is intermediate between $M$ and $N$ (see Lemma 3.3).

If now $F$ is an intermediate space between $M$ and $N$, then we may consider the inverse image $P^{-1}F$ of $F$ and introduce the norm $\|x\|_{E_{P^{-1}}(F)} = \max\{\|x\|_{A+B}, \|Px\|_F\}$ in it. As we saw in subsection 3, the Banach space $E_{P^{-1}}(F)$ thus obtained is intermediate between $A$ and $B$.

Now let $E$ be an interpolation space between $A$ and $B$. Denote by $C_E$ the interpolation constant in the inequality

$$\|T\|_{E\to E} \leqslant C_E\|T\|_{L(AB,AB)}.$$

Since $P \in L(AB, AB)$, the operator $P$ is bounded in $E$ and

$$\|P\|_{E\to E} \leqslant C_E\|P\|,$$

where for brevity $\|P\|$ denotes $\|P\|_{L(AB,AB)}$. In this case the image $PE$ is a subspace in $E$ and the norm in $PE$ as a subspace of $E$ is equivalent to the norm in the space $F_P(E)$. Indeed, if $y \in PE \subset E$, then $y = Py$, and therefore $\|y\|_{F_P(E)} \leqslant \|y\|_E$. Next, for $Px = y$ ($x \in E$) we have $\|y\|_E \leqslant \|P\|_{E\to E}\|x\|_E$, and consequently

$$\|y\|_{F_P(E)} \leqslant \|y\|_E \leqslant \inf_{Px=y}(\|P\|_{E\to E}\|x\|_E) \leqslant C_E\|P\|\,\|y\|_{F_P(E)}. \quad (4.17)$$

Thus, up to isomorphism we may assume that $F_P(E)$ is a subspace of $E$.

LEMMA 4.5. *If $E$ is an interpolation space between $A$ and $B$, then $F_P(E)$ is an interpolation space between $M$ and $N$, and $(A, B, E)$ is an interpolation triple relative to $(M, N, F_P(E))$.*

PROOF. If $T \in L(MN, MN)$, then $TP \in L(AB, AB)$. In view of the interpolation property of $E$ for $y \in F_P(E) \subset E$ we have $TPy = Ty \in E$. The range of $T$ lies in $M + N$, and therefore $Ty = PTy \subset PE = F_P(E)$ and

$$\begin{aligned}\|Ty\|_{F_P(E)} &\leqslant \|Ty\|_E = \|TPy\|_E \leqslant C_E\|TP\|_{L(AB,AB)}\|y\|_E \\ &\leqslant C_E\|T\|_{L(MN,MN)}\|P\|\,\|y\|_E. \quad (4.18)\end{aligned}$$

Taking account of (4.17), we finally obtain that

$$\|Ty\|_{F_P(E)} \leqslant (C_E\|P\|)^2\|T\|_{L(MN,MN)}\|y\|_{F_P(E)}.$$

Now let $T \in L(AB, MN)$; then $T \in L(AB, AB)$, and so, if $x \in E$, then $Tx \in E$. Next, $PTx = Tx \in PE = F_P(E)$ and

$$\|Tx\|_{F_P(E)} \leqslant \|Tx\|_E \leqslant C_E\|T\|_{L(AB,MN)}\|x\|_E. \quad (4.19)$$

COROLLARY. *The space $F_P(E)$ is isomorphic to the space $F(E)$ constructed from the triples $(A, B, E)$ according to Theorem 4.2.*

Indeed, by the construction of $F(E)$ the membership relation $P \in L(AB, MN)$ implies that $F_P(E)$ is imbedded in $F(E)$. On the other hand, since $(A, B, E)$ is an interpolation triple relative to $(M, N, F_P(E))$, by Theorem 4.2 we have $F(E) \subset F_P(E)$. Thus, up to isomorphism, $F(E) = F_P(E)$.

If $F$ is an interpolation space between $M$ and $N$, then in general $E_{P^{-1}}(F)$ is not an interpolation space between $A$ and $B$. We explain this via an example. We set $A = L_1(0, 1)$ and $B = L_\infty(0, 1)$. Then $A + B = L_1(0, 1)$. As $M$ and $N$ we take the subspaces of $L_1(0, 1)$ and $L_\infty(0, 1)$ consisting of all functions vanishing for $t \in [1/2, 1]$. The couple $M, N$ is a complemented subcouple of $A, B$. Let $F$ be an interpolation space between $M$ and $N$, and let $F$ not coincide with $M + N$. The inverse image $P^{-1}F$ consists of all functions on $[0, 1]$ whose restriction to $[0, 1/2]$ belongs to $F$ and whose restriction to $[1/2, 1]$ belongs to $L_1(1/2, 1)$. Let the function $y_0(t)$ from $M + N$ not belong to $F$. We define the following function: $x_0(t) = 0$ for $0 < t \leqslant 1/2$, and $x_0(t) = y_0(1 - t)$ for $1/2 < t \leqslant 1$. It is obvious that $x_0 \in P^{-1}F$. Denote by $T$ the operator defined by the equality $Tz(t) = z(1 - t)$ $(z \in L_1(0, 1))$. This operator has norm 1 in $L_1(0, 1)$ and in $L_\infty(0, 1)$. Next, $PTx_0 = y_0 \in F$, and consequently $Tx_0 \notin P^{-1}F$. Thus, $P^{-1}F$ is not an interpolation space.

We may consider the part of the inverse image $P^{-1}F$ studied in subsection 3, namely, the intersection $E(F)$ of the inverse images $T^{-1}F$ of all operators $T$ in $\pi_1(AB, MN)$. As is shown in the remark to Theorem 4.3, this is an interpolation space between $A$ and $B$. We recall that the norm in this space can be introduced by (4.15); namely,

$$\|x\|_{E(F)} = \sup \|Tx\|_F. \tag{4.20}$$

We note that the interpolation constant for the space $E(F)$ is equal to 1 (see p. 27). If $F$ is an interpolation space between $M$ and $N$, then it is imbedded in $E(F)$. Indeed, if $T \in \pi_1(AB, MN)$, then $T \in \pi_1(MN, MN)$, and so for $x \in F$ we have

$$\|x\|_{E(F)} \leqslant C_F \|x\|_F,$$

where $C_F$ is the interpolation constant in the inequality

$$\|T\|_{F \to F} \leqslant C_F \|T\|_{L(MN,MN)}. \tag{4.21}$$

We also note that $E(F) \subset P^{-1}(F)$, and therefore $PE(F) = F$.

It will sometimes be convenient to introduce an equivalent norm in $E(F)$ by the formula

$$\|x\|^*_{E(F)} = \sup_{T \in \pi_1(AB,AB)} \|PTx\|_F. \tag{4.22}$$

If $T \in \pi_1(AB, AB)$, then $\|PT\|_{L(AB,MN)} \leqslant \|P\| \|T\|_{L(AB,AB)} \leqslant \|P\|$, and therefore $\|P\|^{-1}PT \in \pi_1(AB, MN)$. From this we obtain that

$$\|x\|_{E(F)} \geqslant \sup_{T \in \pi_1(AB,AB)} \| \, \|P\|^{-1}PTx\|_F = \|P\|^{-1}\|x\|^*_{E(F)}.$$

On the other hand, if $T \in \pi_1(AB, MN)$, then $PT = T \in \pi_1(AB, AB)$, and therefore

$$\|x\|^*_{E(F)} \geqslant \sup_{T \in \pi_1(AB,MN)} \|PTx\|_F = \|x\|_{E(F)}.$$

Thus,

$$\|x\|_{E(F)} \leqslant \|x\|^*_{E(F)} \leqslant \|P\| \, \|x\|_{E(F)}. \tag{4.23}$$

The interpolation inequality for $E(F)$ in the new norm has the form

$$\begin{aligned} \|Tx\|^*_{E(F)} &\leqslant \|P\| \, \|Tx\|_{E(F)} \leqslant \|P\| \, \|T\|_{L(AB,AB)} \|x\|_{E(F)} \\ &\leqslant \|P\| \, \|T\|_{L(AB,AB)} \|x\|^*_{E(F)}. \end{aligned} \tag{4.24}$$

Now let $E$ be an interpolation space between $A$ and $B$. By Lemma 4.5 the subspace $F = PE$ is an interpolation space between $M$ and $N$. From it we can construct the space $E(F)$. It turns out that $E$ is imbedded in $E(F)$. Indeed, if $T \in \pi_1(AB, MN)$, then by (4.19) we have $\|Tx\|_F \leqslant C_E \|x\|_E$ and hence $\|x\|_{E(F)} \leqslant C_E \|x\|_E$.

We summarize our results.

THEOREM 4.7. *Let $M$, $N$ be a complemented subcouple of the Banach couple $A$, $B$, and let $P$ be the corresponding projection operator from $A$, $B$ onto $M$, $N$. The operator $P$ projects every interpolation space between $A$ and $B$ onto a subspace $PE$ which is an interpolation space between $M$ and $N$. The space $PE$ is the smallest among the spaces $F$ for which $(A, B, E)$ is an interpolation triple relative to the triple $(M, N, F)$ ($PE$ is isomorphic to $F(E)$).*

*From every interpolation space $F$ between $M$ and $N$ a space $E(F)$ can be constructed which is an interpolation space between $A$ and $B$ and is the largest among all interpolation spaces $E$ between $A$ and $B$ for which $PE = F$.*

*$(A, B, E(F))$ is the optimal interpolation triple relative to $(M, N, F)$.*

The last assertion follows from the corollary of Lemma 4.5 and Theorem 4.4.

We specified the constants in (4.17), (4.18), (4.23), and (4.24) in order to make the following useful remark.

REMARK. If $\|P\|_{L(AB,AB)} = 1$ and the interpolation constant $C_E$ is equal to 1, then $F_P(E)$ is isometric to the subspace $PE$, the interpolation constant in the inequality of type (4.21) for the space $F = PE$ is equal to 1, the norms (4.20) and (4.22) coincide, and the interpolation constant in (4.24) is equal to 1.

We mention still another assertion.

LEMMA 4.6. *If the triples $(A, B, E)$ and $(M, N, F)$ are interpolation triples relative to each other, then $F$ is isomorphic to $PE$.*

PROOF. Since $P \in L(AB, MN)$, we have $PE \subset F$. The operator $j$ imbedding $M + N$ in $A + B$ belongs to $L(MN, AB)$, and therefore $jF \subset E$, i.e. $F \subset E$. If we take into account that $PF = F$, our assertion follows from the above inclusions.

**5. *Right invertible mappings of Banach couples.*** Let $R$ be a bounded linear operator defined on a Banach space $E$, mapping into a Banach space $F$ and having a right inverse, i.e. a linear operator $\Pi$ from $F$ to $E$ such that $R\Pi = I_F$. Since $R\Pi y = y$ for any $y \in F$, the range of $R$ coincides with the whole space $F$. The operator $P = \Pi R$ is a bounded projection in the space $E$: $P^2 = \Pi R \Pi R = \Pi R = P$. We denote by $E_1$ the subspace onto which $E$ is projected by $P$.

This subspace coincides with the range of $\Pi$. Indeed, if $x = \Pi y$, then $Px = \Pi R \Pi y = \Pi y = x$. The complementary projection $I - P$ maps $E$ onto the kernel $N(R)$ of $R$. Indeed, every element $z$ of the kernel can be represented in the form $z = z - \Pi Rz = (I - P)z$. Conversely, every element $(1 - P)z$ belongs to $N(R)$: $R(I - \Pi R)z = (R - R)z = \theta$. Thus, $E$ decomposes into the direct sum of the subspaces $N(R)$ and $E_1$. The restriction $R_1$ of $R$ to $E_1$ effects an isomorphic mapping of $E_1$ onto $F$, and $\Pi = R_1^{-1}$.

Conversely, if the range of $R$ coincides with the space $F$, and the kernel $N(R)$ is a complemented subspace of $E$, then this operator is right invertible. Indeed, if $E = N(R) \dot{+} E_1$, then we may set $\Pi = R_1^{-1}$, where $R_1$ is the restriction of $R$ to $E_1$.

Now let $A$, $B$ and $C$, $D$ be two Banach couples, and let $R \in L(AB, CD)$. Suppose that there exists an operator $\Pi$ defined on $C + D$ whose restrictions to $C$ and $D$ are right inverses of the restrictions of $R$ to $A$ and $B$, respectively. Then we say that $R$ is a *right invertible mapping of the couple $A$, $B$ onto the couple $C$, $D$.* If the restrictions of $R$ to $A$ and $B$ have right inverses $\Pi_1$ and $\Pi_2$, then for the existence of the operator $\Pi$ discussed above, which coincides with $\Pi_1$ on $C$ and with $\Pi_2$ on $D$, it is necessary and sufficient that the operators $\Pi_1$ and $\Pi_2$ coincide on $C \cap D$.

We denote by $M$ and $N$ the images of $C$ and $D$ under the mapping $\Pi$. As was explained above, these images are subspaces of $A$ and $B$ isomorphic to $C$ and $D$, respectively. The operator $P = \Pi R$ projects $A$ onto $M$ and $B$ onto $N$, and so $M$, $N$ is a complemented subcouple of the couple $A$, $B$. The restriction $R_1$ of $R$ to $M + N$ effects an isomorphism of this space and the space $C + D$. To every space $F$ intermediate between $M$ and $N$ this isomorphism assigns a space $G$ intermediate between $C$ and $D$. Moreover, a natural correspondence is thus established between the interpolation properties of $(M, N, F)$ and $(C, D, G)$. Then Theorem 4.7 immediately implies the following theorem.

THEOREM 4.8. *Let $R$ be a right invertible mapping of the Banach couple $A, B$ onto the Banach couple $C, D$.*

*To every interpolation space $E$ between $A$ and $B$ corresponds an interpolation space $G_R(E)$ between $C$ and $D$, with norm*

$$\|y\|_{G_R(E)} = \inf_{Rx=y} \|x\|_E.$$

*The space $G_R(E)$ is the smallest among all spaces $G$ for which $(A, B, E)$ is an interpolation triple relative to the triple $(C, D, G)$.*

*From every interpolation space $G$ between $C$ and $D$ an interpolation space $E(G)$ between $A$ and $B$ can be constructed that is the largest among all interpolation spaces $E$ between $A$ and $B$ for which $RE = G$. An equivalent norm can be introduced in $E(G)$ by the formula*

$$\|x\|_{E(G)} = \sup_{T\in\pi_1(AB,AB)} \|RTx\|_G.$$

*$(A, B, E(G))$ is the optimal interpolation triple relative to $(C, D, G)$.*

Similarly, from Lemma 4.6 follows

LEMMA 4.7. *If $(A, B, E)$ and $(C, D, G)$ are interpolation triples relative to each other, then the space $G$ is isomorphic to $G_R(E)$.*

We note that formally Theorem 4.7 and Lemma 4.6 are special cases of Theorem 4.8 and Lemma 4.7. The projection $P$ induces a right invertible mapping of the couple $A, B$ onto the couple $M, N$, where the role of the right invertible mapping $\Pi$ is played by the operator imbedding $M + N$ naturally in $A + B$.

Another special case can be obtained if we consider the quotients of the spaces of the couple $A, B$ modulo the spaces of a complemented subcouple $M, N$. We consider the quotient spaces $(A + B)/(M + N)$, $A/M$ and $B/N$. To every coset $x + M$ $(x \in A)$ of $A$ modulo $M$ corresponds the coset $x + M + N$ of $A + B$ modulo $M + N$, which contains $x + M$. Under this mapping distinct cosets in $A/M$ are mapped into distinct cosets in $(A + B)/(M + N)$. Indeed, if $x + M$ and $y + M$ are contained in the same class $z + M + N$, then $x - y \in M + N$, and consequently $x - y = P(x - y)$. Since $x - y \in A$, we then have $x - y \in M$ and the classes $x + M$ and $y + M$ coincide. Finally,

$$\|x + M + N\|_{(A+B)/(M+N)} = \inf_{u\in M+N} \|x + u\|_{A+B}$$
$$\leqslant \inf_{u\in M} \|x + u\|_A = \|x + M\|_{A/M}.$$

Thus, $A/M$ and $B/N$ are imbedded in $(A + B)/(M + N)$, and so they form a Banach couple. The natural projection

$$\pi: (A + B) \to (A + B)/(M + N)$$

induces a right invertible mapping of the couple $A$, $B$ onto the couple $A/M$, $B/N$. Its right inverse is the isomorphism which is naturally established between $A/M$ and $B/N$ and the subspaces $(I - P)A$ and $(I - P)B$ complementary to $M$ and $N$.

The reader can easily reformulate Theorem 4.8 and Lemma 4.7 for this special case.

**6. *The interpolation functor.*** Let two categories $\mathfrak{C}$ and $\mathfrak{C}_1$ be given. A *covariant functor* $\mathfrak{X}$ acting from $\mathfrak{C}$ to $\mathfrak{C}_1$ is, by definition, a pair of mappings, each of which is usually denoted by the same letter $\mathfrak{X}$. One of the mappings is defined on objects and to every object $\mathfrak{A}$ of $\mathfrak{C}$ assigns an object $\mathfrak{X}(\mathfrak{A})$ of $\mathfrak{C}_1$; the other one is defined on morphisms and to every morphism $\gamma \in \operatorname{Mor}(\mathfrak{A}, \mathfrak{A}')$ assigns the morphism $\mathfrak{X}(\gamma) \in \operatorname{Mor}(\mathfrak{X}(\mathfrak{A}), \mathfrak{X}(\mathfrak{A}'))$. This pair of mappings must satisfy the following conditions:

$$\mathfrak{X}(1_{\mathfrak{A}}) = 1_{\mathfrak{X}(\mathfrak{A})} \quad (\mathfrak{A} \in \mathfrak{C}) \quad \text{and} \quad \mathfrak{X}(\beta\gamma) = \mathfrak{X}(\beta)\mathfrak{X}(\gamma),$$

where $\beta\gamma$ is defined in $\mathfrak{C}$.

As $\mathfrak{C}$ we take the category of Banach couples $A$, $B$ in which the morphisms are the operators from $L(AB, CD)$ ($A$, $B$ is the first object, and $C$, $D$ is the second object). Denote by $\mathfrak{C}_1$ the category of Banach spaces $E$ in which the morphisms are the bounded linear operators from $L(E, F)$.

DEFINITION 4.7. An *interpolation functor* is a functor $\mathfrak{F}$ acting from the category of Banach couples into the category of Banach spaces and assigning to every Banach couple $A$, $B$ a Banach space $\mathfrak{F}(A, B)$ intermediate between $A$ and $B$: $A \cap B \subset \mathfrak{F}(A, B) \subset A + B$ and to every operator $T \in L(AB, CD)$ assigning its restriction $\mathfrak{F}(T)$ to $\mathfrak{F}(A, B)$.

If the above correspondence is a functor, then the operator $\mathfrak{F}(T)$ belongs to the set of morphisms $\operatorname{Mor}(\mathfrak{F}(A, B), \mathfrak{F}(C, D))$, i.e. is a bounded linear operator from $\mathfrak{F}(A, B)$ into $\mathfrak{F}(C, D)$. In other words, the triples $(A, B, \mathfrak{F}(A, B))$ and $(C, D, \mathfrak{F}(C, D))$ must be interpolation triples for any Banach couples $A$, $B$ and $C$, $D$.

Simple examples of functors from the category of Banach couples into the category of Banach spaces are the functors of the sum and intersection of spaces in a Banach couple. By Lemma 4.1 these are interpolation functors.

An interpolation functor $\mathfrak{F}$ is said to be *normalized* if the interpolation constant for any triples $(A, B, \mathfrak{F}(A, B))$ and $(C, D, \mathfrak{F}(C, D))$ is not greater than 1. The functors of sum and intersection are normalized.

We say that an interpolation functor $\mathfrak{F}$ has *type* (*normal type*) $\alpha$ if any triple $(A, B, \mathfrak{F}(A, B))$ is a (normalized) interpolation triple of type $\alpha$ relative to any triple $(C, D, \mathfrak{F}(C, D))$.

The study in subsection 3 enables us to give some constructions of general interpolation functors. Let $A$, $B$ be a fixed Banach couple and $E$ a space intermediate between $A$ and $B$. For any other Banach couple $C$, $D$ (which may coincide with $A$, $B$) a space $F(E)$ intermediate between $C$ and $D$ has been constructed in subsection 3. We show that the correspondence $\mathfrak{F}(C, D) = F(E)$ induces an interpolation functor. $C^0$, $D^0$, and $C^1$, $D^1$ are two Banach couples and $F^0(E)$ and $F^1(E)$ are the corresponding intermediate spaces. We recall that $F^i(E)$, $i = 0, 1$, is the sum of the spaces $F_T^i(E)$, where $T$ runs over the set $\pi_1(AB, C^iD^i)$. Let $S \in \pi_1(C^0D^0, C^1D^1)$ and $z \in F^0(E)$. The latter means that $z = \Sigma y_k$, where $y_k \in F_{T_k}^0(E)$ and $T_k \in \pi_1(AB, C^0D^0)$. Then $Sz = \Sigma Sy_k$, since $\Sigma y_k$ converges in $C^0 + D^0$, and the operator $S$ acts continuously from $C^0 + D^0$ to $C^1 + D^1$. Next, $y_k = T_k x_k$ $(x_k \in E)$, and consequently, $Sy_k = ST_k x_k$, where $ST_k \in \pi_1(AB, C^1D^1)$. To prove that $Sz \in F^1(E)$ it is sufficient to verify that

$$\sum \|Sy_k\|_{F^1_{ST_k}(E)} < \infty.$$

We choose $x_k \in E$ so that $y_k = T_k x_k$ and $\|y_k\|_{F^0_{T_k}(E)} + \varepsilon/2^k \geq \|x_k\|_E$. Then

$$\sum \|Sy_k\|_{F^1_{ST_k}(E)} \leq \sum \|x_k\|_E \leq \sum \|y_k\|_{F^0_{T_k}(E)} + \varepsilon.$$

Using the arbitrariness of $\varepsilon$ and then taking the infimum over all representations $z = \Sigma y_k$, we obtain

$$\|Sz\|_{F^1(E)} \leq \|z\|_{F^0(E)}.$$

Thus we have constructed a normalized interpolation functor, which we denote by $\mathfrak{F}_{(A,B,E)}$.

If we apply $\mathfrak{F}_{(A,B,E)}$ to the couple $A$, $B$, we obtain a space which is in general different from $E$. Now let $E$ be an interpolation space between $A$ and $B$. We construct the space $\mathfrak{F}_{(A,B,E)}(A, B)$. It consists of elements of the form $z = \Sigma y_k$, where $Y_k = T_k x_k$, $T_k \in \pi_1(AB, AB)$, $x_k \in E$, and $\Sigma \|y_k\|_{F_{T_k}(E)} < \infty$. If $y = Tx$ $(T \in \pi_1(AB, AB))$, then by the interpolation property of $E$ we have $\|y\|_E \leq C_E \|x\|_E$. Therefore

$$\|y\|_{F_T(E)} = \inf_{Tx=y} \|x\|_E \geq C_E^{-1} \|y\|_E.$$

Then from the convergence of $\Sigma \|y_k\|_{F_{T_k}(E)}$ it follows that $\Sigma \|y_k\|_E$ converges, and hence $z \in E$. Besides,

$$\|z\|_{F(E)} = \inf \sum \|y_k\|_{F_{T_k}(E)} \geq C_E^{-1} \inf \sum \|y_k\|_E = C_E^{-1} \|z\|_E.$$

Finally, we consider the representation $z = Iz$ and conclude that $E$ is contained in $F(E)$ and $\|z\|_{F(E)} \leqslant \|z\|_E$. Thus $E$ and $\mathfrak{F}_{(A,B,E)}(A, B)$ are isomorphic, and if $C_E = 1$, then they coincide.

We have arrived at the following important assertion.

THEOREM 4.9. *If a Banach space $E$ is an interpolation space between the spaces $A$ and $B$ of a Banach couple, then there exists a normalized interpolation functor $\mathfrak{F}$ such that $E$ is isomorphic to the space $\mathfrak{F}(A, B)$. If the interpolation constant $C_E$ is equal to* 1, *then $E = \mathfrak{F}(A, B)$.*

In subsection 3, from the intermediate space $F$ between $C$ and $D$ in the fixed Banach couple we have constructed for every other Banach couple $A, B$ an intermediate space $E(F)$. The correspondence $\mathfrak{F}(A, B) = E(F)$ also induces an interpolation functor, which we denote by $\mathfrak{E}_{(C,D,F)}$. If $F$ is an interpolation space between $C$ and $D$, then it is isomorphic to $\mathfrak{E}_{(C,D,F)}(C, D)$ and, if $C_F = 1$, then $F = \mathfrak{E}_{(C,D,F)}(C, D)$. Thus, we obtain yet another functor having the properties described in Theorem 4.9.

It turns out that the functors $\mathfrak{F}_{(A,B,E)}$ and $\mathfrak{E}_{(A,B,E)}$ are extreme cases in a certain sense: for any functor having the properties indicated in Theorem 4.9 the imbedding

$$\mathfrak{F}_{(A,B,E)}(C, D) \subset \mathfrak{F}(C, D) \subset \mathfrak{E}_{(A,B,E)}(C, D)$$

is valid.

The proof of the above facts is left to the reader (see the Notes on the Literature).

Lemma 4.7 immediately implies the following theorem.

THEOREM 4.10. *Let $R$ be a right invertible mapping of the Banach couple $A, B$ onto the Banach couple $C, D$. For every interpolatio functor $\mathfrak{F}$ the space $\mathfrak{F}(C, D)$ is isomorphic to $F_R(\mathfrak{F}(A, B))$ (the image of $\mathfrak{F}(A, B)$ under the mapping $R$).*

The hypotheses of Lemma 4.7 are satisfied, since $(A, B, \mathfrak{F}(A, B))$ and $(C, D, \mathfrak{F}(C, D))$ are interpolation triples relative to each other.

## CHAPTER II

# INTERPOLATION IN SPACES OF MEASURABLE FUNCTIONS

### Introduction

1. ***Measurable functions on a measure space.*** Let $\mathfrak{M}$ be a measurable space and $\mu$ a measure defined on its $\sigma$-algebra of measurable sets $\mathfrak{M}$ (see [16]). In what follows we always assume that this measure is $\sigma$-finite, i.e. the space $\mathfrak{M}$ is the union of countably many sets of finite measure.*

By $\chi_e(t)$ we always denote the characteristic function of a measurable set $e \subset \mathfrak{M}$. A function $x(t)$ is said to be simple if it assumes only finitely many nonzero finite values, each on a measurable set, and strictly simple if the requirement of finiteness of the measure on these sets is imposed.** Every simple function can be written in the form $x(t) = \sum_1^n x_i \chi_{e_i}(t)$, where the $e_i$ are measurable sets.

Every measurable function is the limit of a sequence of simple functions, and if the function is nonnegative, the sequence can be chosen increasing. If the function is bounded, the sequence can be chosen uniformly convergent.

EGOROV'S THEOREM. *If the measure of $\mathfrak{M}$ is finite, and the sequence of almost everywhere finite measurable functions $x_n(t)$ almost everywhere converges to an almost everywhere finite function, then for every $\varepsilon$ there exists a measurable subset $e$ with measure $\mu(e) > \mu(\mathfrak{M}) - \varepsilon$ on which the sequence $x_n(t)$ converges to $x(t)$ uniformly.*

We say that a sequence $x_n(t)$ *converges* to a function $x(t)$ *in measure* if the measure of the set where $|x_n(t) - x(t)| \geqslant \varepsilon$ converges to zero as $n \to \infty$ for any $\varepsilon > 0$.

---

* *Editor's note.* Actually the measure is totally $\sigma$-finite in Halmos' terminology ([10], §7).

** *Editor's note.* The Russian terms used by the authors are Конечнозначная функция (literally: finitely valued function) for strictly simple function, and обобщенно конечнозначная функция (literally: generalized finitely valued function) for simple function. The terminology adopted for the translation is consistent with that of Halmos ([10], §20).

If a sequence converges almost everywhere on a space with a finite measure, then it converges in measure. If a sequence converges in measure, then it contains a subsequence which converges almost everywhere.

The space $S(\mathfrak{M}, \mu)$ of all real-valued measurable functions (more precisely, of classes of equivalent functions) on $\mathfrak{M}$ is linear. In this space a metric can be introduced in the following way: Let $\nu(t)$ be a positive function integrable with respect to $\mu$, and set

$$\rho(x, y) = \int \frac{|x(t) - y(t)|}{1 + |x(t) - y(t)|} \nu(t) \, d\mu.$$

The space $S(\mathfrak{M}, \mu)$ becomes a complete linear metric space with this metric. Convergence in this space is equivalent to convergence in measure on every set of finite measure.

The Banach space of all functions (more precisely, of all classes of equivalent functions) integrable on $\mathfrak{M}$ with respect to $\mu$ is denoted by $L_1(\mathfrak{M}, \mu)$.

LEBESGUE'S THEOREM. *If a sequence of integrable functions $x_n(t)$ converges almost everywhere to a function $x(t)$ and $|x_n(t)| \leqslant |y(t)|$ almost everywhere, where $y(t)$ is some integrable function, then $x(t)$ is integrable and the sequence $\{x_n\}$ converges to $x$ in the norm of $L_1(\mathfrak{M}, \mu)$.*

FATOU'S LEMMA. *If $x_n(t)$ is a sequence of nonnegative measurable functions, then*

$$\int \underline{\lim}_{n \to \infty} x_n(t) \, d\mu \leqslant \underline{\lim}_{n \to \infty} \int x_n(t) \, d\mu.$$

B. LEVI'S THEOREM. *If an increasing sequence of measurable nonnegative functions $x_n(t)$ converges to a function $x(t)$ almost everywhere, then*

$$\lim_{n \to \infty} \int x_n(t) \, d\mu = \int x(t) \, d\mu.$$

**2. *Ideal Banach lattices.*** A Banach space $E$ which is a nonzero linear manifold of $S(\mathfrak{M}, \mu)$ is called a *Banach function space.*

In $S(\mathfrak{M}, \mu)$ there exists a natural partial ordering: $x \leqslant y$ $(x, y \in S(\mathfrak{M}, \mu))$ is equivalent to the inequality $x(t) \leqslant y(t)$ for almost all $t \in \mathfrak{M}$.

DEFINITION 1.1. A Banach function space $E$ is said to be an *ideal Banach lattice*, or briefly an *ideal lattice*, if the condition $|x| \leqslant |y|$, where $x(t)$ is a measurable function and $y \in E$, implies that $x \in E$ and $\|x\|_E \leqslant \|y\|_E$.(1)

(1) In the literature such spaces are also called "Banach lattices", and "function lattices".

It follows from this definition that $\| |x| \|_E = \|x\|_E$ for $x \in E$.

We cite the simplest properties of ideal lattices.

1°. *If $x_n \to x$ in $E$ and the sequence $x_n(t)$ is increasing, then $x_n(t)$ converges to $x(t)$ almost everywhere.*

For the proof we first note that $x(t) - x_n(t) \geqslant 0$. If we had $x(t) - x_{n_0}(t) < -\gamma$ ($\gamma > 0$), on a set $e$ of positive measure, then for all $n > n_0$ we would also have $x(t) - x_n(t) < -\gamma$ ($t \in e$) and

$$\|x - x_n\|_E \geqslant \|(x - x_n)\chi_e\|_E > \gamma\|\chi_e\|_E,$$

which contradicts the convergence of $x_n$ to $x$ in $E$. We write $z(t) = \lim_{n\to\infty} x_n(t)$. Then $x(t) - x_n(t) \geqslant x(t) - z(t) \geqslant 0$, and therefore the function $x(t) - z(t)$, and so $z(t)$ as well, belongs to $E$. Next, $\|x - z\|_E \leqslant \|x - x_n\|_E \to 0$, and so $x(t) = z(t)$ almost everywhere.

2°. *If the sequence $\{y_n\}$ ($y_n \in E$) is increasing and tends to $\infty$ on a set $e$ of positive measure, then $\|y_n\|_E \to \infty$.*

Indeed, routine arguments show that there exists a set $e' \subset e$ with $\mu(e') > 0$ on which the sequence $\{y_n(t)\}$ uniformly tends to $\infty$. Then the characteristic function $\chi_{e'}(t)$ belongs to $E$, and

$$\|\chi_{e'}(t)\|_E \min_{t\in e'} y_n(t) \leqslant \|y_n\|_E \to \infty.$$

3°. *Let $\mu(\mathfrak{M}) < \infty$, $e_k \in \mathfrak{M}$ ($k = 1, 2, \ldots$) and $\mu(e_k) > \delta$. Then $\|\sum_1^N \chi_{e_k}\| \to \infty$ as $N \to \infty$.*

We write $y_N(t) = \sum_{k=1}^N \chi_{e_k}(t)$. The sequence $y_N$ is increasing. If we assume that its limit is finite almost everywhere, then because of the finiteness of the measure of $\mathfrak{M}$ there exists a set $e' \subset \mathfrak{M}$ with measure arbitrarily close to that of $\mathfrak{M}$ (for example, $\mu(\mathfrak{M} - e') < \delta/2$) on which the sequence $y_N(t)$ is uniformly bounded. However,

$$\int_{e'} y_N(t)\, d\mu = \sum_{k=1}^N \int_{e'} \chi_{e_k}(t)\, d\mu = \sum_{k=1}^N \mu(e_k \cap e') \geqslant N\delta/2,$$

since $\mu(e' \cap e_k) > \delta/2$, and this contradicts the boundedness of the sequence $y_N(t)$ on $e'$. Consequently the limit of $y_N(t)$ is infinite on a set of positive measure, and by 2° we have $\|y_N\|_E \to \infty$ as $N \to \infty$.

THEOREM 1. *An ideal lattice is imbedded in $S(\mathfrak{M}, \mu)$.*

PROOF. It is sufficient to show that the condition $\|x_n\|_E \to 0$ implies the convergence of $x_n$ to zero in $S(\mathfrak{M}, \mu)$, or, which is the same, the convergence of $x_n(t)$ to zero in measure on any set $\Omega$ of finite measure. Assume the contrary: Let there exist a set $\Omega$, $0 < \mu(\Omega) < \infty$, and a positive number $\varepsilon > 0$ such that for some sequence $x_{n_k}(t)$ the inequality $|x_{n_k}(t)| > \varepsilon$ is satisfied on a set $\Omega_k \subset \Omega$ with measure $\mu(\Omega_k) > \delta$, where $\delta$ is some fixed number. Then $\varepsilon\chi_{\Omega_k}(t) \leqslant |x_{n_k}(t)|$, and so $\varepsilon\|\chi_{\Omega_k}\|_E \leqslant \|x_{n_k}\|_E$. Without loss of generality we may assume that the sequence $n_k$ is chosen so that $\sum_{k=1}^{\infty} \|x_{n_k}\|_E < \infty$. From this we obtain that

$$\varepsilon\left\|\sum_{k=1}^{N} \chi_{\Omega_k}\right\|_E \leqslant \varepsilon \sum_{k=1}^{N} \|\chi_{\Omega_k}\|_E \leqslant \sum_{k=1}^{N} \|x_{n_k}\|_E \leqslant \sum_{k=1}^{\infty} \|x_{n_k}\|_E < \infty,$$

which contradicts property 3°.

COROLLARY 1. *Any two ideal lattices (on the same set $\mathfrak{M}$ and with the same measure $\mu$) forms a Banach couple.*

COROLLARY 2. *If $\sum_1^\infty \|x_n\|_E < \infty$ and $\sum_1^N x_n \to x$ in $E$ as $N \to \infty$, then $\sum_1^\infty x_n(t)$ converges absolutely to $x(t)$ almost everywhere.*

Indeed, $\sum_1^\infty |x_n|$ converges in $E$, and by 1° it converges almost everywhere. Therefore $\sum_1^\infty x_n(t)$ also converges almost everywhere. Since $E$ is imbedded in $\mathbb{S}(\mathfrak{M}, \mu)$, the latter series converges to $x(t)$ in $\mathbb{S}(\mathfrak{M}, \mu)$, and consequently $\sum_1^\infty x_n(t) = x(t)$ almost everywhere.

4°. *If $0 \leqslant x \leqslant y$ and $y_n \to y$ in $E$, then the functions* $\min(x, y_n) = \bar{x}_n$ *converge to $x$ in $E$.*

Indeed,

$$0 \leqslant x(t) - \bar{x}_n(t) = [x(t) - y_n(t)]_+ \leqslant |y(t) - y_n(t)|.\text{(}^2\text{)}$$

Then

$$\lim_{n\to\infty} \|x - \bar{x}_n\|_E \leqslant \lim_{n\to\infty} \|\,|y - y_n|\,\|_E = \lim_{n\to\infty} \|y - y_n\|_E = 0.$$

5°. *If $x, z \in E$, then the functions $|x(t)|^\theta |z(t)|^{1-\theta}$ also belong to $E$ for* $0 \leqslant \theta \leqslant 1$.

This property follows from the inequalities

$$|x(t)|^\theta |z(t)|^{1-\theta} \leqslant \max(|x(t)|, |z(t)|) \leqslant |x(t)| + |z(t)|.$$

6°. *If $x, z \in E$ and for almost all $t$ the inequality*

$$|y(t)| \leqslant |x(t)|^\theta |z(t)|^{1-\theta} \tag{0.1}$$

*holds for some $\theta \in [0, 1]$, then*

$$\|y\|_E \leqslant \|x\|_E^\theta \|z\|_E^{1-\theta}, \tag{0.2}$$

*and for every positive functional from $E'$ we have*

$$f(|y|) \leqslant [f(|x|)]^\theta [f(|z|)]^{1-\theta}. \tag{0.3}$$

($^2$) $z^+(t) = \max\{z(t), 0\}$.

Inequality (0.1) is equivalent to the following:

$$|y(t)| \leqslant \theta^{\theta-1}|x(t)| + (1-\theta)\varepsilon^{\theta}|z(t)|$$

for all $\varepsilon > 0$. Therefore,

$$\|y\|_E \leqslant \theta\varepsilon^{\theta-1}\|x\|_E + (1-\theta)\varepsilon^{\theta}\|z\|_E$$

and

$$f(|y|) \leqslant \theta\varepsilon^{\theta-1}f(|x|) + (1-\theta)\varepsilon^{\theta}f(|z|).$$

By minimizing the right sides with respect to $\varepsilon$, we obtain (0.2) and (0.3).

A special case of (0.2) is the inequality

$$\| \, |x|^{\theta}|z|^{1-\theta}\|_E \leqslant \|x\|_E^{\theta}\|z\|_E^{1-\theta}, \tag{0.4}$$

which becomes the Hölder inequality in the case where $E = L_1$.

7°. *Let $x, y, x_n$ and $y_n$ be positive elements of an ideal lattice $E$, and assume that $x_n \to x$ and $y_n \to y$ in $E$. Then*

$$x_n^{\theta}y_n^{1-\theta} \to x^{\theta}y^{1-\theta} \quad \textit{in } E.$$

Using (0.4), we obtain

$$\begin{aligned}\|x^{\theta}y^{1-\theta} - x_n^{\theta}y_n^{1-\theta}\|_E &\leqslant \|x^{\theta}(y^{1-\theta} - y_n^{1-\theta})\|_E + \|y_n^{1-\theta}(x^{\theta} - x_n^{\theta})\|_E \\ &\leqslant \|x\|_E^{\theta}\left\| \, |y^{1-\theta} - y_n^{1-\theta}|^{1/(1-\theta)}\right\|_E^{1-\theta} \\ &\quad + \|y_n\|_E^{1-\theta}\left\| \, |x^{\theta} - x_n^{\theta}|^{1/\theta}\right\|_E^{\theta}.\end{aligned}$$

The function $|x^{\theta} - x_n^{\theta}|^{1/\theta}$ belongs to $E$, since $|x^{\theta} - x_n^{\theta}|^{1/\theta} \leqslant |x - x_n|$. Similarly, $|y^{1-\theta} - y_n^{1-\theta}|^{1/(1-\theta)} \leqslant |y - y_n|$. Consequently,

$$\|x^{\theta}y^{1-\theta} - x_n^{\theta}y_n^{1-\theta}\|_E \leqslant \|x\|_E^{\theta}\|y - y_n\|_E^{1-\theta} + \|y_n\|_E^{1-\theta}\|x - x_n\|_E^{\theta} \to 0.$$

We consider some operations which allow us to construct ideal lattices from other ideal lattices.

*Restriction of an ideal lattice.* If $\mathfrak{M}_1$ is a measurable subset of $\mathfrak{M}$ of positive measure and $\chi_1(t)$ is the characteristic function of $\mathfrak{M}_1$, then the functions $\chi_1(t)x(t)$ $(x \in E)$ form a subspace $E_1$ of $E$. Since $\|\chi_1 x\|_E \leqslant \|x\|_E$, multiplication by $\chi_1$ is a projection onto this subspace, bounded by unity. On the other hand, the elements of the Banach space $E_1$ can be identified with restrictions of functions in $E$ to $\mathfrak{M}_1$ in a natural manner. Thus, this space can be considered as a Banach function space on $\mathfrak{M}_1$ with measure $\mu_1$ which is the restriction of $\mu$ to $\mathfrak{M}_1$. In what follows we often speak of $E_1$ as "the space $E$ on $\mathfrak{M}_1$".

*Weighted ideal lattices.* Let $\rho(t)$ be an almost everywhere positive measurable function (weight) on the space $\mathfrak{M}$. The collection of all measurable functions (classes) $x(t)$ for which the functions $\rho(t)x(t)$ belong to an ideal lattice $E$ and which is equipped with the norm

$$\|x\|_{E_{\rho(t)}} = \|\rho x\|_E$$

is obviously an ideal lattice on $\mathfrak{M}$, to be denoted by $E_{\rho(t)}$.

*Sum and intersection of ideal lattices.* If $E_0$ and $E_1$ are two ideal lattices on the same space $\mathfrak{M}$ with the same measure $\mu$, their intersection $E_0 \cap E_1$ is obviously also an ideal lattice. We show that the sum $E_0 + E_1$ also has this property.

Let $y \in E_0 + E_1$ and $|x| \leqslant |y|$. Given $\varepsilon > 0$, we choose $y_0 \in E_0$ and $y_1 \in E_1$ so that $\|y_0\|_{E_0} + \|y_1\|_{E_1} \leqslant \|y\|_{E_0+E_1} + \varepsilon$ and $y = y_0 + y_1$. We set $h(t) = x(t)/y(t)$ if $y(t) \neq 0$ and $h(t) = 0$ if $y(t) = 0$. Then $x = y_0 h + y_1 h$, and since $|y_i h| \leqslant |y_i|$, $i = 0, 1$, we have $y_i h \in E_i$. Therefore $x \in E_0 + E_1$ and

$$\begin{aligned} \|x\|_{E_0+E_1} &\leqslant \|y_0 h\|_{E_0} + \|y_1 h\|_{E_1} \\ &\leqslant \|y_0\|_{E_0} + \|y_1\|_{E_1} \leqslant \|y\|_{E_0+E_1} + \varepsilon. \end{aligned}$$

Since $\varepsilon$ is arbitrary, we have $\|x\|_{E_0+E_1} \leqslant |y|_{E_0+E_1}$.

*Discretization of an ideal lattice.* Let $\mathfrak{M} = \bigcup_1^\infty \mathfrak{M}_n$ be a decomposition of $\mathfrak{M}$ into disjoint sets $\mathfrak{M}_n$ of finite positive measure $\mu_n$. Assume that $\chi_{\mathfrak{M}_n} \in E$. Such a decomposition always exists if $E$ contains an almost everywhere positive function. Denote by $E^c$ the collection of all functions in $E$ constant on each set $\mathfrak{M}_n$. This collection is a subspace of $E$. On the set $\mathbf{N}_+$ of natural numbers we introduce a measure $\mu'$ by the formula $\mu'(\{n\}) = \mu_n$. Then the space $E^c$ can be identified with the space $dE$ of all functions $x(n)$ on $\mathbf{N}_+$ with finite norm

$$\|x\|_{dE} = \left\| \sum_{n=1}^{\infty} x(n)\chi_{\mathfrak{M}_n} \right\|_E .$$

The space $dE$ is an ideal lattice. We shall say that the lattice $dE$ is obtained by *discretization* of the lattice $E$.

From the decomposition $\bigcup_1^\infty \mathfrak{M}_n$ we may construct an averaging operator

$$Tx = \sum_{n=1}^{\infty} \frac{1}{\mu_n} \int_{\mathfrak{M}_n} x(s)\, d\mu\, \chi_{\mathfrak{M}_n}(t).$$

If this operator is bounded in $E$, then it maps $E$ onto $E^c$. The operator

$$T^d x = \left\{ \frac{1}{\mu_n} \int_{\mathfrak{M}_n} x(s)\, d\mu \right\}$$

then maps $E$ onto $dE$. The imbedding operator $dE \to E^c$ is its right inverse.

**3.** ***Special classes of ideal lattices.*** We say that an ideal lattice has the *Fatou property* if the condition that a sequence of functions $x_n(t) \in E$ converges to a function $x(t)$ almost everywhere and is bounded in $E$ ($\|x_n\| \leqslant M$) implies that $x \in E$ and $\|x\|_E \leqslant \underline{\lim}\|x_n\|_E$.

A norm in an ideal lattice $E$ is said to be *absolutely continuous* if for any $x \in E$ and any decreasing sequence of measurable sets $e_n$ with empty intersection we have $\|\chi_{e_n} x\|_E \to 0$ as $n \to \infty$. A lattice with an absolutely continuous norm is said to be *regular*.

The property of absolute continuity of the norm is closely related to the separability of the space $E$. The measure $\mu$ is said to be *separable* if there exists a countable system $\Gamma$ of measurable sets in $\mathfrak{M}$ such that for every $\varepsilon > 0$ and every measurable set $e \in \mathfrak{M}$ there exists a set $e' \subset \Gamma$ for which

$$\mu(e \setminus e' \cup e' \setminus e) < \varepsilon.$$

It turns out that an ideal lattice $E$ is separable if and only if the measure $\mu$ is separable, and the norm in $E$ is absolutely continuous.

The *support* $\mathfrak{N}$ of an ideal lattice $E$ is, by definition, the smallest measurable set outside which all functions in $E$ are equal to zero. In an ideal lattice there is always a function which is positive at all points of the support.

The *associated space* $E^1$ of an ideal lattice $E$ is, by definition, the collection of all measurable functions $y(t)$ on $\mathfrak{M}$ whose support is contained in the support of $E$ and for which

$$\|y\|_{E^1} = \sup_{\|x\|_E = 1} \int x(t)y(t)\, dt < \infty.$$

The associated space is itself an ideal lattice. It is contained in the dual $E'$ of $E$ and is a subspace of it. The space $E^1$ coincides with the whole dual if and only if the norm of $E$ is absolutely continuous.

From the space $E^1$ we construct the second associated space $E^{11} = (E^1)^1$. The space $E$ can be imbedded in $E^{11}$ in a natural way; moreover, we have $\|x\|_{E^{11}} \leqslant \|x\|_E$ for all $x \in E$. If the space $E$ is separable, then $\|x\|_{E^{11}} = \|x\|_E$ for $x \in E$; that is, $E$ is isometrically imbedded in $E^{11}$. If $E$ has the Fatou property, then $E^{11} = E$.

The proofs of the above assertions can be found in [29] and [49].

We note still another generalization of the Minkowski inequality. Let $u(t, \tau)$ be a function of two variables $t, \tau \in \mathfrak{M}$ such that for almost every fixed $t$ it belongs to $E$ as a function of $\tau$, the function $\|u(t, \tau)\|_E$ is measurable, and

$$\int \|u(t, \tau)\|_E\, dt < \infty.$$

Then

$$\left\| \int u(t, \tau)\, dt \right\|_{E^{11}} \leqslant \int \|u(t, \tau)\|_E\, dt. \tag{0.5}$$

Indeed, for $z \in E^1$ we obtain

$$\int |u(t, \tau)|\, |z(\tau)|\, d\tau \leqslant \|u(t, \tau)\|_E \|z\|_{E^1},$$

and therefore, by Fubini's theorem,

$$\left\| \int u(t, \tau)\, dt \right\|_{E^{11}} \leqslant \sup_{\|z\|_{E^1} \leqslant 1} \int \left[ \int |u(t, \tau)|\, |z(\tau)|\, dt \right] d\tau$$

$$= \sup_{\|z\|_{E^1} \leqslant 1} \int \left[ \int |u(t, \tau)|\, |z(\tau)|\, d\tau \right] dt = \int \|u(t, \tau)\|_E\, dt.$$

**4.** ***Lebesgue spaces.*** Two spaces $\mathfrak{M}_1$ and $\mathfrak{M}_2$ with measures $\mu_1$ and $\mu_2$ are said to be *isomorphic* if, after deleting sets of measure zero from them, we can establish a one-to-one correspondence between the remaining parts which preserves the class of measurable sets and the measure of sets.

The spaces which are isomorphic to an interval $(0, a)$ of the real axis (with finite or infinite $a$) with Lebesgue measure are called here *Lebesgue spaces.** There is an axiomatic definition of Lebesgue spaces. It requires the separability of the measure, its continuity (the measure of any point is equal to zero), and a certain completeness property which will not be formulated here (see [324]). An important property of Lebesgue spaces is the fact that every measurable subset of finite measure is also a Lebesgue space. The space $\mathbf{R}^n$ with Lebesgue measure is obviously a Lebesgue space, and consequently any measurable subset of finite measure in $\mathbf{R}^n$ is a Lebesgue space. From this it follows in particular that for two sets in the spaces $\mathbf{R}^n$ and $\mathbf{R}^m$, respectively, with equal nonzero finite measures there exists a one-to-one mapping of one set onto the other preserving the measure of all measurable subsets.

## §1. Positive functions on a semiaxis and their dilation functions

**1.** ***Classes of positive functions on the semiaxis*** $(0, \infty)$. A function $\psi(t)$ given on an interval $I$ of the real line is said to be *concave* (*convex*) if

$$\psi\left(\frac{t_1 + t_2}{2}\right) \geqslant \frac{1}{2}[\psi(t_1) + \psi(t_2)];$$

$$\left(\psi\left(\frac{t_1 + t_2}{2}\right) \leqslant \frac{1}{2}[\psi(t_1) + \psi(t_2)]\right) \qquad t_1, t_2 \in I. \tag{1.1}$$

The properties of concave (convex) functions are well known, and we only recall some of them without proof, referring the reader to the books [18] and [22].

---

* *Editor's note.* Actually these are Lebesgue spaces with continuous measure, in Rohlin's terminology (see [305], §2.4).

If a concave function is measurable, then it is continuous. If it is bounded below on some interval in the interior of an open interval $I$, then it is also continuous on $I$. In particular, a nonnegative function which is concave on an interval is continuous there. For a continuous concave function, (1.1) implies the Jensen inequality

$$\psi((1-\alpha)t_1+\alpha t_2) \geqslant (1-\alpha)\psi(t_1)+\alpha\psi(t_2). \tag{1.2}$$

A continuous concave function is absolutely continuous; it has a derivative everywhere except at countably many points, and this derivative is a decreasing function.([3])

In what follows, we shall be interested in positive concave functions on the semiaxis $(0,\infty)$. As we mentioned above, such a function is always continuous. Inequality (1.2) implies that $\psi((1-\alpha)t_1+\alpha t_2) > \alpha\psi(t_2)$. Letting $t_1$ tend to zero, we obtain that $\psi(\alpha t_2) \geqslant \alpha\psi(t_2)$. This says that if we set $\psi(0)=0$, then the concavity inequality (1.2) will be satisfied for all points of the closed semiaxis $[0,\infty)$. If in (1.2) we write $t=(1-\alpha)t_1+\alpha t_2$, then it takes the form

$$\psi(t) \geqslant \frac{t_2-t}{t_2-t_1}\psi(t_1)+\frac{t-t_1}{t_2-t_1}\psi(t_2) \qquad (t_1 \leqslant t \leqslant t_2). \tag{1.3}$$

From this it is obvious that $\psi(t)$ is increasing. Indeed, if we assume that $\psi(t)<\psi(t_1)$ $(t_1<t)$, then for sufficiently large $t_2$ (1.3) implies that $\psi(t_2)<0$. This contradicts the hypothesis.

Inequality (1.3) can be written also in the form

$$(\psi(t)-\psi(t_1))/(t-t_1) \geqslant (\psi(t_2)-\psi(t_1))/(t_2-t_1). \tag{1.4}$$

Thus, the concavity inequality (1.3) is equivalent to the condition that the function $(\psi(t)-\psi(t_1))/(t-t_1)$ is decreasing for any $t_1 \in [0,\infty)$. In particular, if we set $t_1=0$, we obtain that $\psi(t)/t$ is decreasing.

It is easy to verify that the infimum of any set of positive concave functions has the property (1.2), and consequently it will be either a positive concave function or identically zero. If now $\varphi(t)$ is a positive function having a concave majorant $\psi_0(t)$: $\varphi(t) \leqslant \psi_0(t)$ $(0<t<\infty)$, then the infimum of all concave majorants of $\varphi(t)$ will be the smallest concave majorant of $\varphi(t)$. We denote this majorant by $\tilde{\varphi}(t)$. We give its constructive description. From (1.2) we obtain by induction that

$$\psi\left(\sum_{i=1}^{n}\lambda_i t_i\right) \geqslant \sum_{i=1}^{n}\lambda_i\psi(t_i)$$

([3]) Following Bourbaki, we do not use the term nondecreasing (nonincreasing) function, but speak of increasing (decreasing) functions and strictly increasing (strictly decreasing) functions.

for any $n \geqslant 1$ and $\lambda_i \geqslant 0$, $\sum_{i=1}^{n} \lambda_i = 1$. Therefore, if we write $t = \sum_1^n \lambda_i t_i$, we have

$$\tilde{\varphi}(t) \geqslant \sum_{i=1}^{n} \lambda_i \tilde{\varphi}(t_i) \geqslant \sum_{i=1}^{n} \lambda_i \varphi(t_i).$$

We write

$$\psi(t) = \sup \sum_{i=1}^{n} \lambda_i \varphi(t_i)$$

$$\left(n = 1, 2, \ldots; \lambda_i \geqslant 0; \sum_1^n \lambda_i = 1; \sum_1^n \lambda_i t_i = t\right).$$

It follows from the above that $\varphi(t) \leqslant \psi(t) \leqslant \tilde{\varphi}(t)$. The function $\psi(t)$ is concave. Indeed, if $t', t'' \in (0, \infty)$, then there exist the $\lambda_i$, $\mu_i$, $t_i'$ and $t_i''$ such that

$$\sum_1^{n'} \lambda_i \varphi(t_i') \geqslant \psi(t') - \varepsilon, \qquad \sum_1^{n''} \mu_i \varphi(t_i'') \geqslant \psi(t'') - \varepsilon.$$

Then we have

$$(1 - \alpha) \sum_1^{n'} \lambda_i \varphi(t_i') + \alpha \sum_1^{n''} \mu_i \varphi(t_i'') \geqslant (1 - \alpha)\psi(t') + \alpha\psi(t'') - \varepsilon.$$

Since

$$\sum_1^{n''} (1 - \alpha)\lambda_i + \sum_1^{n''} \alpha\mu_i = 1$$

and

$$\sum_1^{n'} (1 - \alpha)\lambda_i t_i' + \sum_1^{n''} \alpha\mu_i t_i'' = (1 - \alpha)t' + \alpha t'',$$

it follows that

$$\psi((1 - \alpha)t' + \alpha t'') \geqslant (1 - \alpha)\psi(t') + \alpha\psi(t'') - \varepsilon.$$

Since $\varepsilon$ is arbitrary, it follows from this that the function is concave, and so by the definition of the smallest concave majorant we have $\psi(t) = \tilde{\varphi}(t)$. Thus,

$$\tilde{\varphi}(t) = \sup \sum_1^{n} \lambda_i \varphi(t_i)$$

$$\left(n = 1, 2, \ldots; \lambda_i \geqslant 0; \sum_1^n \lambda_i = 1; \sum_1^n \lambda_i t_i = t\right). \tag{1.5}$$

Two positive functions on $(0, \infty)$ are said to be *equivalent* if there exist positive constants $C_1$ and $C_2$ such that

$$C_1\psi(t) \leqslant \varphi(t) \leqslant C_2\psi(t)$$

for all $t > 0$.

THEOREM 1.1. *A positive function* $\varphi(t)$ *is equivalent to a positive concave function if and only if for any positive* $t_0$ *and* $t_1$ $(t_0 < t_1)$

$$\varphi(t_0) \leqslant k\varphi(t_1), \qquad \varphi(t_1)/t_1 \leqslant k\varphi(t_0)/t_0 \tag{1.6}$$

*with some constant* $k$. *Moreover, as equivalent concave function we may take the smallest concave majorant* $\tilde{\varphi}(t)$ *of* $\varphi(t)$.

PROOF. *Necessity*. From the definition of equivalent functions we obtain

$$\varphi(t_0) \leqslant C_2\psi(t_0) \leqslant C_2\psi(t_1) \leqslant (C_2/C_1)\varphi(t_1)$$

and

$$\varphi(t_1)/t_1 \leqslant C_2\psi(t_1)/t_1 \leqslant C_2\psi(t_0)/t_0 \leqslant (C_2/C_1)(\varphi(t_0)/t_0),$$

i.e., (1.6) is satisfied with $k = C_2/C_1$.

*Sufficiency*. By definition, $\tilde{\varphi}(t) \geqslant \varphi(t)$. Next, by (1.5) and (1.6) we have

$$\begin{aligned}\tilde{\varphi}(t) &= \sup \sum_1^n \lambda_i\varphi(t_i) \leqslant \sup \sum_1^n \lambda_i k \max\left\{1, \frac{t_i}{t}\right\}\varphi(t)\\ &\leqslant k \sup\left\{\sum_1^n \lambda_i + \sum_1^n \lambda_i \frac{t_i}{t}\right\}\varphi(t) = 2k\varphi(t).\end{aligned}$$

Hence $\tilde{\varphi}(t)/2k \leqslant \varphi(t) \leqslant \tilde{\varphi}(t)$.

DEFINITION 1.1. A function $\varphi(t)$ on the semiaxis $[0, \infty)$ is said to be *quasiconcave* if 1) $\varphi(0) = 0$, 2) $\varphi(t)$ is positive and increasing for $t > 0$, and 3) $\varphi(t)/t$ is decreasing for $t > 0$.

From the above it follows that a positive concave function on $(0, \infty)$, extended by zero to the origin is quasiconcave on $[0, \infty)$. The converse is not true in general. For example, the function defined by the equalities

$$\varphi(t) = \begin{cases} 0 & \text{for } t = 0,\\ \frac{1}{2} & \text{for } 0 < t \leqslant 1,\\ \frac{1}{2}x & \text{for } t > 1,\end{cases}$$

is quasiconcave but not concave.

Theorem 1.1 has the following corollary.

COROLLARY. *Every quasiconcave function* $\varphi(t)$ *is equivalent to its smallest concave majorant* $\tilde{\varphi}(t)$, *and*

$$\tfrac{1}{2}\tilde{\varphi}(t) \leqslant \varphi(t) \leqslant \tilde{\varphi}(t). \tag{1.7}$$

Every quasiconcave function is continuous for $t > 0$.

Indeed, if we had $\varphi(t_0 - 0) < \varphi(t_0 + 0)$ for some $t_0 > 0$, then for points $t'$ and $t''$, $t' < t_0 < t''$, sufficiently close to $t_0$ we would have $\varphi(t')/t' \leqslant \varphi(t'')/t''$, which contradicts Definition 1.1.

LEMMA 1.1. *Let $\Delta_i(t) = (t_i', t_i'')$, $i = 1, \ldots, n$, be a system of disjoint intervals on the semiaxis, and let $\sum_1^n |\Delta_i| = d$. Then, for any quasiconcave function $\varphi(t)$,*

$$\sum_{i=1}^{n} [\varphi(t_i'') - \varphi(t_i')] \leqslant 2\varphi(d). \tag{1.8}$$

*In particular, if $\varphi(+0) = 0$, then $\varphi(t)$ is absolutely continuous.*

PROOF. We split the sum on the left side of (1.8) into two: $\Sigma'$ and $\Sigma''$, where $\Sigma'$ is taken over those intervals which lie to the left of $d$ and $\Sigma''$ over those intervals which lie to the right of $d$ (if $d$ belongs to one of the intervals, then the right-hand part of it is taken into $\Sigma''$, and the left into $\Sigma'$). Then by the monotonicity of $\varphi(t)$ we have

$$\sum{}' [\varphi(t_i'') - \varphi(t_i')] \leqslant \varphi(d).$$

Next, for $d \leqslant t_i'$ we have

$$\begin{aligned}\varphi(t_i'') - \varphi(t_i') &= \frac{\varphi(t_i'')}{t_i''} t_i'' - \frac{\varphi(t_i')}{t_i'} t_i' \\ &\leqslant \frac{\varphi(t_i')}{t_i'}(t_i'' - t_i') \leqslant \frac{\varphi(d)}{d}(t_i'' - t_i')\end{aligned}$$

and consequently

$$\sum{}'' [\varphi(t_i'') - \varphi(t_i')] \leqslant \frac{\varphi(d)}{d} \sum{}'' (t_i'' - t_i') \leqslant \varphi(d).$$

Inequality (1.8) is established, and the second assertion of the lemma immediately follows from it. The lemma is proved.

We note that (1.8) is a sharp inequality. If we consider the example of the quasiconcave function on p. 49, then for the intervals $(0, \delta)$ and $(1, 2 - \delta)$ with arbitrarily small $\delta > 0$ we have $d = 1$ and

$$\varphi(\delta) - \varphi(0) + \varphi(2 - \delta) - \varphi(1) = \tfrac{1}{2} + \tfrac{1}{2}(1 - \delta) = 2\varphi(1) - \tfrac{1}{2}\delta.$$

If $\varphi(+0) > 0$, then the quasiconcave function is absolutely continuous on every ray $[a, \infty)$ $(a > 0)$. In order to see this it is sufficient to consider the quasiconcave function

$$\varphi_1(t) = \begin{cases} \frac{t}{a}\varphi(a) & \text{for } 0 \leqslant t \leqslant a, \\ \varphi(t) & \text{for } t \geqslant a. \end{cases}$$

The sum of two quasiconcave functions is obviously quasiconcave. If the supremum of a family of quasiconcave functions is finite at at least one point different from zero, then it is a quasiconcave function. The infimum has an analogous property if it is not identically zero.

DEFINITION 1.2. A positive function $u(t)$ given on an interval $I$ of the real axis is said to be *logarithmically convex* if the function $\ln u(t)$ is convex on $I$.

It follows from (1.1) that this definition is equivalent to the requirement that

$$u^2((t_1 + t_2)/2) \leqslant u(t_1)u(t_2). \tag{1.9}$$

If $u(t)$ is continuous, then the inequality (1.3) for a convex function implies that

$$u(t) \leqslant [u(t_1)]^{(t_2-t)/(t_2-t_1)}[u(t_2)]^{(t-t_1)/(t_2-t_1)}. \tag{1.10}$$

We note the following important property of logarithmically convex functions: *The sum of two logarithmically convex functions is logarithmically convex.*

For the proof we note that (1.9) is equivalent to the condition that the quadratic form

$$u(t_1)\xi^2 + 2u((t_1 + t_2)/2)\xi\eta + u(t_2)\eta^2$$

is nonnegative. Since the sum of two nonnegative quadratic forms is nonnegative, it is clear that the property of logarithmic convexity is preserved by addition.

Clearly the product of two logarithmically convex functions is also logarithmically convex.

DEFINITION 1.3. A function $\nu(t)$ defined on the axis $(-\infty, \infty)$ or the semiaxis $(0, \infty)$ is said to be *subadditive* if

$$\nu(t_1 + t_2) \leqslant \nu(t_1) + \nu(t_2). \tag{1.11}$$

The properties of subadditivity are expounded in detail in [18].

We note that, for example, every quasiconcave function on $(0, \infty)$ is subadditive on $(0, \infty)$. In fact,

$$\begin{aligned}\nu(t_1 + t_2) &= t_1\frac{\nu(t_1 + t_2)}{t_1 + t_2} + t_2\frac{\nu(t_1 + t_2)}{t_1 + t_2}\\ &\leqslant t_1\frac{\nu(t_1)}{t_1} + t_2\frac{\nu(t_2)}{t_2} = \nu(t_1) + \nu(t_2).\end{aligned}$$

If a subadditive function is different from $+\infty$ at every point, then it is bounded from above on any finite interval lying in its domain; if it is also different from $-\infty$, then it is bounded on every such interval. Hence an everywhere finite subadditive function can become unbounded only when approaching the endpoints of the interval where it is defined. Its behavior near the endpoints is described by the following important theorem, which we give with proof.

THEOREM 1.2. *If a function $\nu(t)$ is everywhere finite and subadditive on the whole axis $(-\infty, \infty)$, and if*

$$\inf_{t>0} \frac{\nu(t)}{t} = \beta \quad \textit{and} \quad \sup_{t<0} \frac{\nu(t)}{t} = \alpha, \tag{1.12}$$

*then*

$$\lim_{t\to+\infty} \frac{\nu(t)}{t} = \beta \quad \textit{and} \quad \lim_{t\to-\infty} \frac{\nu(t)}{t} = \alpha. \tag{1.13}$$

*Here* $-\infty < \alpha \leqslant \beta < +\infty$.

PROOF. Since $\nu(t)$ is everywhere finite, it follows that $\beta$ is either finite or equal to $-\infty$, and $\alpha$ is either finite or equal to $+\infty$. We assume that $\beta$ is finite. Then there exists $t_0 > 0$ such that $\nu(t_0)/t_0 < \beta + \varepsilon$. Now let $t > t_0$. There exists an integer $n$ such that $(n+1)t_0 \leqslant t \leqslant (n+2)t_0$. We have

$$\begin{aligned}\beta &\leqslant \frac{\nu(t)}{t} \leqslant \frac{\nu(nt_0)}{t} + \frac{\nu(t-nt_0)}{t} \leqslant \frac{n\nu(t_0)}{t} + \frac{\nu(t-nt_0)}{t} \\ &\leqslant \frac{nt_0}{(n+1)t_0}(\beta+\varepsilon) + \frac{\nu(t-nt_0)}{t}.\end{aligned}$$

The quantity $t - nt_0$ varies on the interval $[t_0, 2t_0]$ on which the function $\nu(t)$ is bounded. Therefore, the right side converges to $\beta + \varepsilon$ as $t \to \infty$. From this it follows that $\lim_{t\to\infty}(\nu(t)/t) = \beta$. This fact can be proved similarly in the case where $\beta = -\infty$. The second limit in (1.13) can be considered in the same way.

Next, we note that $\nu(0) \leqslant \nu(0) + \nu(0)$, and so $\nu(0) \geqslant 0$. Using (1.11) again, we obtain that $0 \leqslant \nu(0) \leqslant \nu(t) + \nu(-t)$. Dividing by $t$ and passing to the limit as $t \to \infty$, we obtain the inequality $\alpha \leqslant \beta$. Then $-\infty < -\nu(-1) \leqslant \alpha \leqslant \beta \leqslant \nu(1) < \infty$, and the theorem is completely proved.

The theorem just proved says that an everywhere finite subadditive function on the whole axis has the same asymptotic behavior at $+\infty$ and $-\infty$ as some linear functions. We note that the behavior of a subadditive function on the semiaxis can be more complicated (see [18]).

DEFINITION 1.4. A positive everywhere finite function $v(t)$ on $(0, \infty)$ is said to be *submultiplicative* if

$$v(t_1 t_2) \leqslant v(t_1)v(t_2). \tag{1.14}$$

It is obvious that if we introduce the variable $s = \ln t$, then the function $\nu(s) = \ln v(e^s)$ is an everywhere finite subadditive function on the whole axis $-\infty < s < \infty$. An application of Theorem 1.2 to $\nu(s)$ immediately leads to the following assertion.

THEOREM 1.3. *For every submultiplicative function $v(t)$ there exist numbers $\alpha$ and $\beta$ such that $-\infty < \alpha \leqslant \beta < \infty$ and*

$$v(t) \geqslant t^{\beta} \quad \text{for } t > 1 \qquad \text{and} \qquad v(t) \geqslant t^{\alpha} \quad \text{for } t < 1. \tag{1.15}$$

*For any $\varepsilon > 0$ we have $v(t) \leqslant t^{\beta+\varepsilon}$ for sufficiently large $t$, and $v(t) \leqslant t^{\alpha-\varepsilon}$ for $t$ sufficiently close to zero. Moreover,*

$$\alpha = \lim_{t\to 0} \frac{\ln v(t)}{\ln t} = \sup_{0<t<1} \frac{\ln v(t)}{\ln t}, \qquad \beta = \lim_{t\to\infty} \frac{\ln v(t)}{\ln t} = \inf_{t>1} \frac{\ln v(t)}{\ln t}.$$

COROLLARY 1. *If $v(t)$ is a submultiplicative function and $v(t_0) < t_0^{\mu_0}$ for some $t_0 > 1$ (for some $t_0 < 1$), then $v(t) < t^{\mu}$ for sufficiently large $t$ (for $t$ sufficiently close to zero), where $\mu$ is a number less than $\mu_0$ (greater than $\mu_0$).*

PROOF. If $v(t_0) < t_0^{\mu_0}$ for $t_0 > 1$, then $\beta \leqslant \ln v(t_0)[\ln t_0]^{-1} < \mu_0$. Choosing $\mu$ so that $\beta < \mu < \mu_0$ and applying Theorem 1.3 with $\varepsilon = \mu - \beta$, we arrive at the required assertion.

The second case can be considered similarly.

COROLLARY 2. *If $v(t)$ is a submultiplicative function and $v(t) = o(t^{\mu_0})$ as $t \to \infty$ (as $t \to 0$), then there exists a number $\mu_1 < \mu_0$ ($\mu_1 > \mu_0$) such that $v(t) = o(t^{\mu_1})$ as $t \to \infty$ ($t \to 0$).*

The proof immediately follows from the preceding corollary.

**2.** ***Dilation functions of positive functions on the semiaxis.*** Let $\psi(t)$ be a positive everywhere finite function on the semiaxis $(0, \infty)$. We introduce the function

$$M_{\psi}(s) = \sup_{0<t<\infty} \frac{\psi(st)}{\psi(t)} \qquad (0 < s < \infty), \tag{1.16}$$

which we call the *dilation function* of $\psi(t)$. Obviously

$$\inf_{0<t<\infty} \frac{\psi(st)}{\psi(t)} = \inf_{0<\tau<\infty} \frac{\psi(\tau)}{\psi(s^{-1}\tau)} = \left[ \sup_{0<\tau<\infty} \frac{\psi(s^{-1}\tau)}{\psi(\tau)} \right]^{-1}$$
$$= \left[ M_{\psi}(s^{-1}) \right]^{-1} \leqslant M_{\psi}(s). \tag{1.17}$$

The function $M_{\psi}(s)$ satisfies the submultiplicativity relation (1.14). Indeed,

$$M_{\psi}(s_1 s_2) = \sup_{0<t<\infty} \frac{\psi(s_1 s_2 t)}{\psi(t)} \leqslant \sup_{0<t<\infty} \frac{\psi(s_1 s_2 t)}{\psi(s_2 t)} \sup_{0<t<\infty} \frac{\psi(s_2 t)}{\psi(t)}$$
$$= M_{\psi}(s_1) M_{\psi}(s_2). \tag{1.18}$$

From this it follows that $M_{\psi}(s)$ is everywhere finite if it is bounded in the neighborhood of one.

If the function $M_\psi(s)$ is everywhere finite, then it is submultiplicative, and by Theorem 1.3 there exist numbers $\gamma_\psi$ and $\delta_\psi$ $(-\infty < \gamma_\psi \leqslant \delta_\psi < \infty)$ such that

$$M_\psi(s) \geqslant s^{\delta_\psi} \quad \text{for } s > 1 \qquad \text{and} \qquad M_\psi(s) \geqslant s^{\gamma_\psi} \quad \text{for } s < 1, \tag{1.19}$$

and, for any $\varepsilon > 0$, for sufficiently large $s$

$$M_\psi(s) \leqslant s^{\delta_\psi + \varepsilon}, \tag{1.20}$$

and for sufficiently small $s$

$$M_\psi(s) \leqslant s^{\gamma_\psi - \varepsilon}. \tag{1.21}$$

The numbers $\gamma_\psi$ and $\delta_\psi$ will be called *lower* and *upper dilation exponents of* $\psi(t)$.

Combining (1.17), (1.20), and (1.21), we may say that for sufficiently large $s$

$$s^{\gamma_\psi - \varepsilon} \leqslant \left[M_\psi(s^{-1})\right]^{-1} \leqslant M_\psi(s) \leqslant s^{\delta_\psi + \varepsilon}. \tag{1.22}$$

If the positive function $\varphi(t)$ is increasing, then $M_\varphi(s) \leqslant 1$ for $s < 1$, and consequently by (1.19) we have $s^{\gamma_\varphi} \leqslant 1$, i.e. $\gamma_\varphi \geqslant 0$. Let the function be quasiconcave. Then $\varphi(st)/st \leqslant \varphi(t)/t$ for $s > 1$, and hence $\varphi(st)/\varphi(t) \leqslant s$. From this it follows that $M_\varphi(s) \leqslant s$, and thus $\delta_\varphi \leqslant 1$. Thus, for a quasiconcave function $\varphi(t)$ we have

$$1 \leqslant \left[M_\varphi(s^{-1})\right]^{-1} \leqslant M_\varphi(s) \leqslant s \quad \text{for } s > 1, \tag{1.23}$$

$$0 \leqslant \gamma_\varphi \leqslant \delta_\varphi \leqslant 1. \tag{1.24}$$

From the definition of dilation exponents it is clear that they coincide for equivalent functions, and therefore, if a function is equivalent to a concave function, then it satisfies (1.24). In testing whether a function is equivalent to a concave function, the following lemma is useful.

LEMMA 1.2. *If for a function $\varphi(t)$ the inequalities $M_\varphi(s_0^{-1}) \leqslant 1$ and $M_\varphi(s_1) \leqslant s_1$ hold for some $s_0$, $s_1 > 1$ and the function $M_\varphi(s)$ is bounded on some interval $[1, a]$, then $\varphi(t)$ is equivalent to its smallest concave majorant.*

PROOF. It is sufficient to verify that for any $t_0 < t_1$ the function $\varphi(t)$ satisfies (1.6) with some constant $k$.

We write $C = \sup_{1 \leqslant s \leqslant a} M_\varphi(s)$. From the submultiplicativity inequality (1.18) it follows that $M_\varphi(s)$ does not exceed the number $C^r$ on $[1, a^r]$. If we choose the integer $r$ so that $a^r \geqslant s_0$, then $M_\varphi(s)$ will not exceed $C^r$ on $[1, s_0]$. This implies that the inequality $\varphi(t) \leqslant \varphi(t_1)M_\varphi(t/t_1) \leqslant C^r\varphi(t_1)$ is satisfied on $[t_1, s_0t_1]$. Let an integer $i$ be chosen so that $s_0^i t_0 \leqslant t_1 \leqslant s_0^{i+1}t_0$. The condition $M_\varphi(s_0^{-1}) \leqslant 1$ implies that $\varphi(t) \leqslant \varphi(s_0t)$. Therefore

$$\varphi(t_0) \leqslant \varphi(s_0^{i+1}t_0) \leqslant C^r\varphi(t_1). \tag{1.25}$$

We now choose an integer $j$ so that $s_1^{-(j+1)}t_1 \leqslant t_0 \leqslant s_1^{-j}t_1$. Then from (1.25), with $t_1$ replaced by $t_0$ and $t_0$ by $s_1^{-(j+1)}t_1$, we obtain

$$\varphi(s_1^{-j-1}t_1) \leqslant C^r\varphi(t_0).$$

The condition $M_\varphi(s_1) \leqslant s_1$ implies that $s_1^{-1}\varphi(t) \leqslant \varphi(t/s_1)$, and therefore

$$\varphi(s_1^{-j-1}t_1) \geqslant s_1^{-j-1}\varphi(t_1) \geqslant s_1^{-1}(t_0/t_1)\varphi(t_1).$$

Combining this inequality with the previous one, we see that

$$\varphi(t_1)/t_1 \leqslant s_1C^r\varphi(t_0)/t_0. \tag{1.26}$$

Inequalities (1.25) and (1.26) imply that (1.6) is satisfied with $k = s_1C^r$.

COROLLARY 1. *If the function $\psi(t)$ is increasing and $M_\psi(s_1) \leqslant s_1$ for some $s_1 > 1$, then it is equivalent to its smallest concave majorant.*

Indeed, because of monotonicity, $M_\psi(s_0^{-1}) < 1$ for any $s_0 > 1$. Moreover, for any $s \leqslant s_1$ we have $\psi(st) \leqslant \psi(s_1t) \leqslant s_1\psi(t)$, i.e. $M_\psi(s) \leqslant s_1$ on $[0, s_1]$.

In this case we may set $a = C = s_1$, and $r = 1$, and then (1.26) becomes

$$\varphi(t_1)/t_1 \leqslant s_1^2\varphi(t_0)/t_0.$$

COROLLARY 2. *If $0 < \gamma_\psi \leqslant \delta_\psi < 1$, then $\psi(t)$ is equivalent to its smallest concave majorant.*

PROOF. If we choose $\varepsilon$ so that $\gamma_\psi - \varepsilon > 0$ and $\delta_\psi + \varepsilon < 1$, then (1.22) implies that

$$1 < s_1^{\gamma_\psi - \varepsilon} \leqslant \left[M_\psi(s_1^{-1})\right]^{-1} \leqslant M_\psi(s_1) \leqslant s_1^{\delta_\psi + \varepsilon} < s_1 \tag{1.27}$$

for sufficiently large $s_1$. Next, for fixed $a > 1$ and $1 \leqslant s \leqslant a$

$$M_\psi(s) \leqslant M_\psi(s_1)M_\psi(s/s_1) \leqslant s_1^{\delta_\psi+\varepsilon}(s/s_1)^{\gamma_\psi-\varepsilon} \leqslant s_1^{\delta_\psi-\gamma_\psi+2\varepsilon}a^{\gamma_\psi-\varepsilon}$$

for sufficiently large $s_1$. Hence, all hypotheses of Lemma 1.2 are satisfied.

REMARK. Under the hypotheses of Corollary 2 the equivalence constant depends on $\gamma_\psi$, $\delta_\psi$, and that $s_1$ for which (1.27) is satisfied.

The assertion of Corollary 2 is in a certain sense unimprovable. For example, the function $\psi_0(t) = \max\{t, t\ln t\}$ is increasing more rapidly than a linear function at infinity. Therefore it is not equivalent to a concave function. We have $M_{\psi_0}(s) = \max\{s, s(1 + \ln s)\}$, and consequently $0 < \gamma_{\psi_0} = \delta_{\psi_0} = 1$. Thus, under the hypothesis of Corollary 2, the inequalities cannot be replaced by equalities.

LEMMA 1.3. *If we have $M_\psi(s_0^{-1}) = 1$ $(s_0 > 1)$, for a positive concave function $\psi$, then $M_\psi(s^{-1}) \equiv 1$ $(s \geqslant 1)$. If we have $M_\psi(s_1) = s_1$ $(s_1 > 1)$, then $M_\psi(s) \equiv s$ $(s \geqslant 1)$.*

PROOF. In consequence of (1.17), the function $[M_\psi(s^{-1})]^{-1}$ is the infimum of the concave functions $\psi(st)/\psi(t)$ as functions of $s$, and therefore it is concave. By (1.23) we have $[M_\psi(s^{-1})]^{-1} \geqslant 1$ for $s > 1$, and if this concave function assumes its minimal value 1 at $s = s_0 > 1$, then it is a constant: $[M_\psi(s^{-1})]^{-1} \equiv 1$ $(s > 1)$.

The function $s^{-1}M_\psi(s) = \sup_{0<t<\infty} (\psi(st)/st) \cdot (t/\psi(t))$ is decreasing because of the concavity of $\psi(t)$. Therefore the equality $s_1^{-1}M_\psi(s_1) = 1$ and (1.23) imply that $s^{-1}M_\psi(s) = 1$ for $1 < s \leqslant s_1$. Now let $1 < s_1 < s$. Then

$$\psi(st) \geqslant \frac{s(s_1 - 1)}{(s-1)s_1}\psi(s_1 t) + \frac{s - s_1}{(s-1)s_1}\psi(ss_1 t).$$

Hence

$$\frac{\psi(s_1 t)}{\psi(ss_1 t)} \leqslant \frac{s_1(s-1)}{s(s_1-1)}\frac{\psi(st)}{\psi(ss_1 t)} - \frac{s - s_1}{s(s_1 - 1)}.$$

Then

$$\begin{aligned} M_\psi(s) &= \left[\inf_{0<t<\infty} \frac{\psi(s_1 t)}{\psi(ss_1 t)}\right]^{-1} \\ &\geqslant \left[\inf_{0<t<\infty} \left\{\frac{s_1(s-1)}{s(s_1-1)}\frac{\psi(st)}{\psi(ss_1 t)} - \frac{s - s_1}{s(s_1 - 1)}\right\}\right]^{-1} = s. \end{aligned}$$

By (1.23) we have $M_\psi(s) \equiv s$. The lemma is proved.

We note some further properties of the lower and upper dilation exponents. It is obvious that

$$M_{(\psi)^\theta}(s) = [M_\psi(s)]^\theta.$$

Therefore $\gamma_{(\psi)^\theta} = \theta\gamma_\psi$ and $\delta_{(\psi)^\theta} = \theta\delta_\psi$.

Next, $M_{\varphi_0\varphi_1}(s) \leqslant M_{\varphi_0}(s)M_{\varphi_1}(s)$, and therefore

$$\gamma_{\varphi_0\varphi_1} = \lim_{t\to 0}\frac{\ln M_{\varphi_0\varphi_1}(t)}{\ln t} \leqslant \lim_{t\to 0}\frac{\ln M_{\varphi_0}(t) + \ln M_{\varphi_1}(t)}{\ln t} = \gamma_{\varphi_0} + \gamma_{\varphi_1}. \tag{1.28}$$

Similarly,

$$\delta_{\varphi_0\varphi_1} \leqslant \delta_{\varphi_0} + \delta_{\varphi_1}. \tag{1.29}$$

**3. *The study of the integral* $\int_0^t \psi(\tau)\tau^{-1}\,d\tau$.** For a positive function $\psi(t)$ on the semiaxis $(0, \infty)$ we wish to compare the behavior of the indicated integral with that of the function $\psi(t)$ itself.

LEMMA 1.4. *If $M_\psi(s_0^{-1}) < 1$ for some $s_0 > 1$, and $M_\psi(s)$ is bounded on some interval $[1, a]$, then the function $\int_0^t \psi(\tau)\tau^{-1}\,d\tau$ is equivalent to $\psi(t)$.*

PROOF. It follows from the proof of Lemma 1.2 that, under our hypotheses, $\psi(\tau) \leqslant C'\psi(t_1)$ for any $t_1$ and $\tau \in [t_1, s_0 t_1]$. Setting $t_1 = s_0^{-j+1}t$ successively in this inequality, we obtain

$$\begin{aligned}\int_0^t \frac{\psi(\tau)}{\tau}\,d\tau &= \sum_{j=1}^{\infty} \int_{s_0^{-j}t}^{s_0^{-j+1}t} \frac{\psi(\tau)}{\tau}\,d\tau \\ &\leqslant C' \sum_{j=1}^{\infty} \psi(s_0^{-j+1}t)\ln s_0 \leqslant C' \ln s_0 \sum_{j=1}^{\infty} \left[M_\psi(s_0^{-1})\right]^{j-1}\psi(t) \\ &= C' \ln s_0\left[1 - M_\psi(s_0^{-1})\right]^{-1}\psi(t) = C_2\psi(t).\end{aligned}$$

On the other hand,

$$\begin{aligned}\int_0^t \frac{\psi(\tau)}{\tau}\,d\tau &\geqslant \int_{ta^{-1}}^t \frac{\psi(\tau)}{\tau}\,d\tau \geqslant \ln a \inf_{ta^{-1}\leqslant\tau\leqslant t} \psi(\tau) \\ &\geqslant \ln a\left[\sup_{1\leqslant s\leqslant a} M_\psi(s)\right]^{-1}\psi(t) = C_1\psi(t).\end{aligned}$$

Thus,

$$C_1\psi(t) \leqslant \int_0^t \frac{\psi(\tau)}{\tau}\,d\tau \leqslant C_2\psi(t). \tag{1.30}$$

COROLLARY 1. *If the function $\psi(t)$ is increasing, and $M_\psi(s_0^{-1}) < 1$ and $M_\psi(s_1) < \infty$ for some $s_0$, $s_1 > 1$, then* (1.30) *holds.*

COROLLARY 2. *If the positive function $\psi(t)$ is concave and $M_\psi(s_0^{-1}) < 1$ for some $s_0 > 1$, then* (1.30) *holds.*

COROLLARY 3. *If $0 < \gamma_\psi \leqslant \delta_\psi < \infty$, then* (1.30) *holds.*

Lemma 1.4 immediately implies the following lemma.

LEMMA 1.5. *If $M_\psi(s_0) < 1$ for some $s_0 > 1$ and $M_\psi(s)$ is bounded on an interval $[1/a, 1]$ $(a > 1)$, then*

$$C_1\psi(\tau) \leqslant \int_t^\infty \frac{\psi(\tau)}{\tau}\,d\tau \leqslant C_2\psi(t). \tag{1.31}$$

Indeed, the changes $\tau = 1/z$, $\theta = 1/t$, $\psi(1/\theta) = \psi_1(\theta)$ reduce (1.31) for $\psi(t)$ to (1.30) for $\psi_1(\theta)$. We also have to take into account that $M_{\psi_1}(s) = M_\psi(s^{-1})$.

Of the three corollaries which follow from this lemma similarly to the preceding one, we mention only one.

COROLLARY. *If $-\infty < \gamma_\psi \leqslant \delta_\psi < 0$, then* (1.31) *holds.*

## §2. Rearrangements of measurable functions

**1.** ***Equimeasurable functions. Rearrangements.*** We shall consider the metric space $S(0, \infty)$ of Lebesgue measurable functions, finite almost everywhere on the semiaxis $(0, \infty)$. For every nonnegative function $x \in S(0, \infty)$ we introduce its *distribution function* $n_x(\tau)$ defined by the formula

$$n_x(\tau) = \operatorname{mes}\{t: x(t) > \tau\}.$$

The distribution function is decreasing, right continuous, and may take on infinite values. The collection of all functions $x(t)$ for which $n_{|x|}(\tau) \not\equiv \infty$, will be denoted by $S_0(0, \infty)$. For every function from $S_0(0, \infty)$ we have $n_{|x|}(\tau) \to 0$ as $\tau \to \infty$. Indeed, if $n_{|x|}(\tau_0) < \infty$, then for $\tau > \tau_0$ the measure of the sets $\{t: |x(t)| > \tau\}$ is finite and their intersection has measure zero. Therefore $\operatorname{mes}\{t: |x(t)| > \tau\} \to 0$ as $\tau \to \infty$. In what follows we consider only functions in $S_0(0, \infty)$.

If a function $x(t)$ is summable on $(0, \infty)$, then

$$\|x\|_{L_1} = \int_0^\infty |x(t)|\, dt = \int_0^\infty n_{|x|}(\tau)\, d\tau. \tag{2.1}$$

DEFINITION 2.1. Two nonnegative functions $x(t)$ and $y(t)$ in $S_0(0, \infty)$ are said to be *equimeasurable* if

$$n_x(\tau) = n_y(\tau).$$

The equimeasurability of $x(t)$ and $y(t)$ does not imply in general the equality

$$\operatorname{mes}\{t: x(t) \geqslant \tau\} = \operatorname{mes}\{t: y(t) \geqslant \tau\}. \tag{2.2}$$

For example, the functions $x(t) = \frac{2}{\pi} \operatorname{arctg} t$ and $y(t) \equiv 1$ are equimeasurable. However, $\operatorname{mes}\{t: x(t) \geqslant 1\} = 0$, and $\operatorname{mes}\{t: y(t) \geqslant 1\} = \infty$.

Equality (2.2) will hold for $\tau$ interior to the interval where $n_x(\tau) < \infty$. Indeed, in this case

$$\operatorname{mes}\{t: x(t) \geqslant \tau\} = \lim_{\varepsilon\to 0} \operatorname{mes}\{t: x(t) > \tau - \varepsilon\} = \lim_{\varepsilon\to 0} n_x(\tau - \varepsilon)$$

and similarly $\operatorname{mes}\{t: y(t) \geqslant \tau\} = \lim n_y(\tau - \varepsilon)$. The condition $n_x(\tau) = n_y(\tau)$ $(0 \leqslant \tau < \infty)$, implies (2.2).

We also note that for two equimeasurable functions $x(t)$ and $y(t)$ the functions $x(t)\chi_{e_a}(t)$ and $y(t)\chi_{e'_a}(t)$, will also be equimeasurable, where $\chi_{e_a}(t)$ and $\chi'_{e_a}(t)$ are the characteristic functions of the Lebesgue sets $e_a = \{t: x(t) > a\}$ and $e'_a = \{t: y(t) > a\}$, respectively. If $n_x(\tau) < \infty$ for $\tau > 0$, then the sets $e_a$ and $e'_a$ can be replaced by $\{t: x(t) \geqslant a\}$ and $\{t: y(t) \geqslant a\}$.

Let $t \to \omega(t)$ be a measurable mapping of the semiaxis $(0, \infty)$ into itself. We shall say that $\omega(t)$ *does not decrease the measure of sets* if $\operatorname{mes} \omega^{-1}(e) \leqslant \operatorname{mes} e$, and that it *preserves the measure of sets* if mes $\omega(e) =$ mes $e$, for any measurable set $e$. Since $\omega(\omega^{-1}(e)) \subset e$, every measure preserving mapping does not decrease measure.

If $\omega(t)$ is a mapping which does not decrease measure, then

$$\operatorname{mes}\{t: y(\omega(t)) > \tau\} \leqslant \operatorname{mes}\{s: y(s) > \tau\},$$

and so for nonnegative functions $y(t)$ and $y_\omega(t) \equiv y(\omega(t))$ we have

$$n_{y_\omega}(\tau) \leqslant n_y(\tau).$$

It is obvious that $\|y_\omega\|_{L_\infty} \leqslant \|y\|_{L_\infty}$, and (2.1) implies that

$$\|y_\omega\|_{L_1} \leqslant \|y\|_{L_1}.$$

DEFINITION 2.2. The *rearrangement* of a nonnegative function $x \in S_0(0, \infty)$ is a decreasing left continuous function $x^*(t)$ equimeasurable with $x(t)$.

We show that the rearrangement is unique. Let $x^*(t)$ and $x_1^*(t)$ be two different rearrangements of the same function $x(t)$. Assume that $x_1^*(t_0) > x^*(t_0)$. Let $\varepsilon > 0$ be such that $x_1^*(t_0) - \varepsilon > x^*(t_0)$. By the left continuity of $x^*(t)$ there exists an interval $[t_1, t_0]$ in which $x^*(t) < x_1^*(t_0) - \varepsilon$. Then

$$\operatorname{mes}\{t: x_1^*(t) > x_1^*(t_0) - \varepsilon\} \geqslant t_0,$$

while

$$\operatorname{mes}\{t: x^*(t) > x^*(t_0) - \varepsilon\} \leqslant t_1 < t_0.$$

The functions $x^*(t)$ and $x_1^*(t)$ are not equimeasurable, which contradicts their equimeasurability with $x(t)$.

Consequently, the rearrangement is unique. It can be defined by the formula

$$x^*(t) = \inf\{\tau: n_x(\tau) < t\}.$$

It is easy to verify that $n_x(x^*(t)) = t$, if $x^*(t)$ is a point of continuity of $n_x(\tau)$, and $n_x(x^*(t)) \leqslant t$, if $x^*(t)$ is a point of discontinuity of $n_x(\tau)$. Similarly, $x^*(n_x(\tau)) = \tau$, if $n_x(\tau)$ is a point of continuity of $x^*(t)$ and $x^*(n_x(\tau)) \geqslant \tau$, if $n_x(\tau)$ is a point of discontinuity of $x^*(t)$.

We denote by $x^*(t)$ the rearrangement of the absolute value of an arbitrary function $x(t)$ in $S_0(0, \infty)$.

It is easy to calculate the rearrangements of strictly simple functions. Every strictly simple function can be written in the form

$$x(t) = \sum_{k=1}^{N} x_k \chi_{e_k}(t),$$

where the $e_k$ are disjoint sets of finite measure and $x_i \neq x_j$ for $i \neq j$. Then

$$x^*(t) = \sum_{j=1}^{N} x_j^* \chi_{(\tau_{j-1}, \tau_j]},$$

where $x_j^*$ is the rearrangement of the numbers $|x_k|$ in decreasing order and $\tau_j = \sum_{|x_i| \geqslant x_j^*} \operatorname{mes} e_i$ $(\tau_0 = 0)$.

It is sometimes convenient to write a nonnegative strictly simple function in the form

$$x(t) = \sum_{k=1}^{N} \Delta_k \chi_{e_k'},$$

where $\Delta_k > 0$ and $e_1' \subset e_2' \subset \cdots \subset e_N'$ are sets of finite measure. For $t > 0$ we then have

$$x^*(t) = \sum_{k=1}^{N} \Delta_k \chi_{(0, \operatorname{mes} e_k']}. \tag{2.3}$$

We introduce the number

$$x^*(\infty) = \lim_{t \to \infty} x^*(t) = \inf\{\tau \colon n_x(\tau) < \infty\}.$$

This number has the property that

$$n_x(\tau) \begin{cases} < \infty & \text{for } \tau > x^*(\infty), \\ \leqslant \infty & \text{for } \tau = x^*(\infty), \\ = \infty & \text{for } \tau < x^*(\infty). \end{cases}$$

LEMMA 2.1. *Let $x(t)$ and $y(t)$ be two nonnegative equimeasurable functions. For any $\varepsilon > 0$ there exists a measure preserving mapping $\omega(t)$ of the semiaxis $(0, \infty)$ into itself such that*

$$\|x_1 - y_\omega\|_{L_\infty \cap L_1} \leqslant \varepsilon, \tag{2.4}$$

*where $x_1(t) = \max\{x(t), x^*(\infty)\}$ and $y_\omega(t) = y(\omega(t))$.*

PROOF. We set $\tau_0 = x^*(\infty) + \varepsilon/8$ and partition the semiaxis into three sets:

$$E_x^1 = \{t \colon \tau_0 < x(t)\}, \qquad E_x^2 = \{t \colon x^*(\infty) < x(t) \leqslant \tau_0\},$$
$$E_x^3 = \{t \colon x(t) \leqslant x^*(\infty)\}.$$

The mapping $\omega(t)$ will be defined separately on each of these sets. For the function $y(t)$ we introduce three similar sets $E_y^i$, $i = 1, 2, 3$, having in mind that $x^*(\infty) = y^*(\infty)$.

*The definition of $\omega(t)$ on $E_x^1$.* The set $E_x^1$ has finite measure. We choose $\delta = \min\{\varepsilon, \varepsilon/(2 \operatorname{mes} E_x^1)\}$ and consider the sets $e_{k,x}^1 = \{t \colon \tau_0 + k\delta < x(t) \leqslant \tau_0 + (k+1)\delta\}$ and $e_{k,y}^1 = \{t \colon \tau_0 + k\delta < y(t) \leqslant \tau_0 + (k+1)\delta\}$ $(k = 0, 1, \ldots)$. They have equal measures for the same $k$, and $E_x^1 = \bigcup_k e_{k,x}^1$ and

$E_y^1 = \bigcup_k e_{k,y}^1$. On every set $e_{k,x}^1$ we define the measure preserving mapping $\omega(t)$ onto $e_{k,y}^1$. Then, by construction,

$$|x(t) - y(\omega(t))| \leqslant \delta \leqslant \varepsilon$$

and

$$\int_{E_x^1} |x(t) - y(\omega(t))|\, dt \leqslant \delta \operatorname{mes} E_x^1 < \varepsilon/2. \tag{2.5}$$

*The definition of* $\omega(t)$ *on* $E_x^2$. If the measure of the set $E_x^2$ is finite, then $\omega(t)$ is defined on it in the same way as on $E_x^1$ with $\tau_0$ replaced by $x^*(\infty)$ and $\varepsilon$ by $\varepsilon/2$. Then (2.5) will be satisfied. The case where $\operatorname{mes} E_x^2 = \operatorname{mes} E_y^2 = \infty$ is more complicated. In this case the sets $E_x^2$ and $E_y^2$ can be partitioned into disjoint sets $e_{k,x}^2$ and $e_{k,y}^2$, $k = 0, 1, \ldots,$ respectively, so that $\operatorname{mes} e_{k,x}^2 = \operatorname{mes} e_{k,y}^2 = 1$,

$$\tau_{k+1} \leqslant x(t) \leqslant \tau_k \qquad \left(t \in e_{k,x}^2\right)$$

and

$$\tau_{k+1} \leqslant y(t) \leqslant \tau_k \qquad \left(t \in e_{k,y}^2\right),$$

where $\tau_k \downarrow x^*(\infty)$ as $k \to \infty$. We choose an infinite sequence of natural numbers $m_1, m_2, \ldots$ so that

$$\sum_{i=1}^{\infty} \left(\tau_{m_i+1} - x^*(\infty)\right) < \infty/8.$$

Take the union $F_y$ of the sets $e_{m_i,y}^2$. Denote the remaining sets by $e_{n_k,y}^2$ ($k = 0, 1, \ldots;\ n_k < n_{k+1}$). We define the function $\omega(t)$ so that it maps $e_{k,x}^2$ onto $e_{n_k,y}^2$, preserving measure.

Then

$$\int_{E_x^2} |x(t) - y_\omega(t)|\, dt \leqslant \sum_{k=0}^{\infty} \int_{e_{k,x}^2} |x(t) - y_\omega(t)|\, dt$$

$$\leqslant \sum_{k=0}^{\infty} (\tau_k - \tau_{n_k+1}).$$

We estimate the last sum. For any $N > 0$ we have

$$\sum_{k=0}^{N} (\tau_k - \tau_{n_k+1}) = \sum_{k=0}^{N} \tau_k - \sum_{n_k+1 \leqslant N} \tau_{n_k+1} - \sum_{n_k+1 > N,\, k \leqslant N} \tau_{n_k+1}$$

$$= \tau_0 + \sum_{m_i+1 \leqslant N} \tau_{m_i+1} - \sum_{\substack{k \leqslant N \\ n_k+1 > N}} \tau_{n_k+1}$$

$$\leqslant \tau_0 - x^*(\infty) + \sum_{i=1}^{\infty} \left(\tau_{m_i+1} - x^*(\infty)\right) < \varepsilon/4.$$

Thus

$$|x(t) - y_\omega(t)| \leqslant \tau_k - \tau_{n_k+1} \leqslant \tau_{0^-} - x^*(\infty) \leqslant \varepsilon \qquad (t \in E_x^2) \tag{2.6}$$

and

$$\int_{E_x^2} |x(t) - y_\omega(t)|\, dt < \frac{\varepsilon}{4}.$$

*The definition of* $\omega(t)$ *on* $E_x^3$. On the set $E_x^3$ we have $x_1(t) \equiv x^*(\infty)$, and so the mapping $\omega(t)$ has to send $E_x^3$ into parts of the semiaxis where $y(t)$ is sufficiently close to $x^*(\infty) = y^*(\infty)$.

If mes $E_x^2 = \infty$, then for this purpose we use the set $F_y$ chosen above. If mes $E_x^3 < \infty$, then in $F_y$ we choose a set $e_y$ of the same measure so that

$$x^*(\infty) \leqslant y^*(t) \leqslant x^*(\infty) + \min\{\varepsilon, \varepsilon/(4 \operatorname{mes} E_x^3)\}$$

on $e_y$, and we map $E_x^3$ onto $e_y$ in a measure preserving way. Then

$$|x^*(\infty) - y(\omega(t))| \leqslant \varepsilon$$

on $E_x^3$ and

$$\int_{E_x^3} |x^*(\infty) - y(\omega(t))|\, dt \leqslant \frac{\varepsilon}{4}. \tag{2.7}$$

If, however, mes $E_x^3 = \infty$, then we partition $E_x^2$ into sets $e_{k,x}^3$ ($k = 1, 2, \ldots$) of measure one in an arbitrary way, and map them with preservation of measure onto sets $e_{m_{i_k},y}^2$ for which $\tau_{m_{i_k}} \leqslant x^*(\infty) + \varepsilon/2^{k+2}$. Then (2.7) will also be satisfied.

It remains to consider the case where mes $E_x^2 =$ mes $E_y^2 < \infty$. But then mes $E_x^3 =$ mes $E_y^3 = \infty$, and for any $\tau < x^*(\infty)$ we have $\operatorname{mes}\{t\colon \tau < y(t) \leqslant x^*(\infty)\} = \infty$. Therefore the set $E_y^3$ may now play the same role as $F_y$ did above. Repeating the preceding constructions, we obtain a mapping with the properties (2.7).

Thus, we have defined a measure preserving mapping $\omega(t)$ on the whole semiaxis, and by (2.5)–(2.7) we have

$$|x_1(t) - y(\omega(t))| \leqslant \varepsilon \quad \text{and} \quad \int_0^\infty |x_1(t) - y(\omega(t))|\, dt < \varepsilon.$$

The lemma implies the following important theorem.

THEOREM 2.1. *If for two measurable functions* $x(t)$ *and* $y(t)$ *the inequality* $x^*(t) \leqslant y^*(t)$ *holds for all* $t > 0$, *then for any* $\varepsilon > 0$ *there exist a measure preserving mapping* $\omega(t)$ *of the semiaxis* $(0, \infty)$ *into itself and a measurable function* $\beta(t)$ *with* $|\beta(t)| \leqslant 1$ *such that*

$$\|x - \beta y_\omega\|_{L_\infty \cap L_1} \leqslant \varepsilon \qquad (y_\omega(t) = y(\omega(t))). \tag{2.8}$$

PROOF. The function $y^*(t)$ is equimeasurable with $|y(t)|$, and therefore by Lemma 2.1 and the monotonicity of $y^*(t)$ there exists a measure preserving mapping $\omega_1(t)$ such that $\|y^* - |y|_{\omega_1}\|_{L_\infty \cap L_1} < \varepsilon/2$. We write $\alpha_1(t) = x^*(t)[y^*(t)]^{-1}$ (assuming that $\alpha(t) = 0$ if $y^*(t) = 0$); then $0 \leqslant \alpha_1(t) \leqslant 1$ and

$$\|x^* - \alpha_1|y|_{\omega_1}\|_{L_\infty \cap L_1} = \|\alpha_1(y^* - |y|_{\omega_1})\|_{L_\infty \cap L_1} < \varepsilon/2. \tag{2.9}$$

Similarly, by the equimeasurability of $x^*$ and $|x|$, there exists a measure preserving mapping $\omega_2(t)$ such that

$$\|x_1 - (x^*)\omega_2\|_{L_\infty \cap L_1} < \varepsilon/2,$$

where $x_1(t) = \max\{|x(t)|, x^*(\infty)\}$. Write $\alpha_2(t) = |x(t)|[x_1(t)]^{-1}$. Then $0 \leqslant \alpha_2(t) \leqslant 1$ and

$$\| \, |x| - \alpha_2(x^*)\omega_2\|_{L_\infty \cap L_1} < \varepsilon/2. \tag{2.10}$$

We choose $\omega(t) = \omega_1(\omega_2(t))$ and

$$\beta(t) = \alpha_2(t)\alpha_1(\omega_2(t))\operatorname{sgn} x(t)\operatorname{sgn} y_\omega(t).$$

Then (2.9) and (2.10) imply that

$$\begin{aligned}\|x - \beta y_\omega\|_{L_\infty \cap L_1} &= \| \, |x| - \alpha_2(\alpha_1)_{\omega_2}|y|_{\omega_1 \circ \omega_2}\|_{L_\infty \cap L_1} \\ &\leqslant \| \, |x| - \alpha_2(x^*)_{\omega_2}\|_{L_\infty \cap L_1} + \|\alpha_2[(x^*)_{\omega_2} - (\alpha_1)_{\omega_2}|y|_{\omega_1 \circ \omega_2}]\|_{L_\infty \cap L_1} \\ &< \frac{\varepsilon}{2} + \|x^* - \alpha_1|y|_{\omega_1}\|_{L_\infty \cap L_1} < \varepsilon.\end{aligned}$$

**2. *The properties of rearrangements.***

1°. *The equimeasurability of $|x(t)|$ and $x^*(t)$ implies that*

$$\int_0^\infty |x(t)|\, dt = \int_0^\infty x^*(t)\, dt = \int_0^\infty n_{|x|}(\tau)\, d\tau, \tag{2.11}$$

*where both sides may be infinite.*

2°. *If $|x(t)| \leqslant |y(t)|$, then $x^*(t) \leqslant y^*(t)$.*

3°. *If $x(t) = y(t)$ everywhere except for a set of measure not exceeding a number $\gamma < \infty$, then $x^*(t) \leqslant y^*(t - \gamma)$ and $y^*(t) \leqslant x^*(t - \gamma)$ $(t > \gamma)$.*

Indeed, let $x^*(t_0) = \tau_0$ $(t_0 > \gamma)$. For any $\varepsilon > 0$ there exists a set $e$ of measure $t_0$ on which $|x(s)| > \tau_0 - \varepsilon$. Then we can construct a set $e'$ with $\operatorname{mes} e' \geqslant t_0 - \gamma$ on which $|y(s)| > \tau_0 - \varepsilon$. From the definition of $y^*$ it follows that $y^*(t_0 - \gamma) > \tau_0 - \varepsilon = x^*(t_0) - \varepsilon$. Due to the arbitrariness of $\varepsilon$ and the equimeasurability of $x$ and $y$ the assertion is proved.

4°. *If $x(t) \geqslant 0$ and $\varepsilon > 0$, then $(x + \varepsilon)^*(t) = x^*(t) + \varepsilon$.*

5°. *If $|x(t) - y(t)| \leqslant \varepsilon$ for almost all $t$, then $|x^*(t) - y^*(t)| \leqslant \varepsilon$.*

Indeed, $|x(t)| \leqslant |y(t)| + \varepsilon$, and hence $x^*(t) \leqslant (|y| + \varepsilon)^*(t) = y^*(t) + \varepsilon$. Similarly, $y^*(t) \leqslant x^*(t) + \varepsilon$.

6°. *The inequality*

$$\int_e |x(s)|\,ds \leqslant \int_0^{\operatorname{mes} e} x^*(s)\,ds \tag{2.12}$$

*holds.*

We denote by $\chi_e(s)$ the characteristic function of the measurable set $e$. Then $n_{|x|\chi_e}(\tau) \leqslant \operatorname{mes} e$ and $n_{|x|\chi_e}(\tau) \leqslant n_{|x|}(\tau)$; therefore $n_{|x|\chi_e}(\tau) \leqslant \min\{\operatorname{mes} e,\, n_{|x|}(\tau)\}$. Integrating this inequality and using (2.11), we obtain

$$\begin{aligned}\int_e |x(s)|\,ds &= \int_0^\infty n_{|x|\chi_e}(\tau)\,d\tau \\ &\leqslant \int_0^\infty \min\{\operatorname{mes} e,\, n_{|x|}(\tau)\}\,d\tau = \int_0^{\operatorname{mes} e} x^*(s)\,ds.\end{aligned}$$

7°. *If* $x^*(\theta) > x^*(\infty)$, *then there exists a set* $e_\theta(x)$ *with* $\operatorname{mes} e_\theta(x) = \theta$ *for which*

$$\int_0^\theta x^*(s)\,ds = \int_{e_\theta(x)} |x(s)|\,ds. \tag{2.13}$$

The condition $x^*(\theta) > x^*(\infty)$ implies that in the neighborhood of $x^*(\theta)$ we have $n_{|x|}(\tau) < \infty$, and therefore the sets of those $s$ at which $x^*(s) \geqslant x^*(\theta)$ and $|x(s)| \geqslant x^*(\theta)$ have the same measure, which is at least $\theta$. If we denote by $e'_\theta$ the set of those $s$ where $|x(s)| > x^*(\theta)$ and by $e''_\theta$ the set of those $s$ where $|x(s)| \geqslant x^*(\theta)$, then $\operatorname{mes} e'_\theta \leqslant \theta \leqslant \operatorname{mes} e''_\theta$. The set $e_\theta(x)$ can be chosen so that $e'_\theta \subset e_\theta(x) \subset e''_\theta$ and $\operatorname{mes} e_\theta(x) = \theta$. The restrictions of $|x(t)|$ and $x^*(t)$ to $e_\theta(x)$ and $(0, \theta]$, respectively, are equimeasurable, and so (2.13) is satisfied.

If $x^*(\theta) = x^*(\infty)$, then there is no set $e_\theta(x)$ on which equality (2.13) is attained. The function $x(t) = \frac{2}{\pi}\operatorname{arctg} t$ considered above serves as such an example. However, we have the following assertion.

8°. *The equality*

$$\int_0^\theta x^*(s)\,ds = \sup_{\operatorname{mes} e = \theta} \int_e |x(s)|\,ds \tag{2.14}$$

*holds.*

From (2.12) we immediately obtain the inequality

$$\int_0^\theta x^*(s)ds \geqslant \sup_{\operatorname{mes} e = \theta} \int_e |x(s)|\,ds. \tag{2.15}$$

Therefore, in the case where $x^*(\theta) > x^*(\infty)$, (2.13) and (2.15) imply (2.14).

If $x^*(\theta) = x^*(\infty)$, then by $e'_\theta$ we again denote the set of those $s$ for which $|x(s)| > x^*(\infty)$, and by $e''_\theta$ the set of those $s$ for which $|x(s)| > x^*(\theta) - \varepsilon/\theta$, where $\varepsilon > 0$ is an arbitrarily small number. Then by the equimeasurability of $x^*(s)$ and $|x(s)|$ we have $\operatorname{mes} e'_\theta \leqslant \theta \leqslant \operatorname{mes} e''_\theta = \infty$. Therefore there exists a set $e_\theta^\varepsilon$ such that $e'_\theta \subset e_\theta^\varepsilon \subset e''_\theta$ and $\operatorname{mes} e_\theta^\varepsilon = \theta$. The restrictions of $|x(t)|$ and $x^*(t)$ to the sets $e'_\theta$ and $(0, \operatorname{mes} e'_\theta]$ are equimeasurable. Therefore

$$\int_{e'_\theta} |x(s)|\, ds = \int_0^{\operatorname{mes} e'_\theta} x^*(s)\, ds.$$

Then

$$\begin{aligned}\int_{e_\theta^\varepsilon} |x(s)|\, ds &= \int_{e'_\theta} |x(s)|\, ds + \int_{e_\theta^\varepsilon - e'_\theta} |x(s)|\, ds \\ &\geqslant \int_0^{\operatorname{mes} e'_\theta} x^*(s)\, ds + \int_{e^\varepsilon - e'_\theta} \left[x^*(\infty) - \varepsilon/\theta\right] ds \\ &\geqslant \int_0^{\operatorname{mes} e'_\theta} x^*(s)\, ds + x^*(\infty)(\theta - \operatorname{mes} e'_\theta) - \varepsilon.\end{aligned}$$

On the interval $(\operatorname{mes} e'_\theta, \theta)$ the function $x^*(s)$ is equal to $x^*(\infty)$, and so

$$\int_{e_\theta^\varepsilon} |x(s)|\, ds \geqslant \int_0^\theta x^*(s)\, ds - \varepsilon. \tag{2.16}$$

Inequalities (2.15) and (2.16) imply (2.14).

We note that (2.14) implies the important inequality

$$\int_0^\theta (x + y)^*(s)\, ds \leqslant \int_0^\theta x^*(s)\, ds + \int_0^\theta y^*(s)\, ds. \tag{2.17}$$

REMARK. If $n_{|x|}(\tau) < \infty$ for all $\tau > 0$, i.e. if $x^*(\infty) = 0$, then the sets $e_\theta(x)$ constructed in 7° are defined for all $\theta$ for which $x^*(\theta) > 0$. We write

$$e(x) = \bigcup_{x^*(\theta) > 0} e_\theta(x). \tag{2.18}$$

Then $e(x)$ coincides with the support of $x(t)$.

9°. *For two locally summable functions $x(t)$ and $y(t)$ with $n_x(\tau)$, $n_y(\tau) < \infty$ the equality*

$$(x + y)^*(t) = x^*(t) + y^*(t) \tag{2.19}$$

*holds for all $\tau > 0$ if and only if $x(t)$ and $y(t)$ are of constant sign almost everywhere and have a common system of sets $e_\theta$: $e_\theta(x) = e_\theta(y)$ $(0 < \theta < \infty)$.*

PROOF. *Necessity*. Let $(x + y)^* = x^* + y^*$. By (2.14) we have

$$\begin{aligned}\int_{e_\theta(x+y)} |x + y|\, dt &= \int_0^\theta (x + y)^*\, dt = \int_0^\theta x^*\, dt + \int_0^\theta y^*\, dt \\ &= \sup_{\operatorname{mes} e = \theta} \int_e |x|\, dt + \sup_{\operatorname{mes} e = \theta} \int_e |y|\, dt.\end{aligned} \tag{2.20}$$

If we assume that

$$\int_{e_\theta(x+y)} |x|\,dt < \sup_{\operatorname{mes} e=\theta} \int_e |x|\,dt$$

or

$$\int_{e_\theta(x+y)} |y|\,dt < \sup_{\operatorname{mes} e=\theta} \int_e |y|\,dt,$$

then

$$\int_{e_\theta(x+y)} |x+y|\,dt \leqslant \int_{e_\theta(x+y)} |x|\,dt + \int_{e_\theta(x+y)} |y|\,dt$$
$$< \sup_{\operatorname{mes} e=\theta} \int_e |x|\,dt + \sup_{\operatorname{mes} e=\theta} \int_e |y|\,dt,$$

which contradicts (2.20). Therefore,

$$\int_{e_\theta(x+y)} |x|\,dt = \sup_{\operatorname{mes} e=\theta} \int_e |x|\,dt \quad\text{and}\quad \int_{e_\theta(x+y)} |y|\,dt = \sup_{\operatorname{mes} e=\theta} \int_e |y|\,dt, \tag{2.21}$$

and consequently the system of the sets $e_\theta(x+y)$ can serve as the system $e_\theta(x)$ and $e_\theta(y)$. From (2.20) and (2.21) it follows that

$$\int_{e_\theta(x+y)} |x+y|\,dt = \int_{e_\theta(x+y)} |x|\,dt + \int_{e_\theta(x+y)} |y|\,dt,$$

which implies that

$$|x(t)+y(t)| = |x(t)| + |y(t)| \tag{2.22}$$

at almost all points of $e_\theta(x+y)$.

By the remark following (2.17) this equality holds almost everywhere on the whole semiaxis. The necessity is proved.

*Sufficiency*. If $e_\theta$ is a common system of sets for the functions $x(t)$ and $y(t)$, and these functions satisfy (2.22) almost everywhere, then

$$\int_0^\theta x^*\,dt + \int_0^\theta y^*\,dt = \int_{e_\theta} |x|\,dt + \int_{e_\theta} |y|\,dt$$
$$= \int_{e_\theta} |x+y|\,dt \leqslant \int_0^\theta (x+y)^*\,dt.$$

Then it follows from (2.17) that

$$\int_0^\theta x^*\,dt + \int_0^\theta y^*\,dt = \int_0^\theta (x+y)^*\,dt.$$

Differentiating with respect to $\theta$, we obtain (2.19).

10°. *The inequality*

$$(xy)^*(t_1 + t_2) \leqslant x^*(t_1)y^*(t_2)$$

*is true.*

For the proof we write $x^*(t_1 - \varepsilon) = a$ and $y^*(t_2 - \varepsilon) = b$, where $\varepsilon > 0$ is sufficiently small. We have the inclusion

$$\{t\colon |x(t)y(t)| > ab\} \subset \{t\colon |x(t)| > a\} \cup \{t\colon |y(t)| > b\},$$

which implies that

$$n_{|xy|}(ab) \leqslant n_{|x|}(a) + n_{|y|}(b) \leqslant t_1 + t_2 - 2\varepsilon < t_1 + t_2.$$

Hence

$$(xy)^*(t_1 + t_2) \leqslant ab = x^*(t_1 - \varepsilon)y^*(t_2 - \varepsilon).$$

This inequality implies the required one by the left continuity of rearrangements.

COROLLARY. *The following inequality holds*:

$$(x + y)^*(t_1 + t_2) \leqslant x^*(t_1) + y^*(t_2). \tag{2.23}$$

Indeed,

$$\begin{aligned} e^{(x+y)^*(t_1+t_2)} &= (e^{|x+y|})^*(t_1 + t_2) \leqslant (e^{|x|}e^{|y|})^*(t_1 + t_2) \\ &\leqslant (e^{|x|})^*(t_1)(e^{|y|})^*(t_2) \leqslant e^{x^*(t_1)}e^{y^*(t_2)} = e^{x^*(t_1)+y^*(t_2)}, \end{aligned}$$

which implies (2.23).

11°. *If a sequence $x_n(t)$ converges to a function $x(t)$ in measure, then $x_n^*(t) \to x^*(t)$ at all points of continuity of $x^*(t)$.*

If $\operatorname{mes}(t\colon |x(t) - x_n(t)| > \varepsilon) < \tau$ for $n \geqslant n_0(\varepsilon, \tau)$, then $(x - x_n)^*(\tau) \leqslant \varepsilon$, as $n \to \infty$. Then by the corollary of property 10° we have

$$x^*(t + \eta) \leqslant x_n^*(t) + (x - x_n)^*(\eta)$$

and

$$x_n^*(t) \leqslant x^*(t - \eta) + (x - x_n)^*(\eta).$$

From this and the preceding it follows that for any $\delta > 0$ and $\eta > 0$ for sufficiently large $n$

$$x^*(t + \eta) - \delta \leqslant x_n^*(t) \leqslant x^*(t - \eta) + \delta$$

holds.

If $x^*(t)$ is continuous at $t$, this inequality implies that $\lim_{n\to\infty} x_n^*(t) = x^*(t)$.

12°. *Let $n_{|y|}(\tau) < \infty$ for all $\tau > 0$, and $|x_m(t)| \leqslant y(t)$. If $x_m(t) \to x(t)$ almost everywhere, then $(x - x_m)^*(\tau) \to 0$ for all $\tau > 0$.*

For every $\varepsilon > 0$ the semiaxis can be divided into two sets $e$ and $e'$ so that $|y(t)| < \varepsilon/2$ on $e$ and $e'$ has finite measure. On $e$ we then have $|x(t) - x_m(t)| < \varepsilon$, and on $e'$ the sequence $x_m(t)$ converges to $x(t)$ in measure. Therefore $\text{mes}\{t\colon |x(t) - x_m(t)| > \varepsilon\} \to 0$ as $m \to \infty$. Thus, on the whole semiaxis we have $x_m - x \to 0$ in measure. Property 11° implies that $(x - x_m)^*(\tau) \to 0$ for all $\tau > 0$.

We note that the restrictions imposed on the sequence of the functions $x_m(t)$ are essential. For example, for the sequence of the functions $\chi_{[m,m+1]}(t)$ the condition $n_{\chi_{[m,m+1]}}(\tau) \leqslant 1$ is satisfied; it converges to the function $x(t) \equiv 0$ everywhere, but $\chi^*_{[m,m+1]}(\tau) = \chi_{[0,1]}(\tau)$.

COROLLARY. *If under the preceding hypotheses the function $y(t)$ is summable on the whole semiaxis, then, by Lebesgue's theorem,*

$$\lim_{n\to\infty} \int_e x_n^*(t)\,dt = \int_e x^*(t)\,dt \tag{2.24}$$

*for any measurable set $e$.*

13°. *The following inequality holds*:

$$\int_0^\infty x(s)y(s)\,ds \leqslant \int_0^\infty x^*(s)y^*(s)\,ds. \tag{2.25}$$

PROOF. Without loss of generality we may assume that the functions $x$ and $y$ are nonnegative. We construct an increasing sequence of strictly simple functions $y_n(t)$ converging to $y(t)$ almost everywhere. We represent the functions $y_n(t)$ in the form

$$y_n(t) = \sum_{k=1}^{N_n} y_{n,k}\chi_{e_{n,k}}(t),$$

where the $y_{n,k}$ are positive numbers, and $e_{n,1} \subset e_{n,2} \subset \cdots \subset e_{n,N_n}$. Then 6° implies that

$$\int_0^\infty x(s)y_n(s)\,ds = \sum_{k=1}^{N_n} y_{n,k}\int_{e_{n,k}} x(s)\,ds \leqslant \sum_{k=1}^{N_n} y_{n,k}\int_0^{\text{mes}\,e_{n,k}} x^*(s)\,ds$$

$$= \int_0^\infty x^*(s)\sum_{k=1}^{N_n} y_{n,k}\chi_{[0,\text{mes}\,e_{n,k}]}(s)\,ds.$$

The function $\sum_{k=1}^{N_n} y_{n,k}\chi_{[0,\text{mes}\,e_{n,k}]}(t)$ coincides with $y_n(t)$ by (2.3). Thus

$$\int_0^\infty x(s)y_n(s)\,ds \leqslant \int_0^\infty x^*(s)y_n^*(s)\,ds \leqslant \int_0^\infty x^*(s)y^*(s)\,ds.$$

Passing to the limit as $n \to \infty$, by B. Levi's theorem we obtain (2.25).

Inequality (2.25) is sharp in the following sense.

14°. *The equality*

$$\sup_u \int_0^\infty u(s)y(s)\,ds = \int_0^\infty x^*(s)y^*(s)\,ds \tag{2.26}$$

*holds, where the supremum is taken over all functions $u(s)$ for with $|u(s)|$ is equimeasurable with $x^*(s)$.*

By virtue of (2.25) it is sufficient to establish the inequality (2.27).

Without loss of generality we may assume that $y(t) \geqslant 0$, and then $u(t) \geqslant 0$. We also assume that $y^*(t) > y^*(\infty)$ for $t \in (0, \theta)$ and $y^*(t) = y^*(\infty)$ for $t > \theta$ ($\theta = \infty$ is allowed). We partition the semiaxis into intervals of length $\delta$ so that $N\delta = \theta$ ($N \leqslant \infty$). Reasoning in the same way as in 7° and 8°, we can construct sets $e_1, \ldots, e_N$ of measure $\delta$ such that $y^*((i-1)\delta) \geqslant y(t) \geqslant y^*(i\delta)$ for $t \in e_i$, and sets $e_{N+1}, e_{N+2}, \ldots$ of measure $\delta$ such that $y^*(\infty) \geqslant y(t) \geqslant y^*(\infty) - \varepsilon$ for $t \in e_i$, where $\varepsilon > 0$ is arbitrarily small. We map every set $e_i$ one-to-one onto the interval $((i-1)\delta, i\delta]$ in a measure preserving manner by means of the functions $\omega_i(t)$, and set

$$u_\delta(t) = \sum_{i=1}^\infty x^*(\omega_i(t))\chi_{e_i}(t).$$

The function $u_\delta(t)$ is equimeasurable with $x^*(t)$. Next,

$$\sup_u \int_0^\infty u(s)y(s)\,ds \geqslant \int_0^\infty u_\delta(s)y(s)\,ds = \sum_{i=1}^\infty \int_{e_i} x^*(\omega_i(s))y(s)\,ds$$
$$\geqslant \delta \sum_{i=1}^N x^*(i\delta)y^*(i\delta) + \sum_{i=N+1}^\infty \int_{e_i} x^*(\omega_i(s))[y^*(\infty) - \varepsilon]\,ds.$$

We have

$$\sum_{i=N+1}^\infty \int_{e_i} x^*(\omega_i(s))\,ds = \int_\theta^\infty x^*(s)\,ds.$$

If this integral is infinite, then (2.27) is true, since its left side is infinite. If, however, the last integral is finite, then

$$\sup_u \int_0^\infty u(s)y(s)\,ds \geqslant \delta \sum_{i=1}^\infty x^*(i\delta)y^*(i\delta) - \varepsilon \int_\theta^\infty x^*(s)\,ds.$$

The function $x^*(t)y^*(t)$ is decreasing. Therefore, it is Riemann integrable on any interval $[1/a, a]$. From this it follows that the limit as $\delta \to 0$ of the first term on the right side is equal to the integral of $x^*(t)y^*(t)$ over $(0, \infty)$. Thus,

$$\sup_u \int_0^\infty u(s)y(s)\,ds \geqslant \int_0^\infty x^*(s)y^*(s)\,ds - \varepsilon \int_\theta^\infty x^*(s)\,ds,$$

which implies (2.27) in view of the arbitrariness of $\varepsilon$, and hence (2.26).

REMARK. It can be seen from the proof that for $y(t) \geqslant 0$ the supremum can be taken over functions $u(t)$ of the form $u(t) = x^*(\omega(t))$, where $\omega(t)$ is a one-to-one measure preserving mapping of $(0, \infty)$ onto itself.

COROLLARY. *The following inequality holds*:

$$\int_0^\infty x^*(s)[y(s) + z(s)]^* \, ds \leqslant \int_0^\infty x^*(s)y^*(s) \, ds + \int_0^\infty x^*(s)z^*(s) \, ds.$$

15°. *The equality*

$$\int_\theta^\infty x^*(s) \, ds = \inf \int_{e'} |x(s)| \, ds \tag{2.28}$$

*holds, where the infimum is taken over all measurable sets $e'$ whose complement has measure $\theta$.*

First we establish that

$$\int_\theta^\infty x^*(s) \, ds \leqslant \int_{e'} |x(s)| \, ds. \tag{2.29}$$

If $x(s)$ is summable on $(0, \infty)$, this inequality follows from (2.11) and (2.12). Let $x(s)$ not be summable on $(0, \infty)$. For the proof of the inequality it is of interest to consider only those sets $e'$ for which $\int_{e'} |x(s)| \, ds < \infty$. The function $y_\theta(s) = \min\{|x(s)|, x^*(\theta)\}$ is bounded on the complement of $e'$, which has finite measure $\theta$, and does not exceed $|x(s)|$ on $e'$. Therefore it is summable on $(0, \infty)$. Next, $y_\theta^*(s) = \min\{x^*(s), x^*(\theta)\}$; therefore

$$\int_\theta^\infty x^*(s) \, ds = \int_\theta^\infty y_\theta^*(s) \, ds \leqslant \int_{e'} y_\theta(s) \, ds \leqslant \int_{e'} |x(s)| \, ds.$$

The required inequality (2.29) is proved.

In the proof of (2.28) we may assume that $\int_\theta^\infty x^*(s) \, ds < \infty$. Then $n_{|x|}(\tau) < \infty$ for all $\tau > 0$, and therefore the set $e_\theta(x)$ of measure $\theta$ constructed in 7° exists. We write $e_\theta'(x) = (0, \infty) \setminus e_\theta(x)$. The functions $x^*(s)$ and $|x(s)|$ are equimeasurable on the sets $(\theta, \infty)$ and $e_\theta'(x)$, respectively, and therefore

$$\int_\theta^\infty x^*(s) \, ds = \int_{e_\theta'(x)} |x(s)| \, ds.$$

COROLLARY. *If $x(s) \equiv 0$ for $s > a$, then*

$$\int_\theta^a x^*(s) \, ds \leqslant \int_\theta^a |x(s)| \, ds. \tag{2.30}$$

16°. *If $x(t) = y(t) = 0$ for $t > a$, then*

$$\int_0^a x^*(s)y^*(a - s) \, ds \leqslant \int_0^a |x(s)y(s)| \, ds. \tag{2.31}$$

PROOF. Assuming that $x$ and $y$ are nonnegative, we construct an increasing sequence of strictly simple functions $y_n(t)$ converging to $y(t)$ almost everywhere. The functions $y_n(t)$ can be written in the form

$$y_n(t) = \sum_{k=1}^{N_n} y_{n,k}\chi_{e_{n,k}}(t),$$

where the $y_{n,k}$ are positive numbers and $e_{n,1} \subset e_{n,2} \subset \cdots \subset e_{n,N_n}$. By (2.29) we then have

$$\begin{aligned}\int_0^a x(s)y(s)\,ds &\geqslant \int_0^a x(s)y_n(s)\,ds \\ &= \sum_{k=1}^{N_n} y_{n,k}\int_{e_{n,k}} x(s)\,ds \geqslant \sum_{k=1}^{N_n} y_{n,k}\int_{a-\operatorname{mes} e_{n,k}}^a x^*(s)\,ds \\ &= \int_0^a x^*(s)\sum y_{n,k}\chi_{[a-\operatorname{mes} e_k,\, a]}(s)\,ds \\ &= \int_0^a x^*(s)y_n^*(a-s)\,ds. \end{aligned} \tag{2.32}$$

The sequence $y_n^*(a-s)$ is increasing, and by property 11° it converges to $y^*(a-s)$ almost everywhere. B. Levi's theorem allows us to obtain (2.31) from (2.32) by passing to the limit.

17°. *If $x(t) = y(t) = 0$ for $t > a$, then*

$$\int_0^a x^*(s)y^*(a-s)\,ds = \inf_u \int_0^a |u(s)y(s)|\,ds, \tag{2.33}$$

*where the infimum is taken over all functions $u(t)$ such that $|u(t)|$ is equimeasurable with $x^*(t)$.*

It is sufficient to establish the inequality

$$\int_0^a x^*(s)y^*(a-s)\,ds \geqslant \inf_u \int_0^a |u(s)y(s)|\,ds. \tag{2.34}$$

On the left there is a Riemann integral. For an arbitrary $\varepsilon > 0$ we construct a sequence $\{s_i\}_{-\infty}^{+\infty}$ of points having the following properties:

1) $s_{-i} + s_i = a$,

2) $y^*(a - s_i + 0) \leqslant (1+\varepsilon)y^*(a - s_{i-1})$, and

3)

$$\sum_{i=-\infty}^{\infty} x^*(s_{i-1}+0)y^*(a-s_{i-1})(s_i - s_{i-1}) \leqslant (1+\varepsilon)\int_0^a x^*(s)y^*(a-s)\,ds.$$

Properties 2) and 3) imply the inequality

$$\sum_{i=-\infty}^{\infty} x^*(s_{i-1}+0)y^*(a-s_i+0)(s_i-s_{i-1}) \leqslant (1+\varepsilon)^2\int_0^a x^*(s)y^*(a-s)\,ds.$$

We construct disjoint sets $e_i$ of measure $s_i - s_{i-1}$ on which

$$y^*(a-s_{i-1}) \leqslant |y(t)| \leqslant y^*(a-s_i+0) \qquad (t \in e_i).$$

We map every set $e_i$ onto $[s_{i-1}, s_i]$ with preservation of measure by means of the function $\omega_i(t)$, and set

$$v(t) = \sum_{ti=-\infty}^{\infty} x^*(\omega_i(t))\chi_{e_i}(t).$$

The function $v(t)$ is equimeasurable with $x^*(t)$. We have

$$\begin{aligned}\inf\int_0^a |u(s)y(s)|\,ds &\leqslant \int_0^a v(s)|y(s)|\,ds = \sum_{i=-\infty}^{\infty}\int_{e_i} x^*(\omega_i(s))|y(s)|\,ds\\ &\leqslant \sum_{i=-\infty}^{\infty} x^*(s_{i-1}+0)y^*(a-s_i+0)(s_i-s_{i-1})\\ &\leqslant (1+\varepsilon)^2\int_0^a x^*(s)y^*(a-s)\,ds.\end{aligned}$$

In view of the arbitrariness of $\varepsilon$, the last inequality implies (2.34), and with it (2.33).

18°. *Assume that for the locally summable functions $x(t)$ and $y(t)$*

$$\int_0^\theta x(s)\,ds \leqslant \int_0^\theta y(s)\,ds \tag{2.35}$$

*holds and $b(t)$ is a decreasing nonnegative function on $(0, \infty)$ such that the functions $x(t)b(t)$ and $y(t)b(t)$ are locally summable. Then*

$$\int_0^\infty x(s)b(s)\,ds \leqslant \int_0^\infty y(s)b(s)\,ds. \tag{2.36}$$

*If $x(t)$ and $y(t)$ are nonnegative, then the condition of summability of $xb$ and $yb$ can be omitted.*

PROOF. First consider the case where $b(t)$ is strictly simple:

$$b(t) = \sum_{k=1}^{N} \Delta_k\chi_{(0,\theta_k]}(t),$$

where $\Delta_k > 0$ and $\theta_1 < \theta_2 < \cdots < \theta_N$. Then

$$\int_0^\infty x(s)b(s)\,ds = \sum_{k=1}^N \Delta_k \int_0^{\theta_k} x(s)\,ds$$
$$\leqslant \sum_{k=1}^N \Delta_k \int_0^{\theta_k} y(s)\,ds = \int_0^\infty y(s)b(s)\,ds,$$

i.e. (2.36) is satisfied. If $b(t)$ is a bounded function, then it can be approximated uniformly by a sequence of decreasing strictly simple functions $b_N(s)$ on every finite interval $[0, a]$, and then

$$\int_0^a x(s)b(s)\,ds = \lim_{N\to\infty} \int_0^a x(s)b_N(s)\,ds$$
$$\leqslant \lim_{N\to\infty} \int_0^a y(s)b_N(s)\,ds = \int_0^a y(s)b(s)\,ds,$$

from which we also obtain (2.36).

Finally, for any $b(t)$ we write $b^{(n)}(t) = \min\{b(t), n\}$. Then in the inequality

$$\int_0^a x(s)b^{(n)}(s)\,ds \leqslant \int_0^a y(s)b^{(n)}(s)\,ds, \tag{2.37}$$

already proved, we can pass to the limit in view of the absolute continuity of the integrals of summable functions. Thus, we again arrive at (2.36).

If $x(t)$ and $y(t)$ are nonnegative, then in the case where $y(t)b(t)$ is not summable, (2.36) is obvious, and in the case where it is summable, the passage to the limit in (2.37) is possible in view of Fatou's lemma.

REMARK 1. If the support of $b(t)$ is contained in $(0, N]$, it is sufficient to require that (2.35) be satisfied on $(0, N]$.

REMARK 2. If $b(t)$ is increasing, then the inequality

$$\int_0^\infty |x(s)|b(s)\,ds \leqslant \int_0^\infty |y(s)|b(s)\,ds$$

follows from the inequality

$$\int_\theta^\infty |x(s)|\,ds \leqslant \int_\theta^\infty |y(s)|\,ds.$$

19°. *The inequality*

$$\int_0^\theta x^*(s)\,ds \leqslant \int_0^\theta y^*(s)\,ds$$

*for all* $\theta > 0$ *implies the inequality*

$$\int_0^\infty x(s)a(s)\,ds \leqslant \int_0^\infty y^*(s)a^*(s)\,ds \tag{2.38}$$

*for any measurable function* $a(s)$.

Indeed, this follows from properties 18° and 13° of rearrangements.

20°. *If $b(t)$ is an increasing nonnegative function, then*

$$\int_0^\infty x^*(t)b(t)\,dt = \inf_\omega \int_0^\infty x(\omega(t))b(t)\,dt, \tag{2.39}$$

*where $\omega(t)$ is a one-to-one measure preserving mapping of* $(0, \infty)$ *onto itself.*

PROOF. The functions $|x(\omega(t))|$ and $x^*(t)$ are equimeasurable. Therefore (2.29) applied to $x(\omega(t))$ and Remark 2 after property 18° imply the inequality

$$\int_0^\infty x^*(s)b(s)\,ds \leqslant \int_0^\infty |x(\omega(s))|b(s)\,ds.$$

Thus it is sufficient to prove the reverse inequality under the assumption that the left side is finite. Given $\varepsilon > 0$, consider the sequence of numbers $(1+\varepsilon)^k$ $(-\infty < k < \infty)$. Assuming that $b(-0) = 0$, we denote by $\tau_k$ a point on the semiaxis at which $b(\tau_k - 0) \leqslant (1+\varepsilon)^k \leqslant b(\tau_k + 0)$. The limit of $\tau_k$ as $k \to -\infty$ will be denoted by $\tau_{-\infty}$. If $b(t)$ is bounded, then among the $\tau_k$ there exists a last one $\tau_{k_0 - 1}$, and we set $\tau_{k_0} = \infty$. The numbers $\tau_k$ have the properties

$$b(\tau_k - 0) \leqslant (1+\varepsilon)b(\tau_{k-1} + 0) \qquad (-\infty < k < \infty).$$

The finiteness of the integral on the left side and the increasing property of $b(t)$ imply that $x^*(\infty) = 0$, and therefore $n_{|x|}(\tau) < \infty$ for all $\tau > 0$. Hence, reasoning in the same way as in 7° and 8°, we can construct sets $e_k$ with measure $\tau_k - \tau_{k-1}$ so that the functions $x^*(t)\chi_{(\tau_{k-1}, \tau_k]}(t)$ are equimeasurable with the functions $|x(t)|\chi_{e_k}(t)$ for all numbers $\tau_k \leqslant \infty$ $(k > -\infty)$. If $\tau_{-\infty} > 0$, then we also construct a set $e_{-\infty}$ so that the function $x^*(t)\chi_{(0, \tau_{-\infty}]}(t)$ is equimeasurable with the function $|x(t)|\chi_{e_{-\infty}}(t)$. Next, we define a one-to-one measure preserving mapping $\omega(t)$ of $(0, \infty)$ onto itself which carries the intervals $(\tau_{k-1}, \tau_k]$ onto the sets $e_k$ and the interval $(0, \tau_{-\infty}]$ onto $e_{-\infty}$. Then the functions $x^*(t)\chi_{(\tau_{k-1}, \tau_k]}(t)$ and $|x(\omega(t))|\chi_{(\tau_{k-1}, \tau_k]}(t)$, and also $x^*(t)\chi_{(0, \tau_{-\infty}]}(t)$ and $|x(\omega(t))|\chi_{(0, \tau_{-\infty}]}(t)$, respectively, are equimeasurable, and therefore

$$\begin{aligned}
\int_0^\infty x(\omega(t))b(t)\,dt &= \sum_k \int_{\tau_{k-1}}^{\tau_k} x(\omega(t))b(t)\,dt \\
&\leqslant \sum_k b(\tau_k - 0)\int_{\tau_{k-1}}^{\tau_k} x(\omega(t))\,dt = \sum b(\tau_k - 0)\int_{\tau_{k-1}}^{\tau_k} x^*(t)\,dt \\
&\leqslant (1+\varepsilon)\sum b(\tau_{k-1} + 0)\int_{\tau_{k-1}}^{\tau_k} x^*(t)\,dt \leqslant (1+\varepsilon)\int_0^\infty x^*(t)b(t)\,dt.
\end{aligned}$$

By the arbitrariness of $\varepsilon$ the required inequality, and with it (2.39) as well, are proved.

COROLLARY. *If $b(t)$ is an increasing nonnegative function, then*

$$\int_0^\infty (x+y)^*(t)b(t)\,dt) \geq \int_0^\infty x^*(t)b(t)\,dt + \int_0^\infty y^*(t)b(t)\,dt.$$

21°. *If $\alpha > -1$ and $x(t)$ is a decreasing nonnegative function, then*

$$C_1\int_0^\infty x(t)t^{\gamma+\alpha}\,dt \leq \int_0^\infty (x(t)t^\gamma)^* t^\alpha\,dt \leq C_2\int_0^\infty x(t)t^{\gamma+\alpha}\,dt, \quad (2.40)$$

*where $C_1$ and $C_2$ do not depend on the choice of $x(t)$.*

For $\gamma \leq 0$ the integrals on all sides of the inequalities coincide. Therefore $\gamma > 0$ is the only interesting case.

For $\chi_{(0,h]}(t)$ we have

$$\begin{aligned}\int_0^\infty (\chi_{(0,h]}(t)t^\gamma)^* t^\alpha\,dt &= \int_0^h (\chi_{(0,h]}(t)t^\gamma)^* t^\alpha\,dt = \int_0^h (h-t)^\gamma t^\alpha\,dt\\ &= h^{\alpha+\gamma+1}\int_0^1 (1-s)^\gamma s^\alpha\,ds\\ &= (\alpha+\gamma+1)\int_0^1 (1-s)^\gamma s^\alpha\,ds \int_0^h t^{\gamma+\alpha}\,dt\\ &= C_{\alpha\gamma}\int_0^\infty \chi_{(0,h]}(t)t^{\gamma+\alpha}\,dt. \qquad (2.41)\end{aligned}$$

1) Let $\alpha > 0$. We represent the decreasing nonnegative strictly simple function $x(t)$ in the form

$$x(t) = \sum_{k=1}^N \Delta_k \chi_{(0,h_k]}(t) \qquad (\Delta_k > 0;\ h_1 > h_2 > \cdots > h_n). \quad (2.42)$$

The function $t^\alpha$ is increasing, and so in view of 20°

$$\begin{aligned}\int_0^\infty (x(t)t^\gamma)^* t^\alpha\,dt &= \inf_\omega \int_0^\infty x(\omega(t))[\omega(t)]^\gamma t^\alpha\,dt\\ &\geq \sum_{k-1}^N \Delta_k \inf_\omega \int_0^\infty \chi_{(0,h_k]}(\omega(t))[\omega(t)]^\gamma t^\alpha\,dt\\ &= \sum_{k=1}^N \Delta_k \int_0^\infty (\chi_{(0,h_k]}(t)t^\gamma)^* t^\alpha\,dt\\ &= C_{\alpha\gamma}\sum_{k=1}^N \Delta_k \int_0^\infty \chi_{(0,h_k]}(t)t^{\gamma+\alpha}\,dt\\ &= C_{\alpha\gamma}\int_0^\infty x(t)t^{\gamma+\alpha}\,dt.\end{aligned}$$

(Here $\omega(t)$ is a one-to-one measure preserving mapping of $(0,\infty)$ onto itself.) We represent an arbitrary decreasing nonnegative function $x(t)$ as the limit of an increasing sequence of nonnegative step functions $x_n(t)$. Then

$$\int_0^\infty (x(t)t^\gamma)^* t^\alpha\,dt \geqslant \int_0^\infty (x_n(t)t^\gamma)^* t^\alpha\,dt \geqslant C_{\alpha\gamma}\int_0^\infty x_n(t)t^{\gamma+\alpha}\,dt.$$

Passing to the limit as $n \to \infty$ on the right side, we obtain

$$\int_0^\infty (x(t)t^\gamma)^* t^\alpha\,dt \geqslant C_{\alpha\gamma}\int_0^\infty x(t)t^{\gamma+\alpha}\,dt.$$

The left side of (2.40) is proved. The right side is satisfied with $C_2 = 1$ in view of (2.39).

2) Let $-1 < \alpha < 0$. The function $t^\alpha$ is decreasing and the left side of (2.40) is satisfied with $C_1 = 1$ in view of 13°.

For a function of the form (2.42), we obtain from (2.26) and (2.41)

$$\begin{aligned}\int_0^\infty (x(t)t^\gamma)^* t^\alpha\,dt &= \sup \int_0^\infty x(t)t^\gamma u_\alpha(t)\,dt\\ &\leqslant \sum_{k=1}^N \Delta_k \sup \int_0^\infty \chi_{(0,h_k]}(t)t^\gamma u_\alpha(t)\,dt\\ &= \sum_{k=1}^N \Delta_k \int_0^\infty (\chi_{(0,h_k]}(t)t^\gamma)^* t^\alpha\,dt\\ &= C_{\alpha\gamma}\sum_{k=1}^N \Delta_k \int_0^\infty \chi_{(0,h_k]}(t)t^{\gamma+\alpha}\,dt = C_{\alpha\gamma}\int_0^\infty x(t)t^{\gamma+\alpha}\,dt,\end{aligned}$$

where the $u_\alpha(t)$ are functions equimeasurable with $t^\alpha$.

We represent any decreasing nonnegative function $x(t)$ as the limit of an increasing sequence $x_n(t)$ of the form (2.42). Then

$$\int_0^\infty x_n(t)t^\gamma u_\alpha(t)\,dt \leqslant C_{\alpha\gamma}\int_0^\infty x_n(t)t^{\gamma+\alpha}\,dt \leqslant C_{\alpha\gamma}\int_0^\infty x(t)t^{\gamma+\alpha}\,dt.$$

Passing to the limit as $n \to \infty$, we obtain

$$\int_0^\infty x(t)t^\gamma u_\alpha(t)\,dt \leqslant C_{\alpha\gamma}\int_0^\infty x(t)t^{\gamma+\alpha}\,dt.$$

Taking the supremum over all functions $u_\alpha(t)$ equimeasurable with $t^\alpha$, by virtue of (2.26) we arrive at the inequality

$$\int_0^\infty (x(t)t^\gamma)^* t^\alpha\,dt \leqslant C_{\alpha\gamma}\int_0^\infty x(t)t^{\gamma+\alpha}\,dt.$$

22°. *Let* $\alpha > -1$ *and* $\alpha + \gamma < -1$, *and let* $x(t)$ *be a nonnegative increasing function. Then* (2.40) *is satisfied.*

Let $x(t) = \chi_{(h,\infty)}(t)$. Then

$$\int_0^\infty (\chi_{(h,\infty)}(t)t^\gamma)^* t^\alpha\, dt = \int_0^\infty (t+h)^\gamma t^\alpha\, dt = h^{\gamma+\alpha+1}\int_0^\infty (1+s)^\gamma s^\alpha\, ds$$
$$= |\gamma+\alpha+1| \int_0^\infty (1+s)^\gamma s^\alpha\, ds \int_h^\infty t^{\gamma+\alpha}\, dt$$
$$= C_{\alpha\gamma}\int_0^\infty \chi_{(h,\infty)}(t) t^{\gamma+\alpha}\, dt.$$

(Here $\int_0^\infty (1+s)^\gamma s^\alpha\, ds < \infty$ in view of the conditions imposed on $\alpha$ and $\gamma$.)

The remaining reasoning is analogous to that carried out in the proof of 21°. The increasing function $x(t)$ is approximated by a sequence of simple functions of the form

$$x(t) = \sum_{k=1}^{N} \Delta_k \chi_{(h_k,\infty)}(t).$$

## §3. Operators in the Banach couple $L_1(0, \infty), L_\infty(0, \infty)$

**1.** ***The spaces*** $L_1 \cap L_\infty$ ***and*** $L_1 + L_\infty$. The space $L_1 \cap L_\infty$ consists of all bounded summable functions on $(0, \infty)$ with norm

$$\|x\|_{L_1\cap L_\infty} = \max\left\{\operatorname{ess\,sup}|x(t)|, \int_0^\infty |x(t)|\, dt\right\}.$$

The set of strictly simple functions is dense in $L_1 \cap L_\infty$. Indeed, let $x \in L_1 \cap L_\infty$ and $x(t) \geqslant 0$. We partition the interval $[0, \|x\|_{L_\infty}]$ into $N$ equal parts and set

$$x_N(t) = \sum_{k=0}^{N-1} \frac{k}{N}\|x\|_{L_\infty}\chi_{e_k}(t),$$

where $e_k = \{t\colon (k/N)\|x\|_{L_\infty} \leqslant x(t) < ((k+1)/N)\|x\|_{L_\infty}\}$. The condition $x \in L_1$ implies that $\operatorname{mes} e_k < \infty$ for $1 \leqslant k \leqslant N-1$, and consequently the functions $x_N(t)$ are strictly simple. It is obvious that $\|x - x_N\|_{L_\infty} \leqslant \frac{1}{N}\|x\|_{L_\infty} \to 0$ as $N \to \infty$. Next,

$$\|x - x_N\|_{L_1} = \int_0^\infty [x(t) - x_N(t)]\, dt \to 0$$

by Lebesgue's theorem, since $x(t) - x_N(t) \leqslant x(t)$.

For any function $x \in L_1 \cap L_\infty$ the assertion follows from the representation $x = x_+ - x_-$, where $x_+ = \max\{x(t), 0\}$.

The space $L_1 + L_\infty$ consists of functions which are sums of bounded measurable and summable functions on $(0, \infty)$. The norm in this space is defined by

$$\|x\|_{L_1+L_\infty} = \inf_{\substack{x(t)=u(t)+v(t)\\ u\in L_1,\ v\in L_\infty}} (\|u\|_{L_1} + \|v\|_{L_\infty}). \tag{3.1}$$

It is easy to see that $L_1 + L_\infty$ consists of all locally summable functions $x(t)$ for which $n_{|x|}(\tau) \not\equiv \infty$. Indeed, if $x \in L_1 + L_\infty$, then $x(t) = u(t) + v(t)$, where $u \in L_1$ and $v \in L_\infty$, and so $x(t)$ is locally summable. Next, if $\tau >$ ess $\sup|v(t)| = c$, then $n_{|x|}(\tau) \leqslant n_{|u|}(\tau - c) < \infty$. Conversely, let $x(t)$ be locally summable and let $n_{|x|}(\tau_0) < \infty$. Then for the set $e_0 = \{t\colon |x(t)| > \tau_0\}$ we have mes $e_0 < \infty$. Setting $x(t) = x(t)\chi_{e_0}(t) + x(t)(1 - \chi_{e_0}(t))$, we obtain that the first term on the right belongs to $L_1$, and the second to $L_\infty$, i.e. $x \in L_1 + L_\infty$.

An explicit formula can be given for the calculation of the norm of a function $x(t)$ in the space $L_1 + L_\infty$. Let mes $e = 1$. We have

$$\|x\|_{L_1+L_\infty} \geqslant \|x\chi_e\|_{L_1+L_\infty} = \inf_{u(t)+v(t)=x(t)\chi_e(t)} \left(\|u\|_{L_1} + \|v\|_{L_\infty}\right).$$

It is obvious that for the calculation of the infimum it is sufficient to take functions $u$ and $v$ with supports on the set $e$, and then $\|v\|_{L_1} \leqslant \|v\|_{L_\infty}$. Therefore

$$\|x\|_{L_1+L_\infty} \geqslant \inf\left(\|u\|_{L_1} + \|v\|_{L_1}\right) \geqslant \|x\chi_e\|_{L_1}.$$

By (2.14) we have

$$\|x\|_{L_1+L_\infty} \geqslant \sup_{me=1} \|x\chi_e\|_{L_1} = \int_0^1 x^*(s)\,ds. \tag{3.2}$$

For the proof of the reverse inequality we introduce the functions

$$v(t) = \operatorname{sgn} x(t)\min(|x(t)|, x^*(1))$$

and $u(t) = x(t) - v(t)$. Then $|u(t)| = [|x(t)| - x^*(1)]_+$, and since the functions $[|x(t)| - x^*(1)]_+$ and $[x^*(t) - x^*(1)]_+$ are equimeasurable, we have

$$\|u\|_{L_1} = \int_0^\infty |\,|x(s)| - x^*(1)|_+\,ds = \int_0^1 \left[x^*(s) - x^*(1)\right] ds.$$

On the other hand, $\|v\|_{L_\infty} = x^*(1)$, and consequently

$$\|u\|_{L_1} + \|v\|_{L_\infty} = \int_0^1 x^*(s)\,ds. \tag{3.3}$$

From (3.1)–(3.3) we obtain the basic formula

$$\|x\|_{L_1+L_\infty} = \int_0^1 x^*(s)\,ds. \tag{3.4}$$

This formula in particular implies that $x^*(t) \in L_1 + L_\infty$ for any $x \in L_1 + L_\infty$.

In what follows we will be interested in the functional defined on $L_1 + L_\infty$ by the formula

$$K(t, x) = \inf_{x(s)=u(s)+v(s)} \left(\|u\|_{L_1} + t\|v\|_{L_\infty}\right) \qquad (x \in L_1 + L_\infty, L > 0).$$

For fixed $t$ the functional $K(t, x)$ is equivalent to the norm of the space $L_1 + L_\infty$.

The expression for $K(t, x)$ can be written in the form

$$K(t, x) = t \inf_{x(t\sigma) = u(t\sigma) + v(t\sigma)} \left[ \int_0^\infty |u(t\sigma)| d\sigma + \sup_{0 < \sigma < \infty} |v(t\sigma)| \right]$$

and, by the above (see (3.4)), we have

$$K(t, x) = t \int_0^1 [x(t\sigma)]^* \, d\sigma = \int_0^t x^*(s) \, ds. \tag{3.5}$$

Here we have used the obvious equality $[x(t\sigma)]^* = x^*(t\sigma)$.

We note that the spaces $L_1 \cap L_\infty$ and $L_1 + L_\infty$ are not separable. Indeed, in $L_1 \cap L_\infty$ for the continuous family of functions $\chi_{[0,\tau]}$ the distance between any two different functions is equal to one. In $L_1 + L_\infty$ the same property is possessed by the continuous family of functions $\chi_{\bigcup_1^\infty [n_k, n_k+1]}$, where $\{n_k\}$ is an arbitrary increasing sequence of natural numbers.

We note that the set $L_1$ is not dense in $L_1 + L_\infty$. The closure $\overline{L_1}$ in $L_1 \cap L_\infty$ consists of all locally summable functions for which $n_{|x|}(\tau) < \infty$ for all $\tau > 0$. Indeed, if $x(t)$ has these properties, we write $e_k = \{t\colon |x| > 1/k\}$. Then $\operatorname{mes} e_k < \infty$, and so the function $x(t)\chi_{e_k}(t)$ belongs to $L_1$. Next,

$$\|x - x\chi_{e_k}\|_{L_1 + L_\infty} \leqslant \|x - x\chi_{e_k}\|_{L_\infty} \leqslant 1/k$$

and hence $x \in \overline{L_1}$. Conversely, if $n_{|x|}(\tau_0) = \infty$ $(\tau_0 > 0)$ and $y(t)$ is any function in $L_1$, then $|y(t)| > \varepsilon$ on a set of finite measure, and so $|x(t) - y(t)| \geqslant \tau_0 - \varepsilon$ on a set of infinite measure. By (3.4) we have $\|x - y\|_{L_1 + L_\infty} \geqslant \tau_0 - \varepsilon$, i.e. $x \notin \overline{L_1}$.

It is easy to see that $L_\infty$ is dense in $L_1 + L_\infty$.

LEMMA 3.1. *For every function $x(t)$ from $L_1 + L_\infty$ there exists a sequence of sets $e_n$ of finite measure such that the functions $[x(t)\chi_{e_n}(t)]^*$ converge to $x^*(t)$ almost everywhere.*

PROOF. We consider the sets $e'_n = \{t\colon |x(t)| > x^*(\infty) + 1/n\}$. Then $[x(t)\chi_{e'_n}(t)]^* = x^*(t)\chi_{[0, \operatorname{mes} e'_n]}(t)$. If the measures of the sets $e'_n$ tend to $\infty$, we use them as sets $e_n$, and the assertion is obvious.

If $\operatorname{mes} e'_n \leqslant M < \infty$, then the set $e = \bigcup_1^\infty e'_n$ is of finite measure. The function $x^*(t)$ is equal to $x^*(\infty)$ on the interval $(\operatorname{mes} e, \infty)$. We denote by $f_n$ a set of measure $n$ on which $x^*(\infty) \geqslant |x(t)| > x^*(\infty) - 1/n$ holds, and write $e_n = e \cup f_n$. Then the function $[x(t)\chi_{e_n}(t)]^*$ coincides with $x^*(t)$ on $[0, \operatorname{mes} e]$, and differs from it by less than $1/n$ on the interval $[\operatorname{mes} e, \operatorname{mes} e + n]$.

REMARK. If we replace the system $\{e_n\}$ of sets by a larger system $\{e'_n\}$ $(e'_n \supset e_n)$, then the lemma remains valid for the system $\{e'_n\}$.

The spaces $L_1 \cap L_\infty$ and $L_1 + L_\infty$ are dual in the sense that for $x \in L_1 + L_\infty$ and $y \in L_1 \cap L_\infty$ we have

$$\left|\int_0^\infty x(t)y(t)\,dt\right| \leqslant \|x\|_{L_1+L_\infty}\|y\|_{L_1\cap L_\infty}. \tag{3.6}$$

Indeed, if $x = u + v$, where $u \in L_1$ and $v \in L_\infty$, then

$$\left|\int_0^\infty x(t)y(t)\,dt\right| \leqslant \|u\|_{L_1}\|y\|_{L_\infty} + \|v\|_{L_\infty}\|y\|_{L_1}$$
$$\leqslant \|y\|_{L_1\cap L_\infty}\left(\|u\|_{L_1} + \|v\|_{L_\infty}\right).$$

Inequality (3.6) then follows from the definition of the norm in $L_1 + L_\infty$. Next,

$$\sup_{\|x\|_{L_1+L_\infty}\leqslant 1}\left|\int_0^\infty x(t)y(t)\,dt\right| = \|y\|_{L_1\cap L_\infty}. \tag{3.7}$$

Indeed,

$$\sup_{\|x\|_{L_1+L_\infty}\leqslant 1}\left|\int_0^\infty x(t)y(t)\,dt\right|$$
$$\geqslant \max\left[\sup_{\|x\|_{L_1}\leqslant 1}\left|\int_0^\infty x(t)y(t)\,dt\right|,\ \sup_{\|x\|_{L_\infty}\leqslant 1}\left|\int_0^\infty x(t)y(t)\,dt\right|\right]$$
$$= \max\left[\|y\|_{L_1}, \|y\|_{L_\infty}\right] = \|y\|_{L_1\cap L_\infty}. \tag{3.8}$$

Formulas (3.6) and (3.8) imply (3.7).

Similarly,

$$\sup_{\|y\|_{L_1\cap L_\infty}\leqslant 1}\left|\int_0^\infty x(t)y(t)\,dt\right| = \|x\|_{L_1+L_\infty}. \tag{3.9}$$

Indeed, (3.4) and (2.14) imply that

$$\|x\|_{L_1+L_\infty} = \int_0^1 |x(t)|^*\,dt \leqslant \int_{e_1^\varepsilon} |x(t)|\,dt + \varepsilon,$$

where $e_1^\varepsilon$ is a set of measure 1. Then

$$\sup_{\|y\|_{L_1\cap L_\infty}\leqslant 1}\left|\int_0^\infty x(t)y(t)\,dt\right| \geqslant \int_0^\infty x(t)\chi_{e_1^\varepsilon}(t)\operatorname{sgn} x(t)\,dt \geqslant \|x\|_{L_1+L_\infty} - \varepsilon.$$

From what was said on p. 45 and from (3.7) and (3.9) it follows that the spaces $L_1 \cap L_\infty$ and $L_1 + L_\infty$ are associated with each other.

**2. *Averaging operator.*** Let the semiaxis $(0, \infty)$ be partitioned into a countable number of measurable sets $e_i$ $(i = 1, 2, \ldots)$ of finite measure. From the system $\Delta = \{e_i\}_1^\infty$ we construct an averaging operator defined for every locally summable function by

$$T_\Delta x(t) = \sum_{i=1}^{\infty} \frac{1}{\operatorname{mes} e_i} \int_{e_i} x(\tau)\, d\tau \chi_{e_i}(t) = \sum_{i=1}^{\infty} x_i \chi_{e_i}(t). \tag{3.10}$$

It is obvious that $T_\Delta \chi_{e_i}(t) = \chi_{e_i}(t)$. The operator $T_\Delta$ is bounded in $L_\infty$ and has norm equal to one. Next, for $x \in L_1$ we have

$$\|T_\Delta x\|_{L_1} = \sum_{i=1}^{\infty} \left| \int_{e_i} x(\tau)\, d\tau \right| \leqslant \|x\|_{L_1}.$$

This inequality becomes equality for the functions $\chi_{e_i}(t)$. Therefore $\|T_\Delta\|_{L_1} = 1$. From what was said it follows that $T_\Delta$ is bounded in $L_1 + L_\infty$, and $\|T_\Delta\|_{L_1+L_\infty} \leqslant 1$. By (3.4) for the norm in $L_1 + L_\infty$, this is equivalent to the inequality

$$\int_0^1 (T_\Delta u)^*(\tau)\, d\tau \leqslant \int_0^1 u^*(\tau)\, d\tau.$$

Applying the same reasoning to the norm in $L_1 + L_\infty$ determined by the functional $K(t, x)$, from (3.5) we obtain

$$\int_0^t (T_\Delta u)^*(\tau)\, d\tau \leqslant \int_0^t u^*(\tau)\, d\tau. \tag{3.11}$$

3. ***An additional property of rearrangements.*** We consider the finite-dimensional Banach space $E_n$ of vectors $x = (\xi_1, \dots, \xi_n)$. We shall say that this space is *symmetric* if the norm of a vector does not change under arbitrary permutations and changes of sign of its components.

If $x = (\xi_1, \dots, \xi_n)$ and $y = (\eta_1, \dots, \eta_n)$, and $|\xi_i| \geqslant |\eta_i|$ $(i = 1, \dots, n)$, then the point $y$ belongs to the parallelepiped with vertices at the points $(\pm\xi_1, \dots, \pm\xi_n)$. The radius vectors of these vertices have the same norm equal to $\|x\|$. Therefore, all points of the cube have norm not exceeding $\|x\|$, and, in particular, $\|y\| \leqslant \|x\|$.

By $x^*$ we denote the vector with components $\xi_1^* = |\xi_{i_1}|, \dots, \xi_n^* = |\xi_{i_n}|$, where $(i_1, \dots, i_n)$ is a permutation of the indices $(1, \dots, n)$ such that $|\xi_{i_1}| \geqslant |\xi_{i_2}| \geqslant \cdots \geqslant |\xi_{i_n}|$.

LEMMA 3.2. *In a space with symmetric norm*

$$\|x^* - y^*\| \leqslant \|x - y\|. \tag{3.12}$$

PROOF. The left and right sides of the inequality will not change if we carry out the same permutations and changes of sign of the components in both $x$ and $y$. Therefore in the proof we may assume that $x = x^*$. Under this assumption, by the monotonicity property of the norm described above, we have $\|x - |y|\,\| \leqslant \|x - y\|$, $|y| = (|\eta_1|, \ldots, |\eta_n|)$. Consequently, (3.12) will be proved if we establish it for vectors $y$ with nonnegative components. Thus, let $x = x^*$ and $y \geqslant 0$. Let $\eta_i < \eta_j$ for some $i$ and $j$ with $i < j$. This means that the points $x$ and $y$ lie on different sides of the plane $l$ with equation $\zeta_i = \zeta_j$ ($x$ can lie on this plane). We consider the point $\bar{y}$ symmetric to $y$ with respect to $l$ and the point $z$ symmetric to $y$ with respect to the plane parallel to $l$ and passing through $x$. It is obvious that $\bar{y}$ lies on the line segment connecting $y$ and $z$. Next, $\|x - y\| = \|x - z\|$ in view of the symmetricity of the space, and consequently $\|x - \bar{y}\| \leqslant \|x - y\|$. Thus, the transposition of a pair of components of $y$ in decreasing order does not increase $\|x - y\|$. Applying this property successively, we arrive at (3.12).

We leave it to the reader to prove that the quantity

$$\|x\| = \sum_{i=1}^{l} \xi_i^*$$

has the properties of the norm of a symmetric Banach space for every fixed $l \leqslant n$ (see (2.16)). Then the lemma will imply the inequality

$$\sum_{i=1}^{l} (\xi_i^* - \eta_i^*)^* \leqslant \sum_{i=1}^{l} (\xi_i - \eta_i)^* \qquad (1 \leqslant l \leqslant n). \tag{3.13}$$

We now return to rearrangements of functions.

THEOREM 3.1. *For any functions $x, y \in L_1 + L_\infty$*

$$\int_0^\tau [x^*(t) - y^*(t)]^*\,dt \leqslant \int_0^\tau [x(t) - y(t)]^*\,dt. \tag{3.14}$$

PROOF. First we establish the following auxiliary assertion: Assume that for the sequences of pairs of functions $x_n(t), y_n(t) \in L_1 + L_\infty$ we have already shown the validity of (3.14). If a) $x_n^*(t) \to x^*(t)$ and $y_n^*(t) \to y^*(t)$ almost everywhere, and b)

$$\int_0^\tau [x_n(t) - y_n(t)]^*\,dt \leqslant \int_0^\tau [x(t) - y(t)]^*\,dt,$$

then (3.14) will be true for the two functions $x$ and $y$, as well. Indeed, (3.14) and condition b) imply that

$$\int_0^\tau [x_n^*(t) - y_n^*(t)]^*\,dt \leqslant \int_0^\tau [x(t) - y(t)]^*\,dt.$$

From property 7° it follows that for any $e$ with $\operatorname{mes} e = \tau$ we have

$$\int_e |x_n^*(t) - y_n^*(t)|\,dt \leqslant \int_0^\tau [x(t) - y(t)]^*\,dt.$$

By Fatou's lemma, this inequality and property a) imply that

$$\int_e |x^*(t) - y^*(t)|\, dt \leqslant \int_0^\tau [x(t) - y(t)]^*\, dt.$$

Taking the supremum over $e$ on the left side, from 7° we obtain (3.14).

The above assertion enables us to enlarge the class of functions for which (3.14) holds.

We first consider step functions of the form

$$x(t) = \sum_{k=1}^{n} x_k \chi_{[(k-1)a,\, ka)}(t) \quad \text{and} \quad y(t) = \sum_{k=1}^{n} y_k \chi_{[(k-1)a, ka)}(t).$$

For such functions we have

$$x^*(t) = \sum_{k=1}^{n} x_k^* \chi_{[(k-1)a, ka)}(t) \quad \text{and} \quad y^*(t) = \sum_{k=1}^{n} y_k^* \chi_{[(k-1)a, ka]}(t).$$

Inequality (3.13) immediately leads to the inequality

$$\int_0^{la} [x^*(t) - y^*(t)]^*\, dt \leqslant \int_0^{la} [x(t) - y(t)]^*\, dt \qquad (0 \leqslant l \leqslant n). \quad (3.15)$$

The functions of $\tau$ on the left and right sides of (3.14) are piecewise linear in our case, and so (3.15) implies (3.14).

Now let $x, y \in L_1(0, \infty)$. We introduce the step functions

$$x_n(t) = \sum_{k=1}^{n^2} n \int_{(k-1)/n}^{k/n} x(s)\, ds \chi_{[(k-1)/n,\, k/n)}(t) = T_n x(t),$$

$$y_n(t) = \sum_{k=1}^{n^2} n \int_{(k-1)/n}^{k/n} y(s)\, ds \chi_{[(k-1)/n,\, k/n)}(t) = T_n y(t).$$

By the above, (3.14) holds for these functions. We verify that the sequences $x_n$ and $y_n$ satisfy conditions a) and b). By the properties of the integral of a summable function the sequences $T_n x(t)$ and $T_n y(t)$ converge to $x(t)$ and $y(t)$ almost everywhere. Since $x$ and $y$ belong to $L_1$, the indicated sequences converge in measure. By property 11° of rearrangements this implies that $x_n^*(t) \to x^*(t)$ and $y_n^*(t) \to y^*(t)$ at all points of continuity of $x^*(t)$ and $y^*(t)$, and consequently, almost everywhere. Property a) is satisfied. Property b) follows from (3.11):

$$\int_0^\tau [x_n(t) - y_n(t)]^*\, dt = \int_0^\tau [T_n(x - y)]^*(t)\, dt \leqslant \int_0^\tau (x - y)^*(t)\, dt.$$

Now let $x, y \in L_1 + L_\infty$. According to Lemma 3.1 we construct the sets $e_n$ and $f_n$ for $x(t)$ and $y(t)$, respectively, and then take their unions $\tilde{e}_n = e_n \cup f_n$. The functions $x_n(t) = x(t)\chi_{\tilde{e}_n}(t)$ and $y_n(t) = y(t)\chi_{\tilde{e}_n}(t)$ belong to $L_1(0, \infty)$.

Next, by the remark to Lemma 3.1, we have $x_n^*(t) \to x^*(t)$ and $y_n^*(t) \to y^*(t)$ almost everywhere. Finally, by property 2° of rearrangements we have

$$\int_0^T [x_n(t) - y_n(t)]^* \, dt = \int_0^T [(x - y)(t)\chi_{\tilde{e}_n}(t)]^* \, dt \leqslant \int_0^T [x(t) - y(t)]^* \, dt.$$

Properties a) and b) are satisfied. The theorem is proved.

**4.** ***The contraction semigroup operators.*** We consider the set $\Sigma$ of linear operators in $L_1 + L_\infty$ for which the restrictions to $L_1$ and $L_\infty$ act in each of these spaces and are contractions:

$$\Sigma = \{ T\colon \|T\|_{L_1 \to L_1} \leqslant 1, \|T\|_{L_\infty \to L_\infty} \leqslant 1 \}.$$

In accordance with the notation of the first chapter, $\Sigma = \pi_1(L_1 L_1, L_\infty L_\infty)$; the set $\Sigma$ is convex and forms a semigroup with respect to the composition of operators.

The operators in $\Sigma$ have the following important property:

$$\int_0^\theta (Tx)^*(s) \, ds \leqslant \int_0^\theta x^*(s) \, ds \qquad (0 < \theta < \infty). \tag{3.16}$$

Indeed, by (3.5) this inequality is equivalent to $K(\theta, Tx) \leqslant K(\theta, x)$. The functional $K(\theta, x)$ is the norm in the sum of the spaces $L_1$ and $\theta L_\infty$ (with norm $\theta \|x\|_{L_\infty}$), and the above inequality follows from the fact that this sum is an interpolation space between $L_1$ and $\theta L_\infty$ with interpolation constant 1 (see Lemma 4.2 in Chapter I).

It is easy to see that, conversely, (3.16) implies that $T \in \Sigma$.

We give examples of operators in $\Sigma$. It is obvious that $\Sigma$ contains the operators of multiplication by a measurable function not exceeding one in absolute value.

By (3.11) the set $\Sigma$ contains all averaging operators.

Finally, $\Sigma$ contains the operators of the form $Tx(t) = x(\omega(t))$, where $\omega(t)$ is a transformation of $(0, \infty)$ into itself that does not decrease measure.

We introduce an operator $V$ which .s the adjoint of the restriction of $T \in \Sigma$ to $L_1$: $V = (T|L_1)^*$. The operator $V$ acts in $L_\infty$ and has norm $\leqslant 1$. We show that $V$ can be continuously extended from $L_1 \cap L_\infty$ to a contraction in $L_1$. By definition,

$$\int_0^\infty Tx(t)y(t) \, dt = \int_0^\infty x(t) \, Vy(t) \, dt \tag{3.17}$$

for all $x \in L_1$ and $y \in L_\infty$.

For $y \in L_1 \cap L_\infty$ we have

$$\begin{aligned}\|Vy\|_{L_1} &= \int_0^\infty |Vy(t)|\, dt \\ &= \sup_{0<N<\infty} \int_0^\infty \chi_{[0,N]}(t) \operatorname{sgn} Vy(t)\, Vy(t)\, dt \\ &= \sup_{0<N<\infty} \int_0^\infty T\left[\chi_{[0,N]}(t) \operatorname{sgn} Vy\right](t) y(t)\, dt \\ &\leqslant \|T\|_{L_\infty \to L_\infty} \|y\|_{L_1} \leqslant \|y\|_{L_1}.\end{aligned}$$

Since $L_1 \cap L_\infty$ is dense in $L_1$, the last inequality implies our assertion.

The operator $V$ can be extended to $L_1 + L_\infty$ by linearity. Then it becomes a member of $\Sigma$ and is denoted by $T^0$.

If $x \in L_1 \cap L_\infty$, on passing to the limit we obtain from (3.17) that

$$\int_0^\infty Tx(t)y(t)\, dt = \int_0^\infty x(t) T^0 y(t)\, dt \\ (x \in L_1 \cap L_\infty, y \in L_1 + L_\infty). \tag{3.18}$$

Any operator of multiplication is selfadjoint in the sense that $T = T^0$. The averaging operators have the same property, which follows from the equality

$$\int_0^\infty Tx(t)y(t)\, dt = \sum \frac{1}{\operatorname{mes} e_j} \int_{e_j} x(s)\, ds \int_{e_j} y(s)\, ds.$$

We can give an example of a nonzero operator $T_1 \in \Sigma$ for which $T_1^0 = 0$. On $L_\infty$ we construct a linear functional $f_0(x)$ with norm equal to one, vanishing on $L_1 \cap L_\infty$ and not identically zero. Then the operator $T_1 x = f_0(x)\chi_{(0,\infty)}$ belongs to $\Sigma$ (on $L_1$ the operator $T_1$ is identically zero). Then it follows from (3.18) that

$$\int_0^\infty x(t) T_1^0 y(t)\, dt = 0$$

for all $x \in L_1$ and $y \in L_\infty$, and so $T_1^0 y = 0$.

From this example it follows in particular that $T^{00} \neq T$ in general.

By $\Sigma_0$ we denote the collection of all operators $T \in \Sigma$ for which $T^{00} = T$. For the operators $T \in \Sigma_0$ it follows from (3.18) that

$$\int_0^\infty x(t) T^0 y(t)\, dt = \int_0^\infty T^{00} x(t) y(t)\, dt = \int_0^\infty Tx(t)y(t)\, dt \\ (x \in L_1 + L_\infty, y \in L_1 \cap L_\infty).$$

By (3.18) the equality

$$\int_0^\infty Tx(t)y(t)\,dt = \int_0^\infty x(t)T^{\circ}y(t)\,dt$$
$$(x \in L_1 + L_\infty, y \in L_1 \cap L_\infty) \tag{3.19}$$

is a characteristic property distinguishing $\Sigma_0$ in $\Sigma$.

In $L_1 + L_\infty$ we may introduce the weak topology $\sigma(L_1 + L_\infty, L_1 \cap L_\infty)$. An operator $T \in \Sigma$ belongs to $\Sigma_0$ if and only if it is continuous in this topology. Indeed, if $T \in \Sigma_0$ and $x_n$ converges to $x$ weakly, then for $y \in L_1 \cap L_\infty$ we have

$$\lim \int_0^\infty Tx_n(t)y(t)\,dt = \lim \int_0^\infty x_n(t)T^{\circ}y(t)\,dt$$
$$= \int_0^\infty x(t)T^{\circ}y(t)\,dt = \int_0^\infty Tx(t)y(t)\,dt.$$

Conversely, if $T \in \Sigma$ is continuous, then, using the fact that for any $x \in L_1 + L_\infty$ we have $x(t)\chi_{[0,N]}(t) \in L_1$ and this sequence converges to $x(t)$ in the topology $\sigma(L_1 + L_\infty, L_1 \cap L_\infty)$ as $N \to \infty$, from (3.17) for $y \in L_1 \cap L_\infty$ we obtain

$$\int_0^\infty Tx(t)y(t)\,dt = \lim \int_0^\infty Tx_N(t)y(t)\,dt$$
$$= \lim \int_0^\infty x_N(t)T^{\circ}y(t)\,dt = \int_0^\infty x(t)T^{\circ}y(t)\,dt,$$

which proves that $T$ belongs to $\Sigma_0$.

From the last property it follows in particular that $\Sigma_0$ is a semigroup with respect to composition of operators.

Let $\omega(t)$ be a one-to-one measure preserving mapping of the semiaxis into itself. We show that the operator $T_\omega x = x(\omega(t))$ belongs to $\Sigma_0$. For this we denote by $S$ the image of $(0, \infty)$ under the mapping $\omega$. It is easy to verify that for any functions $x \in L_1 + L_\infty$ and $y \in L_1 \cap L_\infty$ we have

$$\int_0^\infty x(\omega(t))y(t)\,dt = \int_S x(\tau)y(\omega^{-1}(\tau))\,d\tau.$$

If we introduce the function

$$z(\tau) = \begin{cases} y(\omega^{-1}(\tau)) & \text{for } \tau \in S, \\ 0 & \text{for } \tau \notin S, \end{cases}$$

we obtain

$$\int_0^\infty x(\omega(t))y(t)\,dt = \int_0^\infty x(\tau)z(\tau)\,d\tau.$$

It is obvious that $z \in L_1 \cap L_\infty$. Therefore from the weak convergence of $x_n$ to $x$ it follows that $x_n(\omega(t))$ converges to $x(\omega(t))$ weakly, and consequently $T_\omega \in \Sigma_0$.

THEOREM 3.2. *If $u \in L_1 \cap L_\infty$ and $v \in L_1 + L_\infty$, then*

$$\sup_{T\in\Sigma_0} \int_0^\infty Tu(t)v(t)\,dt = \sup_{T\in\Sigma_0} \int_0^\infty u(t)Tv(t)\,dt = \int_0^\infty u^*(t)v^*(t)\,dt. \tag{3.20}$$

PROOF. By property 13°, (3.16), and property 18° of rearrangements we have

$$\int_0^\infty Tu(t)v(t)\,dt \leqslant \int_0^\infty (Tu)^*(t)v^*(t)\,dt \leqslant \int_0^\infty u^*(t)v^*(t)\,dt. \tag{3.21}$$

Similarly,

$$\int_0^\infty u(t)Tv(t)\,dt \leqslant \int_0^\infty u^*(t)v^*(t)\,dt. \tag{3.22}$$

We prove that

$$\sup_{T\in\Sigma_0} \int_0^\infty Tu(t)v(t)\,dt \geqslant \int_0^\infty u^*(t)v^*(t)\,dt. \tag{3.23}$$

Since $\Sigma_0$ contains the operators of multiplication by a function not exceeding one in modulus, we may assume that $u(t) \geqslant 0$, $v(t) \geqslant 0$ and $Tu(t) \geqslant 0$.

According to Lemma 2.1, we construct operators $T_{\omega_1}$ and $T_{\omega_2}$ belonging to $\Sigma_0$ such that

$$T_{\omega_1}u(t) = u^*(t) + \alpha(t), \qquad T_{\omega_2}v(t) = v^*(t) + \beta(t)$$

and $\|\alpha\|_{L_1\cap L_\infty} < \varepsilon$, $\|\beta\|_{L_1\cap L_\infty} < \varepsilon$. Then

$$\int_0^\infty T^0_{\omega_2}T_{\omega_1}u(t)v(t)\,dt = \int_0^\infty T_{\omega_1}u(t)T_{\omega_2}v(t)\,dt = \int_0^\infty u^*(t)v^*(t)\,dt + \gamma,$$

where

$$\begin{aligned} |\gamma| &\leqslant \|\alpha\|_{L_1\cap L_\infty}\|v^* + \beta\|_{L_1+L_\infty} + \|u\|_{L_1\cap L_\infty}\|\beta\|_{L_1+L_\infty} \\ &\leqslant \varepsilon(\|u\|_{L_1\cap L_\infty} + \|v\|_{L_1+L_\infty} + \varepsilon) \to 0 \end{aligned} \tag{3.24}$$

as $\varepsilon \to 0$.

Thus

$$\begin{aligned} \sup_{T\in\Sigma_0} \int_0^\infty Tu(t)v(t)\,dt &\geqslant \int_0^\infty T^0_{\omega_2}T_{\omega_1}u(t)v(t)\,dt \\ &\geqslant \int_0^\infty u^*(t)v^*(t)\,dt + |\gamma|, \end{aligned}$$

from which (3.23) follows by (3.24).

Similarly, by means of the operator $T^0_{\omega_1}T_{\omega_2}$ we may prove the inequality

$$\sup_{T\in\Sigma_0}\int_0^\infty u(t)Tv(t)\,dt \geqslant \int_0^\infty u^*(t)v^*(t)\,dt,$$

which, together with (3.21) and (3.22), gives (3.20).

With every operator $T \in \Sigma$ we associate the function

$$\rho_T(x,y) = \int_0^\infty Tx(t)y(t)\,dt$$

on the product $L_\infty \times (L_1 \cap L_\infty)$. It is obvious that

$$|\rho_T(x,y)| \leqslant \|Tx\|_{L_\infty}\|y\|_{L_1} \leqslant \|x\|_{L_\infty}\|y\|_{L_1}. \tag{3.25}$$

It can be verified easily that the mapping $T \to \rho_T$ is one-to-one.

By means of the functions $\rho_T$ we introduce a topology on $\Sigma$ in which a basis of neighborhoods of the point $T_0$ is determined by all possible finite collections of functions $\{x_i\} \in L_\infty$ and $\{y_i\} \in L_1 \cap L_\infty$. A neighborhood consists of all those $T$ for which $|\rho_T(x_i,y_j) - \rho_{T_0}(x_i,y_j)| < 1$. In this topology the mapping $T \to \rho_T$ is a continuous imbedding of $\Sigma$ in the compactum $Q$ which is the product of the intervals $I_{xy} = \{\alpha: |\alpha| \leqslant \|x\|_{L_\infty}\|y\|_{L_1}\}$ $(x \in L_\infty, y \in L_1 \cap L_\infty)$.

We show that the image of $\Sigma$ is closed, whence it will follow that it is compact. Let $\tilde\rho(x,y)$ be a limit point of the image of $\Sigma$. Then $\tilde\rho(x,y)$ is a bilinear functional of $x$ and $y$. From (3.25) for all $T \in \Sigma$ we obtain the inequality

$$|\tilde\rho(x,y)| \leqslant |\ \|x\|_{L_\infty}|\ \|y\|_{L_1}. \tag{3.26}$$

Thus, $\tilde\rho(x,y)$ is bounded in $L_1$ for fixed $x \in L_\infty$ and so it can be written in the form

$$\tilde\rho(x,y) = \int_0^\infty \tilde z(t)y(t)\,dt,$$

where $\tilde z \in L_\infty$. We set $\tilde Tx = \tilde z$. From the linearity of $\tilde\rho(x,y)$ in $x$ and the density of $L_1 \cap L_\infty$ in $L_1$ it follows that the operator $\tilde T$ is linear. The set $L_1 \cap L_\infty$ as a subset of $L'_\infty$ is normative (see Chapter I, §2) for $L_\infty$. Therefore (3.26) implies that $\|\tilde T\|_{L_\infty} \leqslant 1$.

Since for the functionals $\rho_T(x,y)$ for $x \in L_\infty \cap L_1$ we have

$$|\rho_T(x,y)| \leqslant \|x\|_{L_1}\|y\|_{L_\infty},$$

the same inequality holds for $\tilde\rho(x,y)$ as well, i.e.

$$\left|\int_0^\infty \tilde Tx(t)y(t)\,dt\right| \leqslant \|x\|_{L_1}\|y\|_{L_\infty} \qquad (x \in L_\infty \cap L_1),$$

whence $\|\tilde Tx\|_{L_1} \leqslant \|x\|_{L_1}$. Since $L_1 \cap L_\infty$ is dense in $L_1$, the operator $\tilde T$ admits a continuous extension to the whole space $L_1$, and then to $L_1 + L_\infty$

by linearity; here $\|\tilde{T}\|_{L_1} \leqslant 1$. Thus, $\tilde{T} \in \Sigma$. Consequently $\rho(x, y) = \rho_{\tilde{T}}(x, y)$, and the image of $\Sigma$ is closed in $Q$.

We denote by $O_x$ the orbit of the point $x \in L_1 + L_\infty$ with respect to the semigroup $\Sigma$: $O_x = \{y: y = Tx,\ T \in \Sigma\}$.

THEOREM 3.3. *The orbit of any point $x \in L_1 + L_\infty$ is closed in the weak topology $\sigma(L_1 + L_\infty, L_1 \cap L_\infty)$.*

PROOF. Let $x \in L_1 + L_\infty$. For the proof of the assertion it suffices to show that the mapping $T \to Tx$ of the compactum $\Sigma$ into $L_1 + L_\infty$ is continuous. Let $T_\alpha \to T_0$ in $\Sigma$. Then

$$\rho_{T_\alpha}(z, y) = \int_0^\infty T_\alpha z(t) y(t)\, dt \to \int_0^\infty T_0 z(t) y(t)\, dt$$

for any $z \in L_\infty$ and $y \in L_1 \cap L_\infty$. We choose a sequence $z_n \in L_\infty$ so that $z_n \to x$ in $L_1 + L_\infty$. Then

$$\begin{aligned} &\left| \int_0^\infty T_\alpha x(t) y(t)\, dt - \int_0^\infty T_0 x(t) y(t)\, dt \right| \\ &\qquad \leqslant \left| \int_0^\infty T_\alpha(x - z_n) y\, dt \right| + \left| \int_0^\infty T_0(x - z_n) y\, dt \right| \\ &\qquad\quad + \left| \int_0^\infty (T_\alpha z_n - T_0 z_n) y\, dt \right| \\ &\qquad \leqslant 2\|x - z_n\|_{L_1 + L_\infty} \|y\|_{L_1 \cap L_\infty} + \left| \int_0^\infty (T_\alpha z_n - T_0 z_n) y\, dt \right|. \end{aligned}$$

Choosing $n$ and then $\alpha$ sufficiently large, we obtain

$$\int_0^\infty T_\alpha x(t) y(t)\, dt \to \int_0^\infty T_0 x(t) y(t)\, dt$$

for any $y \in L_1 \cap L_\infty$.

In what follows, the next theorem plays an important role.

THEOREM 3.4. *Let $x \in L_1 + L_\infty$. A measurable function $y$ belongs to the orbit $O_x$ if and only if*

$$\int_0^\theta y^*(t)\, dt \leqslant \int_0^\theta x^*(t)\, dt \qquad (0 < \theta < \infty). \tag{3.27}$$

PROOF. The necessity of (3.27) immediately follows from (3.16).

We prove the sufficiency indirectly. For the function $y(x)$ let (3.27) be satisfied and assume that $y \notin O_x$. We observe that (3.27) implies that $y \in L_1 + L_\infty$. The orbit $O_x$ is convex and closed in the weak topology

$$\sigma(L_1 + L_\infty, L_1 \cap L_\infty).$$

By general duality theorems (see [6], Chapter II) the space $L_1 \cap L_\infty$ is dual to the locally convex linear space $L_1 + L_\infty$ with the topology $\sigma(L_1 + L_\infty, L_1 \cap L_\infty)$. Using the separability theorem ([6], loc. cit.), we may assert that there exists a function $a(t) \in L_1 \cap L_\infty$ such that

$$\int_0^\infty a(t)y(t)\,dt > \sup_{T\in\Sigma} \int_0^\infty a(t)Tx(t)\,dt.$$

From (2.25) and Theorem 3.2 we obtain

$$\int_0^\infty a^*(t)y^*(t)\,dt \geqslant \int_0^\infty a(t)y(t)\,dt > \int_0^\infty a^*(t)x^*(t)\,dt. \tag{3.28}$$

By property 18° of rearrangements, inequality (3.28) contradicts (3.27).

## §4. Symmetric spaces. Interpolation between $L_1$ and $L_\infty$

**1.** ***Definition of symmetric spaces.*** A Banach function space on the semiaxis $(0, \infty)$ with Lebesgue measure is said to be *symmetric* if it satisfies the following conditions:

1) If $y \in E$ and $|x(t)| \leqslant |y(t)|$ almost everywhere on $(0, \infty)$, then $x \in E$ and $\|x\|_E \leqslant \|y\|_E$ (i.e., $E$ is an ideal lattice).

2) If $y \in E$ and $|x(t)|$ is equimeasurable with $|y(t)|$, then $x \in E$ and $\|x\|_E = \|y\|_E$.

Conditions 1) and 2) are equivalent to the following single condition.

1′) If $y \in E$ and $x^*(t) \leqslant y^*(t)$ for all $t \in (0, \infty)$, then $x \in E$ and $\|x\|_E \leqslant \|y\|_E$.

Let the Banach function space $E$ satisfy condition 1). It contains at least one function $x_0(t) \not\equiv 0$. Then for some $\varepsilon > 0$ there exists a set $e_0$ of positive measure mes $e_0 = \mu_0$, such that $|x_0(t)| \geqslant \varepsilon$ for $t \in e_0$. This implies that $\varepsilon\chi_{e_0}(t)$, and hence $\chi_{e_0}(t)$ as well, belongs to $E$.

LEMMA 4.1. *If an ideal lattice $E$ (condition 1)) contains all functions $\chi_e(t)$ with mes $e = \mu_0$ and $\|\chi_e\|_E \leqslant C$, where $C$ does not depend on the choice of $e$, then the space $L_1 \cap L_\infty$ is imbedded in $E$.*

PROOF. Let $x \in L_1 \cap L_\infty$. Without loss of generality we may assume that $x(t) \geqslant 0$. Write $\lambda_j = x^*(j\mu_0)$ $(j = 0, 1, \dots)$. There exist disjoint sets $e_j$ such that mes $e_j = \mu_0$ and $\lambda_{j+1} \leqslant x(t) \leqslant \lambda_j$ $(j = 0, 1, \dots)$. If

$$\bar{x}(t) = \sum_{j=0}^\infty \lambda_j\chi_{e_j}(t) \quad\text{and}\quad \underline{x}(t) = \sum_{j=0}^\infty \lambda_{j+1}\chi_{e_j}(t),$$

then $\underline{x}(t) \leqslant x(t) \leqslant \bar{x}(t)$. Hence

$$\mu_0 \sum_{j=0}^\infty \lambda_{j+1} = \|\underline{x}\|_{L_1} \leqslant \|x\|_{L_1} < \infty. \tag{4.1}$$

We consider the functions $x_N(t) = \sum_0^N \lambda_j \chi_{e_j}(t)$. The sequence $x_N$ is fundamental in $E$. Indeed,

$$\|x_{N+p} - x_N\|_E \leq \sum_{j=N+1}^{N+p} \lambda_j \|\chi_{e_j}\|_E \leq C \sum_{j=N+1}^{N+p} \lambda_j \to 0$$

as $N \to \infty$, because the series (4.1) converges. Since $x_N \to \bar{x}$ in $S(0, \infty)$, we have $x_N \to \bar{x}$ in $E$, and consequently $\bar{x} \in E$. By 1) we also have $x \in E$ and

$$\|x\|_E \leq \|\bar{x}\|_E \leq C \sum_{j=0}^{\infty} \lambda_j = C\left(\lambda_0 + \sum_{j=0}^{\infty} \lambda_{j+1}\right)$$

$$\leq C\left(\|x\|_{L_\infty} + \frac{1}{\mu_0}\|x\|_{L_1}\right) \leq C\left(1 + \frac{1}{\mu_0}\right)\|x\|_{L_1 \cap L_\infty}.$$

REMARK. The hypothesis of the lemma and 1) imply that the space $E$ contains the characteristic function of any measurable set of finite measure. Therefore we may set $\mu_0 = 1$, and the preceding inequality takes the form

$$\|x\|_E \leq 2 \sup_{\text{mes } e = 1} \|\chi_e\| \, \|x\|_{L_1 \cap L_\infty}.$$

THEOREM 4.1. *Any symmetric space is intermediate between the spaces* $L_1(0, \infty)$ *and* $L_\infty(0, \infty)$, *i.e.* $L_1 \cap L_\infty \subset E \subset L_1 + L_\infty$.

PROOF. The space $E$ obviously satisfies the hypotheses of Lemma 4.1, and $\|\chi_e\|_E = \|\chi_{(0,1)}\|_E$ if mes $e = 1$. Therefore $L_1 \cap L_\infty$ is imbedded in $E$, and

$$\|x\|_E \leq 2\|\chi_{(0,1)}\|_E \|x\|_{L_1 \cap L_\infty}. \tag{4.2}$$

We pass to the proof of the second imbedding. Let $x \in E$ and $x(t) = x^*(t)$. By (3.4) we have

$$\|x\|_{L_\infty + L_1} = \int_0^1 x(t)\, dt, \tag{4.3}$$

and therefore it is sufficient to estimate the norm in $L_1$ of functions belonging to $E$ with support in $[0, 1]$. First we consider a step function $y_0(t)$ on $[0, 1]$, assuming the values $x_k \geq 0$ on sets $e_k$ $(k = 0, 1, \ldots, N-1)$ in $[0, 1]$ of the same measure mes $e_k = 1/N$. Then

$$y_0(t) = \sum_{k=0}^{N-1} x_k \chi_{e_k}(t).$$

We set

$$y_j(t) = \sum_{k=0}^{N-1} x_{k+j (\text{mod } N)} \chi_{e_k}(t).$$

It is obvious that the functions $y_0(t)$ and $y_j(t)$ are equimeasurable. Therefore $\|y_j\|_E = \|y_0\|_E$. Next,

$$\sum_{j=0}^{N-1} y_j(t) \equiv \sum_{k=0}^{N-1} x_k \chi_{[0,1]},$$

and hence

$$\left(\sum_{k=0}^{N-1} x_k\right)\|\chi_{[0,1]}\|_E = \left\|\sum_{j=0}^{N-1} y_j\right\|_E \leqslant \sum_{j=0}^{N-1} \|y_j\|_E = N\|y_0\|_E.$$

From this we obtain

$$\|y_0\|_E \geqslant \|\chi_{[0,1]}\|_E \frac{1}{N} \sum_{k=0}^{N-1} x_k = \|\chi_{[0,1]}\|_E \|y_0\|_{L_1}. \tag{4.4}$$

An arbitrary decreasing nonnegative function $z \in E$ which is equal to zero for $t > 1$ can be approximated by an increasing sequence of step functions

$$z_N(t) = \sum_{k=0}^{N-1} z\left(\frac{k+1}{N}\right)\chi_{[k/N,\,(k+1)/N]}(t) \qquad (N = 2^n),$$

which converges to $z(t)$ almost everywhere. Since $z_N(t) \leqslant z(t)$, we have $\lim_{N\to\infty} \|z_N\|_E \leqslant \|z\|_E$. On the other hand, by B. Levi's theorem we have $\|z\|_{L_1} = \lim_{N\to\infty} \|z_N\|_{L_1}$. Applying (4.4) to the functions $z_N$ and using the last relations, we obtain

$$\|z\|_{L_1} \leqslant \frac{1}{\|\chi_{[0,1]}\|_E} \|z\|_E. \tag{4.5}$$

Combining this with (4.3), we arrive at the final inequality

$$\|x\|_{L_\infty + L_1} \leqslant \frac{1}{\|\chi_{[0,1]}\|_E} \|x\chi_{[0,1]}\|_E \leqslant \frac{1}{\|\chi_{[0,1]}\|_E} \|x\|_E.$$

Thus, $E \subset L_\infty + L_1$. The theorem is proved.

REMARK. In the sequel we need the inequality

$$\|x^*\chi_{[0,h]}\|_{L_1} \leqslant \frac{h}{\|\chi_{[0,h]}\|_E} \|x^*\chi_{[0,h]}\|_E. \tag{4.6}$$

LEMMA 4.2. *The set of elementary* functions is dense in any symmetric space $E$.*

PROOF. By Theorem 4.1 every function $x \in E$ can be represented as the sum of a summable and a bounded function, and so its restriction to any

---

* *Editor's note.* The Russian term used by the author is счетнозначная (literally: countably valued). The term adopted for the translation is that used by Halmos in [10].

finite interval belongs to $L_1$. Then the function $x\chi_{(n-1,n]} \in L(n-1, n)$ can be uniformly approximated by an elementary function $z_n(t)$ $(n-1 \leqslant t \leqslant n)$ of smaller modulus with accuracy up to $\varepsilon/2^n$. This means that if $z_n(t)$ is extended to be equal to zero outside $(n-1, n]$, then

$$\|x\chi_{(n-1,n]} - z_n\|_{L_1 \cap L_\infty} < \varepsilon/2^n. \tag{4.7}$$

Then $z(t) = \sum_1^\infty z_n(t)$ is elementary, and $|z(t)| \leqslant |x(t)|$. Therefore $z \in E$. Next, by (4.7) we have

$$\|x - z\|_{L_1 \cap L_\infty} \leqslant \sum_1^\infty \|x\chi_{(n-1,n]} - z_n\|_{L_1 \cap L_\infty} \leqslant \varepsilon. \tag{4.8}$$

By (4.2) we have $\|x - z\|_E \leqslant 2\|\chi_{(0,1]}\|_E \varepsilon$.

LEMMA 4.3. *Under condition* 1) *of the definition of a symmetric space, condition* 2) *is equivalent to the following*: *For any mapping* $t \to \omega(t)$ *of* $(0, \infty)$ *into itself which does not increase measure and for any* $y \in E$ *we have* $y_\omega(t) = y(\omega(t)) \in E$ *and*

$$\|y_\omega\|_E \leqslant \|y\|_E. \tag{4.8$'$}$$

PROOF. As was shown above, condition 1) implies that $E$ contains a function of the form $\chi_{e_0}(t)$ with mes $e_0 = \mu_0 > 0$. If $e_1$ is another measurable set of the same measure, then we can construct a mapping $\omega(t)$ of $(0, \infty)$ onto itself which maps $e_1$ onto $e_0$ and preserves measure. Then $\chi_{e_1}(t) = \chi_{e_0}(\omega(t))$, and therefore $\chi_{e_1} \in E$ and $\|\chi_{e_1}\|_E \leqslant \|\chi_{e_0}\|_E$. Thus, the hypotheses of Lemma 4.1 are satisfied, and $L_1 \cap L_\infty$ is imbedded in $E$.

Now let $y \in E$ and let $|x(t)|$ be equimeasurable with $|y(t)|$. Then by Theorem 2.1 for any $\varepsilon > 0$ there exist a measure preserving mapping $\omega(t)$ of $(0, \infty)$ into itself and a function $\beta(t)$ with $|\beta(t)| \leqslant 1$ such that $\|x - \beta y_\omega\|_{L_1 \cap L_\infty} < \varepsilon$. By Lemma 4.1 and (4.2) we have $x - \beta y_\omega \in E$ and

$$\|x - \beta y_\omega\|_E \leqslant 2\|\chi_{[0,1]}\|_E \varepsilon.$$

By property 1) and the hypothesis of the lemma we have $\beta y_\omega \in E$, and hence $x \in E$ and

$$\begin{aligned}\|x\|_E &\leqslant \|\beta y_\omega\|_E + 2\|\chi_{[0,1]}\|_E \varepsilon \\ &\leqslant \|y_\omega\|_E + 2\|\chi_{[0,1]}\|_E \varepsilon \leqslant \|y\|_E + 2\|\chi_{[0,1]}\|_E \varepsilon.\end{aligned}$$

Since $\varepsilon$ is arbitrary and $x$ and $y$ occur symmetrically, we obtain that $\|x\|_E = \|y\|_E$.

Conversely, if $\omega(t)$ is a measure nondecreasing mapping of $(0, \infty)$ into itself, then $y_\omega^*(t) \leqslant y^*(t)$. In a symmetric space the condition $y \in E$ then implies that $y_\omega \in E$, and (4.8$'$) is satisfied. The lemma is proved.

A linear subset $F$ of a symmetric space $E$ is said to be *symmetric* if the condition that $y$ belongs to $F$ implies that $x$ belongs to $F$ whenever $x^*(t) \leqslant y^*(t)$.

LEMMA 4.4. *The closure in $E$ of a symmetric linear subset $F$ is a symmetric space with respect to the norm of $E$.*

PROOF. Let $y \in \bar{F}$, and let the sequence $y_n \in F$ converge to $y$ in $E$. If $|x| < |y|$, then the sequence $\min(|x|, |y_n|) \leqslant |y_n|$ belongs to $F$, and by property 4° of ideal lattices (see p. 00) it converges to $|x|$ in $E$; consequently $\operatorname{sgn} x \cdot \min(|x|, |y_n|) \in F$, and this sequence converges to $x$ in $E$, i.e. $x \in \bar{F}$. Thus, property 1) is satisfied. If now $\omega(t)$ is a measure nondecreasing transformation of the semiaxis into itself, then $\|(y_n)_\omega - y_\omega\|_E \leqslant \|y_n - y\|_E \to 0$. Since $(y_n)^*_\omega \leqslant y^*_n \in F$, we have $(y_n)_\omega \in F$, and thus $y_\omega \in \bar{F}$. On the basis of Lemma 4.3 this implies property 2) for the space $\bar{F}$.

LEMMA 4.5. *The intersection and the sum of two symmetric spaces are symmetric spaces.*

PROOF. Let $E_1$ and $E_2$ be symmetric spaces. The fact that $E_1 \cap E_2$ is symmetric is obvious. Consider the sum $E = E_1 + E_2$. Let $y \in E$ and $x^*(t) \leqslant y^*(t)$. By Theorem 2.1, for every $\varepsilon > 0$ there exist a measure preserving transformation of $[0, \infty)$ into itself and a measurable function $\beta(t)$ with $|\beta(t)| \leqslant 1$ such that

$$x(t) = \beta(t)y(\omega(t)) + z(t),$$

where $\|z\|_{L_1 \cap L_\infty} < \varepsilon$. If now $y(t) = u(t) + v(t)$, $u \in E_1$, $v \in E_2$, then $x(t) = \beta(t)u(\omega(t)) + \beta(t)v(\omega(t)) + z(t)$, where by Lemma 4.3 we have $\beta u_\omega \in E_1$ and $\beta v_\omega \in E_2$. From this it follows that $x \in E$ and

$$\|x\|_E \leqslant \|\beta u + z\|_{E_1} + \|\beta v\|_{E_1} \leqslant \|u\|_{E_1} + \|v\|_{E_2} + 2\|\chi_{[0,1]}\|_{E_1}\varepsilon.$$

Taking the infimum over all representations $y = u + v$, $u \in E_1$, $v \in E_2$, we obtain that $\|x\|_E \leqslant \|y\|_E + 2\|\chi_{[0,1]}\|_E\varepsilon$. Since $\varepsilon$ is arbitrary, this means that $\|x\|_E \leqslant \|y\|_E$, and so property 1′) is satisfied in $E_1 + E_2$.

THEOREM 4.2. *Let $E_0$ and $E_1$ be two symmetric spaces, and let $E$ be an interpolation space between them with interpolation constant one, i.e.*

$$\|T\|_{E\to E} \leqslant \max(\|T\|_{E_0\to E_0}, \|T\|_{E_1\to E_1}) \tag{4.9}$$

*for all $T \in \pi(E_0, E_1; E_0, E_1)$. Then $E$ is a symmetric space.*

PROOF. Let $\alpha(t)$ be a function with $|\alpha(t)| \leqslant 1$ and $\omega(t)$ a measure nondecreasing mapping of $(0, \infty)$ into itself. Property 1) and Lemma 4.3 imply that the operator $\alpha T_\omega y(t) = \alpha(t)y(\omega(t))$ acts in $E_0$ and $E_1$, and its norm there does not exceed one. Then it acts in $E$ as well, and $\|\alpha T_\omega y\|_E \leqslant \|y\|_E$. For $\omega(t) \equiv t$ this implies that condition 1) is satisfied in $E$, while for $\alpha \equiv 1$ condition 2) is satisfied by Lemma 4.3.

REMARK. If instead of (4.9) we have

$$\|T\|_{E\to E} \leqslant C\max(\|T\|_{E_0\to E_0}, \|T\|_{E_1\to E_1}), \tag{4.10}$$

then from what was said on p. 20 it follows that we may introduce an equivalent norm in $E$ with respect to which $E$ becomes symmetric.

**2. *The main interpolation theorem.*** If $E$ is an interpolation space between $L_1$ and $L_\infty$ and

$$\|T\|_{E\to E} \leqslant \max(\|T\|_{L_1\to L_1}, \|T\|_{L_\infty\to L_\infty}), \tag{4.11}$$

then $E$ is symmetric by Theorem 4.2.

In what follows we need

LEMMA 4.6. *In every interpolation space $E$ between $L_1$ and $L_\infty$ for which* (4.11) *holds,*

$$\|x^* - y^*\|_E \leqslant \|x - y\|_E \qquad (x, y \in E).$$

Indeed, (3.14) for any functions belonging to $L_1 + L_\infty$ and Theorem 3.4 imply that $x^* - y^* = T(x - y)$ for some $T$ in $\Sigma$. By (4.11) we then have

$$\|x^* - y^*\|_E = \|T(x - y)\|_E \leqslant \|T\|_{E\to E}\|x - y\|_E \leqslant \|x - y\|_E.$$

The lemma is proved.

There arises the problem of describing all symmetric spaces that are interpolation spaces between $L_1$ and $L_\infty$ and have the property (4.11). The answer to this question is given by the following theorem.

THEOREM 4.3. *A Banach space $E$ intermediate between $L_1$ and $L_\infty$ is an interpolation space with interpolation constant one if and only if the following condition is satisfied: If $y \in E$, $x \in L_0 + L_1$ and*

$$\int_0^t x^*(\tau)\,d\tau \leqslant \int_0^t y^*(\tau)\,d\tau \tag{4.12}$$

*for all $t \geqslant 0$, then $x \in E$ and $\|x\|_E \leqslant \|y\|_E$.*

PROOF. *Necessity.* If the functions $x$ and $y$ are connected by the relation (4.12), then by Theorem 3.4 there exists an operator $T \in \Sigma$ such that $x = Ty$. Then the interpolation property of $E$ implies that $x \in E$, and (4.11) implies that

$$\|x\|_E = \|Ty\|_E \leqslant \|T\|_{E\to E}\|y\|_E \leqslant \|y\|_E.$$

*Sufficiency*. Let $T$ be an arbitrary operator in $\Sigma$ and let $y \in E$. By (3.16) we have

$$\int_0^t (Ty)^*(\tau)\, d\tau \leqslant \int_0^t y^*(\tau)\, d\tau.$$

From the property of $E$ we have assumed to be satisfied it follows that $Ty \in E$ and $\|Ty\|_E \leqslant \|y\|_E$.

In connection with Theorem 4.3 it is natural to ask whether there exist symmetric spaces which are not interpolation spaces between $L_1$ and $L_\infty$. It turns out that such spaces exist; they will be given in §5.7.

REMARK. If $E$ is such that (4.12) implies that $\|x\|_E \leqslant C\|y\|_E$, then an equivalent norm can be introduced in $E$ by the formula $\|x\|_E^* = \sup\|u\|_E$, where the supremum is taken over all $u \in L_1 + L_\infty$ for which

$$\int_0^t u^*(\tau)\, d\tau \leqslant \int_0^t x^*(\tau)\, d\tau.$$

The space $E$ with the new norm satisfies the hypotheses of Theorem 4.3, and so it is an interpolation space between $L_1$ and $L_\infty$ with property (4.11). Since $\|x\|_E \leqslant \|x\|_E^* \leqslant C\|x\|_E$, inequality (4.10) is satisfied for the original norm in $E$ with the same constant $C$.

**3.** ***Dilation operators in a symmetric space.*** In the space $S(0, \infty)$ we consider dilation operators defined by

$$\sigma_\tau x(t) = x(\tau^{-1}t) \qquad (\tau > 0).$$

It is obvious that the operators $\sigma_\tau$ are continuous in $S(0, \infty)$. These operators form a group with unit element equal to $\sigma_1 = I$. It is easy to see that the operators $\sigma_\tau$ commute with the operation of rearrangement: $(\sigma_\tau x)^*(t) = \sigma_\tau(x^*)(t) = x^*(\tau^{-1}t)$.

THEOREM 4.4. *The operators $\sigma_\tau$ are bounded in any symmetric space $E$.*

PROOF. If $\tau_1 < \tau_2$, then $x^*(\tau_1^{-1}t) \leqslant x^*(\tau_2^{-1}t)$, and consequently $(\sigma_{\tau_1}x)^*(t) \leqslant (\sigma_{\tau_2}x)^*(t)$. This implies that for $x \in E$ the function $\|\sigma_\tau x\|_E$ is increasing in $\tau$. In particular, for $\tau \leqslant 1$ the operator $\sigma_\tau$ is bounded in $E$ and $\|\sigma_\tau\|_E \leqslant 1$ $(\tau \leqslant 1)$.

Now let $\tau > 1$ and $x \in E$. By what was said earlier, to prove the boundedness of the operators $\sigma_\tau$ it suffices to prove the boundedness of the operators $\sigma_m$, $m = 2, 3, \ldots$. We first assume that the function $x \in E$ is elementary, and

$$x(t) = \sum_1^\infty x_j \chi_{e_j}(t), \tag{4.13}$$

where $e_i \cap e_j = \varnothing$ for $i \neq j$, and mes $e_j < \infty$. We divide every set $e_j$ into $m$ disjoint parts $e_{j1}, \ldots, e_{jm}$ of equal measure and set

$$y_s(t) = \sum_{j=1}^{\infty} x_j \chi_{e_{js}}(t).$$

Then

$$x(t) = \sum_1^m y_s(t) \quad \text{and} \quad (\sigma_m y_s)^*(t) = x^*(t).$$

Therefore

$$\|\sigma_m x\|_E \leqslant \sum_1^m \|\sigma_m y_s\|_E = m\|x\|_E. \tag{4.14}$$

Since the elementary functions are dense in $E$ and the operators $\sigma_\tau$ are continuous in $S(0, \infty)$, inequality (4.14) can be established for any $x \in E$ by a passage to the limit.

THEOREM 4.5. *The function* $\|\sigma_\tau\|_E$ *is submultiplicative and quasiconcave.*

PROOF. The first assertion is obvious:

$$\|\sigma_{\tau_1\tau_2}\|_E \leqslant \|\sigma_{\tau_1}\|_E \|\sigma_{\tau_2}\|_E.$$

To prove quasiconcavity we again consider a function of the form (4.13) in $E$ and construct the functions

$$z_s(t) = \sum_{j=1}^{\infty} x_j \chi_{e_j \setminus e_{js}}(t).$$

Then $z_s(t)$ is equimeasurable with $\sigma_{(n-1)/n} x(t)$, and

$$\sum_1^n z_s(t) = (n-1)x(t).$$

Hence

$$(n-1)\|x\|_E = \left\| \sum_1^n z_s \right\|_E \leqslant \sum_1^n \|z_s\|_E = n\|\sigma_{(n-1)/n} x\|_E. \tag{4.15}$$

The inequality can be established for all $x \in E$ by a passage to the limit. Let $y$ be any function in $E$. We set $x = \sigma_{n/m} y$. Then from (4.15) we obtain that

$$\frac{m}{n} \|\sigma_{n/m} y\|_E \leqslant \frac{m}{n-1} \|\sigma_{(n-1)/m} y\|_E.$$

This implies that for any rational numbers

$$\frac{1}{r''} \|\sigma_{r''}\|_E \leqslant \frac{1}{r'} \|\sigma_{r'}\|_E,$$

if $r' < r''$. For arbitrary $\tau_1$ and $\tau_2$ with $\tau_1 < \tau_2$ we construct sequences of rational numbers $r'_n \uparrow \tau_1$ and $r''_n \downarrow \tau_2$. Then

$$\frac{1}{\tau_2}\|\sigma_{\tau_2}\|_E \leqslant \frac{r''_n}{\tau_2}\frac{1}{r''_n}\|\sigma_{r''_n}\|_E$$
$$\leqslant \frac{r''_n}{\tau_2}\frac{1}{r'_n}\|\sigma_{r'_n}\|_E \leqslant \frac{r''_n}{\tau_2}\frac{\tau_1}{r'_n}\frac{1}{\tau_1}\|\sigma_{\tau_1}\|_E.$$

The passage to the limit as $n \to \infty$ gives the inequality

$$\frac{1}{\tau_2}\|\sigma_{\tau_2}\|_E \leqslant \frac{1}{\tau_1}\|\sigma_{\tau_1}\|_E.$$

COROLLARY 1.

$$\|\sigma_\tau\|_E \leqslant \max\{1, \tau\} \qquad (\tau > 0). \tag{4.16}$$

This follows from the fact that $\|\sigma_1\|_E = 1$.

COROLLARY 2. *A symmetric space $E$ has the following property: If $y \in E$ and*

$$n_{|x|}(\tau) \leqslant C n_{|y|}(\tau), \tag{4.17}$$

*then $x \in E$ and $\|x\|_E \leqslant \max\{1, C\}\|y\|_E$.*

Indeed, (4.17) is equivalent to the inequality

$$x^*(t) \leqslant y^*(C^{-1}t) = \sigma_C y^*(t),$$

and the required inequality immediately follows from (4.16).

REMARK. It can be seen from the proof of Theorem 4.5 that the function $\|\sigma_\tau y\|_E$ is quasiconcave for any $y \in E$.

It can be verified immediately that $\|\sigma_\tau\|_{L_p} = \tau^{1/p}$. In §5 we shall study the Lorentz spaces $\Lambda_\psi$ and Marcinkiewicz spaces $M_\psi$. If $\psi(t)$ is a concave function and $\psi(0) = 0$, then

$$\|x\|_{\Lambda_\psi} = \int_0^\infty x^*(t)\, d\psi(t).$$

We calculate

$$\|\sigma_\tau x\|_{\Lambda_\psi} = \int_0^\infty x^*(\tau^{-1}t)\, d\psi(t) = \int_0^\infty x^*(s)\, d\psi(\tau s).$$

By property 18° of rearrangements we have

$$\|\sigma_\tau x\|_{\Lambda_\psi} \leqslant \sup_{s>0}\frac{\psi(\tau s)}{\psi(s)}\int_0^\infty x^*(s)\, d\psi(s) = M_\psi(\tau)\|x\|_{\Lambda_\psi}. \tag{4.18}$$

On the other hand,

$$\|\sigma_\tau\|_{\Lambda_\psi} \geqslant \sup_{t>0}\frac{\|\sigma_\tau \chi_{(0,t)}\|_{\Lambda_\psi}}{\|\chi_{(0,t]}\|_{\Lambda_\psi}} = \sup_{t>0}\frac{\psi(\tau t)}{\psi(t)} = M_\psi(\tau). \tag{4.19}$$

Inequalities (4.18) and (4.19) show that

$$\|\sigma_\tau\|_{\Lambda_\psi} = M_\psi(\tau). \tag{4.20}$$

If $\psi(t)$ is a quasiconcave function and $\psi(0) = 0$, then

$$\|x\|_{\mathrm{M}_\psi} = \sup_{t>0} \frac{1}{\psi(t)} \int_0^t x^*(s)\, ds.$$

A similar calculation shows that

$$\|\sigma_\tau\|_{\mathrm{M}_\psi} = \tau \sup \frac{\psi(t/\tau)}{\psi(t)} = \tau M_\psi\left(\frac{1}{\tau}\right) = M_{\psi_*}(\tau), \tag{4.21}$$

where $\psi_*(t) = t/\psi(t)$.

DEFINITION. The upper and lower dilation exponents of the function $\|\sigma_\tau\|_E$ are called *upper* and *lower dilation exponents of the space* $E$ and will be denoted by $\beta_E$ and $\alpha_E$.

If we write $s(\tau) = \|\sigma_\tau\|_E$, then by the submultiplicativity of this function we have $M_s(\tau) = s(\tau)$. Indeed,

$$M_s(\tau) = \sup_{t>0} \frac{s(t\tau)}{s(t)} \leqslant s(\tau) \quad \text{and} \quad M_s(\tau) \geqslant \frac{s(1\cdot\tau)}{s(1)} = s(\tau).$$

Therefore the formulas for determining the upper and lower dilation exponents take the form

$$\alpha_E = \lim_{\tau\to 0} \frac{\ln \|\sigma_\tau\|_E}{\ln \tau} \quad \text{and} \quad \beta_E = \lim_{\tau\to\infty} \frac{\ln \|\sigma_\tau\|_E}{\ln \tau}. \tag{4.22}$$

REMARK. If $s(\tau)$, $\tau > 0$, is submultiplicative and quasiconcave and $s(1) = 1$, then there exists a symmetric space $E$ for which $s(\tau) = \|\sigma_\tau\|_E$.

Indeed, as $E$ we may take the Marcinkiewicz space $\mathrm{M}_{t/s(t)}$. By (4.21) we have

$$\|\sigma_\tau\|_E = \tau \sup_{t>0} \frac{s(t)}{\tau s(t/\tau)} = M_s(\tau) - s(\tau).$$

Now we are in a position to improve inequality (4.11) under the hypotheses of Theorem 4.3. Let $T$ be an operator acting in $L_1$ and $L_\infty$. Then for any $\tau > 0$ we have

$$\|T\sigma_\tau\|_{L_1\to L_1} \leqslant \tau\|T\|_{L_1\to L_1} \quad \text{and} \quad \|T\sigma_\tau\|_{L_\infty\to L_\infty} \leqslant \|T\|_{L_\infty\to L_\infty}.$$

Using the representation $T = T\sigma_\tau\sigma_{1/\tau}$, we obtain the reverse inequalities, i.e.

$$\|T\sigma_\tau\|_{L_1\to L_1} = \tau\|T\|_{L_1\to L_1} \quad \text{and} \quad \|T\sigma_\tau\|_{L_\infty\to L_\infty} = \|T\|_{L_\infty\to L_\infty}.$$

We choose

$$\tau_0 = \|T\|_{L_\infty\to L_\infty}\|T\|^{-1}_{L_1\to L_1}. \tag{4.23}$$

Then by Theorem 4.3 we have

$$\|T\sigma_{\tau_0}x\|_E \leqslant \|T\|_{L_\infty\to L_\infty}\|x\|_E.$$

Replacing $\sigma_{\tau_0}x$ by $y$, we obtain

$$\|Ty\|_E \leqslant \|T\|_{L_\infty\to L_\infty}\|\sigma_{1/\tau_0}y\|_E,$$

and hence

$$\|T\|_{E\to E} \leqslant \|T\|_{L_\infty\to L_\infty}\|\sigma_{1/\tau_0}\|_E. \tag{4.24}$$

By (4.16) this inequality implies (4.11). Inequality (4.24) is sharp in the sense that it becomes an equality for the operator $\sigma_{1/\tau_0}$.

THEOREM 4.6. *A space $E$ intermediate between $L_1$ and $L_\infty$ is a normalized interpolation space of type $\alpha$ between $L_1$ and $L_\infty$ if and only if it satisfies the hypothesis of Theorem* 4.3 *and the equality* $\|\sigma_\tau\|_E = \tau^\alpha$.

PROOF. If the last equality is satisfied, then by (4.23) and (4.24) we have

$$\|T\|_{E\to E} \leqslant \|T\|^{1-\alpha}_{L_1\to L_1}\|T\|^{\alpha}_{L_\infty\to L_\infty}. \tag{4.25}$$

Conversely, (4.25) applied to $\sigma_\tau$ gives $\|\sigma_\tau\|_E \leqslant \tau^\alpha$. On the other hand, $\|\sigma_\tau\| \geqslant \|\sigma_{1/\tau}\|^{-1} \geqslant (1/\tau)^{-\alpha} = \tau^\alpha$, i.e. $\|\sigma_\tau\|_E = \tau^\alpha$.

In what follows we shall frequently use

LEMMA 4.7. *Let $\psi(t)$ be a continuous increasing function. Then for any function $x \in E$, where $E$ is a symmetric space,*

$$\left\|\int_0^\infty \sigma_\tau x^* d\psi(\tau)\right\|_E \leqslant \int_0^\infty \|\sigma_\tau x\|_E\, d\psi(\tau). \tag{4.26}$$

PROOF. The function $\sigma_\tau x^*(t)$ is increasing in $\tau$, and so for any $\lambda > 1$ we have

$$\int_0^\infty \sigma_\tau x^*(t)\, d\psi(\tau) = \sum_{k=-\infty}^{\infty} \int_{\lambda^k}^{\lambda^{k+1}} \sigma_\tau x^*(t)\, d\psi(\tau)$$
$$\leqslant \sum_{-\infty}^{\infty} \sigma_{\lambda^{k+1}}x^*(t)\left[\psi(\lambda^{k+1}) - \psi(\lambda^k)\right].$$

Since $E$ is complete, we can apply the triangle inequality to infinite sums in it. Therefore

$$\left\|\int_0^\infty \sigma_\tau x^* d\psi(\tau)\right\|_E \leqslant \sum_{-\infty}^{\infty} \|\sigma_{\lambda^{k+1}}x\|_E\left[\psi(\lambda^{k+1}) - \psi(\lambda^k)\right]$$
$$\leqslant \lambda \sum_{-\infty}^{\infty} \|\sigma_{\lambda^k}x\|_E\left[\psi(\lambda^{k+1}) - \psi(\lambda^k)\right]$$

(we have used (4.16)).

On the other hand, by the monotonicity of the function $\|\sigma_\tau x\|_E$ we have

$$\int_0^\infty \|\sigma_\tau x\|_E \, d\psi(\tau) = \sum_{-\infty}^{\infty} \int_{\lambda^k}^{\lambda^{k+1}} \|\sigma_\tau x\|_E \, d\psi(\tau)$$
$$\geqslant \sum_{-\infty}^{\infty} \|\sigma_{\lambda^k} x\|_E [\psi(\lambda^{k+1}) - \psi(\lambda^k)].$$

Thus,

$$\left\| \int_0^\infty \sigma_\tau x^* \, d\psi(\tau) \right\|_E \leqslant \lambda \int_0^\infty \|\sigma_\tau x\|_E \, d\psi(\tau).$$

By the arbitrariness of $\lambda > 1$ this implies (4.26).

REMARK. The integral on the left side of (4.26) is understood as a Stieltjes integral for every $t$. We could not use the theory of abstract integration in obtaining (4.26), since the range of $\sigma_\tau x^*(t)$ as an abstract function of $\tau$ with values in $E$ is in general not separable, and consequently the function is not measurable.

**4. *The fundamental function.*** From the definition of a symmetric space it follows that the norm of the characteristic function $\chi_e(t)$ of a measurable set $e \subset (0, \infty)$ depends only on the measure of $e$. Hence, for a symmetric space $E$ the function $\varphi_E(t)$ $(0 < t < \infty)$ is defined by

$$\|\chi_e\|_E = \varphi_E(\text{mes } e).$$

It is obvious that $\varphi_E(0) = 0$.

The function $\varphi_E(t)$ is called the *fundamental function* of the space $E$.

It is easy to verify that

$$\varphi_{L_p}(t) = t^{1/p}, \qquad \varphi_{L_M^*}(t) = \frac{1}{M^{-1}(1/t)} \qquad (\text{see } [23]). \tag{4.27}$$

For the Lorentz and Marcinkiewicz spaces mentioned in the previous section we have

$$\varphi_{\Lambda_\psi}(t) = \psi(t) \quad \text{and} \quad \varphi_{M_\psi}(t) = \frac{t}{\psi(t)} = \psi_*(t). \tag{4.28}$$

THEOREM 4.7. *A function $\varphi(t)$ is the fundamental function of some symmetric space if and only if it is quasiconcave.*

PROOF. If $E$ is a symmetric space, then

$$\varphi_E(\tau) = \|\chi_{(0,\tau]}\|_E = \|\sigma_\tau \chi_{(0,1]}\|_E,$$

and therefore the quasiconcavity of $\varphi_E(\tau)$ follows from the remark to Theorem 4.5.

If $\varphi(t)$ is a quasiconcave function, then by (4.28) it is fundamental for the space $M_{t/\varphi(t)}$. The theorem is proved.

Between the functions $\|\sigma_\tau\|_E$ and $\varphi_E(\tau)$ the following relation holds:

$$\|\sigma_\tau\|_E \geqslant \sup_t \frac{\|\sigma_\tau \chi_{(0,t]}\|_E}{\|\chi_{(0,t]}\|_E} = \sup_t \frac{\varphi_E(\tau t)}{\varphi_E(t)} = M_{\varphi_E}(\tau). \tag{4.29}$$

For the spaces $L_p$ we have $\|\sigma_\tau\|_{L_p} = \varphi_{L_p}(\tau) = \tau^{1/p}$. As can be seen from (4.20) and (4.28), for Lorentz and Marcinkiewicz spaces (4.29) turns into an equality. There are examples of spaces for which strict inequality holds in (4.29).

Denote by $\gamma_E$ and $\delta_E$ the lower and upper dilation exponents of $M_{\varphi_E}$. From (4.29) it follows that

$$\alpha_E \geqslant \gamma_E \quad \text{and} \quad \beta_E \geqslant \delta_E. \tag{4.30}$$

**5.** ***Separable symmetric spaces.*** Since every symmetric space $E$ is an ideal lattice, from what was said in the Introduction (p. 45) it follows that $E$ is separable if and only if it is regular, i.e. the norm of any element in $E$ is absolutely continuous. The separability of $E$ is also a necessary and sufficient condition for its associated space $E^1$ and the entire conjugate space $E'$ to coincide. In this case $E$ can be isometrically imbedded in a natural way into $E^{11}$.

We note some further properties of separable symmetric spaces.

The function $\chi_{(0,a)}$ belongs to $E$ for any $a > 0$. Therefore, for $\tau < a$ we have

$$\varphi(\tau) = \|\chi_{(0,\tau)}\|_E = \|\chi_{(0,a)}\chi_{(0,\tau)}\|_E \to 0 \quad \text{as } \tau \to 0$$

by the absolute continuity of the norm of $\chi_{(0,a)}$. Thus, in a separable symmetric space we have $\varphi_E(+0) = 0$.

For any measurable function $x(t)$ the function $x^N(t)$ defined by $x^N(t) = x(t)$ if $|x(t)| \leqslant N$ and $x^N(t) = N \operatorname{sgn} x(t)$, if $|x(t)| > N$ is called the *truncation* of $x(t)$. The *right truncation* of $x(t)$ is defined to be the function $x_N(t) = x(t)\chi_{(0,N)}(t)$. If the norm of $x(t) \in E$ is absolutely continuous, then, since $\operatorname{mes} e_N \to 0$, where $e_N = \{t\colon |x(t)| > N\}$, and due to the fact that $x$ belongs to $L_1 + L_\infty$, we obtain

$$\|x - x^N\|_E \leqslant \|x\chi_{e_N}\|_E \to 0 \quad \text{as } N \to \infty.$$

Under the same conditions, we have

$$\|x - x_N\|_E = \|x\chi_{(N,\infty)}\|_E \to 0 \quad \text{as } N \to \infty.$$

Thus, in a separable space $E$ every function is the limit of its truncations and right truncations.

Every bounded function with bounded support can be uniformly approximated by a sequence of strictly simple functions. This sequence will converge in the norm of $L_1 \cap L_\infty$, and hence in the norm of $E$. Since the two-fold truncations* $x_M^N$ of $x \in E$ are bounded functions with bounded support, from the preceding it follows that the strictly simple functions are dense in a separable symmetric space.

Conversely, if the strictly simple functions are dense in $E$, then every function $x \in E$ is the limit of its truncations and right truncations. Indeed, let $y(t)$ be a strictly simple function such that $\|x - y\|_E < \varepsilon$. For $N > \|y\|_{L\infty}$ we have $\|x^N - y\|_E \leqslant \|x - y\|_E$. Therefore $\|x - x^N\|_E < 2\varepsilon$. Next, if $(0, N)$ contains the support of $y(t)$, then $\|x_N - y\|_E = \|(x - y)\chi_{(0,N)}\|_E < \varepsilon$, and consequently $\|x - x_N\|_E < 2\varepsilon$.

Finally, if $\varphi_E(+0) = 0$ and the strictly simple functions are dense in $E$, then the space is separable. Indeed, for every strictly simple function $y(t)$ with support in $(0, N)$ ($N$ is an integer), given $\tau > 0$, by Luzin's theorem one can construct a continuous function $z_\tau(t)$ such that $|z_\tau(t)| \leqslant |y(t)|$ and mes $e_\tau < \tau$, where $e_\tau = \{t: y(t) \neq z_\tau(t)\}$. Then

$$\|y - z_\tau\|_E \leqslant 2 \sup |y(t)| \, \|\chi_{e_\tau}\|_E = 2\|y\|_{L_\infty}\varphi(\tau) \to 0 \quad \text{as } \tau \to 0.$$

The continuous function $z_\tau(t)$ can be approximated by a polynomial with rational coefficients uniformly on $[0, N]$. Thus, the countable set of functions $p(t)\chi_{(0,N)}$, where $p(t)$ is a polynomial with rational coefficients and $N = 1, 2, \ldots$, is dense in $E$.

We have established the following assertion.

THEOREM 4.8. *If $E$ is a separable symmetric space, then the following assertions are true*:

1) $\varphi_E(+0) = 0$.

2) *Every function in $E$ is the limit in $E$ of its truncations and right truncations.*

3) *The strictly simple functions are dense in $E$.*

*If conditions* 1) *and* 2) *or* 1) *and* 3) *are satisfied, then $E$ is separable.*

**6. *Spaces associated with symmetric spaces.*** Let $E$ be a symmetric space and let $E^1$ be the space associated with it. The space $E^1$ consists of all measurable functions $u(t)$ for which

$$\|u\|_{E^1} = \sup_{\|x\|_E \leqslant 1} \int_0^\infty x(t)u(t)\,dt < \infty. \tag{4.31}$$

---

* *Editor's note.* The meaning is that both the range and the domain of $x$ are truncated (by $N$ and $M$ respectively).

By means of B. Levi's theorem it is easy to see that the supremum can be taken only over strictly simple functions $x(t)$ in $E$ of compact support.

Let $u^*(t)$ be the rearrangement of $u(t)$. For any strictly simple function $x(t)$ of compact support we have

$$\int_0^\infty u^*(t)x(t)\,dt \leqslant \int_0^\infty u^*(t)x^*(t)\,dt = \sum_{k=1}^N \alpha_k \int_0^{\theta_k} u^*(t)\,dt.$$

Using the arguments in the proof of property 7° of rearrangements, we may construct a system of measurable sets $e^\varepsilon_{\theta_k}$ such that

a) $\operatorname{mes} e^\varepsilon_{\theta_k} = \theta_k$,

b) $e^\varepsilon_{\theta_1} \subset e^\varepsilon_{\theta_2} \subset \cdots \subset e^\varepsilon_{\theta_N}$,

c) $\int_0^{\theta_k} u^*(t)\,dt \leqslant \int_{e^\varepsilon_{\theta_k}} |u(t)|\,dt + \varepsilon$.

Then

$$\int_0^\infty u^*(t)x(t)\,dt \leqslant \sum_{k=1}^N \alpha_k \int_{e^\varepsilon_{\theta_k}} u(t)\,dt + \varepsilon \sum_{k=1}^N \alpha_k$$

$$= \int_0^\infty u(t)x_\varepsilon(t)\,dt + \varepsilon \sum_{k=1}^N \alpha_k,$$

where $x_\varepsilon(t) = \sum \alpha_k \chi_{e^\varepsilon_{\theta_k}}(t)$ is a function equimeasurable with $x(t)$. Since $\varepsilon$ is arbitrary and $\|x_\varepsilon\|_E = \|x\|_E \leqslant 1$, we have

$$\sup_{\|x\|_E \leqslant 1} \int_0^\infty u^*(t)x(t)\,dt \leqslant \sup_{\|x\|_E \leqslant 1} \int_0^\infty u(t)x(t)\,dt = \|u\|_{E'}.$$

This inequality shows that $u^*(t) \in E^1$ and $\|u^*\|_{E'} \leqslant \|u\|_{E'}$. Conversely, since $\|x^*\|_E = \|x\|_E$, we have

$$\|u\|_{E^1} = \sup_{\|x\|_E \leqslant 1} \int_0^\infty u(t)x(t)\,dt \leqslant \sup_{\|x\|_E \leqslant 1} \int_0^\infty u^*(t)x^*(t)\,dt$$

$$\leqslant \sup_{\|x\|_E \leqslant 1} \int_0^\infty u^*(t)x(t)\,dt = \|u^*\|_{E^1}.$$

This implies that all functions equimeasurable with a $u(t)$ belonging to $E^1$ belong to $E^1$ and have the same norm.

It is obvious that the space $E^1$ has property 1) of the definition of a symmetric space.

We have arrived at the following assertion: *A space associated with a symmetric space is itself symmetric.*

A symmetric space is said to be *maximal symmetric* if $E$ is isometric to $E^{11}$. Since we always have $E^1 = E^{111}$, the class of maximal symmetric spaces coincides with the class of spaces associated with symmetric spaces.

Examples of maximal symmetric spaces are the Orlicz, Marcinkiewicz, and Lorentz spaces. An example of a nonmaximal symmetric space is the space $M_0(\psi)$ (see §5.4).

THEOREM 4.9. *Every maximal symmetric space $E^1$ is an interpolation space between $L_1$ and $L_\infty$ with interpolation constant one.*

PROOF. Let $v \in E^1$, and for $u \in L_\infty + L_1$ let

$$\int_0^t u^*\,ds \leqslant \int_0^t v^*\,ds.$$

By property 18° of rearrangements, for any measurable function $y(t)$ we have

$$\int_0^\infty u^*(t)y^*(t)\,dt \leqslant \int_0^\infty v^*(t)y^*(t)\,dt.$$

In this inequality we take the supremum over all $y \in E$ such that $\|y\|_E \leqslant 1$. We obtain $\|u\|_{E^1} \leqslant \|v\|_{E^1}$. By Theorem 4.3 the space $E^1$ is an interpolation space between $L_1$ and $L_\infty$, and (4.11) is satisfied.

Theorem 4.5 implies the following important theorem.

THEOREM 4.10. *If $L_1 \cap L_\infty$ is dense in a symmetric space $E$, then $E$ is an interpolation space between $L_1$ and $L_\infty$.*

PROOF. Let $x$ and $y$ be measurable functions, $y \in E$, and let

$$\int_0^\tau x^*(t)\,dt \leqslant \int_0^\tau y^*(t)\,dt.$$

In view of Theorem 4.3, to prove the theorem we need to show that $x \in E$ and $\|x\|_E \leqslant \|y\|_E$. From the preceding theorem it follows that $\|x\|_{E^{11}} \leqslant \|y\|_{E^{11}} = \|y\|_E$. It only remains to verify that $x \in E$. By Theorem 3.4 there exists an operator $T_0 \in \Sigma$ such that $x = T_0 y$. We consider a sequence $y_n \in L_1 \cap L_\infty$ converging to $y$ in $E$. Then by the interpolation property of $E^{11}$ we have $T_0 y_n \to T_0 y = x$ in $E^{11}$. The operator $T_0$ acts in $L_1 \cap L_\infty$. Therefore $T_0 y_n \in L_1 \cap L_\infty \subset E$. Since the imbedding of $E$ in $E^{11}$ is isometric, the sequence $T_0 y_n$ is Cauchy in $E$, and so $x \in E$.

COROLLARY. *If a symmetric space is separable, then it is an interpolation space between $L_1$ and $L_\infty$.*

We consider the dilation operators $\sigma_\tau$ in an associated space. From (4.31) we obtain

$$\begin{aligned}\|\sigma_\tau u\|_{E^1} &= \sup_{\|x\|_E \leqslant 1} \int_0^\infty x(t)\sigma_\tau u(t)\,dt \\ &= \sup_{\|x\|_E \leqslant 1} \tau \int_0^\infty \sigma_{1/\tau} x(t) u(t)\,dt \leqslant \tau \|\sigma_{1/\tau}\|_E \|u\|_{E^1}.\end{aligned}$$

Thus,

$$\|\sigma_\tau\|_{E^1} \leqslant \tau \|\sigma_{1/\tau}\|_E. \tag{4.32}$$

Since $E^1 = E^{111}$,

$$\|\sigma_\tau\|_{E^{111}} = \|\sigma_\tau\|_{E^1} \leqslant \tau\|\sigma_{1/\tau}\|_{E^{11}}. \tag{4.33}$$

Applying (4.32) to $E^1$, we obtain

$$\|\sigma_\tau\|_{E^1} = \tau\|\sigma_{1/\tau}\|_{E^{11}}. \tag{4.34}$$

We also need the quantity $\|\sigma_\tau\|_{E\to E^{11}}$. We have

$$\|\sigma_\tau\|_{E\to E^{11}} = \sup_{\|x\|_E\leqslant 1} \|\sigma_\tau x\|_{E^{11}} \leqslant \sup_{\|x\|_{E^{11}}\leqslant 1} \|\sigma_\tau x\|_{E^{11}} = \|\sigma_\tau\|_{E^{11}} = \tau\|\sigma_{1/\tau}\|_{E^1}.$$

On the other hand,

$$\begin{aligned}\tau\|\sigma_{1/\tau}v\|_{E^1} &= \sup_{\|x\|_E\leqslant 1} \tau\int_0^\infty x(t)\sigma_{1/\tau}v(t)\,dt \\ &= \sup_{\|x\|_E\leqslant 1} \int_0^\infty \sigma_\tau x(t)v(t)\,dt \leqslant \|\sigma_\tau\|_{E\to E^{11}}\|v\|_{E^1}.\end{aligned}$$

Thus,

$$\|\sigma_\tau\|_{E\to E^{11}} = \tau\|\sigma_{1/\tau}\|_{E^1} = \|\sigma_\tau\|_{E^{11}}. \tag{4.35}$$

If $E$ is isometrically imbedded in $E^{11}$, this equality implies that

$$\|\sigma_\tau\|_E = \tau\|\sigma_{1/\tau}\|_{E^1} = \|\sigma_\tau\|_{E^{11}}. \tag{4.36}$$

THEOREM 4.11. *The associated space $E^1$ of a symmetric space $E$ satisfies* (4.32) *and the inequalities*

$$\alpha_{E^1} \leqslant 1 - \beta_E, \qquad \beta_{E^1} \leqslant 1 - \alpha_E. \tag{4.37}$$

*If the natural imbedding of $E$ in $E^{11}$ is isometric, then* (4.36) *holds and*

$$\alpha_{E^1} = 1 - \beta_E, \qquad \beta_{E^1} = 1 - \alpha_E. \tag{4.38}$$

Inequalities (4.37) and equalities (4.38) immediately follow from (4.32), (4.36), and formula (4.22) for the calculation of lower and upper dilation exponents.

Now we pass to the calculation of the fundamental function of the associated space. We have

$$\begin{aligned}\varphi_{E^1}(t) = \|\chi_{(0,t]}\|_{E^1} &= \sup_{\|x\|_E\leqslant 1} \int_0^\infty x(s)\chi_{(0,t]}(s)\,ds \\ &\leqslant \sup_{\|x\|_E\leqslant 1} \int_0^\infty x^*(s)\chi_{(0,t]}(s)\,ds = \sup_{\|x\|_E\leqslant 1} \int_0^t x^*(s)\,ds.\end{aligned}$$

We use (4.6). Then we have

$$\varphi_{E^1}(t) \leqslant \sup_{\|x\|_E\leqslant 1} \frac{1}{\|\chi_{(0,t]}\|_E} \|x^*\chi_{(0,t]}\|_E \leqslant \frac{t}{\varphi_E(t)}.$$

On the other hand,

$$\varphi_{E^1}(t) \geq \int_0^\infty \frac{\chi_{(0,t]}(s)}{\|\chi_{(0,t]}\|_E}\, \chi_{(0,t]}(s)\, ds = \frac{t}{\varphi_E(t)}.$$

Thus,

$$\varphi_{E^1}(t) = t/\varphi_E(t). \tag{4.39}$$

Next we compute

$$M_{\varphi_{E^1}}(\tau) = \sup_{t>0} \frac{\varphi_{E^1}(\tau t)}{\varphi_{E^1}(t)} = \sup_{t>0} \frac{\tau\varphi_E(t)}{\varphi_E(\tau t)} = \tau M_{\varphi_E}\left(\frac{1}{\tau}\right). \tag{4.40}$$

THEOREM 4.12. *For the associated space $E^1$ of a symmetric space the equalities* (4.39) *and* (4.40) *hold, and*

$$\gamma_{E^1} = 1 - \delta_E, \qquad \delta_{E^1} = 1 - \gamma_E.$$

We also compare (4.40) with (4.29) and (4.32). We have

$$\tau\|\sigma_{1/\tau}\|_E \geq \|\sigma_\tau\|_{E^1} \geq M_{\varphi_{E^1}}(\tau) = \tau M_{\varphi_E}\left(\frac{1}{\tau}\right).$$

From this it is clear that if (4.29) becomes an equality, then we have equality in (4.32) as well.

## §5. Lorentz and Marcinkiewicz spaces

**1.** ***Lorentz space.*** Let $\psi(t)$ ($\not\equiv 0$) be an increasing concave function on $[0, \infty)$ such that $\psi(0) = 0$, and let $b(t)$ be a nonnegative decreasing function on $(0, \infty)$. We shall consider the integral

$$\int_0^\infty b(t)\, d\psi(t),$$

understanding by it the following: If $\psi(+0) \neq 0$, then

$$\int_0^\infty b(t)\, d\psi(t) = \psi(+0)b(+0) + \int_{+0}^\infty b(t)\, d\psi(t),$$

where the second term is an improper Stieltjes integral or, which is the same,

$$\int_0^\infty b(t)\, d\psi(t) = \psi(+0)b(+0) + \int_0^\infty b(t)\psi'(t)\, dt,$$

where the second term is a Lebesgue integral. If $\psi(+0) = 0$, then

$$\int_0^\infty b(t)\, d\psi(t) = \int_{+0}^\infty b(t)\, d\psi(t) = \int_0^\infty b(t)\psi'(t)\, dt.$$

We now consider the collection $\Lambda_\psi$ of all measurable functions on $(0, \infty)$ for which

$$\int_0^\infty x^*(t)\, d\psi(t) < \infty. \tag{5.1}$$

$\Lambda_\psi$ is called a *Lorentz space*.

If $\psi(+0) \neq 0$, then by the preceding it follows from (5.1) that $x^*(+0) = \|x\|_{L_\infty} < \infty$ and

$$\int_0^\infty x^*(t)\, d\psi(t) = \psi(+0)\|x\|_{L_\infty} + \int_0^\infty x^*(t)\psi'(t)\, dt. \tag{5.2}$$

Inequality (5.1) implies that $x \in L_1 + L_\infty$. Indeed, if $\psi'(t) \equiv 0$, then $\Lambda_\psi = L_\infty \subset L_1 + L_\infty$. If $\psi'(t_0) > 0$ for some $t_0$, then

$$\int_0^{t_0} x^*(t)\, dt \leqslant \frac{1}{\psi'(t_0)} \int_0^{t_0} x^*(t)\psi'(t)\, dt < \infty.$$

Since the function $\int_0^{t_0} x^*(s)\, ds$ is concave, we have

$$\|x\|_{L_1+L_\infty} = \int_0^1 x^*(t)\, dt \leqslant \max\left\{\frac{1}{t_0}, 1\right\} \int_0^{t_0} x^*(t)\, dt < \infty,$$

which implies our assertion.

The behavior of $\psi(t)$ at zero and at infinity manifests itself in the properties of the space $\Lambda_\psi$ in an essential way. From the preceding it follows that $\Lambda_\psi \subset L_\infty$ if $\psi(+0) \neq 0$. Next, if $\psi(\infty) = \lim_{t\to\infty} \psi(t) < \infty$, then every bounded function belongs to $\Lambda_\psi$, i.e. $\Lambda_\psi \supset L_\infty$. Thus, if $\psi(+0) \neq 0$ and $\psi(\infty) < \infty$, then $\Lambda_\psi = L_\infty$.

If $\psi(\infty) = \infty$, then for any function $x \in \Lambda_\psi$ we have $n_{|x|}(\tau) < \infty$ for all $\tau > 0$. Indeed, if for some $\tau_0$ the set $e = \{t\colon |x(t)| > \tau_0\}$ has infinite measure, then $x^*(t) > \tau_0$ for all $t > 0$, and

$$\int_0^\infty x^*(t)\psi'(t)\, dt \geqslant \tau_0 \int_0^\infty \psi'(t)\, dt = \infty.$$

We obtain still another expression for the quantity (5.2). The function $\psi'(t)$ is decreasing, and as the derivative of a concave function it is locally summable. Therefore $\psi' \in L_1 + L_\infty$. By Theorem 3.2 we have

$$\int_0^\infty x^*(t)\psi'(t)\, dt = \sup_{T\in\Sigma_0} \int_0^\infty x(t) T\psi'(t)\, dt.$$

From this and (5.2) it follows that the quantity (5.2) has the properties of a norm. Thus, the space $\Lambda_\psi$ is normed by

$$\|x\|_{\Lambda_\psi} = \int_0^\infty x^*(t)\, d\psi(t).$$

It is obvious that the relations $|y(t)| \leqslant |x(t)|$, $x \in \Lambda_\psi$, imply $y \in \Lambda_\psi$ and $\|y\|_{\Lambda_\psi} \leqslant \|x\|_{\Lambda_\psi}$.

We prove the completeness of the space $\Lambda_\psi$. For this it is sufficient to prove that the inequality $\sum_{k=1}^\infty \|x_k\|_{\Lambda_\psi} = C < \infty$ $(x_k \in \Lambda_\psi)$ implies that $\sum_1^\infty |x_k(t)|$ belongs to $\Lambda_\psi$. Since $\Lambda_\psi$ is imbedded in $L_1 + L_\infty$, we have

$$\sum_{k=1}^\infty \|x_k\|_{L_1+L_\infty} < \infty,$$

and so $\Sigma_1^\infty |x_k(t)|$ converges to a function $x(t) \in L_1 + L_\infty$ almost everywhere. Next, for $T \in \Sigma_0$ Theorem 3.2 and (5.2) imply that

$$\int_0^\infty \sum_{k=1}^N |x_k(t)| T\psi'(t)\,dt \leqslant \sum_{k=1}^N \int_0^\infty x_k^*(t)\psi'(t)\,dt \leqslant C.$$

Then by B. Levi's theorem we have

$$\int_0^\infty x(t)T\psi'(t)\,dt \leqslant C.$$

Taking the supremum over all $T \in \Sigma_0$, we obtain

$$\int_0^\infty x^*(t)\psi'(t)\,dt \leqslant C.$$

If $\psi(+0) \neq 0$, then the finiteness of $\Sigma_1^\infty \|x_k\|_{\Lambda_\psi}$ implies the finiteness of $\Sigma_1^\infty \|x_k\|_{L_\infty}$ and the inequality $\|x\|_{L_\infty} < \infty$. Thus, $x \in \Lambda_\psi$, which proves that $\Lambda_\psi$ is a Banach space.

It is clear from the definition that $\Lambda_\psi$ is symmetric and its fundamental function is equal to $\psi(t)$. By Theorem 4.1 we obtain that

$$L_1 \cap L_\infty \subset \Lambda_\psi \subset L_1 + L_\infty$$

(both imbeddings can be verified immediately).

$\Lambda_\psi$ is an interpolation space between $L_1$ and $L_\infty$. Indeed,

$$\|Tx\|_{\Lambda_\psi} = \psi(+0)\|Tx\|_{L_\infty} + \int_0^\infty (Tx)^*(t)\psi'(t)\,dt,$$

and if $T \in \Sigma$, then by (3.16) and property 18° (see p. 72) we have

$$\|Tx\|_{\Lambda_\psi} \leqslant \psi(+0)\|x\|_{L_\infty} + \int_0^\infty x^*(t)\psi'(t)\,dt = \|x\|_{\Lambda_\psi}.$$

In what follows we need an assertion on passage to the limit in the space $\Lambda_\psi$.

THEOREM 5.1 (on passage to the limit). *Let* $y \in \Lambda_\psi$, $n_{|y|}(\tau) < \infty$ *for all* $\tau > 0$, *and* $|x_n(t)| < |y(t)|$. *Let one of the following conditions be satisfied*:

a) $\psi(+0) = 0$ *and* $\lim_{n\to\infty} x_n(t) = x(t)$ *almost everywhere*.

b) $\psi(+0) \geqslant 0$ *and* $\lim_{n\to\infty} x_n(t) = x(t)$ *uniformly*.

*Then* $\|x - x_n\|_{\Lambda_\psi} \to 0$.

PROOF. It is obvious that $x \in \Lambda_\psi$ and $n_{|x|}(\tau) < \infty$ for $\tau > 0$. From property 12° of rearrangements it follows that $(x - x_n)^*(t) \to 0$ for any $t > 0$ as $n \to \infty$. If $\psi(+0) = 0$, then

$$\|x - x_n\|_{\Lambda_\psi} = \int_0^\infty (x - x_n)^*(t)\psi'(t)\,dt \to 0$$

by the inequality $(x - x_n)^*(t)\psi'(t) \leqslant 2y^*(t)\psi'(t)$ and Lebesgue's theorem.

If $\psi(+0) > 0$, then

$$\|x - x_n\|_{\Lambda_\psi} = \|x - x_n\|_{L_\infty}\psi(+0) + \int_0^\infty (x - x_n)^*(t)\psi'(t)\,dt \to 0$$

by the preceding and the uniform convergence of $x_n(t)$ to $x(t)$.

COROLLARY 1. *If* $\psi(+0) = 0$ *and* $\psi(\infty) = \infty$, *then the relations* $|x_n(t)| \leqslant y(t)$ $(y \in \Lambda_\psi)$ *and* $x_n(t) \to x(t)$ *almost everywhere imply that* $\|x - x_n\|_\psi \to 0$.

Under the hypotheses of Corollary 1 every function in $\Lambda_\psi$ is the limit of its two-fold truncations. These hypotheses are necessary for this property. Indeed, if $\psi(\infty) < \infty$, then $\chi_{[0,\infty)}(t) \in \Lambda_\psi$, and the truncations $\chi_{[0,N]}(t)$ are at a constant distance $\psi(\infty)$ from $\chi_{[0,\infty)}$. If $\psi(+0) > 0$, then for the function $\sum_{n=1}^\infty \chi_{[n,n+1/n^2]}(t)$ the truncations $\sum_{n=1}^N \chi_{[n,n+1/n^2]}(t)$ are at a distance not less than $\psi(+0)$ from this function.

COROLLARY 2. *Every function* $x \in \Lambda_\psi$ *is the limit, in* $\Lambda_\psi$, *of its truncations* $x^N$.

Indeed, if $\psi(+0) > 0$, then $\Lambda_\psi \subset L_\infty$ and $x^N(t) \equiv x(t)$ for sufficiently large $N$.

Now let $\psi(+0) = 0$. There exists $\tau_0 > 0$ such that $n_{|x|}(\tau_0) < \infty$. Then for $N > \tau_0$ we have $|x(t) - x^N(t)| \leqslant |x(t) - x^{\tau_0}(t)|$, and $n_{|x-x^{\tau_0}|}(\tau) < \infty$ for all $\tau > 0$. Theorem 5.1 implies that $\|x - x^N\|_{\Lambda_\psi} \to 0$ as $N \to \infty$.

COROLLARY 3. *If* $\psi(\infty) = \infty$ $(< \infty)$, *then the strictly simple (simple) functions are dense in* $\Lambda_\psi$.

By Corollary 2 any function $x(t) \in \Lambda_\psi$ can be approximated by a bounded function with arbitrary accuracy, and if $\psi(\infty) = \infty$, then the approximating function can be chosen of compact support. Any bounded function of compact support can be represented as the limit of a uniformly convergent sequence of strictly simple functions not exceeding it. Therefore, in the case $\psi(\infty) = \infty$ our assertion follows from part b) of the theorem on the passage to the limit. Every bounded function can be represented as the limit of a uniformly convergent sequence of simple functions. In the case $\psi(\infty) < \infty$ this implies the convergence of the sequence in $\Lambda_\psi$.

LEMMA 5.1. *The space* $\Lambda_\psi$ *is separable if and only if*

$$\psi(+0) = 0 \quad \textit{and} \quad \psi(\infty) = \infty. \tag{5.3}$$

The proof immediately follows from Theorem 4.8 on separable symmetric spaces and the discussion above .

In what follows we need still another expression for the norm in a Lorentz space. We have

$$\|x\|_{\Lambda_\psi} = \int_0^\infty \psi(n_{|x|}(\tau))\,d\tau. \tag{5.4}$$

The left and right sides of this equality do not change if from the function we pass to its absolute value and to a function equimeasurable to it. Therefore we may assume that $x(t) = x^*(t)$. If the function is elementary, say

$$x(t) = \sum_1^\infty \Delta_k \chi_{(0,t_k)} \qquad (\infty = t_1 > t_2 > \cdots > 0,\ \ \Delta_k \geqslant 0), \tag{5.5}$$

then a direct calculation shows that both sides in (5.4) are equal to $\sum_1^\infty \Delta_k \psi(t_k)$. Any function $x \in \Lambda_\psi$ $(x = x^*)$ can be uniformly approximated by an increasing sequence of functions $x_n(t)$ of the form (5.5). By Theorem 5.1 we have $\|x\|_{\Lambda_\psi} = \lim_{n\to\infty} \|x_n\|_{\Lambda_\psi}$. For every $\tau > 0$ the functions $n_{x_n}(\tau)$ converge to $n_x(\tau)$, and consequently $\psi(n_{x_n}(\tau)) \to \psi(n_x(\tau))$ for all $\tau > 0$. By B. Levi's theorem we have

$$\int_0^\infty \psi(n_x(\tau))\,d\tau = \lim_{n\to\infty} \int_0^\infty \psi(n_{x_n}(\tau))\,d\tau.$$

This and the preceding imply (5.4).

*Convex functionals on a Lorentz space.* Let $\Phi(t)$ be a convex functional on $\Lambda_\psi$, i.e. a functional with the properties

$$\Phi(x + y) \leqslant \Phi(x) + \Phi(y), \qquad \Phi(\lambda x) = |\lambda|\Phi(x).$$

We assume that $\Phi(x)$ can take on infinite values.

LEMMA 5.2. *If a convex functional is bounded on all characteristic functions of measurable sets, then it is bounded on all simple functions in* $\Lambda_\psi$.

PROOF. Let

$$\Phi(\chi_e) \leqslant C\|\chi_e\|_{\Lambda_\psi} = C\psi(\text{mes } e).$$

If $x(t)$ is a nonnegative simple function belonging to $\Lambda_\psi$, then it can be represented in the form

$$x(t) = \sum_{k=1}^N \Delta_k \chi_{e_k}(t),$$

where $e_1 \subset e_2 \subset \cdots \subset e_N$ and $\Delta_k > 0$. Then

$$\Phi(x) \leqslant \sum_{k=1}^N \Delta_k \Phi(\chi_{e_k}) \leqslant C \sum_{k=1}^N \Delta_k \psi(\text{mes } e_k). \tag{5.6}$$

On the other hand, as was shown on p. 60, we have

$$x^*(t) = \sum_{k=1}^{N} \Delta_k \chi_{[0,\, \text{mes}\, e_k]}(t),$$

and consequently

$$\|x\|_{\Lambda_\psi} = \int_0^\infty \sum_{k=1}^{N} \Delta_k \chi_{[0,\, \text{mes}\, e_k]}(t)\, d\psi(t) = \sum_{k=1}^{N} \Delta_k \psi(\text{mes}\, e_k). \tag{5.7}$$

Combining (5.6) with (5.7), we obtain $\Phi(x) \leqslant C\|x\|_{\Lambda_\psi}$.

If now the simple function $x(t)$ from $\Lambda_\psi$ has values of arbitrary sign, then

$$\begin{aligned}\Phi(x) &= \Phi(x_+ - x_-) \leqslant \Phi(x_+) + \Phi(x_-)\\ &\leqslant C(\|x_+\|_{\Lambda_\psi} + \|x_-\|_{\Lambda_\psi}) \leqslant 2C\|x\|_{\Lambda_\psi}.\end{aligned}$$

COROLLARY 1. *Let the functional $\Phi(x)$ have the form $\|Ax\|_E$, where $E$ is some Banach space imbedded in a separated topological linear space $\mathfrak{A}$, and let $A$ be a linear operator on $\Lambda_\psi$ continuous in $\mathfrak{A}$ (if $Ax \notin E$, we set $\Phi(x) = \infty$). If the functional $\Phi(x)$ is bounded on all characteristic functions from $\Lambda_\psi$, then $A$ is bounded in $E$.*

Indeed, let $x \in \Lambda_\psi$ and let $x_n$ be a sequence of simple functions approximating $x$. Then by Lemma 5.2 we have $\|A(x_n - x_m)\| \leqslant 2C\|x_n - x_m\|_{\Lambda_\psi}$, and so the sequence $Ax_n$ is Cauchy in $E$. Since it converges to $Ax$ in $\mathfrak{A}$, its limit will be the same $Ax$ in $E$, i.e. $Ax \in E$ and $\|Ax\|_E \leqslant 2C\|x\|_{\Lambda_\psi}$.

**2. *Marcinkiewicz spaces.*** We study the structure of the dual space of a separable space $\Lambda_\psi$.

THEOREM 5.2. *Every linear functional $f$ on a separable space $\Lambda_\psi$ has the form*

$$f(x) = \int_0^\infty x(t)y(t)\, dt, \tag{5.8}$$

*where $y(t)$ is a measurable function on* $[0, \infty)$ *and*

$$\|f\| = \sup_{0<h<\infty} [\psi(h)]^{-1} \int_0^h y^*(s)\, ds < \infty. \tag{5.9}$$

PROOF. We show that if (5.9) is satisfied, then (5.8) defines a linear functional on $\Lambda_\psi$. We note that (5.9) and the condition $\psi(+0) = 0$ imply that $\int_0^h y^*(s)\, ds \leqslant C\psi(h) \to 0$ as $h \to 0$. Let $x(t)$ be a step function. Then

$$x^*(t) = \sum_{k=1}^{N} \Delta_k \chi_{[0,h_k]}, \qquad \Delta_k > 0.$$

We estimate

$$|f(x)| = \left|\int_0^\infty x(t)y(t)\,dt\right| \leqslant \int_0^\infty x^*(t)y^*(t)\,dt = \sum_{k=1}^N \Delta_k \int_0^{h_k} y^*(t)\,dt$$
$$\leqslant \sup_{0<h<\infty} \left\{[\psi(h)]^{-1}\int_0^h y^*(t)\,dt\right\} \Sigma\, \Delta_k \psi(h_k)$$
$$= \sup_{0<h<\infty} \left\{[\psi(h)]^{-1}\int_0^h y^*(t)\,dt\right\} \|x\|_{\Lambda_\psi}.$$

As we saw in the proof of Lemma 5.1, the step functions are dense in the norm of $\Lambda_\psi$ in the set of bounded functions of compact support. Therefore the inequality

$$|f(x)| \leqslant \int_0^\infty |x(t)|\,|y(t)|\,dt \leqslant \sup_{0<h<\infty} \left\{[\psi(h)]^{-1}\int_0^h y^*(t)\,dt\right\} \|x\|_{\Lambda_\psi} \quad (5.10)$$

can be established for any bounded function of compact support by passage to the limit. By the corollaries to Theorem 5.1, for any function $x \in \Lambda_\psi$ the absolute value $|x(t)|$ is the limit, in $\Lambda_\psi$, of its two-fold truncations, and so by Fatou's lemma inequality (5.10) can be carried over to any functions in $\Lambda_\psi$. This implies that $f \in \Lambda_\psi$ and

$$|f(x)| \leqslant \sup_{0<h<\infty} \left\{[\psi(h)]^{-1}\int_0^h y^*(t)\,dt\right\} \|x\|_{\Lambda_\psi}. \quad (5.11)$$

Now let $f$ be an arbitrary linear functional on $\Lambda_\psi$. Then $|f(x)| \leqslant \|f\|\,\|x\|_{\Lambda_\psi}$. If $x = \chi_e$ is the characteristic function of a set $e$ of finite measure, then for $\alpha(e) = f(\chi_e)$ we have

$$|\alpha(e)| \leqslant \|f\|\,\|\chi_e\|_{\Lambda_\psi} = \|f\|\psi(\text{mes}\, e).$$

Since $\psi(+0) = 0$, the function $\alpha(e)$ is absolutely continuous, and by the Radon-Nikodým theorem we have

$$\alpha(e) = f(\chi_e) = \int_0^\infty \chi_e(t)y(t)\,dt, \quad (5.12)$$

where $y(t)$ is a locally summable function. Formula (5.12) implies that

$$f(x) = \int_0^\infty x(t)y(t)\,dt \quad (5.13)$$

for any strictly simple function $x(t)$.

Next,

$$\left\|\text{sgn}\, y(t) \frac{\chi_e(t)}{\psi(\text{mes}\, e)}\right\|_{\Lambda_\psi} = 1,$$

and therefore

$$[\psi(\text{mes}\, e)]^{-1}\int_e |y(t)|\,dt = f\left(\text{sgn}\, y \frac{\chi_e}{\psi(\text{mes}\, e)}\right) \leqslant \|f\|.$$

By property 7° of rearrangements we then have

$$\sup_{0<h<\infty} [\psi(h)]^{-1}\int_0^h y^*(t)\,dt \leqslant \|f\|. \tag{5.14}$$

The extension of (5.13) to all functions in $\Lambda_\psi$ can be carried out by the same arguments as in the proof of the first part of the theorem. Inequalities (5.11) and (5.14) imply (5.9). The theorem is proved.

The space of all measurable functions $y(t)$ for which (5.9) is satisfied is called the *Marcinkiewicz space* $\mathrm{M}_\psi$. We have proved that $\mathrm{M}_\psi$ *is isometric to the space dual to the separable Lorentz space* $\Lambda_\psi$. Similar arguments show that for an arbitrary Lorentz space $\Lambda_\psi$ the space $\mathrm{M}_\psi$ coincides with the space associated with $\Lambda_\psi$. This implies that $\mathrm{M}_\psi$ is a Banach space. It is obviously symmetric.

$\mathrm{M}_\psi$ is an interpolation space between $L_1$ and $L_\infty$ with interpolation constant 1. Indeed, if $T \in \Sigma$, then by (3.16) we have

$$\begin{aligned}\|Tx\|_{\mathrm{M}_\psi} &= \sup[\psi(h)]^{-1}\int_0^h (Tx)^*(t)\,dt \\ &\leqslant \sup[\psi(h)]^{-1}\int_0^h x^*(t)\,dt = \|x\|_{\mathrm{M}_\psi}.\end{aligned}$$

It is sometimes convenient to consider spaces $\mathrm{M}_\psi$ constructed from a quasiconcave function $\psi(t)$. Then for the fundamental function the equality $\varphi_{\mathrm{M}_\psi}(t) = t/\psi(t)$ holds.

Of course, the norm (5.9) makes sense for any function $\psi(t)$ positive for $t > 0$. However, it can be shown that if the space thus obtained has nonzero elements, then the norm coincides with the norm calculated according to (5.9) with some quasiconcave function (the largest quasiconcave minorant of $\psi(t)$).

LEMMA 5.3. *Let $\psi(t)$ be an increasing concave function. In order that $\psi(t)/t \in \mathrm{M}_\psi$ it is necessary and sufficient that $M_\psi(s_0^{-1}) < 1$ for any $s_0 > 1$.*

PROOF. If $M_\psi(s_0^{-1}) = 1$, then by Lemma 1.3 we have $M_\psi(s^{-1}) \equiv 1$ for $s \geqslant 1$. Then for any integer $n > 1$ there exists a point $t_n$ such that $\psi(nt_n) \leqslant (1+\varepsilon)\psi(t_n)$. The function $\psi(t)/t$ is decreasing, and therefore

$$\begin{aligned}\left\|\frac{\psi(t)}{t}\right\|_{\mathrm{M}_\psi} &\geqslant [\psi(nt_n)]^{-1}\int_0^{nt_n}\frac{\psi(\tau)}{\tau}\,d\tau \geqslant [\psi(nt_n)]^{-1}\int_{t_n}^{nt_n}\frac{\psi(\tau)}{\tau}\,d\tau \\ &\geqslant [\psi(nt_n)]^{-1}\psi(t_n)\int_{t_n}^{nt_n}\frac{d\tau}{\tau} \geqslant \frac{1}{1+\varepsilon}\ln n.\end{aligned}$$

Since $n$ is arbitrary, we obtain that $\psi(t)/t \notin \mathrm{M}_\psi$. Thus, for $\psi(t)/t$ to belong to $\mathrm{M}_\psi$ it is necessary that $M_\psi(s_0^{-1}) < 1$ ($s_0 > 1$). Now let this condition be satisfied. Then by Corollary 2 from Lemma 1.4 we obtain

$$\left\| \frac{\psi(t)}{t} \right\|_{M_\psi} = \sup \frac{1}{\psi(t)} \int_0^t \frac{\psi(\tau)}{\tau}\, d\tau \leqslant C.$$

The lemma is proved.

We introduce the functional

$$F(x) = \sup_{0<t<\infty} \frac{t}{\psi(t)} x^*(t).$$

It is obvious that $F(x) > 0$ if $x(t) \not\equiv 0$, and $F(\lambda x) = |\lambda| F(x)$. However, this functional does not in general have the properties of a norm.

We have

$$\|x\|_{M_\psi} = \sup_{0<h<\infty} \frac{\int_0^h x^*(t)\, dt}{\psi(h)} \geqslant \sup_{0<h<\infty} \frac{hx^*(h)}{\psi(h)} = F(x). \tag{5.15}$$

This and Lemma 5.3 imply the following theorem.

THEOREM 5.3. *The functional $F(x)$ is equivalent to the norm $\|x\|_{M_\psi}$ if and only if $M_\psi(s^{-1}) < 1$ for $s > 1$.*

PROOF. *Necessity*. Since $\psi(t)/t$ is decreasing, $F(\psi(t)/t) = 1$. If the functional $F(x)$ is equivalent to the norm of the space $M_\psi$, then this equality implies that $\psi(t)/t \in M_\psi$, and by Lemma 5.3 we have $M_\psi(s^{-1}) < 1$ for $s > 1$.

*Sufficiency*. Let $F(x) \leqslant 1$. Then $x^*(t) \leqslant \psi(t)/t$ and consequently $\|x\|_{M_\psi} \leqslant \|\psi(t)/t\|_{M_\psi} = C < \infty$. From the homogeneity of both functionals it follows that $\|x\|_{M_\psi} \leqslant CF(x)$. This, together with (5.15), proves the theorem.

**3. *The space* $M_\psi^0$.** Assuming that $\psi(+0) = 0$ and $\psi(\infty) = \infty$, we denote by $M_\psi^0$ the set of all $x \in M_\psi$ for which

$$\lim_{h\to 0,\infty} [\psi(h)]^{-1} \int_0^h x^*(t)\, dt = 0. \tag{5.16}$$

It turns out that $M_\psi^0$ is a subspace of $M_\psi$. The linearity of $M_\psi^0$ follows from (2.17). Let $x_n \to x_0$ in $M_\psi$ and let $x_n \in M_\psi^0$. Then the sequence of the continuous functions $\alpha_n(h) = [\psi(h)]^{-1} \int_0^h x_n^*(t)\, dt$ uniformly converges to the function $\alpha_0[h] = [\psi(h)]^{-1} \int_0^h x_0^*(t)\, dt$.

Indeed, by (3.14) we have

$$\left| [\psi(h)]^{-1} \int_0^h x_n^*(t)\, dt - [\psi(h)]^{-1} \int_0^h x_0^*(t)\, dt \right| \leqslant [\psi(h)]^{-1} \int_0^h (x_n - x_0)^*(t)\, dt \leqslant \|x_n - x_0\|_{M_\psi}.$$

Since all functions $\alpha_n(h)$ have the property that $\lim_{h\to 0,\infty} \alpha_n(h) = 0$, so does the limit function $\alpha_0(h)$. Thus, $x_0 \in M_\psi^0$.

We note that for $x \in M_\psi^0$ we have $n_{|x|}(\tau) < \infty$ for all $\tau > 0$. Indeed, if $n_{|x|}(\tau_0) = \infty$, then $x^*(t) \geqslant \tau_0$ and

$$[\psi(h)]^{-1}\int_0^h x^*(t)\,dt \geqslant \tau_0\frac{h}{\psi(h)} \geqslant \frac{\tau_0}{\psi(1)} > 0$$

for $h \geqslant 1$, i.e. $x \notin M_\psi^0$.

If $\lim_{t\to 0}\psi(t)/t = k < \infty$, then $M_\psi^0$ consists only of the element zero. Indeed, if $x \in M_\psi$ and $x \not\equiv 0$, then for some $t_0 > 0$ we have $x^*(t_0) > 0$. Then

$$\lim_{h\to 0}\frac{\int_0^h x^*(t)\,dt}{\psi(h)} \geqslant x^*(t_0)\lim_{h\to 0}\frac{h}{\psi(h)} = \frac{1}{k}x^*(t_0),$$

i.e. $x \notin M_\psi^0$.

LEMMA 5.4. *Under the condition*

$$\lim_{t\to 0}\psi(t)/t = \infty \tag{5.17}$$

*the subspace* $M_\psi^0$ *coincides with the closure of the set of all bounded functions of compact support in* $M_\psi$.

PROOF. It is obvious that under the condition (5.17) any bounded function of compact support belongs to $M_\psi$ (since $\psi(\infty) = \infty$). Let $x_0 \in M_\psi^0$. For any $\varepsilon > 0$ there is a $\delta > 0$ such that

$$[\psi(h)]^{-1}\int_0^h x_0^*(t)\,dt \leqslant \varepsilon$$

for $h \leqslant \delta$ or $h \geqslant 1/\delta$. Since $n_{|x_0|}(\tau) < \infty$ for all $\tau$, by property 12° of rearrangements $(x_0 - x_0^N)^*(t)$ converges to zero for all $t$ as $N \to \infty$. Hence, for sufficiently large $N$,

$$\int_0^{\delta^{-1}}(x_0 - x_0^N)^*(t)\,dt \leqslant \varepsilon\psi(\delta).$$

For these $N$ we have

$$\begin{aligned}
&\|x_0 - x_0^N\|_{M_\psi}\\
&\quad = \max\left\{\sup_{h<\delta,\,h>\delta^{-1}}\frac{\int_0^h (x_0 - x_0^N)^*(t)\,dt}{\psi(h)},\ \sup_{\delta\leqslant h\leqslant\delta^{-1}}\frac{\int_0^h (x_0 - x_0^N)^*(t)\,dt}{\psi(h)}\right\}\\
&\quad \leqslant \max\left\{\sup_{h<\delta,\,h>\delta^{-1}}\frac{\int_0^h x_0^*(t)\,dt}{\psi(h)},\ \frac{\int_0^{\delta^{-1}}(x_0 - x_0^N)^*(t)\,dt}{\psi(\delta)}\right\} \leqslant \varepsilon.
\end{aligned}$$

Obviously the space $M_\psi^0$ is symmetric. From the imbedding Theorem 4.1 it follows that

$$L_1 \cap L_\infty \subset M_\psi^0. \tag{5.18}$$

Any operator $T \in \Sigma$ maps the set of bounded functions of compact support into $L_1 \cap L_\infty$, and consequently into $\mathrm{M}_\psi^0$. Since $L_1 \cap L_\infty$ is dense in $\mathrm{M}_\psi^0$ and $\mathrm{M}_\psi$ is an interpolation space between $L_1$ and $L_\infty$, it follows that $T\mathrm{M}_\psi^0 \subset \mathrm{M}_\psi^0$, i.e. $\mathrm{M}_\psi^0$ is also an interpolation space between $L_1$ and $L_\infty$, with interpolation constant 1.

THEOREM 5.4. *Under the condition* (5.17) *every linear functional $f$ on $\mathrm{M}_\psi^0$ can be represented in the form*

$$f(x) = \int_0^\infty x(t)y(t)\,dt, \tag{5.19}$$

*where $y \in \Lambda_\psi$ and $\|f\| = \|y\|_{\Lambda_\psi}$.*

PROOF. The fact that (5.19) defines a linear functional on $\mathrm{M}_\psi$, and consequently on $\mathrm{M}_\psi^0$, follows from (5.11) if $x$ and $y$ are interchanged there. This also implies that

$$\|f\| \leqslant \|y\|_{\Lambda_\psi}. \tag{5.20}$$

Now let $f$ be an arbitrary linear functional on $\mathrm{M}_\psi^0$. The function $\alpha(e) = f(\chi_e)$ defined on all sets of finite measure is additive and absolutely continuous:

$$|\alpha(e)| = |f(\chi_e)| \leqslant \|f\|\,\|\chi_e\|_{\mathrm{M}_\psi^0} = \|f\|\frac{\operatorname{mes} e}{\psi(\operatorname{mes} e)} \to 0$$

when $\operatorname{mes} e \to 0$ (by virtue of (5.17)). It follows from the Radon-Nikodým theorem that there exists a locally integrable function $y$ such that

$$f(x) = \int_0^\infty x(t)y(t)\,dt \tag{5.21}$$

for any strictly simple function $x(t)$.

We choose $x_e(t) = \operatorname{sgn} y(t)\chi_e(t)$ with $\operatorname{mes} e = 1$. Then $\|\chi_e\|_{\mathrm{M}_\psi^0} = 1/\psi(1)$. Therefore

$$\sup_{\operatorname{mes} e = 1} \int_e |y(t)|\,dt \leqslant \frac{\|f\|}{\psi(1)},$$

i.e. $y \in L_1 + L_\infty$.

Any function $x$ in $L_1 \cap L_\infty$ can be represented as the limit of a sequence of strictly simple functions converging in $L_1 \cap L_\infty$, and consequently, by virtue of the imbedding (5.18), in $\mathrm{M}_\psi^0$ as well. This allows us to extend (5.21) to all functions $x$ in $L_1 \cap L_\infty$ by passing to the limit.

We introduce the functions $\mu_\varepsilon \in L_1 \cap L_\infty$ $(0 < \varepsilon < 1)$, defined by

$$\mu_\varepsilon(t) = \psi'(t)\chi_{[\varepsilon, 1/\varepsilon]}(t).$$

We have $\|\mu_\varepsilon\|_{M^0_\psi} \leqslant \|\psi'\|_{M_\psi} = 1$. For any operator $T \in \Sigma$ the functions $T\mu_\varepsilon$ belong to $L_1 \cap L_\infty$, and by the interpolation property of the space $M^0_\psi$ we have $\|T\mu_\varepsilon\|_{M^0_\psi} \leqslant 1$. Formula (5.21) implies that

$$|f(T\mu_\varepsilon)| = \left|\int_0^\infty T\mu_\varepsilon(t)y(t)\,dt\right| \leqslant \|f\|.$$

From Theorem 3.2 it follows that

$$\int_0^\infty \mu_\varepsilon^*(t)y^*(t)\,dt = \int_0^{1/\varepsilon-\varepsilon} \psi'(t+\varepsilon)y^*(t)\,dt \leqslant \|f\|.$$

Letting $\varepsilon$ tend to zero, we obtain from B. Levi's theorem that

$$\|y\|_{\Lambda_\psi} = \int_0^\infty \psi'(t)y^*(t)\,dt \leqslant \|f\|. \tag{5.22}$$

Thus, $y \in \Lambda_\psi$, and by (5.20) and (5.22) we have $\|y\|_{\Lambda_\psi} = \|f\|$.

The validity of (5.19) for any $x \in M^0_\psi$ can now be established by standard arguments.

**4. *Imbedding of symmetric spaces.*** The space $\Lambda_\psi$ is the smallest among symmetric spaces with the same fundamental function.

THEOREM 5.5. *Let $E$ be a symmetric space with fundamental function $\varphi(t)$. The space $\Lambda_\psi$, where $\psi(t)$ is the smallest concave majorant of $\varphi(t)$, is imbedded in $E$, and $\|x\|_E \leqslant \|x\|_{\Lambda_\psi}$.*

PROOF. For any characteristic function $\chi_e$ we have

$$\|\chi_e\|_E = \varphi(\text{mes } e) \leqslant \psi(\text{mes } e) = \int_0^{\text{mes } e} d\psi(t) = \|\chi_e\|_{\Lambda_\psi}.$$

Let $x(t)$ be an arbitrary nonnegative strictly simple function:

$$x(t) = \sum_{k=1}^N \Delta_k \chi_{e_k}(t),$$

where $\Delta_k > 0$ and $e_1 \subset e_2 \subset \cdots \subset e_N$. Then

$$\|x\|_E \leqslant \sum_{k=1}^N \|\Delta_k \chi_{e_k}\|_E \leqslant \sum_{k=1}^N \Delta_k \|\chi_{e_k}\|_{\Lambda_\psi} = \sum_{k=1}^N \Delta_k \psi(\text{mes } e_k) = \|x\|_{\Lambda_\psi} \tag{5.23}$$

(see (5.7)). Obviously this inequality holds for an arbitrary simple function, as well. By Corollary 3 of Theorem 5.1 any function in $\Lambda_\psi$ can be represented as the limit of a sequence of simple functions. By (5.23) this sequence is Cauchy in $E$, and so it has a limit in $E$. Since $E$ and $\Lambda_\psi$ are imbedded in $L_1 + L_\infty$, this limit coincides with the initial function. Thus, $\Lambda_\psi \subset E$. Inequality (5.23) can be extended to any function in $\Lambda_\psi$ by passing to the limit.

THEOREM 5.6. *Let $E$ be a symmetric space. If the function $\|\sigma_{t^{-1}}\|_{E\to E}$ belongs to the Lorentz space $\Lambda_\psi$, then $E \subset \Lambda_\psi$.*

PROOF. For any $\tau \in (0, 1]$ we have

$$\int_0^\infty x^*(t)\, d\psi(t) \leqslant \int_0^\infty x^*(\tau t)\, d\psi(t).$$

We multiply both sides of this inequality by the characteristic function $\chi_{[0,1]}(\tau)$ and compare the norms of the functions of $\tau$ thus obtained. Then, by Lemma 4.7,

$$\varphi_E(1)\int_0^\infty x^*(t)\, d\psi(t) \leqslant \left\|\chi_{[0,1]}(\tau)\int_0^\infty x^*(\tau t)\, d\psi(t)\right\|_E \leqslant \left\|\int_0^\infty x^*(\tau t)\, d\psi(t)\right\|_E$$
$$\leqslant \int_0^\infty \|\sigma_{1/t}x^*\|_E\, d\psi(t) \leqslant \int_0^\infty \|\sigma_{1/t}\|_E\, d\psi(t)\|x\|_E.$$

Thus, for $x \in E$ we have

$$\|x\|_{\Lambda_\psi} \leqslant [\varphi_E(1)]^{-1}\|\,\|\sigma_{1/t}\|_E\|_{\Lambda_\psi}\|x\|_E.$$

The largest space among symmetric spaces with the same fundamental function is the Marcinkiewicz space.

THEOREM 5.7. *Let $E$ be a symmetric space with fundamental function $\psi(t)$. Then $E \subset M_{\psi_*}$, and*

$$\|x\|_{M_{\psi_*}} \leqslant \|x\|_E, \tag{5.24}$$

*where $\psi_*(t) = t/\psi(t)$.*

PROOF. Let $x$ be an arbitrary function in $E$. By (4.6) we have

$$\frac{\|x^*\chi_{[0,h]}\|_{L_1}}{\|x^*\chi_{[0,h]}\|_E} \leqslant \frac{h}{\psi(h)}.$$

Hence

$$\|x\|_E = \|x^*\|_E \geqslant \|x^*\chi_{[0,h]}\|_E \geqslant \frac{\|x^*\chi_{[0,h]}\|_{L_1}}{h/\psi(h)} = \frac{1}{\psi_*(h)}\int_0^h x^*(t)\, dt.$$

If we take the supremum over $h$ on the right side, then we obtain (5.24). The theorem is proved.

It is worth noting that every interpolation space $E$ between $L_1$ and $L_\infty$ is a union of Marcinkiewicz spaces. Indeed, let $x \in E$. Consider the concave function

$$\psi_x(\tau) = \int_0^\tau x^*(s)\, ds,$$

and the Marcinkiewicz space $M_{\psi_x}$ constructed from it. We have

$$\|x\|_{M_{\psi_x}} = \sup_{0<\tau<\infty} \frac{\int_0^\tau x^*(s)\, ds}{\psi_x(\tau)} = 1,$$

and consequently $x \in M_{\psi_x}$. Next, if $y \in M_{\psi_x}$, then

$$\int_0^T y^*(s)\,ds \leqslant \|y\|_{M_{\psi_x}} \int_0^T x^*(s)\,ds.$$

It then follows from Theorem 4.3 that $y \in E$. Thus, $M_{\psi_x}$ is imbedded in $E$.

If the interpolation constant corresponding to $E$ does not exceed one, then the last inequality implies that

$$\|y\|_E \leqslant \|y\|_{M_{\psi_x}} \|x\|_E.$$

If we set $x = y/\|y\|_E$, the inequality becomes an equality. Therefore we may write

$$\|y\|_E = \inf \|y\|_{M_\psi},$$

where the infimum is taken over all $\psi$ of the form $\psi_x$ with $\|x\|_E = 1$ or, in other words, over all $\psi$ for which the Marcinkiewicz spaces are imbedded in $E$ with imbedding constant not exceeding 1.

**5. *The renorming of a symmetric space.*** Let $E$ be a symmetric space with fundamental function $\psi(t)$ and let $\psi_1(t)$ be a quasiconcave function equivalent to $\psi(t)$:

$$C_1\psi(t) \leqslant \psi_1(t) \leqslant C_2\psi(t). \tag{5.25}$$

We introduce the new norm

$$\|x\|_E^1 = \max\{C_1\|x\|_E, \|x\|_{M_{\psi_{1*}}}\}$$

in $E$. It is obvious that $\|x\|_E^1 \geqslant C_1\|x\|_E$. Next, by (5.25) and Theorem 5.7 we have

$$\|x\|_{M_{\psi_{1*}}} \leqslant C_2\|x\|_{M_{\psi_*}} \leqslant C_2\|x\|_E.$$

Thus, $\|x\|_E^1$ is equivalent to the original norm of $E$. For $t = \operatorname{mes} e$ we have

$$\|\chi_e\|_E^1 = \max\{C_1\|\chi_e\|_E, \|\chi_e\|_{M_{\psi_{1*}}}\} = \max\{C_1\psi(t), \psi_1(t)\} = \psi_1(t).$$

We have arrived at the following assertion.

THEOREM 5.8. *If the function $\psi_1(t)$ is quasiconcave and equivalent to the fundamental function $\psi(t)$ of a symmetric space $E$, then an equivalent norm can be introduced in $E$ so that it remains symmetric and has fundamental function $\psi_1(t)$.*

From Corollary 1 of Theorem 1.1 we obtain the following corollary.

COROLLARY. *In any symmetric space an equivalent norm can be introduced so that the space remains symmetric and its fundamental function becomes concave.*

**6. *Sum and intersection of two Lorentz spaces.*** We study the sum of two Lorentz spaces in the case where the quotient of their fundamental functions is monotone.

THEOREM 5.9. *Let $\varphi_0(s)$ and $\varphi_1(s)$ be two increasing concave functions whose quotient $\varphi_0(s)[\varphi_1(s)]^{-1}$ is decreasing. The space $\Lambda_{\varphi_0} + \lambda_{\varphi_1}$ is isometric to $\Lambda_{\min(\varphi_0,\varphi_1)}$.*

PROOF. Let $x \in \Lambda_{\varphi_0} + \Lambda_{\varphi_1}$. By property 18° of rearrangements we have

$$\begin{aligned}\|x\|_{\Lambda_{\varphi_0}+\Lambda_{\varphi_1}} &= \inf_{u+v=x} \{\|u\|_{\Lambda_{\varphi_0}} + \|v\|_{\Lambda_{\varphi_1}}\} \\ &\geqslant \inf_{u+v=x} \{\|u\|_{\Lambda_{\min(\varphi_0,\varphi_1)}} + \|v\|_{\Lambda_{\min(\varphi_0,\varphi_1)}}\} \geqslant \|x\|_{\Lambda_{\min(\varphi_0,\varphi_1)}}.\end{aligned}$$

Thus, $\Lambda_{\varphi_0} + \Lambda_{\varphi_1} \subset \Lambda_{\min(\varphi_0,\varphi_1)}$. (Here we have not used the monotonicity of $\varphi_0(s)[\varphi_1(s)]^{-1}$.)

If $\varphi_0(s)[\varphi_1(s)]^{-1}$ is greater than 1 for all $s$ or smaller than 1 for all $s$, then one of the functions $\varphi_0(s)$ or $\varphi_1(s)$ is smaller than the other, and the space corresponding to it contains the other space. The sum of the spaces then coincides with the larger space, and the theorem is proved.

Now let

$$\varphi_0(s_0)[\varphi_1(s_0)]^{-1} = 1. \tag{5.26}$$

Then

$$\varphi_0(s)\begin{cases} \geqslant \varphi_1(s) & \text{for } s \leqslant s_0, \\ \leqslant \varphi_1(s) & \text{for } s \geqslant s_0. \end{cases}$$

Let $x \in \Lambda_{\min(\varphi_0,\varphi_1)}$. We write $x_0 = \operatorname{sgn} x(s)\min\{x^*(s_0), |x(s)|\}$ and $x_1 = x(s) - x_0(s)$.

Then $x_0^*(s) = x^*(s_0)$ for $s \leqslant s_0$ and $x_1^*(s) = 0$ for $s > s_0$. Besides, $x^*(s) = x_0^*(s) + x_1^*(s)$. Therefore

$$\begin{aligned}\|x\|_{\Lambda_{\min(\varphi_0,\varphi_1)}} &= \|x_0^*\|_{\Lambda_{\min(\varphi_0,\varphi_1)}} + \|x_1^*\|_{\Lambda_{\min(\varphi_0,\varphi_1)}} \\ &= x^*(s_0)\min(\varphi_0(s_0), \varphi_1(s_0)) + \int_{s_0}^{\infty} x_0^*(s)\, d\varphi_0(s) + \|x_1^*\|_{\Lambda_{\varphi_1}} \\ &= \|x_0^*\|_{\Lambda_{\varphi_0}} + \|x_1^*\|_{\Lambda_{\varphi_1}} \geqslant \|x^*\|_{\Lambda_{\varphi_0}+\Lambda_{\varphi_1}}.\end{aligned}$$

Since the sum of symmetric spaces is symmetric (Lemma 4.5), we have $\|x^*\|_{\Lambda_{\varphi_0}+\Lambda_{\varphi_1}} = \|x\|_{\Lambda_{\varphi_0}+\Lambda_{\varphi_1}}$, and consequently $\|x\|_{\Lambda_{\min(\varphi_0,\varphi_1)}} \geqslant \|x\|_{\Lambda_{\varphi_0}+\Lambda_{\varphi_1}}$.

From Theorem 5.9 we obtain the following corollary.

COROLLARY. *If under the hypotheses of Theorem 5.9 a function $x$ in $\Lambda_{\varphi_0} + \Lambda_{\varphi_1}$ is of compact support, then it belongs to $\Lambda_{\varphi_1}$; if it is bounded, then it belongs to $\Lambda_{\varphi_0}$.*

The case where (5.26) holds is of interest. In this case, by Theorem 5.9 we have

$$\|x\|_{\Lambda_{\varphi_0}+\Lambda_{\varphi_1}} = \int_0^{s_0} x^*(s)\, d\varphi_1(s) + \int_{s_0}^{\infty} x^*(s)\, d\varphi_0(s),$$

which immediately implies our assertion.

REMARK. Instead of the monotonicity of $\varphi_0(s)[\varphi_1(s)]^{-1}$ it is sufficient to require generalized monotonicity:

$$\varphi_0(s)[\varphi_1(s)]^{-1} \leqslant C\varphi_0(t)[\varphi_1(t)]^{-1} \quad \text{for } t < s,$$

where $C$ does not depend on $s$ or $t$. Then in Theorem 5.9 isometry must be changed to isomorphism. The assertion of the corollary remains valid.

We now consider the intersection of Lorentz spaces.

THEOREM 5.10. *Let $\varphi_0(s)$ and $\varphi_1(s)$ be two increasing concave functions. The space $\Lambda_{\varphi_0} \cap \Lambda_{\varphi_1}$ is isomorphic to $\Lambda_{(\max(\varphi_0,\varphi_1))^\sim}$ (where, as above, $(\max(\varphi_0, \varphi_1))^\sim$ denotes the smallest concave majorant of* $\max(\varphi_0, \varphi_1)$).

PROOF. Let $x \in \Lambda_{\varphi_0} \cap \Lambda_{\varphi_1}$. Since $(\max(\varphi_0, \varphi_1))^\sim \leqslant \varphi_0 + \varphi_1$, by property 18° we have

$$\begin{aligned}\|x\|_{\Lambda_{(\max(\varphi_0,\varphi_1))^\sim}} &= \int_0^{\infty} x^*(s)\, d(\max(\varphi_0, \varphi_1))^\sim \\ &\leqslant \int_0^{\infty} x^*(s)\, d\varphi_0 + \int_0^{\infty} x^*(s)\, d\varphi_1 \leqslant 2 \max_{i=0,1} \|x\|_{\Lambda_{\varphi_i}} = 2\|x\|_{\Lambda_{\varphi_0}\cap\Lambda_{\varphi_1}}.\end{aligned}$$

Thus, $\Lambda_{\varphi_0} \cap \Lambda_{\varphi_1} \subset \Lambda_{(\max(\varphi_0,\varphi_1))^\sim}$. Conversely, if $x \in \Lambda_{(\max(\varphi_0,\varphi_1))^\sim}$, then $\|x\|_{\Lambda_{\varphi_i}} \leqslant \|x\|_{\Lambda_{(\max(\varphi_0,\varphi_1))^\sim}}$, which proves the theorem.

**7. *Symmetric spaces which are not interpolation spaces between $L_1$ and $L_\infty$.*** We give an answer to the question raised in §4.2 on the existence of such spaces.

LEMMA 5.5. *Let $\psi(t)$ be an increasing concave continuously differentiable function on $[0, \infty)$ having the properties $\psi(0) = 0$, $\psi(\infty) = \infty$ and*

$$\lim_{t\to 0} \frac{\psi(2t)}{\psi(t)} = 1. \tag{5.27}$$

*The Marcinkiewicz space $M_\psi$ has a proper subspace $G$ which is a symmetric space and contains the function $\psi'(t)\chi_{(0,1]}(t)$.*

PROOF. Denote by $\tilde{G}$ the collection of functions in $M_\psi$ different from zero only on a set of finite measure (depending on the function) and having the property that

$$\sup_{t>0} \frac{x^*(t)}{\psi'(t)} < \infty.$$

We note that for $x \in \tilde{G}$ the function $x^*(t)$ is of compact support.

We show that $\tilde{G}$ is linear. From (5.27) it follows that

$$\lim_{t\to 0} \frac{\psi'(2t)}{\psi'(t)} = \frac{1}{2},$$

and therefore on every interval $[0, N]$ we have $\psi'(2t) \geqslant a_N \psi'(t)$. If $x_1$ and $x_2$ belong to $\tilde{G}$, then $x_i^*(t) \leqslant C_i \psi'(t)$ $(i = 1, 2)$, and by (2.23) we have

$$\begin{aligned}(x_1 + x_2)^*(t) &\leqslant x_1^*\left(\frac{t}{2}\right) + x_2^*\left(\frac{t}{2}\right) \leqslant (C_1 + C_2)\psi'\left(\frac{t}{2}\right) \\ &\leqslant \frac{C_1 + C_2}{a_N}\psi'(t)\end{aligned}$$

for $0 \leqslant t \leqslant N$. If $N$ is chosen so that the support of the function $(x_1 + x_2)^*(t)$ is contained in $[0, N]$, then the last inequality will be valid for all $t > 0$. Thus, $x_1 + x_2 \in \tilde{G}$.

As the subspace $G$ we take the closure of $\tilde{G}$ in $\mathrm{M}_\psi$. Since $\tilde{G}$ is obviously symmetric, by Lemma 4.4 $G$ is symmetric. It is clear that $\psi'\chi_{(0,1]} \in \tilde{G} \subset G$.

It remains to show that $G$ does not coincide with $\mathrm{M}_\psi$. For this we construct a sequence $\tau_n$ in the following way: We set $\tau_0 = 1$ and determine $\tau_{n+1}$ from $\tau_n$ by the two conditions

$$0 < \tau_{n+1} \leqslant \frac{1}{2}\tau_n \quad \text{and} \quad \frac{\psi(\tau_n) - \psi(\tau_{n+1})}{\psi(\tau_n - \tau_{n+1})} \geqslant \frac{2}{3}.$$

We can satisfy the second condition, because $(\psi(\tau_n) - \psi(\tau))/\psi(\tau_n - \tau)$ is continuous and converges to 1 as $\tau \to 0$. We show that the function

$$u(t) = \sum_{n=0}^{\infty} \frac{\psi(\tau_n) - \psi(\tau_{n+1})}{\tau_n - \tau_{n+1}} \chi_{(\tau_{n+1}, \tau_n]}(t)$$

belongs to $\mathrm{M}_\psi \setminus G$. Since $\psi(t)$ is increasing and concave, the function $u(t)$ is decreasing; thus $u(t) = u^*(t)$. Then $\int_0^\tau u^*(t)\,dt$ is linear on the intervals $[\tau_{n+1}, \tau_n]$, and $\int_0^{\tau_n} u^*(t)\,dt = \psi(\tau_n)$. The concavity of $\psi(\tau)$ implies that

$$\int_0^\tau u^*(t)\,dt \leqslant \psi(\tau) \qquad (\tau > 0), \tag{5.28}$$

and consequently $u \in \mathrm{M}_\psi$.

Let $y \in \tilde{G}$. Then $y^*(t) \leqslant C\psi'(t)$. By Lemma 4.6 we have $\|u - y\|_{\mathrm{M}_\psi} \geqslant \|u - y^*\|_{\mathrm{M}_\psi}$. Next,

$$\begin{aligned}\|u - y^*\|_{\mathrm{M}_\psi} &\geqslant \|(u - y^*)\chi_{(\tau_p/2, \tau_p]}\|_{\mathrm{M}_\psi} \\ &\geqslant \|u\chi_{(\tau_p/2, \tau_p]}\|_{\mathrm{M}_\psi} - C\|\psi'\chi_{(\tau_p/2, \tau_p]}\|_{\mathrm{M}_\psi}.\end{aligned}$$

By the definition of the norm in $M_\psi$ and the choice of the numbers $\tau_p$ we have

$$\frac{\tau_p - \tau_{p+1}}{\psi(\tau_p - \tau_{p+1})} = \|\chi_{(\tau_{p+1},\tau_p]}\|_{M_\psi}$$
$$\leqslant \|\chi_{(\tau_{p+1},\tau_p/2]}\|_{M_\psi} + \|\chi_{(\tau_p/2,\tau_p]}\|_{M_\psi} \leqslant 2\|\chi_{(\tau_p/2,\tau_p]}\|_{M_\psi}.$$

Hence

$$\|u\chi_{(\tau_p/2,\tau_p]}\|_{M_\psi} = \frac{\psi(\tau_p) - \psi(\tau_{p+1})}{\tau_p - \tau_{p+1}} \|\chi_{(\tau_p/2,\tau_p]}\|_{M_\psi}$$
$$\geqslant \frac{1}{2} \frac{\psi(\tau_p) - \psi(\tau_{p+1})}{\psi(\tau_p - \tau_{p+1})} \geqslant \frac{1}{3}.$$

Thus,

$$\|u - y\|_{M_\psi} \geqslant 1/3 - C\|\psi'\chi_{(\tau_p/2,\tau_p]}\|_{M_\psi}.$$

If we show that the last term on the right tends to zero as $p \to \infty$, then this inequality will imply that $u \notin G$. We have

$$\|\psi'\chi_{(\tau,2\tau]}\|_{M_\psi} = \sup_{0<h\leqslant\tau} \frac{\int_0^h \psi'(\tau + s)\, ds}{\psi(h)} = \sup_{0<h\leqslant\tau} \frac{\psi(\tau + h) - \psi(\tau)}{\psi(h)}.$$

Since $\psi(t)$ is concave, the function $\psi(\tau + h) - \psi(\tau)$ is decreasing in $\tau$, and therefore $\psi(\tau + h) - \psi(\tau) \leqslant \psi(2h) - \psi(h)$. Then

$$\lim_{\tau\to 0} \|\psi'\chi_{(\tau,2\tau]}\|_{M_\psi} \leqslant \lim_{\tau\to 0} \sup_{0<h\leqslant\tau} \frac{\psi(2h) - \psi(h)}{\psi(h)} = \overline{\lim_{\tau\to 0}} \left(\frac{\psi(2h)}{\psi(h)} - 1\right) = 0.$$

THEOREM 5.11. *The symmetric space $G$ constructed in Lemma 5.5 is not an interpolation space between $L_1$ and $L_\infty$.*

PROOF. We rewrite (5.28) in the form

$$\int_0^\tau u^*(t)\, dt \leqslant \int_0^\tau [\psi'\chi_{(0,1]}]^*(t)\, dt.$$

The function $\psi'\chi_{(0,1]}$ belongs to $G$, while $u \notin G$. Thus, $G$ does not satisfy the necessary condition of Theorem 4.3. $G$ is not an interpolation space.

## §6. Operators of weakened and weak type

**1.** ***The operation*** $^{**}$. For $x \in L_1 + L_\infty$ we introduce the notation

$$x^{**}(t) = \frac{1}{t}\int_0^t x^*(s)\, ds.$$

Obviously $x^{**}(t) \geqslant x^*(t)$. Using property 7° of rearrangements (see (2.13)), we may write

$$x^{**}(t) = \frac{1}{t} \sup_{\operatorname{mes} e = t} \int_e |x(s)|\, ds.$$

This implies that the operation $x^{**}$ satisfies the triangle inequality

$$[x_1 + x_2]^{**}(t) \leqslant x_1^{**}(t) + x_2^{**}(t). \tag{6.1}$$

We note that the operation $x^*$ does not have this property. We show that $x^{**}$ satisfies the triangle inequality for an infinite number of terms, as well. Let the series $\sum x_k^{**}(\tau)$ $(x_k \in L_1 + L_\infty)$ be convergent. By B. Levi's theorem, for a measurable set $e$ with $\operatorname{mes} e = \tau$ the series $\sum x_k(s)$ converges almost everywhere on $e$, and

$$\frac{1}{\tau} \int_e \left|\sum x_k(s)\right| ds \leqslant \frac{1}{\tau} \sum \int_e |x_k(s)|\, ds \leqslant \sum x_k^{**}(\tau).$$

On the left we take the supremum over all $e$ with $\operatorname{mes} e = \tau$. This gives us

$$\left(\sum x_k\right)^{**}(\tau) \leqslant \sum x_k^{**}(\tau).$$

The last property allows us to construct new symmetric spaces from old ones by means of the operation $**$. Let $\delta(t)$ and $\kappa(t)$ be two positive measurable functions on $(0, \infty)$. For a symmetric space $E$ we denote by $E_{x,\delta}$ the space of functions $x(t)$ in $L_1 + L_\infty$ for which the functions $\kappa(t)x^{**}(\delta(t))$ belong to $E$. (The function $x^{**}(\delta(t))$ is measurable, since $x^{**}$ is monotone and $\delta$ is measurable.) With the norm

$$\|x\|_{E_{\kappa,\delta}} = \|\kappa(t)x^{**}(\delta(t))\|_E$$

this will be a Banach space. To prove this it is sufficient to verify that for any sequence $x_k \in E_{\kappa,\delta}$ $(k = 1, 2, \ldots)$ with $\sum \|x_k\|_{E_{\kappa,\delta}} < \infty$ the series $\sum x_k$ is convergent in $E_{\kappa,\delta}$. By definition

$$\sum \|x_k\|_{E_{\kappa,\delta}} = \sum \|\kappa(t)x_k^{**}(\delta(t))\|_E,$$

and, since $E$ is complete, $\sum \kappa(t)x_k^{**}(\delta(t)) \in E$. It then follows from the triangle inequality that the function

$$\kappa(t)\left(\sum x_k\right)^{**}(\delta(t)) \leqslant \kappa(t)\sum x_k^{**}(\delta(t))$$

also belongs to $E$. This is equivalent to the condition that $\sum x_k(t) \in E_{\kappa,\delta}$; moreover,

$$\left\|\sum x_k\right\|_{E_{\kappa,\delta}} \leqslant \sum \|x_k\|_{E_{\kappa,\delta}}.$$

From this it follows by standard arguments that $\sum x_k$ is the limit in $E_{\kappa,\delta}$ of the partial sums $\sum_1^N x_k$.

We establish some connections between the operation ** and the dilation operators. In the case of characteristic functions

$$\chi_{(0,\theta]}^{**}(t) = \begin{cases} 1 & \text{for } 0 < t \leq \theta, \\ \theta/t & \text{for } t > \theta \end{cases}$$

and we consider the averaging operator

$$T_h x(t) = \frac{1}{h} \int_0^h x(s)\, ds\, \chi_{(0,h]}(t).$$

For $\tau > 1$ we compute

$$\frac{1}{\tau\theta} \int_0^{\tau\theta} \chi_{(0,\theta]}^{**}(s)\, ds = \frac{1}{\tau\theta}\left[\int_0^{\theta} \chi_{(0,\theta]}(s)\, ds + \int_\theta^{\tau\theta} \chi_{(0,\theta]}(s)\, ds\right]$$
$$= \frac{1}{\tau}(1 + \ln \tau).$$

Hence

$$(T_{\tau\theta}\chi_{(0,\theta]}^{**})(t) = \frac{1}{\tau}(1 + \ln \tau)\chi_{(0,\tau\theta]}(t) = \frac{1 + \ln \tau}{\tau} \sigma_\tau \chi_{(0,\theta]}(t).$$

The operator $T_{\tau\theta}$ is of norm one in the spaces $L_1$ and $L_\infty$. Therefore it satisfies (3.16). Hence

$$\int_0^s \sigma_\tau \chi_{(0,\theta]}(t)\, dt \leq \frac{\tau}{1 + \ln \tau} \int_0^s \chi_{(0,\theta]}(t)\, dt \qquad (s > 0).$$

For a strictly simple function of the form

$$z(t) = \sum_{i=1}^{N} \Delta_i \chi_{(0,\theta_i]}(t),$$

where $\Delta_i > 0$ and $\theta_1 > \theta_2 > \cdots > 0$, it follows immediately that

$$\int_0^s \sigma_\tau z(t)\, dt \leq \frac{\tau}{1 + \ln \tau} \int_0^s \sum_{i=1}^{N} \Delta_i \chi_{(0,\theta_i]}^{**}(t)\, dt = \frac{\tau}{1 + \ln \tau} \int_0^s z^{**}(t)\, dt.$$

Any decreasing nonnegative function $x(t)$ can be approximated uniformly by an increasing sequence of elementary functions of the form

$$y(t) = \sum_{i=1}^{\infty} \Delta_i \chi_{(0,\theta_i]}(t) + x(\infty),$$

where $\Delta_i > 0$ and $\theta_1 > \theta_2 > \cdots > 0$. The preceding inequality implies that

$$\int_0^s \sigma_\tau \left[\sum_{i=1}^{N} \Delta_i \chi_{(0,\theta_i]}(t) + x(\infty)\right] dt$$
$$\leq \frac{\tau}{1 + \ln \tau} \int_0^s \left[\sum_{i=1}^{N} \Delta_i \chi_{(0,\theta_i]}^{**}(t) + x(\infty)\right] dt \leq \frac{\tau}{1 + \ln \tau} \int_0^s x^{**}(t)\, dt.$$

Passing to the limit twice we arrive at an inequality which can finally be written in the form

$$\int_0^s \sigma_\tau x^*(t)\,dt \leqslant \frac{\tau}{1+\ln\tau}\int_0^s x^{**}(t)\,dt \qquad (\tau \geqslant 1, s > 0). \qquad (6.1')$$

**2.** ***Operators from Lorentz spaces to Marcinkiewicz spaces.*** Let $\psi(t)$ be a quasiconcave function and let $\psi_*(t) = t/\psi(t)$. The norm in the Marcinkiewicz space $\mathrm{M}_\psi$ can be written in the form

$$\|y\|_{\mathrm{M}_{\psi_*}} = \sup_{0<t<\infty} \frac{\psi(t)}{t}\int_0^t y^*(s)\,ds = \sup_{0<t<\infty} [\psi(t)y^{**}(t)].$$

Now if an operator $A$ acts boundedly from $\Lambda_\varphi$ to $\mathrm{M}_{\psi_*}$, this means that

$$\psi(t)(Ax)^{**}(t) \leqslant C\int_0^\infty x^*(s)\,d\varphi(s). \qquad (6.2)$$

If $\varphi(t)$ is a quasiconcave function, we shall consider the space $\Lambda_{\tilde{\varphi}}$, where $\tilde{\varphi}(t)$ is the smallest concave majorant of $\varphi(t)$. By (1.7) the condition that the operator is bounded from $\mathrm{M}_{\psi_*}$ to $\Lambda_\varphi$ can again be written in the form (6.2).

The main interpolation theorem to which the present subsection is devoted concerns operators that act boundedly from the Banach couple $(\Lambda_{\tilde{\varphi}_0}, \Lambda_{\tilde{\varphi}_1})$ to the Banach couple $(\mathrm{M}_{\psi_{0*}}, \mathrm{M}_{\varphi_{1*}})$. Before formulating this theorem, we introduce a class of spaces which are intermediate between Lorentz spaces.

Let $\varphi_0(t)$ and $\varphi_1(t)$ be two quasiconcave functions for which the quotient $\varphi_0(t)[\varphi_1(t)]^{-1}$ is decreasing. In this case, for $\tau > 1$ we have the inequality

$$\varphi_0(\tau t)[\varphi_1(\tau t)]^{-1} \leqslant \varphi_0(t)[\varphi_1(t)]^{-1}$$

or

$$\varphi_0(\tau t)[\varphi_0(t)]^{-1} \leqslant \varphi_1(\tau t)[\varphi_1(t)]^{-1},$$

which implies that $M_{\varphi_0}(\tau) \leqslant M_{\varphi_1}(\tau)$ for $\tau > 1$. For $\tau < 1$ the reverse inequality holds.

We consider a symmetric space $E$ having the property that for the norm of the dilation operator $\sigma_t$ the inequality

$$\int_0^1 \|\sigma_{1/\tau}\|_E\,dM_{\varphi_1}(\tau) + \int_1^\infty \|\sigma_{1/\tau}\|_E\,dM_{\varphi_0}(\tau) < \infty \qquad (6.3)$$

is satisfied. In view of the above inequalities, this condition can be written in the form

$$\int_0^\infty \|\sigma_{1/\tau}\|_E\,d\min\{M_{\varphi_i}(\tau)\} < \infty, \qquad i = 0, 1. \qquad (6.4)$$

As usual, we denote by $\gamma_i$ and $\delta_i$ the lower and upper dilation exponents of $\varphi_i(t)$ and by $\alpha_E$ and $\beta_E$ the lower and upper dilation exponents of $E$. Then

$$\int_1^N \|\sigma_{1/\tau}\|_E \, dM_{\varphi_0}(\tau) \geqslant \int_1^N \tau^{-\alpha_E} \, dM_{\varphi_0}(\tau)$$
$$= \tau^{-\alpha_E} M_{\varphi_0}(\tau)|_1^N + \alpha_E \int_1^N M_{\varphi_0}(\tau)\tau^{-\alpha_E - 1}\, d\tau$$
$$\geqslant \frac{\delta_0}{\delta_0 - \alpha_E} \tau^{\delta_0 - \alpha_E}|_1^N.$$

Condition (6.3) implies that $\alpha_E > \delta_0$. Similarly, from the finiteness of the first term in (6.3) it follows that $\gamma_1 > \beta_E$. Conversely, if $\alpha_E > \delta_0$ and $\gamma_1 > \beta_E$, then $\|\sigma_{1/\tau}\|_E \leqslant C\tau^{-\alpha_E + \varepsilon}$ and $M_{\varphi_0}(\tau) \leqslant C\tau^{\delta_0 + \varepsilon}$ for $\tau > 1$, and $\|\sigma_{1/\tau}\|_E \leqslant C\tau^{-(\beta_E + \varepsilon)}$ and $M_{\varphi_0}(\tau) \leqslant C\tau^{\gamma_1 - \varepsilon}$ for $\tau < 1$. For sufficiently small $\varepsilon$ this implies the convergence of the integrals in (6.3). Thus, (6.3) is equivalent to the inequalities

$$\gamma_1 > \beta_E \geqslant \alpha_E > \delta_0. \tag{6.5}$$

These inequalities are in turn equivalent to the relations

$$\|\sigma_t\|_E = o(t^{\delta_0}) \quad \text{as } t \to 0 \qquad \text{and} \qquad \|\sigma_t\|_E = o(t^{\gamma_1}) \quad \text{as } t \to 0. \tag{6.6}$$

LEMMA 6.1. *If one of the equivalent conditions* (6.3)–(6.6) *is satisfied, then the space $E$ is intermediate between the Lorentz spaces $\Lambda_{\tilde{\varphi}_0}$ and $\Lambda_{\tilde{\varphi}_1}$.*

PROOF. The function $\|\sigma_{1/\tau}\|_E$ is decreasing, and therefore, by property 18° of rearrangements, from (6.4) we obtain that

$$\int_0^\infty \|\sigma_{1/\tau}\|_E \, d\min\{\tilde{\varphi}_i(\tau)\} \leqslant 2\int_0^\infty \|\sigma_{1/\tau}\|_E \, d\min\{\varphi_i(\tau)\}$$
$$\leqslant 2\max\{\varphi_i(1)\} \int_0^\infty \|\sigma_{1/\tau}\|_E \, d\min\{M\varphi_i(\tau)\} < \infty.$$

Theorem 5.6 implies that $E \subset \Lambda_{\min\{\tilde{\varphi}_i\}}$. The quotient $\tilde{\varphi}_0(t)[\tilde{\varphi}_1(t)]^{-1}$ is generalized monotone, and so from the remark to Theorem 5.9 it follows that $\Lambda_{\min\{\tilde{\varphi}_i\}}$ is isomorphic to $\Lambda_{\tilde{\varphi}_0} + \Lambda_{\tilde{\varphi}_1}$. Thus, $E \subset \Lambda_{\tilde{\varphi}_0} + \Lambda_{\tilde{\varphi}_1}$.

To obtain the second imbedding, we note that

$$\lambda = \int_0^\infty \|\sigma_{1/\tau}\|_E \, d\min\{M_{\varphi_i}(\tau)\}$$
$$\geqslant \int_0^t \|\sigma_{1/\tau}\|_E \, d\min\{M_{\varphi_i}(\tau)\} \geqslant \|\sigma_{1/t}\|_E \min\{M_{\varphi_i}(t)\}$$
$$\geqslant \frac{\|\sigma_{1/t}\chi_{(0,1)}\|_E}{\|\chi_{(0,1)}\|_E} \min\left\{\frac{\varphi_i(1)}{\varphi_i(1/t)}\right\}$$
$$\geqslant \frac{\varphi_E(1/t)}{\varphi_E(1)} \min\{\varphi_i(1)\} \min\left\{\frac{1}{\varphi_i(1/t)}\right\}.$$

Hence

$$\varphi_E(\tau) = \frac{\lambda\varphi_E(1)}{\min\{\varphi_i(1)\}} \max \varphi_i(\tau).$$

An analogous inequality holds for $\tilde{\varphi}_E(\tau)$ and $\tilde{\varphi}_i(\tau)$. By Theorem 5.5 we have $E \supset \Lambda_{\tilde{\varphi}_E} \supset \Lambda_{C(\max\{\varphi_i\})^\sim}$, and by Theorem 5.10 the space $\Lambda_{C(\max\{\varphi_i\})^\sim}$ is isomorphic to $\Lambda_{\tilde{\varphi}_0} \cap \Lambda_{\tilde{\varphi}_1}$. Thus, $E \supset \Lambda_{\tilde{\varphi}_0} \cap \Lambda_{\tilde{\varphi}_1}$.

THEOREM 6.1 (main interpolation theorem). *Let four quasiconcave functions $\varphi_0(t)$, $\varphi_1(t)$, $\psi_0(t)$ and $\psi_1(t)$ satisfy the following conditions:*

1) *The function $\varphi_0(s)[\varphi_1(s)]^{-1}$ is decreasing.*

2) *The range of $\psi_0(s)[\psi_1(s)]^{-1}$ for $0 < s < \infty$ contains the range of $\varphi_0(s)[\varphi_1(s)]^{-1}$ $(0 < s < \infty)$.*

*If a linear operator $A$ acts boundedly from the couple $(\Lambda_{\tilde{\varphi}_0}, \Lambda_{\tilde{\varphi}_1})$ to the couple $(M_{\psi_{0*}}, M_{\psi_{1*}})$, then it acts boundedly from every space $E$ satisfying conditions* (6.3)–(6.6) *into the space $E_{\kappa,\delta}$, where $\delta(t)$ is a measurable positive solution of the equation*

$$\psi_0(\delta(t))[\psi_1(\delta(t))]^{-1} = \varphi_0(t)[\varphi_1(t)]^{-1}, \tag{6.7}$$

*and*

$$\kappa(t) = \psi_0(\delta(t))[\varphi_0(t)]^{-1} = \psi_1(\delta(t))[\varphi_1(t)]^{-1}. \tag{6.8}$$

PROOF. By Lemma 6.1 the space $E$ is imbedded in $\Lambda_{\tilde{\varphi}_0} + \Lambda_{\tilde{\varphi}_1}$, and therefore $A$ is uniquely defined on it. Let $x \in E \subset \Lambda_{\tilde{\varphi}_0 + \tilde{\varphi}_1}$. For fixed $t$ we set $x_0(s) = \operatorname{sgn} x(s) \min\{x^*(t), |x(s)|\}$ and $x_1(s) = x(s) - x_0(s)$. Then $x_0^*(s) = x^*(t)$ for $s \leqslant t$ and $x_1^*(s) = 0$ for $s > t$. Besides, $x^*(s) = x_0^*(s) + x_1^*(s)$, and so the functions $x_0^*(s)$ and $x_1^*(s)$ belong to $\Lambda_{\tilde{\varphi}_0 + \tilde{\varphi}_1}$. From the corollary of Theorem 5.9 we then obtain that $x_0^* \in \Lambda_{\tilde{\varphi}_0}$ and $x_1^* \in \Lambda_{\tilde{\varphi}_1}$.

By (6.1) we have

$$(Ax)^{**}(\delta(t)) \leqslant (Ax_0)^{**}(\delta(t)) + (Ax_1)^{**}(\delta(t)). \tag{6.9}$$

According to (6.2), from the conditions imposed on $A$ we obtain

$$\begin{aligned}\psi_0(\delta(t))(Ax_0)^{**}(\delta(t)) &\leqslant C_0 \int_0^\infty x_0^*(s)\, d\varphi_0(s) \\ &= C_0\left[\int_0^t x^*(t)\, d\varphi_0(s) + \int_t^\infty x_0^*(s)\, d\varphi_0(s)\right].\end{aligned}$$

By (6.7), for the first integral on the right we have

$$\begin{aligned}C_0 x^*(t) \int_0^t d\varphi_0(s) &= C_0 x^*(t)\varphi_0(t) \\ &= C_0 x^*(t) \frac{\varphi_1(t)\psi_0(\delta(t))}{\psi_1(\delta(t))} = C_0 \frac{\psi_0(\delta(t))}{\psi_1(\delta(t))} \int_0^t x_0^*(s)\, d\varphi_1(s).\end{aligned}$$

Next,

$$\psi_1(\delta(t))(Ax_1)^{**}(\delta(t)) \leqslant C_1\int_0^\infty x_1^*(s)\,d\varphi_1(s) = C_1\int_0^t x_1^*(s)\,d\varphi_1(s).$$

From the above inequalities it follows that

$$(Ax)^{**}(\delta(t)) \leqslant C\left[\int_0^t x^*(s)\,d\frac{\varphi_1(s)}{\psi_1(\delta(t))} + \int_t^\infty x^*(s)\,d\frac{\varphi_0(s)}{\psi_0(\delta(t))}\right],$$

where $C = \max\{C_0, C_1\}$, and by (6.8)

$$\kappa(t)(Ax)^{**}(\delta(t)) \leqslant C\left[\int_0^t x^*(s)\,d\frac{\varphi_1(s)}{\varphi_1(t)} + \int_t^\infty x^*(s)\,d\frac{\varphi_0(s)}{\varphi_0(t)}\right].$$

By condition 1) we then have

$$\kappa(t)(Ax)^{**}(\delta(t)) \leqslant C\int_0^\infty x^*(\tau t)\,d\min\frac{\varphi_i(\tau t)}{\varphi_i(t)}$$

and by property 18° of rearrangements we have

$$\kappa(t)(Ax)^{**}(\delta(t)) \leqslant C\int_0^\infty x^*(\tau t)\,d\min M_{\varphi_i}(\tau). \tag{6.10}$$

Lemma 4.7 implies that

$$\|\kappa(t)(Ax)^{**}(\delta(t))\|_E \leqslant C\int_0^\infty \|\sigma_{1/\tau}\|_E\,d\min M_{\varphi_i}(\tau)\|x\|_E.$$

The theorem is proved.

If $\varphi(\infty) = \infty$ ($< \infty$), then the strictly simple (simple) functions are dense in $\Lambda_\varphi$, and therefore, if the operator $A$ is defined on the strictly simple functions (simple functions) and satisfies (6.2), then it admits an extension by continuity to a bounded operator from $\Lambda_\varphi$ to $M_{\psi_*}$. Moreover, by Lemma 5.2 it is sufficient to verify condition 5.2 only on characteristic functions from $\Lambda_\psi$.

**3.** ***Operators of strong, weakened, and weak types*** $\{E, F\}$. Let $E$ and $F$ be two symmetric spaces. The bounded linear operators from $E$ to $F$ are often called *operators of strong type* $\{E, F\}$.

DEFINITION 6.1. A linear operator $A$, defined on all strictly simple functions if $\varphi_E(\infty) = \infty$, or on all simple functions if $\varphi_E(\infty) < \infty$, is called an *operator of weakened type* $\{E, F\}$ if it is bounded as an operator from $\Lambda_{\tilde{\varphi}_E}$ to $M_{(\varphi_F)_*}$.

In §5.5 it was shown that for a symmetric space $E$ we have the imbedding

$$\Lambda_{\tilde{\varphi}_E} \subset E \subset M_{(\varphi_E)_*},$$

from which it follows that an operator of strong type $\{E, F\}$ is an operator of weakened type $\{E, F\}$.

If an operator belongs to the semigroup $\Sigma$ (see §4.4), then, as we pointed out on pp. 109 and 114, it acts boundedly in all Lorentz and Marcinkiewicz spaces. Consequently, if an operator $A$ is of weakened type $\{E, F\}$, then so are the operators $\lambda AT$ and $\lambda TA$ for any $T \in \Sigma$ and $\lambda \in \mathbf{R}^1$.

DEFINITION 6.2. A linear operator $A$, defined on all strictly simple (simple) functions if $\varphi_E(\infty) = \infty$ $(< \infty)$, is called an *operator of weak type* $\{E, F\}$ if

$$(Ax)^*(t)\varphi_F(t) \leqslant C\int_0^\infty x^*(s)\,d\varphi_E(s). \tag{6.11}$$

Since $(Ax)^{**} \geqslant (Ax)^*$, every operator of weakened type is an operator of weak type. Consequently, every operator of strong type is an operator of weak type.

LEMMA 6.2. *If* $M_{\varphi_F}(s_0) < s_0$ *for some* $s_0 > 1$, *then any operator of weak type* $\{E, F\}$ *is an operator of weakened type.*

PROOF. The condition $M_{\varphi_F}(s_0) < s_0$ implies that the function $\varphi_{F^\bullet}(t) = t/\varphi_F(t)$ satisfies the condition $M_{\varphi_{F^\bullet}}(s_0^{-1}) < 1$. Therefore by Corollary 2 of Lemma 1.4 we have

$$\frac{1}{\tau}\int_0^\tau \frac{dt}{\varphi_F(t)} = \frac{1}{\tau}\int_0^\tau \frac{\varphi_{F^\bullet}(t)}{t}\,dt \leqslant \frac{C}{\tau}\varphi_{F^\bullet}(\tau) = \frac{C}{\varphi_F(\tau)}.$$

We then obtain from (6.11) that

$$\begin{aligned}(Ax)^{**}(\tau) &\leqslant C\frac{1}{\tau}\int_0^\tau \frac{dt}{\varphi_F(t)}\int_0^\infty x^*(s)\,d\varphi_E(s)\\ &\leqslant \frac{C}{\varphi_F(\tau)}\int_0^\infty x^*(s)\,d\varphi_E(s).\end{aligned} \tag{6.12}$$

The lemma is proved.

As an example of an operator of weak type which is not an operator of weakened type we may take the Hardy-Littlewood operator

$$Hx(t) = \frac{1}{t}\int_0^t x(s)\,ds, \tag{6.13}$$

considered as an operator from $L_1$ to $L_1$.

By (2.12) we have

$$|Hx(t)| = \left|\frac{1}{t}\int_0^t x(s)\,ds\right| \leqslant \frac{1}{t}\int_0^t x^*(s)\,ds.$$

The last function is decreasing, and so

$$(Hx)^*(t)\varphi_{L_1}(t) = t(Hx)^*(t) \leqslant \left(\frac{1}{t}\int_0^t x^*(s)\,ds\right)^* t \leqslant \int_0^\infty x^*(s)\,ds,$$

i.e. $H$ is an operator of weak type. Now let $x(t) = \chi_{[0,h]}(t)$. Assuming that $t > h$, we compute

$$(Hx)^{**}(t)t = \int_0^t \left[ \frac{1}{s} \int_0^s \chi_{[0,h]}(\tau)\, d\tau \right]^* ds$$
$$= \int_0^t \min\left\{1, \frac{h}{s}\right\} ds = h + h \ln \frac{t}{h}.$$

This implies that the inequality

$$\left(H(\chi_{[0,h)})\right)^{**}(t)t = h + h \ln \frac{t}{h} \leqslant C \int_0^\infty \chi_{[0,h)}(s)\, ds = Ch$$

is impossible.

**4. *Interpolation theorems.*** From Theorem 6.1 and what was said following it we obtain

THEOREM 6.1′. *Let the operator $A$ be of weakened types $\{E_0, F_0\}$ and $\{E_1, F_1\}$, and assume that the following conditions hold:*

*1) The function $\varphi_{E_0}(s)[\varphi_{E_1}(s)]^{-1}$ is decreasing.*

*2) The range of the function $\varphi_{F_0}(s)[\varphi_{F_1}(s)]^{-1}$ for $0 < s < \infty$ contains the range of the function $\varphi_{E_0}(s)[\varphi_{E_1}(s)]^{-1}$ $(0 < s < \infty)$.*

*If for a symmetric space $E$ we have*

$$\int_0^1 \|\sigma_{1/\tau}\|_E \, dM_{\varphi_{E_1}}(\tau) + l \int_1^\infty \|\sigma_{1/\tau}\|_E \, dM_{\varphi_{E_0}}(\tau) < \infty, \tag{6.14}$$

*then $A$ admits closure to a bounded operator from $E$ to $E_{\kappa,\delta}$, where $\delta(t)$ is a measurable positive solution of the equation*

$$\varphi_{F_0}(\delta(t))\left[\varphi_{F_1}(\delta(t))\right]^{-1} = \varphi_{E_0}(t)\left[\varphi_{E_1}(t)\right]^{-1}, \tag{6.15}$$

*and*

$$\kappa(t) = \varphi_{F_0}(\delta(t))\left[\varphi_{E_0}(t)\right]^{-1} = \varphi_{F_1}(\delta(t))\left[\varphi_{E_1}(t)\right]^{-1}. \tag{6.16}$$

COROLLARY 1. *If $\gamma_i$ and $\delta_i$ are the lower and upper dilation exponents of the functions $\varphi_{E_i}$, then under the hypotheses of Theorem 6.1′ the operator $A$ is bounded, acting from any symmetric space $E$ for which*

$$\|\sigma_t\|_E = o(t^{\delta_0}) \quad \text{as } t \to 0 \qquad \text{and} \qquad \|\sigma_t\|_E = o(t^{\gamma_1}) \quad \text{as } t \to \infty \tag{6.17}$$

*into the space $E_{\kappa,\delta}$.*

COROLLARY 2. *If under the hypotheses of Theorem* 6.1′ *we have $E_0 = F_0$ and $E_1 = F_1$, then the operator $A$ is bounded, acting from a space $E$ for which* (6.14) *or* (6.17) *is satisfied into a space $E_{i,t}$, i.e. into a space with norm $\|x\|_{E_{1,t}} = \|x^{**}\|_E$.*

REMARK. If under the hypotheses of Theorem 6.1′ the space $E_0$ coincides with $L_\infty$, then condition 1) is satisfied automatically, and (6.14) takes the form

$$\int_0^1 \|\sigma_{1/\tau}\|_E \, dM_{\varphi_1}(\tau) < \infty \qquad \text{or} \qquad \|\sigma_t\| = o(t^{\gamma_1}) \quad \text{as } t \to \infty. \tag{6.18}$$

Indeed, in this case we have $\varphi_{E_0}(t) = \operatorname{sgn} t$ ($\operatorname{sgn} 0 = 0$).

Now we consider an operator $A$ which is of weak types $\{E_0, F_0\}$ and $\{E_1, F_1\}$, and assume that $M_{\varphi_{F_1}}(s_0) < s_0$ for some $s_0 > 1$. Then $A$ is of weakened type $\{E_1, F_1\}$. The operator $A$ satisfies the inequalities

$$(A\chi_e)^*(t)\varphi_{F_i}(t) \leqslant C_i\varphi_{E_i}(\text{mes } e),$$

which imply that for any $\theta \in [0, 1]$

$$(A\chi_e)^*(t)\varphi_{F_0}^{1-\theta}(t)\varphi_{F_1}^{\theta}(t) \leqslant C_0^{1-\theta}C_1^{\theta}\varphi_{E_0}^{1-\theta}(\text{mes } e)\varphi_{E_1}^{\theta}(\text{mes } e). \tag{6.19}$$

We write $\varphi_{E,\theta}(t) = \varphi_{E_0}^{1-\theta}(t)\varphi_{E_1}^{\theta}(t)$ and $\varphi_{F,\theta}(t) = \varphi_{F_0}^{1-\theta}(t)\varphi_{F_1}^{\theta}(t)$. For the function $\varphi_{F,\theta}(t)$ for $\theta \in (0, 1]$ we have

$$M_{\varphi_{F,\theta}}(s_0) \leqslant \left[M_{\varphi_{F_0}}(s_0)\right]^{1-\theta}\left[M_{\varphi_{F_1}}(s_0)\right]^{\theta} < s_0^{1-\theta}s_0^{\theta} = s_0.$$

Repeating the calculation in (6.12), we obtain from (6.19) that

$$(A\chi_e)^{**}(t)\varphi_{F,\theta}(t) \leqslant C_\theta\varphi_{E,\theta}(\text{mes } e) \leqslant C_\theta\tilde{\varphi}_{E,\theta}(\text{mes } e). \tag{6.20}$$

As we noted on p. 130, this inequality implies that the operator $A$ can be extended by linearity and continuity to a bounded linear operator from $\Lambda_{\tilde{\varphi}_{E,\theta}}$ into $\mathrm{M}_{F,\theta}$, i.e. to an operator of weakened type $(\Lambda_{\tilde{\varphi}_{E,\theta}}, \mathrm{M}_{\varphi_{F,\theta}})$. Now we can apply the main interpolation Theorem 6.1 to it.

As a preliminary, we show that every space $E$ satisfying (6.14) satisfies a similar condition with respect to the functions $\varphi_{E,\theta}(t)$ and $\varphi_1(t)$ for sufficiently small $\theta$. If we denote by $\gamma_\theta$ and $\delta_\theta$ the lower and upper dilation exponents of the function $\varphi_{E,\theta}(t)$, then by (1.29) we have $\delta_\theta \leqslant \delta_0(1-\theta) + \delta_1\theta$. By (6.14) we have $\gamma_1 > \beta \geqslant \alpha > \delta_0$, and so for sufficiently small $\theta$ the inequality $\gamma_1 > \beta \geqslant \alpha > \delta_\theta$ will be satisfied.

We may carry out similar arguments in the case where $M_{\varphi_{F_0}}(s_0) < s_0$ for some $s_0 > 1$.

We have arrived at the following assertion.

THEOREM 6.2. *If to the hypotheses of Theorem* 6.1′ *the condition*

$$M_{\varphi_{F_1}}(s_0) < s_0 \quad \textit{or} \quad M_{\varphi_{F_0}}(s_0) < s_0 \tag{6.21}$$

*for some* $s_0 > 1$ *is added, then every operator of weak types* $\{E_0, F_0\}$ *and* $\{E_1, F_1\}$ *admits closure to an operator which acts boundedly from a space* $E$ *satisfying* (6.14) *into the space* $E_{\kappa,\delta}$.

REMARK. In the definitions of operators of weakened and weak types the fundamental functions of the spaces are used. It would have been possible to give similar definitions by replacing the fundamental functions by norms of the dilation operators. For such operators, Theorems 6.1′ and 6.2 remain valid if in their formulation we replace the functions $\varphi_{E_i}(t)$ and $\varphi_{F_i}(t)$ by $\|\sigma_t\|_{E_i}$ and $\|\sigma_t\|_{F_i}$, respectively. We note that in view of the submultiplicativity of the norm of the operator $\sigma_t$ the functions $M_{\varphi_{E_i}}(t)$ are replaced by the functions $\|\sigma_\tau\|_{E_i}$.

5. ***Optimal interpolation triples.*** Let $\varphi(t)$ and $\psi(t)$ be two quasiconcave functions. We introduce the operator

$$Sx(t) = \frac{1}{\psi(t)} \int_0^\infty x(s)\, d\tilde{\varphi}(s).$$

By the concavity of $\tilde{\varphi}(t)$ and the properties of rearrangements we have

$$\begin{aligned}(Sx)^*(t) &= \frac{1}{\psi(t)} \left| \int_0^\infty x(s)\, d\tilde{\varphi}(s) \right| \\ &\leqslant \frac{1}{\psi(t)} \int_0^\infty x^*(s)\, d\tilde{\varphi}(s) \leqslant \frac{2}{\psi(t)} \int_0^\infty x^*(s)\, d\varphi(s).\end{aligned}$$

Thus, an inequality of the form (6.11) is satisfied for the operator $S$. If $M_\varphi(s_0) < s_0$ for some $s_0 > 1$, then according to Lemma 6.2 inequality (6.2) will also be satisfied for $S$.

Now let two pairs $\varphi_0(t)$, $\psi_0(t)$ and $\varphi_1(t)$, $\psi_1(t)$ of quasiconcave functions be given. Consider the operator

$$S_{01}x(t) = \int_0^\infty x(s) d \min_i \left\{ \tilde{\varphi}_i(s) [\psi_i(t)]^{-1} \right\}. \tag{6.22}$$

By property 13° of rearrangements we have

$$|S_{01}x(t)| \leqslant \int_0^\infty x^*(s)\, d \min_i \left\{ \tilde{\varphi}_i(s) [\psi_i(t)]^{-1} \right\}.$$

By 18°, on the right we have a monotone function. Therefore

$$\begin{aligned}(S_{01}x)^*(t) &\leqslant \int_0^\infty x^*(s)\, d \min_i \left\{ \tilde{\varphi}_i(s) [\psi_i(t)]^{-1} \right\} \\ &\leqslant \min_i \left\{ \int_0^\infty x^*(s)\, d\tilde{\varphi}_i(s) [\psi_i(t)]^{-1} \right\} \\ &\leqslant 2 \min_i \left\{ \frac{1}{\psi_i(t)} \int_0^\infty x^*(s)\, d\varphi_i(s) \right\}.\end{aligned} \tag{6.23}$$

Thus, the operator $S_{01}$ satisfies an inequality of the type (6.11) both with respect to the pairs $\varphi_0$, $\psi_0$ and $\varphi_1$, $\psi_1$.

LEMMA 6.3. *Let the functions* $\varphi_0(t)$, $\varphi_1(t)$, $\psi_0(t)$ *and* $\psi_1(t)$ *satisfy the hypotheses of Theorem* 6.1, *and assume that the following conditions hold:*

a) $M_{\varphi_i}(s_0) < s_0$ $(i = 0, 1)$ *for some* $s_0 > 1$.

b) $M_{\varphi_1}(s_1^{-1}) < 1$ *for some* $s_1 > 1$.

*If for any operator A satisfying the hypotheses of Theorem* 6.1 *the function* $\kappa(t)(Ax)^{**}(\delta(t))$ *belongs to E, then* $x \in E$.

PROOF. According to Theorem 3.4 there is an operator $T \in \Sigma$ such that $Tx = x^*$. By (6.23) and condition a) the operator $S_{01}$ acts boundedly from $\Lambda_{\tilde{\varphi}_i}$ into $M_{\psi_i}$, $i = 0, 1$. Therefore, so does the operator $S_{01}T$. By the hypothesis of the lemma the function $\kappa(t)(S_{01}Tx)^{**}(\delta(t))$ then belongs to $E$, and, a fortiori, $\kappa(t)(S_{01}Tx)^*(\delta(t)) \in E$. As we mentioned above, the function $S_{01}Tx = S_{01}x^*$ is decreasing, and consequently $(S_{01}Tx)^* = S_{01}x^*$. According to (6.22) we have

$$\begin{aligned}\kappa(t)(S_{01}x^*)(\delta(t)) &= \kappa(t)\int_0^\infty x^*(s)\, d\min\{\tilde{\varphi}_i(s)[\psi_i(\delta(t))]^{-1}\}\\ &\geqslant \kappa(t)\int_0^\infty x^*(s)\, d\min\{\varphi_i(s)[\psi_i(\delta(t))]^{-1}\}.\end{aligned}$$

From (6.8) it then follows that

$$\begin{aligned}\kappa(t)(S_{01}x^*)(\delta(t)) &\geqslant \int_0^\infty x^*(s)\, d\min\{\varphi_i(s)[\varphi_i(t)]^{-1}\}\\ &\geqslant \int_{t/s_1}^t x^*(s)\, d\min\{\varphi_i(s)[\varphi_i(t)]^{-1}\}\\ &\geqslant \left[1 - \min\{\varphi_i(t/s_1)[\varphi_i(t)]^{-1}\}\right]x^*(t)\\ &\geqslant \left[1 - \min\{M_{\varphi_i}(s_1^{-1})\}\right]x^*(t).\end{aligned}$$

By condition 1) of Theorem 6.1 we have $M_{\varphi_0}(s_1^{-1}) \geqslant M_{\varphi_1}(s_1^{-1})$ and consequently

$$x^*(t) \leqslant \left[1 - M_{\varphi_1}(s_1^{-1})\right]^{-1}\kappa(t)(S_{01}x^*)(\delta(t)) \in E.$$

LEMMA 6.4. *Let the functions* $\varphi_0(t)$, $\varphi_1(t)$, $\psi_0(t)$ *and* $\psi_1(t)$ *satisfy the hypotheses of Theorem* 6.1 *and condition* a) *of Lemma* 6.3. *Assume that the function* $\delta(t)$ *determined from* (6.7) *is continuous and strictly increasing, and write* $\tilde{\varphi}_i(\delta^{-1}(\tau)) = \nu_i(\tau)$. *If* $M_{\nu_i}(\tau^{-1}) < 1$ *for* $\tau > 1$, *then for every function* $y \in E_{\kappa,\delta}$ *there are an operator A acting boundedly from the couple* $\{\Lambda_{\tilde{\varphi}_0}, \Lambda_{\tilde{\varphi}_1}\}$ *into the couple* $\{M_{\psi_{0*}}, M_{\psi_{1*}}\}$, *and a function* $x \in E$ *such that* $Ax = y$.

PROOF. We write $x(t) = \kappa(t)y^{**}(\delta(t))$ and apply the operator $S_{01}$ to this function. Then we obtain

$$S_{01}x(t) = \int_0^\infty \kappa(s)y^{**}(\delta(s))\, d\min\{\tilde{\varphi}_i(s)[\psi_i(t)]^{-1}\}.$$

Using (6.8) for $j = 0, 1$, we obtain for $\tau_0 > 1$ that

$$\begin{aligned} S_{01}x(t) &\geqslant \int_{\delta^{-1}(t)}^{\delta^{-1}(\tau_0 t)} \psi_j(\delta(s))[\varphi_j(s)]^{-1}y^{**}(\delta(s))\, d\min\{\tilde{\varphi}_i(s)[\psi_i(t)]^{-1}\} \\ &\geqslant \psi_j(t)[\tilde{\varphi}_j(\delta^{-1}(\tau_0 t))]^{-1}y^{**}(\tau_0 t)\big[\min\{\tilde{\varphi}_i(\delta^{-1}(\tau_0 t))[\psi_i(t)]^{-1}\} \\ &\qquad - \min\{\tilde{\varphi}_i(\delta^{-1}(t))[\psi_i(t)]^{-1}\}\big]. \end{aligned}$$

By the continuity of the functions $\tilde{\varphi}_i(\delta^{-1}(t))$ we may choose an index $j$ and a number $\tau_0 > 1$ such that

$$\min\{\tilde{\varphi}_i(\delta^{-1}(t))[\psi_i(t)]^{-1}\} \quad \text{and} \quad \min\{\tilde{\varphi}_i(\delta^{-1}(\tau_0 t))[\psi_i(t)]^{-1}\}$$

are attained for this index $j$. Then

$$S_{01}x(t) \geqslant y^{**}(\tau_0 t)\{1 - \tilde{\varphi}_j(\delta^{-1}(t))[\tilde{\varphi}_j(\delta^{-1}(\tau_0 t))]^{-1}\}.$$

Since

$$y^{**}(\tau_0 t) = \frac{1}{\tau_0 t}\int_0^{\tau_0 t} y^*(s)\, ds \geqslant \frac{1}{\tau_0 t}\int_0^t y^*(s)\, ds = \frac{1}{\tau_0}y^{**}(t)$$

and $\nu_j(t)[\nu_j(\tau_0 t)]^{-1} \leqslant M_{\nu_j}(\tau_0^{-1}) < 1$, we have

$$\tau_0[1 - M_{\nu_j}(\tau_0)]^{-1}S_{01}x(t) \geqslant y^{**}(t) \geqslant y^*(t).$$

By Lemma 3.4 there exists an operator $T \in \Sigma$ such that $y = \tau_0[1 - M_{\nu_j}(\tau_0)]^{-1}TS_{01}x$. By condition a) of Lemma 6.3 the operator $\tau_0[1 - M_{\nu_j}(\tau_0)]^{-1}TS_{01}$ acts boundedly from the couple $\{\Lambda_{\tilde{\varphi}_0}, \Lambda_{\tilde{\varphi}_1}\}$ into the couple $\{\mathrm{M}_{\psi_{0*}}, \mathrm{M}_{\psi_{1*}}\}$.

REMARK. If under the hypotheses of Lemma 6.4 the function $\delta(t)$ is strictly decreasing, then the assertion of the lemma remains valid if $M_{\nu_i}(\tau) < 1$ for $\tau > 1$. The proof is analogous.

Lemmas 6.3 and 6.4 lead to the following assertion.

THEOREM 6.3. *Let the functions* $\varphi_0(t)$, $\varphi_1(t)$, $\psi_0(t)$ *and* $\psi_1(t)$ *satisfy the hypotheses of Theorem* 6.1 *and the following conditions*:

1) $M_{\psi_i}(s_0) < s_0$ $(i = 0, 1)$ *for some* $s_0 > 1$.

2) $M_{\varphi_i}(s_1^{-1}) < 1$ *for some* $s_1 > 1$.

3) *The function* $\delta(t)$ *determined from* (6.7) *is continuous and strictly increasing* (*decreasing*), *and* $M_{\nu_i}(\tau^{-1}) < 1$ $(M_{\nu_i}(\tau) < 1)$ *for* $i = 0, 1$ *and* $\tau > 1$, *where* $\nu_i(t) = \varphi_i(\delta^{-1}(t))$.

*Then* $\{\Lambda_{\tilde{\varphi}_0}, \Lambda_{\tilde{\varphi}_1}, E\}$ *is an optimal interpolation triple with respect to the triple* $\{M_{\psi_{0*}}, M_{\psi_{1*}}, E_{\kappa,\delta}\}$ *for any space E satisfying conditions* (6.3)–(6.6).

6. ***Operators of weakened types*** $\{L_\infty, L_\infty\}$ ***and*** $\{L_1, L_1\}$. The fundamental function of the space $L_1$ is equal to $t$, and that of $L_\infty$ is equal to sgn $t$. Therefore every operator of weakened types $(L_\infty, L_\infty)$ and $\{L_1, L_1\}$ is a linear operator, defined for all simple functions, that satisfies the conditions

$$(Ax)^{**}(t) \leqslant C_0 x^*(0) = C_0\|x\|_{L_\infty},$$
$$t(Ax)^{**}(t) \leqslant C_1\int_0^\infty x^*(t)\,dt = C_1\|x\|_{L_1}.$$

From the first condition it follows that $(Ax)^*(t) \leqslant C_0\|x\|_{L_\infty}$ (weak type), which in turn implies that $\|Ax\|_{L_\infty} = (Ax)^*(+0) \leqslant C_0\|x\|_{L_\infty}$, i.e. the operator $A$ is of strong type $\{L_\infty, L_\infty\}$. The second condition gives the inequality $\int_0^t (Ax)^*(t)\,dt \leqslant C_1\|x\|_{L_1}$, which implies that $\|Ax\|_{L_1} \leqslant C_1\|x\|_{L_1}$. Thus, the operators of weakened type $\{L_\infty, L_\infty\}$, $\{L_1, L_1\}$ coincide with the operators of strong type $\{L_\infty, L_\infty\}$, $\{L_1, L_1\}$.

Corollary 2 and the remark to Theorem 6.1′ lead to the following assertion.

THEOREM 6.4. *If a linear operator A acts boundedly in the spaces* $L_\infty$ *and* $L_1$, *then it acts boundedly from any symmetric space E having the property that*

$$\|\sigma_t\|_E = o(t) \quad \text{as } t \to \infty, \tag{6.24}$$

*into any space* $E_{1,t}$, *i.e.*

$$\|(Ax)^{**}\|_E \leqslant C\|x\|_E. \tag{6.25}$$

The conclusion of Theorem 6.4 is stronger than that of the interpolation Theorem 4.3, and for this reason condition (6.24) is imposed on the space $E$. It turns out that condition (6.24) is necessary for the validity of this conclusion.

THEOREM 6.5. *Condition* (6.24) *is necessary for the boundedness of the identity operator as an operator from a symmetric space E into the space* $E_{1,t}$.

PROOF. If the identity operator acts and is bounded from $E$ into $E_1$, then for any function $x \in E$

$$\|x\|_E \leqslant \left\|\frac{1}{s}\int_0^s x^*(t)\,dt\right\|_E \leqslant C\|x\|_E. \tag{6.26}$$

We use (6.1′). Then

$$\|\sigma_\tau x\|_E \leqslant \left\|\frac{1}{s}\int_0^s \sigma_\tau x^*(t)\,dt\right\|_E \leqslant \frac{\tau}{1+\ln\tau}\left\|\frac{1}{s}\int_0^s x^{**}(t)\,dt\right\|_E$$
$$\leqslant \frac{C\tau}{1+\ln\tau}\|x^{**}\|_E \leqslant \frac{C^2\tau}{1+\ln\tau}\|x\|_E. \tag{6.27}$$

Hence $\|\sigma_\tau\|_E = o(\tau)$ as $\tau \to \infty$.

REMARK. If condition (6.24) is satisfied for the space $E$, then it is satisfied in every symmetric subspace of it. Therefore, every subspace of a symmetric space $E$ with condition (6.24) is an interpolation space between $L_1$ and $L_\infty$. In §5.7 we proved that in the spaces $\mathrm{M}_\psi$ with the condition $\lim_{t\to 0}\psi(2t)/\psi(t) = 1$ there exist subspaces that are not interpolation spaces. Now if $\inf \psi(2t)/\psi(t) > 1$, then by (4.21) we have $\|\sigma_2\|_{\mathrm{M}_\psi} < 2$, which implies (6.24), and for such spaces $\mathrm{M}_\psi$ all subspaces are interpolation spaces.

**7. *Hardy-Littlewood and Hilbert operators.*** The operator

$$Hx(t) = \frac{1}{t}\int_0^t x(s)\,ds \tag{6.28}$$

is called the *Hardy-Littlewood operator*. It coincides with the operator $S_{01}$ constructed from the fundamental functions of the couples of spaces $(L_\infty, L_\infty)$ and $(L_1, L_1)$, and is of strong type $\{L_\infty, L_\infty\}$ and weak type $\{L_1, L_1\}$. The space $L_1$ does not satisfy the hypothesis of Lemma 6.2, and, as we saw on p. 132, the operator $H$ is not of weakened type $\{L_1, L_1\}$. However, if a symmetric space satisfies (6.24), then $H$ acts in it. Indeed, by Lemma 4.7 we have

$$\|Hx\|_E = \left\|\int_0^1 x(ts)\,ds\right\|_E \leqslant \int_0^1 \|\sigma_{1/s}\|_E\,ds\|x\|_E, \tag{6.29}$$

and the integral is finite under condition (6.24). This fact, together with Theorem 6.5, leads us to the following important assertion.

THEOREM 6.6. *For the Hardy-Littlewood operator* (6.28) *to act boundedly in a symmetric space E it is necessary and sufficient that condition* (6.24) *be satisfied.*

Inequality (6.29) implies that

$$\|x\|_E \leqslant \|x^{**}\|_E = \|Hx^*\|_E \leqslant \int_0^1 \|\sigma_{1/s}\|\,ds\|x\|_E. \tag{6.30}$$

Now we introduce the adjoint $H_1$ of $H$:

$$H_1y(t) = \int_t^\infty \frac{y(\tau)}{\tau}\,d\tau.$$

For nonnegative functions $x(t)$ and $y(t)$ we have by Fubini's theorem that

$$\int_0^\infty Hx(\tau)y(\tau)\,d\tau = \int_0^\infty \frac{1}{\tau}\int_0^\tau x(s)\,ds\,y(\tau)\,d\tau$$
$$= \int_0^\infty x(s)\int_s^\infty \frac{y(\tau)}{\tau}\,d\tau\,ds = \int_0^\infty x(s)H_1y(s)\,ds. \tag{6.31}$$

We now assume that the operator $H$ acts from a space $E$ into $E^{11}$. Then for $x \in E$ and $y \in E^1$ we obtain

$$\int_0^\infty x(s) H_1 y(s)\, ds \leqslant \|Hx\|_{E^{11}} \|y\|_{E^1} \leqslant \|H\|_{E \to E^{11}} \|x\|_E \|y\|_{E^1}.$$

Hence $\|H_1 y\|_{E^1} \leqslant \|H\|_{E\to E^{11}} \|y\|_{E^1}$. By the positivity of $H$ and $H_1$ it is sufficient to verify their boundedness for nonnegative functions. Therefore, it follows from the last inequality that $H_1$ acts in $E^1$ and $\|H_1\|_{E^1 \to E^1} \leqslant \|H\|_{E\to E^{11}}$.

Conversely, if $H_1$ acts in $E^1$, then (6.31) implies that

$$\int_0^\infty Hx(t) y(t)\, dt \leqslant \|H_1\|_{E^1 \to E^1} \|x\|_E \|y\|_{E^1},$$

from which we obtain that $\|Hx\|_{E^{11}} \leqslant \|H_1\|_{E^1\to E^1} \|x\|_E$, i.e., $H$ acts from $E$ into $E^{11}$, and $\|H\|_{E\to E^{11}} \leqslant \|H_1\|_{E^1 \to E^1}$.

Thus, $H$ acts from $E$ into $E^{11}$ if and only if $H_1$ acts in $E^1$, and then $\|H\|_{E\to E^{11}} = \|H_1\|_{E^1\to E^1}$. (We note that in the proof of this statement we have used no special properties of $H$ and $H_1$; the assertion is therefore true for any two positive integral operators which are adjoints of each other.)

Interchanging the roles of $H$ and $H_1$ and taking account of the fact that $E^1 = E^{111}$, we conclude that $H_1$ acts in $E^1$ if and only if $H$ acts in $E^{11}$. Thus, $H$ acts from $E$ into $E^{11}$ if and only if it acts in $E^{11}$.

THEOREM 6.7. *If the Hardy-Littlewood operator acts boundedly from a space $E$ into $E^{11}$, then*

$$\|\sigma_t\|_{E\to E^{11}} = o(t) \quad \textit{as } t \to \infty. \tag{6.32}$$

PROOF. As was shown above, the operator $H$ acts in $E^{11}$, and therefore, from (6.1′) we obtain

$$\|\sigma_\tau x\|_{E^{11}} \leqslant \frac{C^2\tau}{1 + \ln \tau} \|x\|_E$$

in the same way as (6.27).

COROLLARY. *If $E$ is isometrically imbedded in $E^{11}$, then the condition that the operator $H$ act boundedly from $E$ into $E^{11}$ implies condition* (6.24).

From arguments based on (6.31) it follows that $H_1$ acts from $E$ into $E^{11}$ if and only if $H$ acts in $E^1$, and for this it is necessary and sufficient that

$$\|\sigma_t\|_{E^1} = o(t) \quad \text{as } t \to \infty. \tag{6.33}$$

By (4.35) this condition is equivalent to

$$\|\sigma_\tau\|_{E\to E^{11}} = o(1) \quad \text{as } \tau \to 0. \tag{6.34}$$

We have proved the second part of the following assertion.

THEOREM 6.8. *For the adjoint $H_1$ to act boundedly in a symmetric space $E$ it is necessary and sufficient that*

$$\|\sigma_\tau\|_E = o(1) \quad \text{as } \tau \to 0. \tag{6.35}$$

*For it to act from $E$ into $E^{11}$ it is necessary and sufficient that condition* (6.34) *be satisfied, or, if $E$ is isometrically imbedded in $E^{11}$, that* (6.35) *be satisfied.*

PROOF OF THE FIRST PART. By Lemma 4.7 we have

$$\|H_1 y\|_E = \left\| \int_\tau^\infty y(s)\frac{ds}{s} \right\|_E = \left\| \int_1^\infty y(t\tau)\frac{d\tau}{\tau} \right\|_E \leqslant \int_1^\infty \|\sigma_{1/\tau}\|_E \frac{d\tau}{\tau} \|y\|_E.$$

If (6.35) is satisfied, the integral on the right is finite.

We show the necessity of (6.35). For $\tau < 1$ we have

$$\left\| \int_t^\infty x^*(s)\frac{ds}{s} \right\|_E \geqslant \left\| \int_t^{t/\tau} x^*(s)\frac{ds}{s} \right\|_E \geqslant \left\| x^*\left(\frac{t}{\tau}\right) \int_t^{t/\tau} \frac{ds}{s} \right\|_E = \ln \frac{1}{\tau} \|\sigma_\tau x^*\|_E.$$

Hence

$$\begin{aligned} \|\sigma_\tau x\|_E = \|\sigma_\tau x^*\|_E &\leqslant (\ln(1/\tau))^{-1} \|H_1 x^*\|_E \\ &\leqslant (\ln(1/\tau))^{-1} \|H_1\|_{E\to E} \|x\|_E, \end{aligned}$$

and so (6.35) is satisfied. The theorem is proved.

The *Hilbert operator*

$$\Gamma x(t) = \int_0^\infty \frac{x(s)}{t+s}\, ds \tag{6.36}$$

is closely connected with the Hardy-Littlewood operator and its adjoint. The Hilbert operator has positive kernel. Therefore it is sufficient to study it on nonnegative functions. If $x(t) \geqslant 0$, then

$$\begin{aligned} \frac{1}{t}\int_0^t x(s)\, ds + \int_t^\infty \frac{x(s)\, ds}{s} &\leqslant 2\int_0^t \frac{x(s)\, ds}{t+s} + 2\int_t^\infty \frac{x(s)\, ds}{t+s} \\ &= 2\int_0^\infty \frac{x(s)\, ds}{t+s}. \end{aligned}$$

On the other hand,

$$\int_0^\infty \frac{x(s)\, ds}{t+s} = \int_0^t \frac{x(s)\, ds}{t+s} + \int_t^\infty \frac{x(s)\, ds}{t+s} \leqslant \frac{1}{t}\int_0^t x(s)\, ds + \int_t^\infty \frac{x(s)\, ds}{s}.$$

Thus, for $x(t) \geqslant 0$ we have

$$\tfrac{1}{2}(H + H_1)x \leqslant \Gamma x \leqslant (H + H_1)x.$$

This and Theorems 6.6–6.8 imply

THEOREM 6.9. *For the Hilbert operator* (6.36) *to act boundedly in a symmetric space E it is necessary and sufficient that*

$$\|\sigma_t\|_E = o(1) \quad \text{as } t \to 0 \qquad \text{and} \qquad \|\sigma_t\|_E = o(t) \quad \text{as } t \to \infty. \tag{6.37}$$

The notion of *majorant function* is connected with the Hardy-Littlewood operator. We set

$$\theta_1(t; x) = \max_{0 < \tau < t} \frac{1}{t - \tau} \int_\tau^t x(s)\, ds. \tag{6.38}$$

We establish the following important inequality:

$$\theta_1^*(t; x) \leqslant x^{**}(t) \qquad (x \in L_1 + L_\infty).$$

Since $|\theta_1(t;x)| \leqslant \theta_1(t;|x|)$, it is sufficient to prove the inequality for nonnegative functions.

Let $\theta_1^*(t_0; x) = p > \theta_1^*(\infty; x)$. We choose $\varepsilon < p - \theta_1^*(\infty; x)$ and consider the set $e = \{t\colon \theta_1(t; x) > p - \varepsilon\}$. It has finite measure. According to (6.38) the set $e$ consists of those points $t$ for which there exist points $\tau \in (0, t)$ such that

$$\int_0^t x(s)\, ds - (p - \varepsilon)t > \int_0^\tau x(s)\, ds - (p - \varepsilon)\tau.$$

The function

$$g(t) = \int_0^t x(s)\, ds - (p - \varepsilon)t$$

is continuous. Therefore the set $e$ is open, and consequently it consists of a system of intervals. Let $(\alpha_k, \beta_k)$ be an interval in $e$ whose endpoints do not belong to $e$. We show that on $[\alpha_k, \beta_k]$ the function $g(t)$ attains its minimum at $\alpha_k$. Indeed, if the minimum were attained at $t' \in (\alpha_k, \beta_k)$ and the inequality $g(\alpha_k) > g(t')$ held, then there would not be a point $\tau$ on $(\alpha_k, t')$ where $g(\tau) < g(t')$. There would not be such a point on $(0, \alpha_k]$, either, since the condition $\alpha_k \notin e$ implies that $g(\tau) \geqslant g(\alpha_k) > g(t')$ for $\tau \in (0, \alpha_k]$. Thus, we arrive at a contradiction to the fact that $t' \in e$. In particular, from what has been said it follows that $g(\beta_k) \geqslant g(\alpha_k)$ or

$$(p - \varepsilon)(\beta_k - \alpha_k) \leqslant \int_{\alpha_k}^{\beta_k} x(s)\, ds.$$

Summing these inequalities over all intervals in $e$, we obtain

$$(p - \varepsilon)\operatorname{mes} e \leqslant \int_e x(s)\, ds \leqslant \int_0^{\operatorname{mes} e} x^*(s)\, ds.$$

By the definition of $e$ we have mes $e \geqslant t_0$. Therefore

$$p - \varepsilon \leqslant \frac{1}{\text{mes } e}\int_0^{\text{mes } e} x^*(s)\, ds \leqslant \frac{1}{t_0}\int_0^{t_0} x^*(s)\, ds = x^{**}(t_0).$$

Since $\varepsilon$ is arbitrary, we have

$$\theta_1^*(t_0; x) \leqslant x^{**}(t_0). \tag{6.39}$$

Now let $\theta_1^*(t_0;\ x) = p = \theta_1^*(\infty;\ x) > 0$. The set $e$ defined as above for $\varepsilon \in (0, p)$ has infinite measure and can contain an interval of the form $(\alpha, \infty)$. On this interval $g(t)$ also attains its minimum at $\alpha$; therefore for $t > \alpha$ we have

$$\int_0^t x(s)\, ds - (p - \varepsilon)t \geqslant \int_0^\alpha x(s)\, ds - (p - \varepsilon)\alpha.$$

Dividing both sides by $t$ and letting $t$ tend to $\infty$, we obtain

$$p - \varepsilon \leqslant \lim_{t\to\infty} \frac{1}{t}\int_0^t x(s)\, ds \leqslant \lim_{t\to\infty} \frac{1}{t}\int_0^t x^*(s)\, ds = x^{**}(\infty) \leqslant x^{**}(t_0).$$

Thus, in this case we also obtain (6.39).

We may introduce the function

$$\theta_2(t; x) = \sup_{\tau > t} \frac{1}{\tau - t}\int_t^\tau x(s)\, ds,$$

and obtain the inequality $\theta_2^*(t; x) \leqslant x^{**}(t)$ by similar arguments.

Finally, for the majorant function

$$\theta(t; x) = \sup_{|t-\tau|>0} \frac{1}{t - \tau}\int_\tau^t x(s)\, ds$$

we have $\theta(t; x) = \max\{\theta_1(t; x), \theta_2(t; x)\}$. Therefore

$$\theta^*(t; x) \leqslant \max\{\theta_1^*(t; x), \theta_2^*(t; x)\} \leqslant x^{**}(t).$$

If a symmetric space $E$ has the property that the Hardy-Littlewood operator acts boundedly in it, then for $x \in E$ we have

$$\|\theta(t; x)\|_E = \|\theta^*(t; x)\|_E \leqslant \|x^{**}\|_E \leqslant C\|x\|_E.$$

We note that for a decreasing nonnegative function $x$ we have

$$\theta(t; x) = \frac{1}{t}\int_0^t x(s)\, ds = Hx.$$

Therefore the inequality

$$\|\theta(t; x)\|_E \leqslant C\|x\|_E \tag{6.40}$$

implies that the Hardy-Littlewood operator acts in $E$ and has norm not exceeding $C$.

Using Theorem 6.6, we arrive at the following assertion.

THEOREM 6.10. *Condition* (6.24) *is necessary and sufficient for the inequality* (6.40) *to hold in a symmetric space* $E$.

We still have to study the Hardy-Littlewood operator in symmetric spaces with weight. Let $E$ be a symmetric space and let $\rho(t)$ be a positive function on $(0, \infty)$. Consider the ideal lattice $E_{\rho(t)}$, i.e. the space of measurable functions on $(0, \infty)$ for which

$$\|x\|_{E_{\rho(t)}} = \|x\rho\|_E < \infty.$$

THEOREM 6.11. *If* $\rho(t)$ *is a submultiplicative function and*

$$\int_0^1 \|\sigma_{1/s}\|_E \rho(s^{-1})\, ds < \infty, \tag{6.41}$$

*then the Hardy operator acts boundedly from* $E_{\rho(t)}$ *into* $E^{11}_{\rho(t)}$.

PROOF. We have

$$\left|\frac{\rho(t)}{t}\int_0^t x(\tau)\, d\tau\right| \leqslant \int_0^1 |x(ts)|\rho(t)\, ds$$
$$\leqslant \int_0^1 (\sigma_{1/s}|x|\rho)(t)\ \rho(s^{-})\, ds. \tag{6.42}$$

By the generalized Minkowski inequality (0.5) we have

$$\|\rho Hx\|_{E^{11}} \leqslant \int_0^1 \|(\sigma_{1/s}|x|\rho)(t)\|_E \rho(s^{-1})\, ds$$
$$\leqslant \int_0^1 \|\sigma_{1/s}\|_E \rho(s^{-1})\, ds \|x\rho\|_E. \tag{6.43}$$

The theorem is proved.

REMARK 1. If the weight $\rho(t)$ is decreasing, then the Hardy operator acts boundedly in the space $E_{\rho(t)}$ itself.

Indeed, from (6.42) it follows that

$$|\rho(t)Hx(t)| \leqslant \int_0^1 (\sigma_{1/s}x^*\rho)(t)\rho(s^{-1})\, ds,$$

and then by Lemma 6.7 we have

$$\|\rho Hx\|_E \leqslant \int_0^1 \|\sigma_{1/s}x^*\rho\|_E \rho(s^{-1})\, ds \leqslant \int_0^1 \|\sigma_{1/s}\|_E \rho(s^{-1})\, ds \|\rho x\|_E.$$

REMARK 2. Inequality (6.41) is equivalent to $\delta_E + \beta_\rho < 1$.

REMARK 3. If $E = L_p$ and $\rho(t) = t^\gamma$, then (6.41) assumes the form $\gamma p' < 1$, and (6.42) becomes the famous Hardy inequality

$$\|t^\gamma Hx\|_{L_p} \leqslant \frac{p'}{1 - \gamma p'}\|t^\gamma x\|_{L_p},$$

which is unimprovable (see [17]).

COROLLARY 1. *If $\gamma p' < 1$, then*

$$\int_0^\infty t^{\gamma p}|x^*(t)|^p\,dt \leqslant \int_0^\infty t^{\gamma p}|x^{**}(t)|^p\,dt \leqslant \left(\frac{p'}{1-\gamma p'}\right)^p \int_0^\infty t^{\gamma p}|x^*(t)|^p\,dt.$$

COROLLARY 2. *The Hardy-Littlewood operator of the form*

$$H_l x(t) = \frac{1}{t^{l+1}} \int_0^t s^l x(s)\,ds$$

*is bounded in the spaces $L_p$ with weight $t^\gamma$, provided that $(\gamma - l)p' < 1$.*

Indeed,

$$\|t^\gamma H_l x\|_{L_p} = \|t^{\gamma - l} H(s^l x)\|_{L_p} \leqslant \frac{p'}{1-(\gamma - l)p'} \|t^\gamma x\|_{L_p}.$$

**8. *Operators of weakened and weak types $\{L_{p_i}, L_{q_i}\}$.*** We consider the case where the spaces $E_i$ coincide with the spaces $L_{p_i}$, and the $F_i$ with the spaces $L_{q_i}$ $(i = 0, 1)$. We compute $\varphi_{E_i} = t^{1/p_i}$ and $\varphi_{F_i}(t) = t^{1/q_i}$. From (6.15) and (6.16) we find that

$$\delta(t) = t^{(1/p_1 - 1/p_0)/(1/q_1 - 1/q_0)} = t^\mu \text{ and } \kappa(t) = t^{(1/p_1 q_0 - 1/p_0 q_1)/(1/q_1 - 1/q_0)} = t^\nu.$$

To satisfy condition 1) of Theorem 6.1′ we have to assume that $p_1 < p_0$. Condition 2) is satisfied if $q_0 \neq q_1$. Subject to the satisfaction of this inequality, at least one of the numbers $q_i$ is greater than 1. Therefore condition (6.21) of Theorem 6.2 is satisfied. Because of (6.6), condition (6.14) imposed on the intermediate space $E$ can be written in the form

$$\|\sigma_t\|_E = o(\tau^{1/p_1}) \quad \text{as } t \to \infty \qquad \text{and} \qquad \|\sigma_t\|_E = o(t^{1/p_0}) \quad \text{as } t \to 0. \tag{6.44}$$

Theorem 6.2 immediately implies the following theorem.

THEOREM 6.12. *Let $1 \leqslant p_1 < p_0 \leqslant \infty$, $1 \leqslant q_0, q_1 \leqslant \infty$, $q_0 \neq q_1$ and let $A$ be an operator of weak types $\{L_{p_i}, L_{q_i}\}$, i.e. let the inequalities*

$$t^{1/q_i}(Ax)^*(t) \leqslant c \int_0^\infty \tau^{1/p_i - 1} x^*(\tau)\,d\tau \qquad (i = 0, 1) \tag{6.45}$$

*hold for the characteristic functions of sets of finite measure if $p_0 < \infty$, or for all characteristic functions if $p_0 = \infty$. The operator $A$ admits a bounded linear extension acting from every symmetric space satisfying conditions* (6.44) *into the space $E_{t^\nu, t^\mu}$ with norm*

$$\|x\|_{E_{t^\nu, t^\mu}} = \|t^\nu x^{**}(t^\mu)\|_E.$$

We remark that conditions (6.45) will automatically be satisfied if

$$t^{1/q_i}(Ax)^*(t) \leqslant C\|x\|_{L_{p_i}}$$

(see the imbedding Theorem 5.5).

From Theorem 6.3 we obtain

THEOREM 6.13. *If $1 \leqslant p_1 < p_0 < \infty$, $1 < q_0, q_1 \leqslant \infty$, $q_0 \neq q_1$, and condition (6.44) holds, then $\{\Lambda_{t^{1/p_0}}, \Lambda_{t^{1/p_1}}, E\}$ is an optimal interpolation triple of spaces with respect to the triple $\{M_{t^{1-1/q_0}}, M_{t^{1-1/q_1}}, E_{t^\nu,t^\mu}\}$.*

Now we consider the case where the space $E$ is an $L_p$ with $1/p = (1-\lambda)/p_0 + \lambda/p_1$ $(0 < \lambda < 1)$. Since $\|\sigma_t\|_{L_p} = t^{1/p}$, conditions (6.44) are satisfied. In this case the space $E_{t^\nu,t^\mu}$ consists of all functions in $L_1 + L_\infty$ for which

$$\int_0^\infty (x^{**}(t^\mu)t^\nu)^p\, dt < \infty.$$

We carry out the change of variables $t^\mu = \tau$. Then we obtain

$$\int_0^\infty (x^{**}(\tau))^p \tau^{p/q-1}\, d\tau < \infty,$$

where $1/q = (1-\lambda)/q_0 + \lambda/q_1$. We have arrived at spaces which are usually denoted by $L_{r,p}$. In these spaces the norm is given by

$$\|x\|_{L_{r,p}} = \left\{r^{-2}(r-1)\int_0^\infty [x^{**}(t)]^p t^{p/r-1}\, dt\right\}^{1/p}.$$

Thus, under the hypotheses of Theorem 6.12 the operator $A$ acts from $L_p$ into $L_{q,p}$ $(1/p = (1-\lambda)/p_0 + \lambda/p_1,\ 1/q = (1-\lambda)/q_0 + \lambda/q_1,\ 0 < \lambda < 1)$.

LEMMA 6.5. *If $1 < r < \infty$ and $p < p_1$, then $L_{r,p}$ is imbedded in $L_{r,p_1}$.*

PROOF. Since $x^{**}(t)$ is decreasing, we have

$$(r-1)r^{-2}\int_0^\infty [x^{**}(s)]^p s^{p/r-1}\, ds \geqslant (r-1)r^{-2}\int_0^t [x^{**}(s)]^p s^{p/r-1}\, ds$$
$$\geqslant (r-1)(rp)^{-1}[x^{**}(t)]^p t^{p/r}.$$

Raising to the power $(p_1 - p)/p > 0$, we obtain

$$\|x\|_{L_{r,p}}^{p_1-p} \geqslant \left(\frac{r-1}{rp}\right)^{(p_1-p)/p}[x^{**}(t)]^{p_1-p}t^{(p_1-p)/r},$$

and hence

$$[x^{**}(t)]^{p_1}t^{p_1/r-1} \leqslant \left(\frac{rp}{r-1}\right)^{(p_1-p)/p}[x^{**}(t)]^p t^{p/r-1}\|x\|_{L_{r,p}}^{p_1-p}.$$

Integrating with respect to $t$ from 0 to $\infty$, we finally obtain

$$\|x\|_{L_{r,p_1}} \leqslant \left(\frac{rp}{r-1}\right)^{1/p-1/p_1} \|x\|_{L_{r,p}}.$$

The lemma is proved.

We note that for $p = r > 1$ the space $L_{r,p}$ coincides with $L_p$. Indeed,

$$\|x\|_{L_p} = \left\{\int_0^\infty [x^*(t)]^p\, dt\right\}^{1/p}$$

$$\leqslant \left\{\int_0^\infty [x^{**}(t)]^p\, dt\right\}^{1/p} = \left(\frac{p^2}{p-1}\right)^{1/p} \|x\|_{L_{p,p}}.$$

On the other hand, since $\|\sigma_t\|_{L_p} = t^{1/p}$, condition (6.24) is satisfied, and the operator ** is bounded in $L_p$, i.e. (see (6.30))

$$\|x\|_{L_{p,p}} = \left(\frac{p-1}{p^2}\right)^{1/p} \|x^{**}\|_{L_p}$$

$$\leqslant \left(\frac{p-1}{p^2}\right)^{1/p} \int_0^1 \tau^{-1/p}\, d\tau \|x\|_{L_p} = \frac{p^{1-r/p}}{(p-1)^{1-1/p}} \|x\|_{L_p}.$$

Thus, the norms in $L_{p,p}$ and $L_p$ are equivalent. Lemma 6.5 implies that for $p \leqslant q$ $(1 < q < \infty)$ the space $L_{q,p}$ is imbedded in $L_{q,q} = L_q$.

Returning now to our interpolation Theorem 6.12, we see that for $p < q$ the operator $A$ acts from $L_p$ into $L_{q,p}$, which is smaller than $L_q$.

THEOREM 6.14. *$\{L_{p_0}, L_{p_1}, L_p\}$ is an optimal interpolation triple of spaces with respect to the triple $\{L_{q_0}, L_{q_1}, L_{q,p}\}$, if $1 < p_i \leqslant q_i < \infty$.*

PROOF. Every operator acting boundedly from $L_{p_i}$ into $L_{q_i}$, $i = 0, 1$, is of weakened types $\{L_{p_i}, L_{q_i}\}$. Therefore, as was shown above, such an operator acts from $L_p$ into $L_{q,p}$, and consequently the first triple of spaces is an interpolation triple with respect to the second one.

Since $1 < p_i, q_i < \infty$, there exist two numbers $\lambda_0 < 0$ and $\lambda_1 > 1$ such that

$$0 < \frac{1}{p_j'} = \frac{1-\lambda_j}{p_0} + \frac{\lambda_j}{p_1} < 1 \quad \text{and} \quad 0 < \frac{1}{q_j'} = \frac{1-\lambda_j}{q_0} + \frac{\lambda_j}{q_1} < 1$$
$$(j = 0, 1).$$

Every operator of weak types $\{L_{p_j'}, L_{q_j'}\}$ acts from the $L_{p_i}$ into the $L_{q_i,p_i}$ by Theorem 6.12, and into the $L_{q_i}$ since $p_i \leqslant q_i$. The optimality of the triple $\{L_0, L_1, L_p\}$ with respect to $\{L_{q_0}, L_{q_1}, L_{q,p}\}$ then follows from Theorem 6.13. The theorem is proved.

It is natural to call the operators acting from $L_{p_i}$ into $L_{q_i}$, with $p_i \leqslant q_i$, "improving". Theorem 6.14 refines the classical Riesz-Thorin theorem (see Chapter I, §4.2) for "improving" operators.

We now consider the case where the space $E$ is an $L_{r,p}$ with $p_1 < r < p_0$. A calculation shows that $\|\sigma_t\|_{L_{r,p}} = t^{1/r}$. Therefore (6.44) is satisfied.

In this case the space $E_{t^\nu, t^\mu}$ consists of all functions in $L_1 + L_\infty$ for which

$$\int_0^\infty \left[(x^{**}(t^\mu)t^\nu)^{**}\right]^p t^{p/r-1}\,dt < \infty.$$

Since $(1/r - 1/p)p' < 1$, we can apply Corollary 1 of Theorem 6.11 with $\gamma = 1/r - 1/p$. It implies that the quantity on the left is equivalent to

$$\int_0^\infty \left[(x^{**}(t^\mu)t^\nu)^{*}\right]^p t^{p/r-1}\,dt.$$

If we use properties 21° and 22° of rearrangements, we obtain that the last quantity is equivalent to

$$\int_0^\infty \left[x^{**}(t^\mu)\right]^p t^{p\nu+p/r-1}\,dt = \frac{1}{\mu}\int_0^\infty \left[x^{**}(t)\right]^p t^{p((\nu+1/r)/\mu)-1}\,dt.$$

We have arrived at the following assertion.

THEOREM 6.15. *If the hypotheses of Theorem 6.12 are satisfied, then the operator $A$ admits an extension to a linear operator acting boundedly from the space $L_{r,p}$ $(p_1 < r < p_0)$ into $L_{s,p}$, where*

$$s = \frac{r\mu}{r\nu + 1}, \qquad \mu = \frac{1/p_1 - 1/p_0}{1/q_1 - 1/q_0}, \qquad \nu = \frac{1/(p_1 q_0) - 1/(p_0 q_1)}{1/q_1 - 1/q_0}.$$

**9. *Convolution operator.*** For functions defined on $(0, \infty)$ the convolution operator is defined by

$$x * y(t) = \int_0^t x(s)y(t-s)\,ds. \tag{6.46}$$

It can be verified immediately that this bilinear operator acts boundedly from $L_1 \times L_\infty$ into $L_\infty$ and from $L_1 \times L_1$ into $L_1$. If we fix the first factor $x \in L_1$, then the linear operator thus obtained acts boundedly in the spaces $L_1$ and $L_\infty$, and so it acts in every interpolation space $G$ between $L_1$ and $L_\infty$ (see Theorem 4.3 and what follows). Thus, the convolution operator acts boundedly from $L_1 \times G$ into $G$.

Let $G^1$ be the space associated with $G$. For $x \in G^1$ and $y \in G$ we have

$$|x * y(t)| = \left|\int_0^t x(s)y(t-s)\,ds\right| = \left|\int_0^\infty x(s)z_t(s)\,ds\right| \leqslant \|x\|_{G^1}\|z_t\|_G,$$

where

$$z_t(s) = \begin{cases} y(t-s) & \text{for } s \leqslant t, \\ 0 & \text{for } s > t. \end{cases}$$

Clearly $z_t \in G$ and $\|z_t\|_G \leqslant \|y\|_G$. Therefore

$$\|x*y\|_{L_\infty} \leqslant \|x\|_{G^1}\|y\|_G.$$

This means that the convolution operator acts boundedly from $G^1 \times G$ into $L_\infty$.

We now fix the second factor $y \in G$. Then we obtain a linear operator acting boundedly from $G^1$ into $L_\infty$ and from $L_1$ into $G$. We apply the interpolation Theorem 6.1′ to the operator, putting $E_0 = G^1$, $E_1 = L_1$, $F_0 = L_\infty$ and $F_1 = G$. In this case we have

$$\varphi_{E_0}(s)[\varphi_{E_1}(s)]^{-1} = \varphi_{F_0}(s)[\varphi_{F_1}(s)]^{-1} = [\varphi_G(s)]^{-1},$$

and so, conditions 1) and 2) of this theorem are satisfied. In view of (4.40), condition (6.24) on the intermediate space $E$ takes the form

$$\int_0^1 \|\sigma_{1/\tau}\|_E \, d\tau + \int_1^\infty \|\sigma_{1/\tau}\|_E \, d\left[\tau M_{\varphi_G}\left(\frac{1}{\tau}\right)\right] < \infty.$$

Using the equivalence of (6.3) and (6.6), this condition can be written in the form

$$\|\sigma_t\|_E = o(t^{1-\gamma_{\varphi_G}}) \quad \text{as } t \to 0 \qquad \text{and} \qquad \|\sigma_t\|_E = o(t) \quad \text{as } t \to \infty. \tag{6.47}$$

The functions $\delta$ and $\kappa$ determined from (6.1) and (6.6) are thus $\delta(t) \equiv t$ and $\kappa(t) = t^{-1}\varphi_G(t)$. Consequently the norm in the space $F = E_{\kappa,\delta}$ is defined by

$$\|z\|_F = \left\| z^{**}(t)\frac{\varphi_G(t)}{t} \right\|_E. \tag{6.48}$$

From Theorem 6.1′ it now follows that *the convolution operator acts boundedly from every space $E \times G$ (where $G$ is an interpolation space between $L_\infty$ and $L_1$, and the symmetric space $E$ satisfies* (6.47)) *into the space $F$ with norm* (6.48).

We note that in (6.47) and the expression for the norm (6.48) only the fundamental function of the space $G$ is present. Therefore, for spaces with a given fundamental function $\varphi(t)$ the preceding assertion will be strongest when $G$ is largest. By the imbedding Theorem 5.7, $G$ will be largest if it coincides with the Marcinkiewicz space with the same fundamental function. Then our assertion takes the following form.

THEOREM 6.16. *Let $\varphi(t)$ be an arbitrary quasiconcave function. The convolution operator acts boundedly from $E \times M_{\varphi*}$, where $E$ satisfies the conditions*

$$\|\sigma_t\|_E = o(t^{1-\gamma_\varphi}) \quad \text{as} \quad t \to 0 \qquad \text{and} \qquad \|\sigma_t\|_E = o(t) \quad \text{as } t \to \infty, \tag{6.49}$$

*into the space $F$ with norm*

$$\|z\|_F = \|z^{**}(t)t^{-1}\varphi(t)\|_E. \tag{6.50}$$

*Moreover,*

$$\|x*y\|_F \leqslant C\|x\|_E\|y\|_{M_{\varphi_*}}, \tag{6.51}$$

*where $\varphi_* = t/\varphi(t)$.*

From the assertion formulated earlier we obtain the inequality $\|x*y\|_F \leqslant C(x)\|y\|_{\varphi_*}$, which implies (6.51) in view of the uniform boundedness principle.

We consider the special case where $\varphi(t) = t^{1/q}$ and $E = L_p$. Condition (6.49) is satisfied if $p > 1$ and $1/p + 1/q > 1$. The norm (6.50) can be calculated by the formula

$$\|z\|_F^p = \int_0^\infty [z^{**}(t)]^p t^{p/q-p}\, dt = \int_0^\infty [z^{**}(t)]^p t^{p(1/p+1/q-1)-1}\, dt,$$

and consequently $F$ coincides with $L_{r,p}$, where $1/r = 1/p + 1/q - 1$.

THEOREM 6.17. *If $p > 1$ and $1/p + 1/q - 1 > 0$, then*

$$\|x*y\|_{L_{r,p}} \leqslant C\|x\|_{L_p}\|y\|_{M_{t^{1-1/q}}}, \tag{6.52}$$

*where $1/r = 1/p + 1/q - 1$.*

This assertion is a sharpening, in the case $p > 1$, of Young's famous inequality. Indeed, the space $M_{t^{1-1/q}}$ is larger than the space $L_q$ (see Theorem 5.7) and the space $L_{r,p}$ is imbedded in $L_r$ (see Lemma 6.5), and so, (6.52) implies the Young inequality

$$\|x \times y\|_{L_r} \leqslant C\|x\|_{L_p}\|y\|_{L_q}. \tag{6.53}$$

A classical example of an operator of convolution type is the Riemann-Liouville integral of fractional order

$$\mathcal{J}_\alpha x = \frac{1}{\Gamma(\alpha)}\int_0^t (t-s)^{\alpha-1}x(s)\, ds. \tag{6.54}$$

If $0 < \alpha < 1$, then $t^{\alpha-1}$ belongs to the Marcinkiewicz space $M_{t^\alpha}$. Then Theorem 6.5 implies the following theorem.

THEOREM 6.18. *The operator* (6.54) *of fractional integration acts boundedly from any space E satisfying the conditions* $\|\sigma_t\|_E = o(t^\alpha)$ *as* $t \to 0$ *and* $\|\sigma_t\|_E = o(t)$ *as* $t \to \infty$ *into the space F with norm*

$$\|x\|_F = \|x^{**}(t)t^{-\alpha}\|_E.$$

*In particular, if* $p > 1$ *and* $1/p - \alpha > 0$, *then the operator* $\mathcal{J}_\alpha$ *acts from* $L_p$ *into* $L_{r,p}$, *where* $1/r = 1/p - \alpha$.

## §7. The singular Hilbert operator

In this section we shall consider functions $x(t)$ defined on the whole axis $(-\infty, \infty)$. The *singular Hilbert operator** is defined by

$$Sx(t) = \lim_{\delta \to 0} \int_{|t-s|>\delta} \frac{x(s)}{t-s}\, ds = \text{p.v.} \int_{-\infty}^{\infty} \frac{x(s)}{t-s}\, ds, \tag{7.1}$$

where p.v. means that the integral is taken in the sense of principal value.

If $x(t) \equiv \chi_{[a,b]}(t)$, then

$$S\chi_{[a,b]}(t) = \ln|(t-a)/(t-b)|. \tag{7.2}$$

From this it is clear that the operator acts in neither $L_\infty$ nor $L_1$. The operator $S$ is not defined for the function $\chi_{[a,\infty]}(t)$, since the right side of (7.1) is identically equal to $-\infty$. In connection with this there arises the problem of determining a class of functions for which the integral in (7.1) exists for almost all values of $t$. It turns out that $L_1(-\infty, \infty)$ may serve as such a class. For the proof of this deep result we establish three lemmas.

LEMMA 7.1. *For the function* $x(t) = \Sigma_1^n \mu_j/(t - a_j)$, *where* $a_1 < a_2 < \cdots < a_n$ *and* $\mu_j > 0$, *the following equality holds*:

$$n_{|x|}(\tau) = \frac{2}{\tau} \sum_{j=1}^{n} \mu_j \qquad (\tau > 0).$$

PROOF. We consider the set $e$ of those points where $x(t) > \tau$. The function $x(t)$ is decreasing on each of the intervals $(-\infty, a_1), (a_1, a_2), \ldots, (a_n, \infty)$ and is negative on $(-\infty, a_1)$. Therefore it assumes the value $\tau$ once on each of the intervals $(a_1, a_2), \ldots, (a_n, \infty)$ at the points $\alpha_1, \ldots, \alpha_n$. Thus, mes $e = \Sigma(\alpha_i - a_i) = \Sigma\alpha_i - \Sigma a_j$. The numbers $\alpha_j$ are roots of the equation

$$\sum_{j=1}^{n} \frac{\mu}{t - a_j} = \tau$$

* *Editor's note.* This operator is commonly called the Hilbert transform.

or the equation

$$\tau \prod_{j=1}^{n} (t - a_j) - \sum_{j=1}^{n} \mu_j \prod_{i \neq j} (t - a_i) = 0.$$

By Viète's theorem we have

$$\sum_{i=1}^{n} \alpha_i = \frac{\tau \sum_{j=1}^{n} a_i + \sum_{j=1}^{n} \mu_j}{\tau}.$$

From this we obtain that mes $e = \tau^{-1} \sum_1^n \mu_j$. It can be shown similarly that the measure of the set of points at which $x(t) < -\tau$ is equal to the same number. The lemma is proved.

We write

$$S_\delta x(t) = \int_{|t-s|>\delta} \frac{x(s)}{t-s}\, ds.$$

It is obvious that the operator $S_\delta$ is defined for every function $x \in L_1(-\infty, \infty)$.

LEMMA 7.2. *Let*

$$S_{\delta_i} x(t_i) > \tau > 0 \qquad (i = 1, 2, \dots, n), \tag{7.3}$$

*where the intervals* $I_i = [t_i - \delta_i, t_i + \delta_i]$ *are disjoint. Then*

$$\sum_{i=1}^{n} \delta_i \leqslant \frac{8}{\tau} \|x\|_{L_1}.$$

PROOF. First we assume that $x(t) \geqslant 0$. In the integral

$$S_{\delta_i} x(t_i) = \int_{[s-t_i]>\delta_i} \frac{x(s)}{t_i - s}\, ds$$

we replace the function $1/(t_i - s)$ by a step function. We choose points $s_i^{(1)}, \dots, s_i^{(N_i)}$ outside $I_i$ so that for any system of points $(s^{(1)}, \dots, s^{(N)})$ containing these points we have

$$\left| \sum_{j=1}^{N-1} \frac{1}{t_i - s^{(j)}} \int_{s^{(j)}}^{s^{(j+1)}} x(s)\, ds - S_{\delta_i} x(t_i) \right| < \varepsilon. \tag{7.4}$$

We combine all points $s_i^{(j)}$ $(i = 1, \dots, n, j = 1, \dots, N_i)$ in one system, arranged in increasing order: $\sigma = (s_1, \dots, s_N)$, and denote by $\sigma_k$ the collection of all points in $\sigma$ which are in the interval $(t_k - \delta_k, t_k + \delta_k)$. We consider the functions

$$y_i(t) = \sum_{s_j \notin \sigma_i} \frac{\mu_j}{t - s_j}, \qquad z_j(t) = \sum_{s_j \in \sigma_j} \frac{\mu_j}{t - s_j}, \qquad z(t) = \sum_{j=1}^{N} \frac{\mu_j}{t - s_j},$$

where $\mu_j = \int_{s_j}^{s_{j+1}} x(s)\,ds$. It is obvious that $z(t) = y_i(t) + z_i(t)$. If we choose $\varepsilon$ so that $\varepsilon < S_{\delta_i}(t_i) - \tau$ $(i = 1, \ldots, n)$, then by (7.4) we have

$$y_i(t_i) > S_{\delta_i}x(t_i) - \varepsilon > \tau. \tag{7.5}$$

The function $y_i(t)$ is decreasing on each of the intervals where it is defined, and therefore (7.5) implies that $y_i(t) > \tau$ for $\tau \in [t_i - \delta_i, t_i]$. Then for these $\tau$ we have either $z(t) \geqslant \tau/2$ or $z_i(t) < -\tau/2$. We denote by $e_0$ the set of those $t$ for which $z(t) > \tau/2$, and by $e_i$ the set of those $t$ for which $z_i(t) < -\tau/2$. Then

$$\bigcup_{i=1}^{n} [t_i - \delta_i, t_i] \subseteq \bigcup_{i=0}^{n} e_i.$$

From the proof of Lemma 7.1 we obtain

$$\sum_{i=1}^{n} \delta_i \leqslant \sum_{i=0}^{n} \operatorname{mes} e_i = \frac{2}{\tau} \sum_{j=1}^{N-1} \mu_j + \frac{2}{\tau} \sum_{i=1}^{n} \sum_{j \in \sigma_i} \mu_j$$
$$\leqslant \frac{4}{\tau} \sum_{j=1}^{N-1} \mu_j \leqslant \frac{4}{\tau} \int_{-\infty}^{\infty} x(s)\,ds.$$

Similarly, for a nonpositive $x(t)$ we can show that the inequality $S_\delta x(t_i) < -\tau$ implies that $\sum_{i=1}^{n} \delta_i \leqslant (4/\tau)\|x\|_{L_1}$.

For an arbitrary function $x \in L_1$ we have

$$S_{\delta_i}x(t) = S_{\delta_i}x_+(t) - S_{\delta_i}x_-(t).$$

Inequality (7.3) implies that either $S_{\delta_i}x_+(t_i) > \tau/2$, or $S_{\delta_i}x_-(t_i) < -\tau/2$. If the index $k$ denotes those indices for which the first inequality is satisfied, and $k'$ denotes those for which the second is satisfied, then

$$\sum_{k=1}^{n} \delta_i \leqslant \sum \delta_k + \sum \delta_{k'} \leqslant \frac{8}{\tau}\|x_+\|_{L_1} + \frac{8}{\tau}\|x_-\|_{L_1} = \frac{8}{\tau}\|x\|_{L_1}.$$

REMARK. If at the points $t_i$ we have

$$|S_{\delta_i}x(t_i)| > \tau > 0, \qquad i = 1, 2, \ldots, n,$$

then

$$\sum_{i=1}^{n} \delta_i \leqslant \frac{16}{\tau}\|x\|_{L_1}. \tag{7.6}$$

LEMMA 7.3. *For* $x \in L_1$

$$\operatorname{mes}\left\{t:\ \overline{\lim_{\delta \to 0}}\ |S_\delta x(t)| > \tau\right\} < 64\tau^{-1}\|x\|_{L_1}. \tag{7.7}$$

PROOF. The measure of the set $e$ being considered is the supremum of the measures of the compacta contained in it. Therefore it is sufficient to prove the inequality $\operatorname{mes} K < 64\tau^{-1}\|x\|_{L_1}$ for any compactum $K$ at whose points we have $\overline{\lim}_\delta |S_\delta x(t)| > \tau$. For every $t \in K$ the inequality $|S_\delta x(t)| > \tau$, is satisfied for some $\delta$, and so, the intervals $(t - \delta, t + \delta)$ cover $K$. From this cover we choose a finite cover $(t_1 - \delta_1, t_1 + \delta_1), \ldots, (t_N - \delta_N, t_N + \delta_N)$ so that in this system only consecutive intervals intersect each other. Then the intervals with even and with odd indices form two systems of disjoint intervals. At least one of them (for example the one with even indices) contains a part of $K$ with measure not less than $\frac{1}{2}$ mes $K$. Thus $\frac{1}{2} \operatorname{mes} K \leqslant \sum 2\delta_{2k}$. From (7.6) it follows that $\operatorname{mes} K \leqslant 64\tau^{-1}\|x\|_{L_1}$.

From Lemma 7.3 it follows that

$$\operatorname{mes}\left\{t\colon \overline{\lim_{\delta,\,\delta'\to 0}} |S_\delta x(t) - S_{\delta'}x(t)| > \tau\right\} \leqslant \operatorname{mes}\left\{t\colon \overline{\lim_{\delta\to 0}} |S_\delta x(t)| > \frac{\tau}{2}\right\}$$

$$\leqslant \frac{128}{\tau}\|x\|_{L_1}.$$

THEOREM 7.1. *If $f \in L_1$, then the limit in* (7.1) *exists almost everywhere. The operator $S$ is of weak type $\{L_1, L_1\}$, i.e.*

$$t(Sx)^*(t) \leqslant C\|x\|_{L_1}. \tag{7.8}$$

PROOF. We prove the first part of the theorem. We represent the function $x(t)$ as the sum of a step function $x_1(t)$ of compact support and a function $x_2(t)$ with sufficiently small norm in $L_1$. Then

$$\overline{\lim_{\delta,\delta'\to 0}} |S_\delta x(t) - S_{\delta'}x(t)|$$

$$\leqslant \overline{\lim_{\delta,\delta'\to 0}} |S_\delta x_1(t) - S_{\delta'}x_1(t)| + \overline{\lim_{\delta,\delta'\to 0}} |S_\delta x_2(t) - S_{\delta'}x_2(t)|.$$

At the points of continuity of $x_1(t)$ the first term equals zero, since $S_\delta x_1(t) = S_{\delta'}x_1(t)$ for sufficiently small $\delta$ and $\delta'$. The measure of the set where the second term is greater than $\tau$ does not exceed the number $128\tau^{-1}\|x_2\|_{L_1}$. Thus, the left side does not exceed $\tau$ anywhere outside a set of measure $128\tau^{-1}\|x_2\|_{L_1}$. Since $\|x_2\|_{L_1}$ can be made as small as we wish, and $\pi$ is arbitrary we have

$$\overline{\lim_{\delta,\delta'\to 0}} |S_\delta x(t) - S_{\delta'}x(t)| = 0$$

almost everywhere. The Cauchy test implies the first assertion of the theorem.

Inequality (7.7) implies that

$$\operatorname{mes}\{t: |Sx(t)| > \tau\} < 64\tau^{-1}\|x\|_{L_1}$$

almost everywhere, and this is equivalent to (7.8).

Inequality (7.8) implies the following corollary.

COROLLARY 1. *If a sequence of functions $x_n$ converges to a function $x$ in $L_1$, then $Sx_n$ converges to $Sx$ in measure.*

COROLLARY 2. *If $x \in L_p$, $1 \leqslant p < \infty$, then the limit in* (7.1) *exists almost everywhere.*

In fact, $x(t) = x(t)\chi_{(-a,a)}(t) + x(t)(1 - \chi_{(-a,a)}(t))$. The first term belongs to $L_1$, and therefore $Sx\chi_{(-a,a)}(t)$ is defined almost everywhere. For the second term with $t \in (-a, a)$ the corresponding integral is absolutely convergent. Consequently, $S_\delta x(t)$ is defined for almost all $t$ in $(-a, a)$, and thus for almost all $t$ in $(-\infty, \infty)$.

LEMMA 7.4. *If a sequence of functions $x_n(t)$ converges to a function $x(t)$ in $L_p$, then $Sx_n(t)$ converges to $Sx(t)$ in measure on every set of finite measure.*

PROOF. We have $x_n(t) = x_n(t)\chi_{(-2a,2a)}(t) + x_n(t)(1 - \chi_{(-2a,2a)}(t))$. The sequence of the first terms converges to $x(t)\chi_{(-2a,2a)}(t)$ in $L_1$, and consequently it converges in measure according to Corollary 1. The second terms converge uniformly on $(-a, a)$:

$$\begin{aligned}|Sx_n(1 - \chi_{(-2a,2a)})(t) - Sx(1 - \chi_{(-2a,2a)})(t)| &= \left|\int_{|s|>2a} \frac{x_n(s) - x(s)}{t - s}\, ds\right| \\ &\leqslant \left\{\int_{|s|>2a} |x_n(s) - x(s)|^p\, ds\right\}^{1/p}\end{aligned}$$

$$\left\{\int_{|s|>2a} \frac{ds}{|t - ds|^{p'}}\right\}^{1/p'} \leqslant \left\{\int_a^\infty \frac{ds}{s^{p'}}\right\}^{1/p'} \|x_n - x\|_{L_p}.$$

LEMMA 7.5. *Let $e$ be a measurable set of finite measure. Then*

$$(S\chi_e)^*(t) = \operatorname{arcsinh}\frac{2 \operatorname{mes} e}{t}. \tag{7.9}$$

PROOF. By Corollary 1 of Theorem 7.1 and property 11° of rearrangements it is sufficient to prove (7.9) for a set $e$ consisting of a finite number of disjoint intervals $[a_i, b_i]$ $(a_1 < b_1 < a_2 < b_2 < \cdots < a_n < b_n)$. Then by (7.2) we have

$$S\chi_e(t) = \sum_{j=1}^{n} \ln\left|\frac{t-a_j}{t-b_j}\right|.$$

We consider the equation

$$S\chi_e(t) = \tau > 0. \tag{7.10}$$

This equation has at least one root in every interval $(a_j, b_j)$ and $(b_j, a_{j+1})$ and in $(b_n, \infty)$. Thus, it has at least $2n$ roots. For $t \in e$ our equation is equivalent to

$$\prod_{j=1}^{n} \frac{t-a_j}{t-b_j} = -e^{\tau},$$

and for $t \notin e$ to

$$\prod_{j=1}^{n} \frac{t-a_j}{t-b_j} = e^{\tau}.$$

Each of these equations has $n$ roots, and so our equation has exactly $2n$ roots. A similar calculation applies to the case $\tau < 0$. Hence it follows that the function $S\chi_e(t)$ is monotone in every interval where it is continuous. Denote by $\alpha_j$ the root of (7.10) in the interval $(a_j, b_j)$ and by $\beta_j$ the root in $(b_j, a_{j+1})$ $(a_{n+1} = \infty)$. For $\tau > 0$ we have

$$\operatorname{mes}(t\colon S\chi_e(t) > \tau) = \sum_{j=1}^{n} (\beta_j - \alpha_j) = \sum_{j=1}^{n} \beta_j - \sum_{j=1}^{n} \alpha_j.$$

By Viète's theorem we have

$$\sum_{j=1}^{n} \alpha_j = \frac{\Sigma a_j + e^{\tau}\Sigma b_j}{1+e^{\tau}} \quad \text{and} \quad \sum_{j=1}^{n} \beta_j = \frac{\Sigma a_j - e^{\tau}\Sigma b_j}{1-e^{\tau}}.$$

Hence

$$\operatorname{mes}(t\colon S\chi_e(t) > \tau) = \frac{2(\Sigma a_j - \Sigma b_j)e^{\tau}}{1-e^{2\tau}} = \frac{\operatorname{mes} e}{\sinh \tau}.$$

Similarly,

$$\operatorname{mes}(t\colon S\chi_e(t) < -\tau) = \operatorname{mes} e/\sinh \tau.$$

Thus,

$$n_{|S\chi_e|}(\tau) = 2 \operatorname{mes} e/\sinh \tau,$$

which is equivalent to (7.9).

THEOREM 7.2. *For the singular Hilbert operator to act boundedly in a symmetric space E it is necessary and sufficient that*

$$\|\sigma_t\|_E = o(1) \quad \text{as } t \to 0 \qquad \text{and} \qquad \|\sigma_t\|_E = o(t) \quad \text{as } t \to \infty. \tag{7.11}$$

PROOF. For every $p \in (1, \infty)$ inequality (7.9) implies that

$$t^{1/p}(S\chi_e)^*(t) = t^{1/p} \operatorname{arcsinh} \frac{2 \operatorname{mes} e}{t}$$
$$\leqslant (\operatorname{mes} e)^{1/p} \sup_{0<\tau<\infty} \tau^{1/p} \operatorname{arcsinh} \frac{2}{\tau} = C_p \int_0^\infty \chi_e^*(t)\, dt^{1/p}.$$

Thus, $S$ is an operator of weak type $\{L_p, L_p\}$ $(1 < p < \infty)$, and hence also an operator of weakened type $\{L_p, L_p\}$ according to Lemma 6.4. From (7.11) and Theorem 1.3 on submultiplicative functions it follows that there exist numbers $\nu_0 > 1/p_0$ and $\nu_1 > 1 - 1/p_1$ such that $\|\sigma_t\|_E = o(t^{\nu_0})$ as $t \to 0$ and $\|\sigma_t\|_E = o(t^{1-\nu_1})$ as $t \to \infty$, where $p_0$ and $p_1$ are numbers from $(1, \infty)$. Then (6.5) is satisfied for the space $E$ and the spaces $E_0 = L_{p_0}$ and $E_1 = L_{p_1}$. Theorem 6.1 implies that the restriction of $S$ to strictly simple functions admits a bounded extension to an operator acting in $E$. By Lemma 7.4 this extension will coincide with the operator $S$ itself.

We now show that the hypotheses of the theorem are necessary for the operator $S$ to act in the symmetric space $E$. We consider an arbitrary function $x \in E$ equal to zero on the negative semiaxis. Then the function $x(-t)$ belongs to $E$, as well:

$$\text{p.v.} \int_{-\infty}^\infty \frac{x(-s)}{t-s}\, ds = \text{p.v.} \int_0^\infty \frac{x(s)}{t+s}\, ds \in E.$$

If we restrict ourselves to $t > 0$, we see that

$$\chi_{(0,\infty)}(t) \int_0^\infty \frac{x(s)}{t+s}\, ds \in E.$$

If we introduce the symmetric space $E(0, \infty)$ of functions on the semiaxis, consisting of the restrictions of the functions from $E$ to $(0, \infty)$, then it is clear from the preceding that the Hilbert operator $\Gamma$ acts in this space. By Theorem 6.8 the space $E(0, \infty)$, and consequently the space $E$ as well, has properties (7.11).

## §8. Interpolation theorems for spaces with different measures

**1.** ***The operators*** $\Pi$ ***and*** $R$. Let $\mathfrak{M}$ be a space with $\sigma$-finite measure $\mu$. Similarly to §2, we may introduce the notion of distribution function for any nonnegative measurable function $u(m)$, $m \in \mathfrak{M}$, as follows:

$$n_u(\tau) = \mu(m\colon u(m) > \tau) \qquad (\tau > 0).$$

We assume that a one-to-one continuous linear mapping $\Pi$ is defined from $S(\mathfrak{M}, \mu)$ into $S(0, \infty)$ having the following property:

$$n_{|u|}(\tau) = n_{|\Pi u|}(\tau) \qquad (\tau > 0). \tag{8.1}$$

In the same way as in §2, the rearrangement $u^*(t)$ of a function $u(m)$ is defined to be the function

$$u^*(t) = \inf\{\tau : n_{|u|}(\tau) < t\}.$$

This function is defined on $(0, \infty)$. If the operator $\Pi$ has property (8.1), then

$$(\Pi u)^*(t) = u^*(t) \qquad (0 < t < \infty). \tag{8.2}$$

(We note that on the right and left sides the operation * is applied in function spaces of different nature.)

It is easy to verify that

$$\int |u(m)|\, d\mu = \int_0^\infty u^*(t)\, dt \quad \text{and} \quad \operatorname*{ess\,sup}_{m \in \mathfrak{M}} |u(m)| = \operatorname*{ess\,sup}_{0<t<\infty} u^*(t). \tag{8.3}$$

Let $E(0, \infty)$ be a symmetric space of functions on $(0, \infty)$. Denote by $E(\mathfrak{M})$ the inverse image of $E(0, \infty)$ under $\Pi$. Obviously $E(\mathfrak{M})$ is a linear manifold in $S(\mathfrak{M}, \mu)$. We may introduce a norm in it by the formula

$$\|u\|_{E(\mathfrak{M})} = \|\Pi u\|_{E(0,\infty)} = \|(\Pi u)^*\|_{E(0,\infty)} = \|u^*\|_{E(0,\infty)}. \tag{8.4}$$

With this definition the operator $\Pi$ will realize an isometric imbedding of $E(\mathfrak{M})$ in $E(0, \infty)$. From (8.3) it follows immediately that the spaces $L_1(\mathfrak{M})$ and $L_\infty(\mathfrak{M})$ coincide with the ordinary spaces $L_1$ and $L_\infty$ on measure spaces.

The space $E(\mathfrak{M})$ is Banach. Indeed, if a sequence $u_n$ is Cauchy in $E(\mathfrak{M})$, then the sequence $\Pi u_n$ is Cauchy in $E(0, \infty)$, and consequently converges to an element $v$. By Theorem 4.1 the sequence $\Pi u_n$ converges to $v$ in $L_1(0, \infty) + L_\infty(0, \infty)$, and so $u_n$ converges to some element $u$ in $L_1(\mathfrak{M}) + L_\infty(\mathfrak{M})$. The continuity of $\Pi$, which we assumed, implies that $v = \Pi u$, and consequently $v \in E(\mathfrak{M})$. Since $\Pi u_n \to \Pi u$ in $E(0, \infty)$, we have $u_n \to u$ in $E(\mathfrak{M})$.

It is useful to note that to construct the space $E(\mathfrak{M})$ it is sufficient to know only that the operator $\Pi$ with the indicated properties exists, without knowing its concrete form. Indeed, $E(\mathfrak{M})$ may be defined as the collection of all measurable functions $u$ on $\mathfrak{M}$ for which $u^* \in E(0, \infty)$, with norm $\|u\|_{E(\mathfrak{M})} = \|u^*\|_{E(0,\infty)}$. In fact, if $u^* \in E(0, \infty)$, then by (8.2) we have $(\Pi u)^* \in E(0, \infty)$, and, since the space $E(0, \infty)$ is symmetric, $\Pi u \in E(0, \infty)$, i.e. $u \in \Pi^{-1}E(0, \infty) = E(\mathfrak{M})$. Conversely, if $u \in E(\mathfrak{M})$, then by (8.4) we have $u^* \in E(0, \infty)$ and the norm of $u$ is defined by the above formula.

We now assume that the restriction of $\Pi$ to $L_1(\mathfrak{M}) + L_\infty(\mathfrak{M})$ has a left inverse $R$ that maps $L_1(0, \infty) + L_\infty(0, \infty)$ onto $L_1(\mathfrak{M}) + L_\infty(\mathfrak{M})$, acting with norm 1 from $L_1(0, \infty)$ into $L_1(\mathfrak{M})$ and from $L_\infty(0, \infty)$ into $L_\infty(\mathfrak{M})$.

We give the most important examples of measure spaces and the operators $\Pi$ and $R$.

1) *One-to-one image of the semiaxis.* We assume that there exists a one-to-one measure preserving mapping $\omega(m)$ of the measure space $\mathfrak{M}$ with measure $\mu$ onto the semiaxis with Lebesgue measure. Then $\Pi$ is defined as $\Pi u(t) = u(\omega^{-1}(t))$. This operator has all the necessary properties and has the inverse operator $Rx(m) = x(\omega(m))$.

In particular, if $\mathfrak{M} = (-\infty, \infty)$ with Lebesgue measure, then as $\omega$ we may take the mapping shifting the intervals $[k, k+1)$ $(k \geqslant 0)$ to the intervals $[2k, 2k+1)$, and the intervals $[-k, -k+1)$ $(k \geqslant 1)$ to the intervals $[2k-1, 2k)$.

If $\mathfrak{M}$ is the semiaxis $(0, \infty)$ with the measure $\mu$ $(d\mu = ds/s)$, invariant with respect to dilations, then the function $\omega(\ln s)$, where $\omega(t)$ is the same as above, has the desired properties.

2) *The interval* $(0, 1)$ *with Lebesgue measure.* Denote by $\Pi$ the operator which maps a function $u \in S(0, 1)$ onto the function

$$\Pi u(t) = \begin{cases} u(t), & 0 < t < 1, \\ 0, & t \geqslant 1, \end{cases}$$

and by $R$ the operator of restricting functions defined on the semiaxis to the interval $(0, 1)$. All properties are obviously satisfied.

3) *Sequence spaces.* Let $\mathfrak{M}$ consist of countably many points with measure equal to one. The functions on $\mathfrak{M}$ are identified with sequences $\{u_m\}_1^\infty$ in a natural way. To a sequence $u = \{u_m\}$ we assign the step function $\Pi u(t) = u_m$ for $m-1 \leqslant t < m$. The operator $\Pi u$ has the required properties. Its left inverse is the averaging operator

$$Rx = \left\{ \int_{m-1}^{m} x(s)\, ds \right\}.$$

It is obvious that $R$ has norm equal to one as an operator from $L_1(0, \infty)$ $(L_\infty(0, \infty))$ to $l_1$ $(l_\infty)$, which coincides with $L_1(\mathfrak{M})$ $(L_\infty(\mathfrak{M}))$ in our case.

**2.** ***Interpolation theorem.*** Let $\mathfrak{M}$ and $\mathfrak{M}'$ be two measure spaces for which there exist the operators $\Pi$, $R$ and $\Pi'$, $R'$ with the properties indicated above. If the operator $V$ maps functions on $\mathfrak{M}$ onto functions on $\mathfrak{M}'$, then the operator $T = \Pi' V R$ takes functions on $(0, \infty)$ into functions on $(0, \infty)$. If $T = \Pi' V R$ acts boundedly from a symmetric space $E(0, \infty)$ into a symmetric space $F(0, \infty)$, then $V$ acts boundedly from $E(\mathfrak{M})$ into $F(\mathfrak{M}')$, and has smaller norm. Indeed, the inequality $\|\Pi' V R x\|_{F(0,\infty)} \leqslant C\|x\|_{E(0,\infty)}$ implies that for $u \in E(\mathfrak{M})$ and $x = \Pi u$ we have

$$\begin{aligned} \|Vu\|_{F(\mathfrak{M}')} &= \|\Pi' V u\|_{F(0,\infty)} = \|\Pi' V R \Pi u\|_{F(0,\infty)} \\ &\leqslant C\|\Pi u\|_{E(0,\infty)} = C\|u\|_{E(\mathfrak{M})}. \end{aligned}$$

The converse will be true if the operator $R$ has norm one as an operator from $E(0, \infty)$ to $E(\mathfrak{M})$. In fact, then we have

$$\|\Pi' VRx\|_{F(0,\infty)} = \|VRx\|_{F(\mathfrak{M}')} \leqslant c\|Rx\|_{E(\mathfrak{M})} \leqslant c\|x\|_{E(0,\infty)}.$$

Denote by $\Sigma(\mathfrak{M}, \mathfrak{M}')$ the collection of all linear operators acting boundedly, with norm not exceeding one, from $L_1(\mathfrak{M})$ to $L_1(\mathfrak{M}')$ and from $L_\infty(\mathfrak{M})$ to $L_\infty(\mathfrak{M}')$.

From the interpolation Theorem 4.3 we obtain

THEOREM 8.1. *For the spaces* $(L_1(\mathfrak{M}), L_\infty(\mathfrak{M}), E)$ *to form an interpolation triple with respect to the triple* $(L_1(\mathfrak{M}'), L_\infty(\mathfrak{M}'), F)$ *with interpolation constant equal to one it is necessary and sufficient that the inequality*

$$\int_0^t v^*(\tau)\,d\tau \leqslant \int_0^t u^*(\tau)\,d\tau \qquad (0 < t < \infty) \tag{8.5}$$

*for* $u \in E$ *and* $v \in L_1(\mathfrak{M}') + L_\infty(\mathfrak{M}')$ *imply that* $v \in F$ *and* $\|v\|_F \leqslant \|u\|_E$.

PROOF. *Necessity*. (8.5) and (8.2) imply that

$$\int_0^t (\Pi' v)^*(\tau)\,d\tau \leqslant \int_0^t (\Pi u)^*(\tau)\,d\tau.$$

By Theorem 3.4 we then have $\Pi' v = T\Pi u$, where $T \in \Sigma$. Then $v = R' T\Pi u$. The operator $R' T\Pi$ obviously belongs to $\Sigma(\mathfrak{M}, \mathfrak{M}')$, and therefore from the interpolation properties of $E$ and $F$ it follows that $v \in F$ and $\|v\|_F = \|R' T\Pi u\|_F \leqslant \|u\|_E$.

*Sufficiency*. Let $V \in \Sigma(\mathfrak{M}, \mathfrak{M}')$. We have $\Pi' VR \in \Sigma$, and by (3.16) for $x \in L_1(0, \infty) + L_\infty(0, \infty)$ we have

$$\int_0^t (\Pi' VRx)^*(\tau)\,d\tau \leqslant \int_0^t x^*(\tau)\,d\tau.$$

For $u \in E$ we set $x = \Pi u$. Then

$$\int_0^t (\Pi' Vu)^*(\tau)\,d\tau \leqslant \int_0^t (\Pi u)^*(\tau)\,d\tau,$$

or, by (8.2),

$$\int_0^t (Vu)^*(\tau)\,d\tau \leqslant \int_0^t u^*(\tau)\,d\tau.$$

From the properties of $E$ and $F$ we obtain that $Vu \in F$ and $\|Vu\|_F \leqslant \|u\|_E$.

**3. *Correspondence between interpolation spaces.*** From the general discussion in §4.6 of Chapter I it follows that if $E(0, \infty)$ is an interpolation space between $L_1(0, \infty)$ and $L_\infty(0, \infty)$ with interpolation constant equal to one, then so is $E(\mathfrak{M})$ such a space between $L_1(\mathfrak{M})$ and $L_\infty(\mathfrak{M})$. (Indeed, if $V \in \Sigma(\mathfrak{M}, \mathfrak{M})$, then $\Pi VR \in \Sigma$, and so it acts in $E(0, \infty)$; and then $V$ acts in $E(\mathfrak{M})$ with not greater norm.)

In §4.6 of Chapter I a scheme is given according to which from every interpolation space $G$ between $L_1(\mathfrak{M})$ and $L_\infty(\mathfrak{M})$ an interpolation space $E_G$ (with the notation of subsection 6, a space $E(G)$) can be constructed between $L_1(0, \infty)$ and $L_\infty(0, \infty)$. This space consists of all functions $x$ for which $RTx \in G$ for any operator $T \in \Sigma$. The norm in $E_G$ is introduced by the formula

$$\|x\|_{E_G} = \sup_{T \in \Sigma} \|RTx\|_G. \tag{8.6}$$

Under a certain assumption on the operator $R$ this norm can be calculated explicitly. We shall say that $R$ *satisfies condition* A) if there exists a set $e_R \subset (0, \infty)$ having the following two properties:

1) For $t \in e_R$ and $x \in L_1(0, \infty) + L_\infty(0, \infty)$

$$\int_0^t (Rx^*)^*(\tau)\, d\tau = \int_0^t x^*(\tau)\, d\tau. \tag{8.7}$$

2) If for $u, v \in L_1(\mathfrak{M}) + L_\infty(\mathfrak{M})$ the inequality

$$\int_0^t u^*(\tau)\, d\tau \leqslant \int_0^t v^*(\tau)\, d\tau$$

holds when $t \in e_R$, then it holds for all $t \in (0, \infty)$ as well.

Assuming that condition A) is satisfied, we pass to the calculation of the norm (8.6). By (3.16) for $T \in \Sigma$ we have

$$\int_0^t (Tx)^*(\tau)\, d\tau \leqslant \int_0^t x^*(\tau)\, d\tau. \tag{8.8}$$

The projection operator $\Pi R$ also belongs to $\Sigma$, and therefore

$$\int_0^t (\Pi Ry)^*(\tau)\, d\tau \leqslant \int_0^t y^*(\tau)\, d\tau. \tag{8.9}$$

Setting $y = Tx$ and combining (8.8) and (8.9), we obtain

$$\int_0^t (\Pi RTx)^*(\tau)\, d\tau \leqslant \int_0^t (Tx)^*(\tau)\, d\tau \leqslant \int_0^t x^*(\tau)\, d\tau.$$

From this by (8.2) and (8.7) for $t \in e_R$ we obtain

$$\int_0^t (RTx)^*(\tau)\, d\tau \leqslant \int_0^t (Rx^*)^*(\tau)\, d\tau.$$

By condition (2) this inequality is satisfied for $t > 0$. Without loss of generality we may assume that the interpolation constant of $G$ is one. Then Theorem 8.1 implies that

$$\|RTx\|_G \leqslant \|Rx^*\|_G.$$

On the other hand, by Theorem 3.4 there exists an operator $T_0 \in \Sigma$ such that $x^* = T_0 x$, and therefore $RT_0 x = Rx^*$. From this and the preceding it follows that

$$\|x\|_{E_G} = \|Rx^*\|_G. \tag{8.10}$$

In passing we have proved that the space $E_G$ consists of all functions from $L_1(0, \infty) + L_\infty(0, \infty)$, for which $Rx^* \in G$.

We note that (8.7) applied to $x = \Pi u$ implies, by (8.2), that for $t \in e_R$

$$\int_0^t (Ru^*)^*(\tau)\, d\tau = \int_0^t u^*(\tau)\, d\tau,$$

and so this equality is satisfied for all $t$. Then by Theorem 8.1

$$\|Ru^*\|_G = \|u\|_G.$$

From this it follows that the space $E_G(\mathfrak{M})$ coincides isometrically with $G$. Indeed,

$$\|u\|_{E_G(\mathfrak{M})} = \|u^*\|_{E_G} = \|Ru^*\|_G = \|u\|_G. \tag{8.11}$$

We consider the following situation, as well: Let $E(0, \infty)$ be an interpolation space between $L_1(0, \infty)$ and $L_\infty(0, \infty)$ with interpolation constant one. From it we construct the space $E(\mathfrak{M}) = G$, and then the space $E_G$. We calculate the norm in $E_G$:

$$\|x\|_{E_G} = \|Rx^*\|_G = \|Rx^*\|_{E(\mathfrak{M})} = \|\Pi Rx^*\|_{E(0,\infty)}.$$

Since $\Pi R \in \Sigma$, we have $\|\Pi Rx^*\|_{E(0,\infty)} \leqslant \|x\|_{E(0,\infty)}$, and so in this case $E_G$ is larger than the original space $E(0, \infty)$.

In examples 1)–3) condition A) on the operator $R$ is satisfied. In example 1) we have $e_R = (0, \infty)$ and $(Rx^*)^*(t) = x^*(t)$. In example 2) we have $e_R = (0, 1)$ and $(Rx^*)^*(t) = x^*(t)$ for $t \in (0, 1)$. Property 2) is obviously satisfied. In example 3) the set $e_R$ consists of all natural numbers, $(Rx^*)^*(t) \equiv \Pi Rx^*(t)$, and

$$\int_0^n \Pi Rx^*(\tau)\, d\tau = \int_0^n \sum_{m=1}^{\infty} \int_{m-1}^{m} x^*(s)\, ds\, \chi_{[m-1,m)}(\tau)\, d\tau$$
$$= \sum_{m=1}^{n} \int_{m-1}^{m} x^*(s)\, ds = \int_0^n x^*(\tau)\, d\tau.$$

Property 2) is also satisfied.

**4. *Operators of weakened type.*** If $E(\mathfrak{M})$ and $F(\mathfrak{M}')$ are symmetric spaces, then a linear operator $V$, defined on all strictly simple functions on $\mathfrak{M}$ if $\varphi_{E(0,\infty)}(\infty) = \infty$, or on all simple functions if $\varphi_{E(0,\infty)}(\infty) < \infty$, is called an *operator of weakened type* $\{E(\mathfrak{M}), F(\mathfrak{M}')\}$, if $Vu \in S(\mathfrak{M}')$ and

$$(Vu)^{**}(t)\varphi_{F(0,\infty)}(t) \leqslant C\int_0^\infty u^*(t)\, d\varphi_{E(0,\infty)}(t). \tag{8.12}$$

We consider the operator $\Pi' VR$, assuming that the operator $R$ takes strictly simple (simple) functions on $(0, \infty)$ into strictly simple (simple) functions of $\mathfrak{M}$. For a strictly simple (simple) function $x(t)$ on $(0, \infty)$ we then obtain

$$\begin{aligned}(\Pi' VRx)^{**}(t)\varphi_{F(0,\infty)}(t) &= \frac{1}{t}\int_0^t (\Pi' VRx)^*(\tau)\, d\tau\varphi_{F(0,\infty)}(t)\\ &= \frac{1}{t}\int_0^t (VRx)^*(\tau)\, d\tau\varphi_{F(0,\infty)}(t)\\ &\leqslant C\int_0^\infty (Rx)^*(t)\, d\varphi_{E(0,\infty)}(t)\\ &= C\int_0^\infty (\Pi Rx)^*(t)\, d\varphi_{E(0,\infty)}(t).\end{aligned}$$

The operator $\Pi R$ belongs to $\Sigma$. Therefore (3.16) is satisfied for $\Pi R$, and by property 18° of rearrangements we then have

$$(\Pi' VRx)^{**}(t)\varphi_{F(0,\infty)}(t) \leqslant C\int_0^\infty x^*(t)\, d\varphi_{E(0,\infty)}(t).$$

Thus, $\Pi' VR$ is an operator of weakened type $\{E(0, \infty), F(0, \infty)\}$. Then Theorem 6.1′ allows us to obtain the following assertion.

THEOREM 8.2. *Let $E_0$, $F_0$, $E_1$ and $F_1$ be symmetric spaces of functions on the semiaxis for which the hypotheses of Theorem* 6.1′ *are satisfied, and let $V$ be an operator of weakened types $\{E_0(\mathfrak{M}), F_0(\mathfrak{M}')\}$ and $\{E_1(\mathfrak{M}), F_1(\mathfrak{M}')\}$. The operator $V$ admits closure to a bounded operator acting from $E(\mathfrak{M})$ to $E_{\kappa,\delta}(\mathfrak{M}')$, where $E(0, \infty)$ and $E_{\kappa,\delta}(0, \infty)$ are the spaces described in Theorem* 6.1′.

PROOF. The hypotheses of Theorem 6.1′ are satisfied for the operator $\Pi' VR$, and therefore it admits closure to a bounded operator from $E(0, \infty)$ to $E_{\kappa,\delta}(0, \infty)$. Then, as described above, $V$ admits closure to a bounded operator from $E(\mathfrak{M})$ to $E_{\kappa,\delta}(\mathfrak{M}')$. The theorem is proved.

For the symmetric space $E(\mathfrak{M})$ we can introduce the notion of a fundamental function by setting $\varphi_{E(\mathfrak{M})}(\mu(e)) = \|\chi_e\|_{E(\mathfrak{M})}$. By the definition of $E(\mathfrak{M})$, from this it follows that $\varphi_{E(\mathfrak{M})}(\mu(e)) = \|\chi_{(0,\mu(e))}\|_{E(0,\infty)} = \varphi_{E(0,\infty)}(\mu(e))$.

Thus, the functions $\varphi_{E(\mathfrak{M})}$ and $\varphi_{E(0,\infty)}$ coincide at all values assumed by the measure $\mu$ on $\mathfrak{M}$. Denote by $e_\mu$ the closure on $(0, \infty)$ of the set of all values of $\mu$, and assume that the function $\varphi_{E(\mathfrak{M})}$ is extended by continuity to $e_\mu$. It then coincides with the restriction of $\varphi_{E(0,\infty)}$ to $e_\mu$. The distribution function of any measurable function on $\mathfrak{M}$ does not assume values in the intervals complementary to $e_\mu$, and therefore the rearrangement of any measurable function is constant on these intervals. This enables us to modify condition (8.12) a little.

Let $(t_1, t_2)$ be an interval complementary to the set $e_\mu$ and let the inequalities

$$(Vu)^{**}(t_i)\varphi_{F(0,\infty)}(t_i) \leqslant Q \qquad (i = 1, 2)$$

or

$$\int_0^{t_i} (Vu)^*(\tau)\, d\tau \leqslant \frac{t_i}{\varphi_{F(0,\infty)}(t_i)} Q \leqslant \psi(t_i)Q \tag{8.13}$$

hold, where $\psi(t)$ is the smallest concave majorant of the quasiconcave function $t/\varphi_{F(0,\infty)}(t)$. From the preceding it follows that the function $\int_0^t (Vu)^* d\tau$ is linear on $[t_1, t_2]$, and so (8.13) implies that it is smaller than the concave function $Q\psi(t)$ on the whole interval. Consequently, by (1.7) we have

$$\int_0^t (Vu)^*(\tau)\, d\tau \leqslant \psi(t)Q \leqslant 2\frac{t}{\varphi_{F(0,\infty)}(t)} Q \qquad (t_1 \leqslant t \leqslant t_2).$$

Applying these arguments to (8.12), we conclude that it will be satisfied if the inequality

$$(Vu)^{**}(t)\varphi_{F(\mathfrak{M}')}(t) \leqslant C_1 \int_0^\infty u^*(\tau)\, d\varphi_{E(\mathfrak{M})}(\tau) \qquad (t \in e_{\mu'}) \tag{8.14}$$

is, where $C_1 = C/2$ and the function $\varphi_{E(\mathfrak{M})}$ in the integral is arbitrarily (for example, linearly) extended to the intervals complementary to $e_\mu$. Thus, the condition that an operator has weakened type $\{E(\mathfrak{M}), F(\mathfrak{M}')\}$, needs to be verified only at the points $t \in e_{\mu'}$ and can be expressed in terms of the fundamental functions of the indicated spaces.

Theorem 8.2 is formulated in terms of spaces of functions on the semiaxis. We obtain a somewhat different assertion formulated in terms of spaces of functions on $\mathfrak{M}$ and $\mathfrak{M}'$. In this connection, we relax the requirements on $E$ somewhat, but obtain a cruder inequality for operators of weakened type. Let $G$ be an interpolation space between $L_1(\mathfrak{M})$ and $L_\infty(\mathfrak{M})$ with interpolation constant 1, and let $E_G$ be the space constructed in subsection 3. From the proof of Theorem 6.1 we can see that under its hypotheses any strictly simple (simple) function satisfies (6.10), which takes the form

$$\kappa(t)(\Pi' VRx)^{**}(\delta(t)) \leqslant C\int_0^\infty x^*(t\tau)\, d\min\{M_{\varphi_{E_i}}(\tau)\}$$

in our case.

For the function $x = \Pi u$ this implies that

$$\kappa(t)(\Pi' Vu)^{**}(\delta(t)) \leqslant C\int_0^\infty \sigma_{1/\tau}u^*(t)\, d\min\{M_{\varphi_{E_i}}(\tau)\}.$$

By Lemma 4.7 and (8.10) we then have

$$\|\kappa(t)(\Pi' Vu)^{**}(\delta(t))\|_{E_G} \leqslant C\int_0^\infty \|\sigma_{1/\tau}u^*\|_{E_G}\, d\min\{M_{\varphi_{E_i}}(\tau)\}$$
$$= C\int_0^\infty \|R\sigma_{1/\tau}u^*\|_G\, d\min\{M_{\varphi_{E_i}}(\tau)\}.$$

We now make the following assumption: For $u \in L_1(\mathfrak{M}) + L_\infty(\mathfrak{M})$

$$\Pi R u^* = u^*. \tag{8.15}$$

This property is equivalent to the condition that the function $u^*$ belongs to the range of the mapping $\Pi$.

If (8.15) is satisfied, then by (8.11)

$$\|R\sigma_{1/\tau}u^*(t)\|_G = \|R\sigma_{1/\tau}\Pi R u^*\|_G$$
$$\leqslant \|R\sigma_{1/\tau}\Pi\|_G\, \|Ru^*\|_G = \|R\sigma_{1/\tau}\Pi\|_G\, \|u\|_G.$$

Thus,

$$\|\kappa(t)(\Pi' Vu)^{**}(\delta(t))\|_{E_G} \leqslant C\int_0^\infty \|R\sigma_{1/\tau}\Pi\|_G\, d\min\{M_{\varphi_{E_i}}(\tau)\}\,\|u\|_G, \tag{8.16}$$

and if

$$\int_0^1 \|R\sigma_{1/\tau}\Pi\|_G\, dM_{\varphi_{E_1}} + \int_1^\infty \|R\sigma_{1/\tau}\Pi\|_G\, dM_{\varphi_{E_0}} < \infty, \tag{8.17}$$

then $V$ acts boundedly from $G$ to $(E_G)_{\kappa,\delta}(\mathfrak{M}')$.

We consider the operator $R\sigma_{1/\tau}\Pi$ in the examples 1)–3). In example 1) this operator has the form $R\sigma_{1/\tau}\Pi u(m) = u(\omega^{-1}(\tau\omega(m)))$. The transformation $\omega^{-1}(\tau\omega(m))$ may have quite complicated structure. However, it has the property that the measure of the image of any measurable set is $\tau$ times greater than the measure of the set itself. Since the space $G$ is symmetric, any one-to-one mapping $\zeta_\tau(m)$ of $\mathfrak{M}$ onto itself having the same property induces an operator $\tilde{\sigma}_{1/\tau}u(m) = u(\zeta_\tau(m))$, having the same norm in $G$. Therefore the operator $R\sigma_{1/\tau}\Pi$ in (8.17) can be replaced by the operator $\tilde{\sigma}_{1/\tau}$. For example, for $\mathfrak{M} = (-\infty, \infty)$ with Lebesgue measure it is natural to set $\tilde{\sigma}_{1/\tau}u(m) = u(\tau m)$. In the case of the semiaxis with measure $d\mu = ds/s$ we may set $\tilde{\sigma}_{1/\tau}u(s) = u(s^\tau)$. Using this, we obtain

1) *Condition* (8.17) *for* $\mathfrak{M} = (-\infty, \infty)$ *with Lebesgue measure is satisfied if*

$$\int_0^1 \|\tilde{\sigma}_{1/\tau}\|_G\, dM_{\varphi_{E_1}}(\tau) + \int_1^\infty \|\tilde{\sigma}_{1/\tau}\|_G\, dM_{\varphi_{E_0}}(\tau) < \infty, \tag{8.18}$$

*where* $\tilde{\sigma}_{1/\tau}u(m) = u(\tau m)$.

The operators $\tilde{\sigma}_t$ form a semigroup, and so the function $\|\tilde{\sigma}_t\|_G$ is submultiplicative. Therefore (8.18) is equivalent to the following condition:

$$\|\tilde{\sigma}_t\|_G = o(t^{\beta_0}) \quad \text{as } t \to 0 \qquad \text{and} \qquad \|\tilde{\sigma}_t\|_G = o(t^{\alpha_1}) \quad \text{as } t \to \infty, \tag{8.19}$$

where $\alpha_1$ is the lower dilation exponent of $\varphi_{E_1}$ and $\beta_0$ is the upper dilation exponent of $\varphi_{E_0}$.

In example 2) the operator $\tilde{\sigma}_{1/\tau} = R\sigma_{1/\tau}\Pi$ is given by

$$\tilde{\sigma}_{1/\tau}u(t) = \begin{cases} u(\tau t) & \text{if } 0 \leqslant \tau t \leqslant 1, \\ 0 & \text{if } \tau t > 1 \end{cases} \qquad (0 \leqslant t \leqslant 1). \tag{8.20}$$

The operators $\tilde{\sigma}_t$ do not always commute in this case. However, if $t_1 < t_2$, then $\tilde{\sigma}_{t_1 t_2} = \tilde{\sigma}_{t_2}\tilde{\sigma}_{t_1}$, and so $\|\tilde{\sigma}_t\|_G^1$ is again submultiplicative. Thus,

2) *Condition* (8.17) *for* $\mathfrak{M} = (0, 1)$ *with Lebesgue measure is satisfied if the operator* (8.20) *satisfies* (8.18) *or* (8.19).

In example 3) the operators $\tilde{\sigma}_{1/\tau} = R\sigma_{1/\tau}\Pi$ are described in a natural way for $\tau = \ldots, 1/3, 1/2, 1, 2, 3, \ldots$. It is easy to verify that for $u = \{u_k\}$ we have

$$\tilde{\sigma}_n u = \{\underbrace{u_1, \ldots, u_1}, \underbrace{u_2, \ldots, u_2}, \ldots, \underbrace{u_k, \ldots, u_k}, \ldots\}, \tag{8.21}$$

$$\tilde{\sigma}_{1/n} u = \left\{ \frac{u_1 + \cdots + u_n}{n}, \frac{u_{n+1} + \cdots + u_{2n}}{n}, \ldots, \frac{u_{(k-1)n+1} + \cdots + u_{kn}}{n}, \ldots \right\}. \tag{8.22}$$

The quantity $\|R\sigma_{1/\tau}\Pi\|_G$ is a decreasing function of $\tau$, and so by relaxing the condition (8.17) we arrive at the following conclusion:

3) *Condition* (8.17) *for* $\mathfrak{M} = \{1, 2, \ldots\}$ *with* $\mu(n) = 1$ *is satisfied if*

$$\sum_{n=1}^{\infty} \|\tilde{\sigma}_{n+1}\|_G \left[ M_{\varphi_{E_1}}\left(\frac{1}{n}\right) - M_{\varphi_{E_1}}\left(\frac{1}{n+1}\right) \right] + \sum_{n=1}^{\infty} \|\tilde{\sigma}_{1/n}\|_G \left[ M_{\varphi_{E_0}}(n+1) - M_{\varphi_{E_0}}(n) \right] < \infty, \tag{8.23}$$

*where the operators* $\tilde{\sigma}_n$ *and* $\tilde{\sigma}_{1/n}$ *are defined by* (8.21) *and* (8.22).

We pass to the description of the space $(E_G)_{\kappa,\delta}(\mathfrak{M}')$ into which the operator $V$ maps. According to the definitions, we have

$$\|v\|_{(E_G)_{\kappa,\delta}(\mathfrak{M}')} = \|\kappa(t)(\Pi' v)^{**}(\delta(t))\|_{E_G} = \|R[\kappa(t)(\Pi' v)^{**}(\delta(t))]^*\|_G. \tag{8.24}$$

It is sometimes convenient to replace the inequality

$$\|Vu\|_{(E_G)_{\kappa,\delta}(\mathfrak{M}')} \leqslant C\|u\|_G \tag{8.25}$$

by a cruder one, using the following arguments: If we can replace the function $\kappa(t)(\Pi' v)^{**}(\delta(t))$ by a smaller function of the form $\Pi\tilde{v}$, where $\tilde{v} \in L_1(\mathfrak{M}) + L_\infty(\mathfrak{M})$, then by (8.11) and (8.2) we obtain

$$\|\tilde{v}\|_G = \|R\tilde{v}^*\|_G = \|R(\Pi\tilde{v})^*\|_G = \|\Pi\tilde{v}\|_{E_G} \leqslant \|v\|_{(E_G)_{\kappa,\delta}(\mathfrak{M}')}.$$

Thus, (8.25) implies the inequality

$$\|\widetilde{V_u}\|_G \leqslant C\|u\|_G. \tag{8.26}$$

We illustrate this by the case where $\mathfrak{M} = \{1, 2, \dots\}$. We write $\delta_n = \sup_{n-1 \leqslant t \leqslant n} \delta(t)$ and $\kappa_n = \inf_{n-1 \leqslant t \leqslant n} \kappa(t)$. Then, using the fact that the function $(\Pi' v)^{**}(t)$ is monotonic, we can replace $\kappa(t)(\Pi' v)^{**}(\delta(t))$ by a smaller function assuming the constant values $\kappa_n(\Pi' v)^{**}(\delta_n)$ on the intervals $(n-1, n]$. Inequality (8.26) then takes the following form:

$$\|\{\kappa_n(\Pi' Vu)^{**}(\delta_n)\}\|_G \leqslant C\|\{u_n\}\|_G. \tag{8.27}$$

If $\mathfrak{M}' = \{1, 2, \dots\}$, then $\Pi' v$ is constant on the intervals $(n-1, n]$, and so is $(\Pi' v)^*$. Denote by $v_n^*$ the value of this function on $(n-1, n]$. It is more convenient to also replace the function $(\Pi' v)^{**}$ by a smaller one assuming the value

$$v_n^{**} = \frac{1}{n} \sum_1^n v_k^*$$

on $(n-1, n]$.

If $\mathfrak{M} = \mathfrak{M}' = \{1, 2, \dots\}$, then (8.27) implies the inequality

$$\|\{\kappa_n v_{\nu(n)}^{**}\}\|_G \leqslant C\|\{u_n\}\|_G, \tag{8.28}$$

where $\{v_n\} = V\{u_n\}$ and $\nu(n)$ is an integer no smaller than $\delta_n$.

We note that the expression (8.24) for the norm becomes essentially simpler if $E_0 = F_0$ and $E_1 = F_1$. Then $\kappa(t) \equiv 1$ and $\delta(t) \equiv t$. Therefore

$$\|v\|_{(E_G)_{\kappa,\delta(\mathfrak{M}')}} = \|R(\Pi' v)^{**}\|_G. \tag{8.29}$$

We shall not formulate the theorems which can be obtained from Theorem 8.2 in the various cases. The above arguments enable us to apply these theorems in concrete situations.

5. ***Spaces of measurable functions on* (0, ∞), *with a measure that is invariant under dilation.*** We consider the measure on $(0, \infty)$ defined by the differential form $dt/t$. This measure is invariant under dilation. We denote by $L_{p,*}$ the spaces $L_p((0, \infty), dt/t)$, constructed from this measure (for $p = \infty$ we obviously have $L_{p,*} = L_p$, but we keep the *, for the sake of uniformity). We shall be interested in interpolation spaces $E$ between $L_{1,*}$ and $L_{\infty,*}$. From what we said in subsection 2 it follows, for instance, that $L_{p,*}$ are such spaces.

We consider the simplest operators acting in interpolation spaces $E$.

1) The dilation operator $\sigma_\tau \varphi(\tau) = \varphi(t/\tau)$ acts in the spaces $L_{1,*}$ and $L_{\infty,*}$ and has norm 1 there. Therefore

$$\|\sigma_\tau \varphi\|_E \leqslant k_E \|\varphi\|_E,$$

where $k_E$ is the interpolation constant of $E$ relative to the couple $L_{1,*}$, $L_{\infty,*}$.

2) The *power-type transformation* $T_\lambda \varphi(t) = \varphi(t^\lambda)$ acts in $L_{\infty,\bullet}$ with norm 1 and in $L_{1,\bullet}$ with norm $|\lambda|^{-1}$, and therefore

$$\|T_\lambda \varphi\|_E = \|\varphi(t^\lambda)\|_E \leqslant k_E \max\{1, |\lambda|^{-1}\}\|\varphi\|_E. \tag{8.30}$$

3) *Multiplicative convolution transformation* (cf. §6.9). If $\psi \in L_{1,\bullet}$, then for $\varphi \in L_{\infty,\bullet}$ we have

$$\left\| \int_0^\infty \psi\left(\frac{t}{s}\right)\varphi(s)\frac{ds}{s} \right\|_{L_{\infty,\bullet}} \leqslant \|\psi\|_{L_{1,\bullet}}\|\varphi\|_{L_{\infty,\bullet}},$$

while if $\varphi \in L_{1,\bullet}$, then

$$\left\| \int_0^\infty \psi\left(\frac{t}{s}\right)\varphi(s)\frac{ds}{s} \right\|_{L_{1,\bullet}} \leqslant \|\psi\|_{L_{1,\bullet}}\|\varphi\|_{L_{1,\bullet}}.$$

Therefore we have the inequality

$$\left\| \int_0^\infty \psi\left(\frac{t}{s}\right)\varphi(s)\frac{ds}{s} \right\|_E \leqslant k_E\|\psi\|_{L_{1,\bullet}}\|\varphi\|_E, \tag{8.31}$$

which we shall call *Young's inequality*.

4) The operators of Hardy-Littlewood type

$$H_l\varphi(t) = \frac{1}{t^{1+l}}\int_0^t s^l\varphi(s)\,ds \tag{8.32}$$

can be expressed in terms of convolution:

$$H_l\varphi(t) = \int_0^\infty \left(\frac{t}{s}\right)^{-(l+1)} \chi_{(1,\infty)}\left(\frac{t}{s}\right)\varphi(s)\frac{ds}{s}. \tag{8.33}$$

The function $t^{-(l+1)}\chi_{(1,\infty)}(t)$ belongs to $L_{1,\bullet}$ for $l > -1$, and its norm in $L_{1,\bullet}$ is equal to $1/(1+l)$. Therefore (8.31) implies *Hardy's inequality*

$$\left\| \frac{1}{t^{1+l}}\int_0^t s^l\varphi(s)\,ds \right\|_E \leqslant \frac{k_E}{1+l}\|\varphi\|_E \tag{8.34}$$

for $l > -1$.

We construct an averaging operator by the formula

$$T_{\text{avg}} = \sum_{-\infty}^{\infty} \chi_{(e^n, e^{n+1}]}(t)\int_{e^n}^{e^{n+1}} \varphi(s)\frac{ds}{s}. \tag{8.35}$$

It is easy to see that this operator acts, and has norm 1, in the spaces $L_{1,\bullet}$ and $L_{\infty,\bullet}$ (similarly to the situation in §3.2). Therefore its norm in $E$ does not exceed $k_E$.

In what follows we need spaces with power weights generated by interpolation spaces between $L_{1,\bullet}$ and $L_{\infty,\bullet}$. We shall denote by $E_{t^\alpha}$, or, for short, $E^\alpha$, the space with norm $\|x\|_{E^\alpha} = \|t^\alpha x\|_E$. We consider how the above operators act in these spaces.

*Dilation operators.*

$$\|\sigma_\tau\varphi\|_{E^\alpha} = \|t^\alpha\varphi(\tau^{-1}t)\|_E = \tau^\alpha\|(\tau^{-1}t)^\alpha\varphi(\tau^{-1}t)\|_E \leqslant k_E\tau^\alpha\|\varphi\|_{E^\alpha}. \quad (8.36)$$

*Power-type transformation.* By (8.30) we have

$$\|t^{\alpha\lambda}\varphi(t^\lambda)\|_E \leqslant k_E \max\{1, |\lambda|^{-1}\}\|t^\alpha\varphi(t)\|_E,$$

and therefore

$$\|T_\lambda\varphi\|_{E^{\alpha\lambda}} \leqslant k_E \max\{1, |\lambda|^{-1}\}\|\varphi\|_{E^\alpha}. \quad (8.37)$$

*Multiplicative convolution transformation.* From (8.31) we obtain that

$$\left\|t^\alpha\int_0^\infty \left(\frac{t}{s}\right)\varphi(s)\frac{ds}{s}\right\|_E = \left\|\int_0^\infty \left(\frac{t}{s}\right)^\alpha\psi\left(\frac{t}{s}\right)s^\alpha\varphi(s)\frac{ds}{s}\right\|_E$$

$$\leqslant \|t^\alpha\psi(t)\|_{L_{1,\bullet}}\|\varphi\|_{E^\alpha}. \quad (8.38)$$

Thus, the convolution operator acts in $E^\alpha$ if $\|t^\alpha\psi\|_{L_{1,\bullet}} < \infty$.

*Hardy-Littlewood type operators.* We write the expression for the operator in a still more general form,

$$H_{l,m}\varphi(t) = \frac{1}{t^{1+m}}\int_0^t s^l\varphi(s)\,ds. \quad (8.39)$$

We have

$$t^{m+\alpha-l}H_{l,m}\varphi(t) = \frac{1}{t^{1+l-\alpha}}\int_0^t s^{l-\alpha}s^\alpha\varphi(s)\,ds.$$

Assuming that $l > \alpha - 1$, from (8.34) we obtain

$$\|H_{l,m}\varphi\|_{E^{m+\alpha-l}} \leqslant \frac{k_E}{1+l-\alpha}\|\varphi\|_{E^\alpha}. \quad (8.40)$$

We may consider similarly the adjoint operator

$$H^1_{l,m}\varphi(t) = t^l\int_t^\infty \varphi(s)\frac{ds}{s^{1+m}}.$$

We have

$$t^{m-l+\alpha}H^1_{l,m}\varphi(t) = \int_0^\infty \left(\frac{t}{s}\right)^{\alpha+m}\chi_{(0,1]}\left(\frac{t}{s}\right)s^\alpha\varphi(s)\frac{ds}{s}.$$

The function $t^{\alpha+m}\chi_{(0,1)}(t)$ belongs to $L_{1,\bullet}$ if $\alpha + m > 0$, and its norm in $L_{1,\bullet}$ is equal to $1/(\alpha + m)$. Under this condition we have

$$\|H^1_{l,m}\varphi\|_{E^{m-l+\alpha}} \leqslant \frac{k_E}{\alpha+m}\|\varphi\|_{E^\alpha}. \quad (8.41)$$

*Averaging operator.* If $e^n < t \leqslant e^{n+1}$, then we have

$$|t^\alpha T_{\text{avg}}\varphi(t)| \leqslant \int_{e^n}^{e^{n+1}} \frac{t^\alpha}{s^\alpha}s^\alpha|\varphi(s)|\frac{ds}{s} \leqslant e^{|\alpha|}\int_{e^n}^{e^{n+1}} s^\alpha|\varphi(s)|\frac{ds}{s}.$$

Thus, for all $t$

$$|t^{\alpha}T_{\mathrm{avg}}\varphi(t)| \leqslant e^{|\alpha|}T_{\mathrm{avg}}(t^{\alpha}|\varphi(t)|).$$

Then

$$\|T_{\mathrm{avg}}\varphi\|_{E^{\alpha}} \leqslant k_E e^{|\alpha|}\|\varphi\|_{E^{\alpha}}. \tag{8.42}$$

Without loss of generality (see Chapter I, §4.6) we may assume that the space $E$ is constructed from $L_{1,*}$ and $L_{\infty,*}$ by means of some interpolation functor. This enables us to construct, for any Banach couple $(A, B)$, a space $F$ intermediate between $A$ and $B$ such that $(L_{1,*}, L_{\infty,*}, E)$ is an interpolation triple relative to the triple $(A, B, F)$. We apply this reasoning to the trivial case where $A = B = \mathbf{R}$ (or $\mathbf{C}$). The integrals

$$\int_0^a t^{\beta}\varphi(t)\frac{dt}{t}\ (a > 0),\quad \int_b^{\infty} t^{-\beta}\varphi(t)\frac{dt}{t}\ (b > 0)\quad \text{and}\quad \int_a^b \varphi(t)\,dt$$

are finite for $\beta > 0$ for functions from $L_{1,*}$ and for functions from $L_{\infty,*}$, and therefore they realize bounded linear operators from these spaces into $\mathbf{R}$ (or $\mathbf{C}$). From this it follows that these operators act boundedly from $E$ into $\mathbf{R}$ (or $\mathbf{C}$). In particular, all functions from $E$ are locally summable. If $\varphi \in E^{\alpha}$, then

$$\int_0^a t^{-\beta}\varphi(t)\frac{dt}{t} \leqslant C\|\varphi\|_{E^{\alpha}} \quad \text{for } \beta > \alpha, \tag{8.43}$$

$$\int_b^{\infty} t^{-\beta}\varphi(t)\frac{dt}{t} \leqslant C\|\varphi\|_{E^{\alpha}} \quad \text{for } \beta > -\alpha. \tag{8.44}$$

**6. *Operators in spaces of measurable functions on the interval* [0, 1]. *Conjugation operator.*** We consider the Hardy-Littlewood operator on [0, 1]. It is defined by the same formula as in §6.7:

$$Hx(t) = \frac{1}{t}\int_0^t x(s)\,ds.$$

As in §§6.6 and 6.7, with the aid of (6.1′) and (6.29) we can show that $H$ *acts in a symmetric space $E$ if and only if $\|\tilde{\sigma}_t\|_E = o(t)$ as $t \to \infty$.* The adjoint operator has the form

$$H_1x(t) = \int_t^1 x(s)\frac{ds}{s} \qquad (0 \leqslant t \leqslant 1).$$

In the same way as on p. 139 we can verify that $\|H\|_{E\to E^{11}} = \|H_1\|_{E^1\to E^1}$, and it can be shown that *in the case where $H$ acts from $E$ into $E^{11}$ we have $\|\tilde{\sigma}_t\|_{E\to E^{11}} = o(t)$ as $t \to \infty$.*

*For $H_1$ to act boundedly in a symmetric space $E$ it is necessary and sufficient that $\|\tilde{\sigma}_t\|_E = o(t)$ as $t \to 0$.* We repeat the proof of the necessity. For $0 < t \leqslant \tau \leqslant 1$ we have

$$\int_t^1 x^*(s)\frac{ds}{s} \geqslant \int_t^{t/\tau} x^*(s)\frac{ds}{s} \geqslant \ln\frac{1}{\tau}x^*\left(\frac{t}{\tau}\right).$$

The integral on the right is nonnegative for $t > \tau$, and therefore, using the definition of the operators $\tilde{\sigma}_t$ (p. 164), we may write

$$\int_t^1 x^*(s)\frac{ds}{s} \geqslant \ln\frac{1}{\tau}\tilde{\sigma}_\tau x^*(t).$$

Hence for $\tau < 1$

$$\|\tilde{\sigma}_\tau x\|_E = \|\tilde{\sigma}_\tau x^*\|_E \leqslant \frac{1}{\ln\tau^{-1}}\|H_1 x^*\|_E \leqslant \frac{1}{\ln\tau^{-1}}\|H_1\|_E\|x\|_E.$$

The assertion is proved.

We may prove similarly that the conditions

$$\|\tilde{\sigma}_t\|_E = o(1) \quad \text{as } t \to 0 \qquad \text{and} \qquad \|\tilde{\sigma}_t\|_E = o(t) \quad \text{as } t \to \infty \tag{8.45}$$

are necessary and sufficient for the operator

$$\Gamma x(t) = \int_0^1 \frac{x(s)}{t+s}\,ds \qquad (0 \leqslant t \leqslant 1)$$

to act in a symmetric space $E$.

Now we consider the singular Hilbert operator

$$Sx(t) = \text{p.v.}\int_0^1 \frac{x(s)}{t-s}\,ds \qquad (0 \leqslant t \leqslant 1). \tag{8.46}$$

From (7.9) it follows immediately that

$$(S\chi_e)^*(t) \leqslant \operatorname{arcsinh}\frac{2\operatorname{mes} e}{t} \qquad (0 < t \leqslant 1, e \subset [0, 1]).$$

This inequality implies again that the operator $S$ is of weakened type $\{L_p, L_p\}$ for any $p \in (1, \infty)$. If conditions (8.45) are satisfied, then repeating the arguments in the proof of Theorem 7.2 (using the submultiplicativity of $\|\tilde{\sigma}_t\|_E$), we may apply the interpolation theorem described in §8.2. Then from (8.16) we obtain that $\|Sx\|_E \leqslant C\|x\|_E$.

Now we show that conditions (8.45) are necessary for the boundedness of the Hilbert operator.

We can write (8.46) in the form

$$Sx\left(\frac{\tau+1}{2}\right) = \text{p.v.}\int_{-1}^1 \frac{x((\sigma+1)/2)}{\tau-\sigma}\,d\sigma \qquad (-1 \leqslant \tau \leqslant 1). \tag{8.47}$$

On $[-1, 1]$ we introduce a symmetric space $E(-1, 1)$ of functions with norm $\|y\|_{E(-1,1)} = \|y(2t-1)\|_E$. If the Hilbert operator acts in $E$, then by (8.47) it acts in $E(-1, 1)$ as well. Reasoning in the same way as in the proof of Theorem 7.2, from this we obtain that $\Gamma$ acts in the space $\tilde{E}(0, 1)$ consisting of restrictions of the functions from $E(-1, 1)$ to $(0, 1)$. This implies that conditions (8.45) are satisfied for $\tilde{E}(0, 1)$, and consequently for $E(-1, 1)$ as well. Under the isometric correspondence between the spaces $E(-1, 1)$ and $E$, to a dilation operator in $E(-1, 1)$ there corresponds the dilation operator in $E$ with respect to the center $1/2$ with the same dilation coefficient. It is easy to verify that the latter operator has the same norm as the dilation operator with respect to the center 0. Thus, condition (8.45) is satisfied for $E$.

Thus, *condition* (8.45) *is necessary and sufficient for the singular Hilbert operator to act boundedly in E.*

In the theory of trigonometric series, an important role is played by the conjugation operator, i.e. the operator that passes to the conjugate function (see [2] and [50]). It can be written in the form

$$Qx(t) = \text{p.v.}\ \frac{1}{\pi}\int_0^{2\pi} \frac{x(s)}{2\operatorname{tg}((t-s)/2)}\, ds.$$

Let a symmetric space of functions on $[0, 2\pi]$ have property (8.45). Then the Hilbert operator

$$\frac{1}{\pi} Sx(t) = \text{p.v.}\ \frac{1}{\pi}\int_0^{2\pi} \frac{x(s)}{t-s}\, ds$$

acts boundedly in it.

Consider the subspace $E_1$ of $E$ consisting of functions with support in the interval $[\pi/2, 3\pi/2]$. On such functions the operator $Q$ takes the form

$$Qx(t) = \text{p.v.}\ \frac{1}{\pi}\int_{\pi/2}^{3\pi/2} \frac{x(s)}{2\operatorname{tg}((t-s)/2)}\, ds.$$

The quantity $(t-s)/2$ runs over the interval $[-3\pi/4, 3\pi/4]$. On this interval the function $1/(t-s)$ differs from $1/2\operatorname{tg}((t-s)/2)$ by a bounded function. Therefore $Q$ differs from the Hilbert operator $S/\pi$ on the subspace $E_1$ by an operator mapping $E_1$ into $L_\infty \subset E$. Hence it follows that $Q$ acts from $E_1$ into $E$.

We consider a subspace $E_2$ complementary to $E_1$. It consists of functions with support in $[0, \pi/2] \cup [3\pi/2, 2\pi]$. On $E_2$ we have

$$Qx(t) = \text{p.v.}\ \frac{1}{\pi}\int_0^{\pi/2} \frac{x(s)}{2\operatorname{tg}\frac{t-s}{2}}\, ds + \text{p.v.}\ \frac{1}{\pi}\int_{3\pi/2}^{2\pi} \frac{x(s)}{2\operatorname{tg}\frac{t-s}{2}}\, ds.$$

If we assume that the function $x(t)$ is extended periodically with period $2\pi$, we have

$$Qx(t) = \text{p.v.}\ \frac{1}{\pi}\int_{\pi}^{3\pi/2} \frac{x(z-\pi)}{2\operatorname{tg}\frac{t-z+\pi}{2}}\,dz + \text{p.v.}\ \frac{1}{\pi}\int_{\pi/2}^{\pi} \frac{x(z+\pi)}{2\operatorname{tg}\frac{t-z-\pi}{2}}\,dz$$

$$= \text{p.v.}\ \frac{1}{\pi}\int_{\pi/2}^{3\pi/2} \frac{x(z-\pi)}{2\operatorname{tg}\frac{t-z+\pi}{2}}\,dz = Qy(t+\pi),$$

where the function $y(z) = x(z-\pi)$ has support in $[\pi/2, 3\pi/2]$. Then

$$\|Qx\|_E = \|Qy\|_E \leqslant \|Q\|_{E_1\to E}\|y\|_{E_1} = \|Q\|_{E_1\to E}\|x\|_E.$$

Thus, $Q$ acts boundedly from $E$ into $E$.

Conversely, if $Q$ is bounded, then it acts from $E_1$ into $E$ and from $E_2$ into $E$. From the preceding it follows that the operator $S/\pi$ also acts from $E_1$ into $E$. We consider this operator on $E_2$. Let the support of $x(t)$ be contained in $[0, \pi/2]$. Then

$$Sx(t) = \text{p.v.}\int_0^{\pi/2} \frac{x(s)}{t-s}\,ds = \text{p.v.}\int_{\pi/2}^{\pi} \frac{x(\sigma-\pi/2)}{t-\sigma+\pi/2}\,d\sigma$$

$$= \text{p.v.}\int_{\pi/2}^{\pi} \frac{y(\sigma)}{t-\sigma+\pi/2}\,d\sigma,$$

where $y(\sigma) = x(\sigma - \pi/2)$ for $\sigma \in [\pi/2, \pi]$ and $y(\sigma) = 0$ for $\sigma \notin [\pi/2, \pi]$. The function $y$ is equimeasurable with $x$. For $0 \leqslant t \leqslant 3\pi/2$ we have $Sx(t) = Sy(t+\pi/2)$, and for $3\pi/2 \leqslant t \leqslant 2\pi$ we have

$$|Sx(t)| \leqslant \frac{1}{\pi}\int_{\pi/2}^{\pi} |y(\sigma)|\,d\sigma,$$

therefore

$$\|Sx\|_E \leqslant \|\chi_{[0,3\pi/2]}Sx\|_E + \|\chi_{[3\pi/2,2\pi]}Sx\|_E$$

$$\leqslant \|Sy\|_E + \frac{1}{\pi}\|y\|_{L_1} \leqslant \left(\|S\|_{E_1\to E} + \frac{1}{\pi}\right)\|y\|_E \leqslant C\|x\|_E.$$

This inequality can be established similarly for functions from $E$ with support in $[3\pi/2, 2\pi]$. Thus the Hilbert operator acts in $E$, and hence condition (8.45) is satisfied. We have arrived at the following assertion.

THEOREM 8.3. *For the conjugation operator to act in a symmetric space $E$ it is necessary and sufficient that conditions* (8.45) *be satisfied.*

## §9. Applications to the theory of orthogonal series

**1.** ***The generalized Paley theorem.*** We assume that an orthonormal system of functions $u_n(t)$ is given on the interval $[0, 1]$. We denote by $\{c_n\}$ the Fourier coefficients of a function $x(t) \in L_1(0, 1)$:

$$c_n = \int_0^1 x(t)u_n(t)\,dt. \tag{9.1}$$

By the Bessel inequality

$$\sum_1^\infty c_n^2 \leqslant \|x\|_{L_2}^2$$

the Fourier operator $Vx = \{c_n\}$ is a bounded linear operator from $L_2(0, 1)$ into $l_2$.

Let $H(0, \infty)$ be a symmetric space on the semiaxis and let $H(0, 1)$ be the corresponding space on $[0, 1]$. We assume that, in addition, we know that the system $\{u_n\}$ is bounded in the space $H(0, 1)$: $\|u_n\|_H \leqslant C$. Then we obtain the estimate

$$|c_n| = \left|\int_0^1 x(t)u_n(t)\,dt\right| \leqslant C\|x\|_{H^1(0,1)}$$

for the Fourier coefficients. This implies that the Fourier operator $V$ acts boundedly from $H^1(0, 1)$ into $l_\infty$. We denote by $\varphi(t)$ the fundamental function of $H^1(0, \infty)$ and assume that it has the following properties: $\varphi(1) = 1$, the function $\varphi(t)/\sqrt{t}$ is increasing and $\varphi(t)/\sqrt{t} \to 0$ as $t \to 0$. We are thus able to invoke Theorem 8.2. In our case,

$$\varphi_{E_0}(t) = \sqrt{t}, \quad \varphi_{E_1}(t) = \varphi(t), \quad \psi_{F_0}(t) = \sqrt{t}, \quad \psi_{F_1}(t) = \operatorname{sgn} t.$$

Therefore $\delta(t) = t/\varphi^2(t)$ and $\kappa(t) = 1/\varphi(t)$.

Repeating the reasoning in the proof of Theorem 8.2, we consider the operator $\Pi' VR$, where $R$ is the operator of restriction of a function given on the semiaxis to the interval $[0, 1]$, and $\Pi'\{c_n\}$ is the function equal to $c_n$ on $(n - 1, n]$. The operator $\Pi' VR$ acts boundedly from $L_2(0, \infty)$ to $L_2(0, \infty)$ and from $H^1(0, \infty)$ to $L_\infty(0, \infty)$. Applying Theorem 6.1′ to it, we obtain

$$\left\|\frac{1}{\varphi(t)}(\Pi' VRx)^{**}\left(\frac{t}{\varphi^2(t)}\right)\right\|_{E(0,\infty)} \leqslant C\|x\|_{E(0,\infty)}.$$

If $x(t) \equiv 0$ for $t > 1$, then

$$\left\|\frac{1}{\varphi(t)}(\Pi'\{c_n\})^{**}\left(\frac{t}{\varphi^2(t)}\right)\right\|_{E(0,\infty)} \leqslant C\|x\|_{E(0,1)}.$$

If we denote by $c_n^*$ the rearrangement of the sequence of $|c_n|$ in decreasing order, then $(\Pi'\{c_n\})^*(t) = \sum_1^\infty c_n^* \chi_{(n-1,n]}(t)$. The function $(\Pi'\{c_n\})^{**}(t)$ can be replaced by a smaller function, namely

$$\sum_1^\infty c_n^{**} \chi_{(n-1,n]}(t), \quad \text{where } c_n^{**} = \frac{1}{n} \sum_{k=1}^n c_k^*. \tag{9.2}$$

If we now denote by $a_n$ the solution of the equation $t/\varphi^2(t) = n$ $(n = 1, 2, \ldots)$, then from the last inequality it follows that

$$\left\| \frac{1}{\varphi(t)} \sum_1^\infty c_n^{**} \chi_{(a_n, a_{n-1}]}(t) \right\|_{E(0,\infty)} \leqslant C\|x\|_{E(0,1)}.$$

This inequality may be made even more crude by replacing $1/\varphi(t)$ on the intervals $(a_n, a_{n-1}]$ by $1/\varphi(a_{n-1})$ and the norm in $E(0, \infty)$ by the norm in $E(0, 1)$. Then

$$\left\| \sum_2^\infty \frac{c_n^{**}}{\varphi(a_{n-1})} \chi_{(a_n, a_{n-1}]}(t) \right\|_{E(0,1)} \leqslant C\|x\|_{E(0,1)}. \tag{9.3}$$

The functions $\chi_{(a_n, a_{n-1}]}(t)$ form an orthonormal system on $[0, 1]$. It is sometimes convenient to normalize them, i.e. divide them by $\sqrt{a_{n-1} - a_n}$.

Since $a_n/\varphi^2(a_n) = n$, we have

$$(n-1)a_{n-1} - na_n = \frac{a_{n-1}^2}{\varphi^2(a_{n-1})} - \frac{a_n^2}{\varphi^2(a_n)} \geqslant 0$$

in view of the quasiconcavity of $\varphi(t)$. Hence

$$\frac{\sqrt{a_{n-1} - a_n}}{\varphi(a_{n-1})} = \sqrt{\frac{(n-1)(a_{n-1} - a_n)}{a_{n-1}}}$$

$$= \sqrt{\frac{n-1}{n}} \sqrt{\frac{n(a_{n-1} - a_n)}{a_{n-1}}} \geqslant \frac{1}{\sqrt{2}}.$$

Therefore (9.3) implies the inequality

$$\left\| \sum_2^\infty c_n^{**}(a_{n-1} - a_n)^{-1/2} \chi_{(a_n, a_{n-1}]}(t) \right\|_{E(0,1)} \leqslant C\|x\|_{E(0,1)}. \tag{9.4}$$

We have arrived at the following assertion.

THEOREM 9.1. *Let $\{u_n\}$ be an orthonormal sequence on* $[0, 1]$ *that is bounded in the norm of a symmetric space $H(0, \infty)$. Assume that the fundamental function $\varphi_H(t)$ of $H(0, \infty)$ has the following properties*:

1) $\varphi_H(1) = 1$.

2) *The function $\sqrt{t}/\varphi_H(t)$ is increasing.*

3) $\sqrt{t}/\varphi_H(t) \to 0$ *as* $t \to 0$.

*If $E(0, \infty)$ is a symmetric space for which*

$$\|\sigma_t\|_E = o(\sqrt{t}\,) \quad \text{as } t \to 0 \qquad \text{and} \qquad \|\sigma_t\|_E = o(t^{1-\delta_H}) \quad \text{as } t \to \infty, \tag{9.5}$$

*where $\delta_H$ is the upper dilation exponent of $\varphi_H(t)$, then (9.4) holds, where the $c_n^{**}$ are calculated according to (9.2) and the $a_n$ are the solutions of the equations $\varphi_H^2(t)/t = n$.*

In the formulation we have used relation (4.39): $\varphi(t) = t/\varphi_H(t)$.

We consider special cases of Theorem 9.1. Assume that $E = L_p$. Conditions (9.5) will be satisfied if $1/2 < p^{-1} < 1 - \delta_H$. Then we obtain

$$\left\| \sum_2^\infty c_n^{**}(a_{n-1} - a_n)^{-1/2} \chi_{(a_n, a_{n-1}]} \right\|_{L_p} = \left\{ \sum_2^\infty (c_n^{**})^p (a_{n-1} - a_n)^{1-p/2} \right\}^{1/p} \leqslant C\|x\|_{L_p}. \tag{9.6}$$

In the case of a Lorentz space $\Lambda_\varphi = E$ conditions (9.5) are satisfied if $1/2 < \gamma_\psi \leqslant \delta_\psi < 1 - \delta_H$, and then

$$\begin{aligned} \left\| \sum_2^\infty c_n^{**}(a_{n-1} - a_n)^{-1/2} \chi_{(a_n, a_{n-1}]} \right\|_{\Lambda_\psi} &\geqslant \int_0^1 \sum_2^\infty c_n^{**}(a_{n-1} - a_n)^{-1/2} \chi_{(a_n, a_{n-1}]}(t)\, d\psi(t) \\ &= \sum_2^\infty c_n^{**}(a_{n-1} - a_n)^{-1/2} [\psi(a_{n-1}) - \psi(a_n)]. \end{aligned}$$

Therefore

$$\sum_2^\infty c_n^{**}(a_{n-1} - a_n)^{-1/2} [\psi(a_{n-1}) - \psi(a_n)] \leqslant C\|x\|_{\Lambda_\psi}. \tag{9.7}$$

Finally, for a Marcinkiewicz space $M_\psi$ we have

$$\begin{aligned} \left\| \sum_2^\infty c_n^{**}(a_{n-1} - a_n)^{-1/2} \chi_{(a_n, a_{n-1}]} \right\|_{M_\psi} &\geqslant \frac{1}{\psi(a_k)} \int_0^{a_k} \sum_2^\infty c_n^{**}(a_{n-1} - a_n)^{-1/2} \chi_{(a_n, a_{n-1}]}(t)\, dt \\ &= \frac{1}{\psi(a_k)} \sum_2^k c_n^{**}(a_{n-1} - a_n)^{1/2}. \end{aligned}$$

Thus for $1/2 < 1 - \delta_\psi \leqslant 1 - \gamma_\psi < 1 - \delta_H$ we obtain

$$\sup_k \frac{1}{\psi(a_k)} \sum_2^k c_n^{**}(a_{n-1} - a_n)^{1/2} \leqslant C\|x\|_{M_\psi}.$$

We consider still another special case, namely where the sequence $u_n(t)$ is uniformly bounded, i.e. bounded in $L_\infty$. Then $\varphi_H(t) = \operatorname{sgn} t$ and $\varphi_H^2(t)/t = 1/t$. Hence $a_n = 1/n$. Inequality (9.4) then takes the form

$$\left\| \sum_2^\infty c_n^{**}[n(n+1)]^{1/2}\chi_{(1/n,1/(n-1))} \right\|_{E(0,1)} \leqslant C\|x\|_{E(0,1)},$$

or the equivalent form

$$\left\| \sum_2^\infty c_n^{**} n\chi_{(1/n,1/(n+1))} \right\|_{E(0,1)} \leqslant C\|x\|_{E(0,1)}.$$

For the spaces $L_p$ with $1 < p \leqslant 2$ we obtain

$$\left\{ \sum_2^\infty (c_n^{**})^p n^{p-2} \right\}^{1/p} \leqslant C\|x\|_{L_p}.$$

If we replace $c_n^{**}$ by the smaller number $c_n^*$, the inequality thus obtained is the content of the classical Paley theorem.

**2. *Generalized Hardy-Littlewood theorem.*** For trigonometric series with decreasing coefficients, the preceding assertions admit essential sharpening.

THEOREM 9.3. *Let $u_n(t) = \cos \pi nt$, and let the space $G(0, 1)$ have the following properties*:

$$\|\tilde{\sigma}_t\|_G = o(1) \quad \textit{as } t \to 0 \qquad \textit{and} \qquad \|\tilde{\sigma}_t\|_G = o(t) \quad \textit{as } t \to \infty.$$

*If the sequence $c_n$ is decreasing and converges to zero, then for the function $x(t) = \sum_1^\infty c_n \cos \pi nt$ to belong to $G$ it is necessary and sufficient that*

$$\left\| \sum_1^\infty kc_k\chi_{(1/(k+1),1/k)} \right\|_G < \infty, \tag{9.8}$$

*or*

$$\left\| \sum_1^\infty c_k\chi_{(0,1/k)} \right\|_G < \infty. \tag{9.9}$$

PROOF. We use some inequalities for the trigonometric system (see [50], vol. II, Chapter XII, §6). The first asserts that

$$c_n \leqslant B\int_0^{1/n} |x(s)|\, ds.$$

Hence, since $x^{**}(t)$ is monotone, we obtain

$$\begin{aligned}\sum_1^\infty kc_k\chi_{(1/(k+1),1/k)}(t) &\leqslant B\sum_1^\infty k\int_0^{1/k}|x(s)|\,ds\,\chi_{(1/(k+1),1/k)}(t)\\ &\leqslant B\sum_1^\infty k\int_0^{1/k}x^*(s)\,ds\,\chi_{(1/(k+1),1/k)}(t)\\ &\leqslant B\frac{1}{t}\int_0^t x^*(s)\,ds.\end{aligned}$$

The Hardy-Littlewood operator acts continuously in $G$ (Theorem 6.4), and therefore

$$\left\|\sum_1^\infty kc_k\chi_{(1/(k+1),1/k)}\right\|_G \leqslant C\|x\|_G.$$

To prove the converse we use the inequality

$$|x(t)| \leqslant B\sum_1^n c_k \quad \text{for } \frac{1}{n+1} \leqslant t \leqslant \frac{1}{n}$$

(see [50], vol. II, Chapter XII, §6). Then

$$|x(t)| \leqslant B\sum_1^\infty\left(\sum_1^n c_k\right)\chi_{(1/(n+1),1/n)}(t). \tag{9.10}$$

We calculate the value of the adjoint $H_1$ of the Hardy-Littlewood operator at the function $y(t) = \sum_1^\infty kc_k\chi_{(1/(k+1),1/k)}(t)$. For $1/(n+1) < t < 1/n$ we have

$$\begin{aligned}H_1y(t) &= \int_t^1 \frac{\sum_1^\infty kc_k\chi_{(1/(k+1),1/k)}(s)}{s}\,ds\\ &\geqslant \int_{1/n}^1 \frac{\sum_1^\infty kc_k\chi_{(1/(k+1),1/k)}(s)}{s}\,ds\\ &= \sum_1^{n-1} kc_k\ln(1+1/k) \geqslant \ln 2\sum_1^{n-1} c_k,\end{aligned}$$

i.e.

$$\begin{aligned}H_1y(t) &\geqslant \ln 2\sum_1^\infty\left(\sum_1^{n-1} c_k\right)\chi_{(1/(n+1),1/n)}(t)\\ &= \ln 2\sum_1^\infty\left(\sum_1^n c_k\right)\chi_{(1/(n+2),1/(n+1))}(t).\end{aligned}$$

Then

$$H_1 y\left(\frac{t}{2}\right) \geqslant \ln 2 \sum_1^\infty \left(\sum_1^n c_k\right)\chi_{(1/(n+2),1/(n+1))}\left(\frac{t}{2}\right)$$
$$\geqslant \ln 2 \sum_1^\infty \left(\sum_1^n c_k\right)\chi_{(1/(n+1),\,1/n)}(t). \tag{9.11}$$

Thus,

$$\|x\|_G \leqslant B\left\|\sum_1^\infty \left(\sum_1^n c_k\right)\chi_{(1/(n+1),1/n)}\right\|_G$$
$$\leqslant \frac{B}{\ln 2}\left\|H_1 y\left(\frac{t}{2}\right)\right\| \leqslant \frac{B}{\ln 2}\|H_1\|_G\|\sigma_2\|_G\|y\|_G.$$

We have established the inequalities

$$\|x\|_G \leqslant B\left\|\sum_1^\infty \left(\sum_1^n c_k\right)\chi_{(1/(n+1),1/n)}\right\|_G$$
$$\leqslant C_1\left\|\sum_1^\infty kc_k\chi_{(1/(k+1),1/k)}\right\|_G \leqslant C_2\|x\|_G.$$

Since

$$\sum_1^\infty \left(\sum_1^n c_k\right)\chi_{(1/(n+1),1/n)}(t) = \sum_1^\infty c_n\chi_{(0,1/n)}(t),$$

the theorem is proved.

**3. *Bases*.** We recall that a *basis* in a separable space $E$ is, by definition, a system $\{u_k\}_1^\infty$ of elements of $E$ such that every $x \in E$ can be represented uniquely in the form of a convergent series:

$$x = \sum_1^\infty c_k u_k.$$

If $\{u_k\}$ is a complete minimal system (see [25]) and $\{f_k\}$ is the system of linear functionals biorthogonal to it, then the bounded operators

$$S_n x = \sum_1^n f_k(x) u_k$$

are defined. For the system $\{u_k\}$ to form a basis it is necessary and sufficient that the operators $S_n$ be uniformly bounded.

A basis is said to be *unconditional* if it remains a basis under any permutation of its elements. For a basis to be unconditional it is necessary and sufficient that operators of the form

$$\sum_{i=1}^{n} f_{k_i}(x)u_{k_i} \qquad (n = 1, 2, \dots)$$

be uniformly bounded.

Thus, determining whether a complete minimal system is a basis or an unconditional basis reduces to estimating norms of certain operators. From this it follows that if $E_0 \supset E_1$, $E$ is an interpolation space between $E_0$ and $E_1$, $E_1$ is complete in $E$ and the elements $\{u_k\}$ of $E_1$ form a basis (unconditional basis) in $E_0$ and in $E_1$, then they form a basis (unconditional basis) in $E$.

On [0, 1] consider the Haar orthonormal system defined by

$$h_0^0(t) \equiv 1; \qquad h_0^{(1)}(t) = \begin{cases} 1, & 0 \leqslant t < 1/2, \\ -1, & 1/2 < t \leqslant 1 \end{cases}$$

and

$$h_n^{(k)}(t) = \begin{cases} 2^{n/2}, & (2k-2)/2^{n+1} \leqslant t < (2k-1)/2^{n+1}, \\ -2^{n/2}, & (2k-1)/2^{n+1} \leqslant t \leqslant 2k/2^{n+1} \\ & \qquad (k = 1, 2, 3, \dots, 2^n), \\ 0 & \text{otherwise.} \end{cases}$$

The functions $h_n^k(t)$ are arranged in a simple sequence $u_m(t)$, $m = 0, 1, \dots,$ in order of increasing $n$ and for fixed $n$ in order of increasing $k$. It is easy to see that the kernels $Q_N(t, s)$ of the operators $S_N$ corresponding to the partial sums of the Fourier series in the functions $h_n^{(k)}$ are nonnegative. Therefore

$$\int_0^1 |Q_N(t, s)|\, ds = \int_0^1 \sum_{m=0}^{N} u_m(t)u_m(s)\, ds = \int_0^1 u_0(t)u_0(s)\, ds = 1.$$

Similarly,

$$\int_0^1 |Q_N(t, s)|\, dt = 1.$$

These equalities say that the operators $S_N$ are uniformly bounded in $L_1$ and $L_\infty$. By the corollary to the interpolation Theorem 4.6 these operators are bounded by 1 in any separable symmetric space $E$. It is well known that the Haar system is a basis in $L_1$. This in particular implies that the system is complete in $E$. Indeed, otherwise there would be a nonzero function in $E' = E^1 \subset L_1$ orthogonal to all functions $u_m(t)$, which contradicts the basis property of the system $\{u_m\}$ in $L_1$.

We have arrived at the following assertion.

THEOREM 9.4. *The Haar system is a basis in any separable symmetric space.*

We now consider the trigonometric system on $[-\pi, \pi]$.

THEOREM 9.5. *For the trigonometric system to be a basis in a separable symmetric space $E$ it is necessary and sufficient that*

$$\|\tilde{\sigma}_\tau\|_E = o(\tau) \quad \textit{as } \tau \to \infty \qquad \textit{and} \qquad \|\tilde{\sigma}_\tau\|_E = o(1) \quad \textit{as } \tau \to 0. \quad (9.12)$$

PROOF. Denote by $S_n x$ the $n$th partial sum of the Fourier series of $x(t)$. For this sum (see [2])

$$S_n x(t) = \frac{1}{\pi}\int_{-\pi}^{\pi} x(s)\frac{\sin n(t-s)}{2\operatorname{tg}\frac{t-s}{2}}\,ds + \frac{1}{2\pi}\int_{-\pi}^{\pi} x(s)\cos n(t-s)\,ds$$

$$= -\cos nt\cdot\frac{1}{\pi}\int_{-\pi}^{\pi}\frac{x(s)\sin ns}{2\operatorname{tg}\frac{t-s}{2}}\,ds + \sin nt\cdot\frac{1}{\pi}\int_{-\pi}^{\pi}\frac{x(s)\cos ns}{2\operatorname{tg}\frac{t-s}{2}}\,ds$$

$$+\frac{1}{2\pi}\int_{-\pi}^{\pi} x(s)\cos n(t-s)\,ds.$$

If conditions (9.12) are satisfied, then the operator of passage to the conjugate function is bounded in $E$ (see Theorem 8.3), and therefore

$$\|S_n x\|_E \leqslant C(\|x(s)\sin ns\|_E + \|x(s)\cos ns\|_E + \|x\|_{L_1})$$
$$\leqslant C_1\|x\|_E.$$

The sufficiency of conditions (9.12) is proved.

Now we assume that the trigonometric system is a basis in $E$. Then $\|S_n x\|_E \leqslant C(E)\|x\|_E$ for any function $x \in E$. First we assume that $x(t)$ is a trigonometric polynomial. Then for the conjugate function $\tilde{x}(t)$ we have the identity

$$S_{2n}\tilde{x}(t) = -2\cos nt\cdot S_n(-x\sin nt) + 2\sin nt\cdot S_n(x\cos nt),$$

which can be verified in an elementary way for the functions $\sin kt$ and $\cos kt$. Now it follows that

$$\|S_{2n}\tilde{x}\|_E \leqslant 4C(E)\|x\|_E.$$

But for large $n$ we have $S_{2n}\tilde{x} = \tilde{x}$, and therefore

$$\|\tilde{x}\|_E \leqslant 4C(E)\|x\|_E. \quad (9.13)$$

Now let $x(t)$ be an arbitrary function in $E$. By assumption, $S_N x \to x$ in $E$ as $N \to \infty$. Then (9.13) implies that the sequence $\widetilde{S_N x} = S_N\tilde{x}$ is Cauchy in $E$, and consequently converges to a function $y \in E$. The relation $x \in E \subset L_1$

implies, as we know (see [2], Chapter VIII, §21), that $S_N\tilde{x} \to \tilde{x}$ in $L_p$ with $p < 1$, and therefore $y = \tilde{x}$. From this we conclude by passing to the limit that (9.13) holds for all $x \in E$. Thus, the conjugation operator acts boundedly in $E$, and, by Theorem 8.3, conditions (9.12) are satisfied.

**4. *Unconditional bases.***

THEOREM 9.6. *For the Haar system to be an unconditional basis in a separable symmetric space $E$ it is necessary and sufficient that conditions* (9.12) *be satisfied.*

PROOF. As we noted above (§6.6), conditions (9.12) imply that $E$ is an interpolation space between some spaces $L_{p_0}$ and $L_{p_1}$ $(1 < p_0 < p_1 < \infty)$. By what was said in subsection 3, the Haar system is an unconditional basis in $E$.

For the proof of the necessity we consider the subsystem of the Haar system consisting of the functions $u_0(t) \equiv 1$ and $u_n(t) = h^{(1)}_{2n-1}(t)$ $(n = 1, 2, \dots)$.

The kernel of the integral operator corresponding to the sum $S_N x$ in the system $\{u_n\}$ is the function

$$K_N(t, s) = \sum_{n=0}^{N} u_n(t) u_n(s).$$

We fix $t$ in the interval $(2^{-2k-1}, 2^{2k})$. Then $\varphi_n(t) = 2^{(2n-1)/2}$ for $n \leqslant k$ and $\varphi_n(t) = 0$ for $n > k$, and

$$K_N(t, s) = \begin{cases} 1 + \sum_{n=1}^{k} 2^{2n-1}, & 0 < s < 2^{-2k}, \\ 1 + \sum_{n=1}^{j-1} 2^{2n-1} - 2^{2j-1}, & 2^{-2j} < s < 2^{-2j+1} \\ & (j = k, k-1, \dots, 1), \\ 1 + \sum_{n=0}^{j} 2^{2n-1}, & 2^{-2j-1} < s < 2^{-2j} \ (j = k-1, k-2, \dots, 1), \\ 0, & 2^{-1} < s < 1. \end{cases}$$

As $s$ changes, the function $K_N(t, s)$ assumes positive and negative values alternately on the intervals $(0, 2^{-2k})$, $(2^{-2k}, 2^{-2k+1}), \dots, (2^{-2}, 2^{-1})$ and is equal to zero on $(2^{-1}, 1)$. Its indefinite integral

$$\int_0^{\tau} K_N(t, s)\, ds$$

with respect to $s$ is a continuous function having local minima at the points $2^{-2j+1}$ $(j = k, \dots, 1)$. The first of them is equal to

$$\left(1 + \sum_{n=1}^{k} 2^{2n-1} + 1 + \sum_{n=1}^{k-1} 2^{2n-1} - 2^{2k-1}\right)2^{-2k} = \frac{2}{3}(1 + 2^{2k-1})2^{-2k}.$$

Moreover, the integral of $K_N(t, s)$ over $(2^{-2j-1}, 2^{-2j+1})$ is equal to

$$\left(1 + \sum_{n=1}^{j} 2^{2n-1}\right)2^{-2j-1} + \left(1 + \sum_{n=1}^{j-1} 2^{2n-1} - 2^{2j-1}\right)2^{-2j} > 0$$
$$(j = k - 1, \dots, 1),$$

and so all local minima are greater than the first one.

We consider the function $\tilde{K}_N(t, s) = \frac{1}{3}(1 + 2^{2k-1})\chi_{(0,2^{-2k+1})}(s)$. By construction,

$$\int_0^{\tau} K_N(t, s)\, ds \geqslant \int_0^{\tau} \tilde{K}_N(t, s)\, ds \qquad (0 < \tau \leqslant 1).$$

By property 18° of rearrangements, for $x = x^*$ we then have

$$S_N x(t) = \int_0^t K_N(t, s)x(s)\, ds \geqslant \int_0^1 K_N(t, s)x(s)\, ds$$
$$= \frac{1}{3}(1 + 2^{2k-1})\int_0^{2^{-2k+1}} x(s)\, ds \geqslant \frac{1}{12}\frac{1}{t}\int_0^t x(s)\, ds$$
$$(-2^{-2k+1} < t < 2^{-2k}).$$

Hence

$$S_N x(t) \geqslant \frac{1}{12}\frac{1}{t}\int_0^t x(s)\, ds\left(\sum_{k=1}^{N} \chi_{(2^{-2k+1},2^{-2k})}(t)\right).$$

By the assumption that the Haar system is an unconditional basis, the partial sums corresponding to a subsystem of it are uniformly bounded: $\|S_N x\|_E \leqslant C\|x\|_E$. Then

$$\left\|Hx \sum_{k=1}^{N} \chi_{(2^{-2k-1},2^{-2k})}\right\|_E \leqslant 12C\|x\|_E,$$

where $H$ is the Hardy-Littlewood operator.

Reasoning similarly to the proof of the theorem, we may pass to the limit as $N \to \infty$ and obtain the inequality

$$\left\|Hx \sum_{k=1}^{\infty} \chi_{(2^{-2k-1},2^{-2k})}\right\|_{E''} \leqslant 12C\|x\|_E.$$

It is obvious that

$$\sigma_2 \sum_{k=1}^{\infty} \chi_{(2^{-2k-1},2^{-2k})} = \sum_{k=1}^{\infty} \chi_{(2^{-2k},2^{-2k+1})}.$$

Next, $Hx(t)$ is decreasing, and therefore

$$Hx(t)\sum_{k=1}^{\infty}\chi_{(2^{-2k},2^{-2k+1})}(t)\leqslant\sigma_2\left[Hx\sum_{k=1}^{\infty}\chi_{(2^{-2k-1},2^{-2k})}\right](t).$$

Hence $\|Hx\chi_{(0,1/2)}\|_{E''}\leqslant 36C\|x\|_E$.
Similarly,

$$Hx\leqslant\sigma_2\left[Hx\chi_{(0,1/2)}\right].$$

Therefore

$$\|Hx\|_{E''}\leqslant 72C\|x\|_E.$$

Since $|Hx|\leqslant Hx^*$, this inequality holds for any $x\in E$. By Theorem 6.7 and §8.6, the condition $\|\sigma_\tau\|_E=o(\tau)$ is satisfied as $\tau\to\infty$. The kernels $K_N(t,s)$ are symmetric, and therefore $\|S_N\|_E=\|S_N\|_{E'}$. This implies that

$$\|\sigma_\tau\|_{E'}=o(\tau)\quad\text{as } \tau\to\infty.$$

Since $E$ is separable, the imbedding of $E$ in $E^{11}$ is isometric. Then by (4.36) we have

$$\tau\|\sigma_{1/\tau}\|_E=o(\tau)\quad\text{as } \tau\to\infty,\quad\text{i.e. } \|\sigma_\tau\|_E=o(1)\quad\text{as } \tau\to 0.$$

The theorem is proved.

In [251] it was announced that if there exists an unconditional basis in a symmetric space then the Haar system forms an unconditional basis in it (the proof is to be found in [33a]). Thus, conditions (9.12) are necessary and sufficient for the existence of an unconditional basis in a symmetric space.

**5. *Series in the Rademacher system.*** The *Rademacher system* on [0, 1] is the orthonormal system of the functions $r_k(t)=\operatorname{sgn}\sin 2^{k-1}\pi t$. For every sequence $u=\{u_n\}\in l_2$ we construct a function

$$v(t)=Vu=\sum_{k=1}^{\infty}u_kr_k(t).\tag{9.14}$$

(We recall that for $\{u_n\}\in l_2$ the series (9.14) converges almost everywhere; otherwise it is divergent almost everywhere–see [50]).

Thus, we obtain an operator from sequence spaces into function spaces.

THEOREM 9.7. *Let $G$ be a symmetric sequence space satisfying the conditions*

$$\|\tilde\sigma_n\|_G=o(n),\quad\|\tilde\sigma_{1/n}\|_G=o(1/\sqrt{n})\quad\textit{as } n\to\infty.\tag{9.15}$$

*If $u=\{u_n\}\in G$, then $\{v^{**}(e^{-n})/n\}\in G$, and*

$$\left\|\left\{\frac{v^{**}(e^{-n})}{n}\right\}\right\|_G\leqslant C\|u\|_G.\tag{9.16}$$

PROOF. Since $|r_k(t)| \leqslant 1$, we have

$$\|Vu\|_{L_\infty} \leqslant \sum_{k=1}^{\infty} |u_k| = \|u\|_{l_1}.$$

If $u \in l_2$, then according to Khintchine's inequality the function $Vu$ belongs to all spaces $L_p$, $1 \leqslant p < \infty$, and, moreover (see [50], Vol. I, Chapter V, §8), for $\|u\|_{l_2} \leqslant 1$ we have

$$\int_0^1 e^{(Vu(t)/\lambda)^2}\,dt \leqslant \sum_{k=0}^{\infty} \frac{k^k}{k!}\left(\frac{2}{\lambda}\right)^{2k}$$

for any $\lambda > 0$. Since $e^k \geqslant k^k/k!$, for $\lambda_0 = \sqrt{8e}$ we have

$$\int_0^1 e^{(Vu(t)/\lambda_0)^2}\,dt \leqslant \sum_{k=0}^{\infty} \left(\frac{4e}{\lambda_0^2}\right)^k = 2.$$

This inequality means that in the Luxemburg norm in the Orlicz space $L_H^*$ constructed from the $N$-function $M(u) = e^{u^2} - 1$ (see [22]) we have

$$\|Vu\|_{L_M^*} \leqslant \sqrt{8e}\,.$$

Thus, the operator $V$ acts boundedly from $l_1$ into $L_\infty$ and from $l_2$ into $L_M^*$. We apply Theorem 8.2. In our case $M_{\varphi_{E_0}}(t) = \sqrt{t}$ and $M_{\varphi_{E_1}}(t) = t$. Therefore from (9.15) it follows that the space $G$ satisfies condition (8.17). According to (4.27) we have

$$\varphi_{L_M^*}(t) = \left[\ln(1 + 1/t)\right]^{-1/2},$$

and $\varphi_{L_\infty} = \operatorname{sgn} t$. Consequently, equations (6.7) and (6.8) for determining $\kappa(t)$ and $\delta(t)$ take the form $1 = \kappa(t)t$ (i.e. $\kappa(t) = 1/t$) and $\delta(t) = 1/(e^t - 1)$. We calculate $\kappa_n = \inf_{n-1\leqslant t\leqslant n} \kappa(t) = n^{-1}$ and

$$\delta_n = \sup_{n-1\leqslant t\leqslant n} (e^t - 1)^{-1} = (e^{(n-1)} - 1)^{-1}.$$

From (8.27) we obtain that

$$\left\|\left\{\frac{v^{**}((e^n - 1)^{-1})}{n+1}\right\}\right\|_G \leqslant C\|\{u_n\}\|_G. \tag{9.17}$$

Since $V^{**}(at) \geqslant V^{**}(t)/a$ for $a \geqslant 1$, from (9.17) we obtain (9.16).

Theorem 9.7 is sharp in the following sense.

THEOREM 9.8. *Under the assumptions* (9.15), *for every function of the form* (9.14)

$$\|\{u_n\}\|_G \leqslant \left\|\left\{\frac{v^{**}(e^{-n})}{n}\right\}\right\|_G.$$

PROOF. We use the fact (without recalling its proof) that the functions

$$v_m(t) = \sum_{k=1}^{m} u_k r_k(t) \quad \text{and} \quad y_m(t) = \sum_{k=1}^{m} u_k^* r_k(t)$$

are equimeasurable, and consequently $v_m^*(t) = y_m^*(t)$. Using property 11° of rearrangements, we obtain that $v^*(t) = y^*(t)$, where $y(t) = \sum_1^\infty u_k^* r_k(t)$. This in particular implies that any function $v(t)$ of the form (9.14) is equimeasurable with the function $-v(t)$, and consequently $\operatorname{mes}\{t\colon v(t) \geqslant 0\} \geqslant 1/2$.

We represent $y(t)$ in the form

$$y(t) = \sum_{k=1}^{i} u_k^* r_k(t) + \sum_{k=i+1}^{\infty} u_k^* r_k(t).$$

The first term is equal to $\sum_1^i u_k^*$ on $(0, 1/2^{i-1})$. The second term is periodic with period $1/2^{i-1}$, and so it is nonnegative on a set of measure not less than $1/2^i$ on $(0, 1/2^{i-1})$. Thus, $y(t) \geqslant \sum_1^i u_k^*$ on a set of measure not less than $1/2^i$. Then

$$v^*(t) = y^*(t) \geqslant \sum_{k=1}^{\infty} u_k^* \chi_{(0,2^{-k}]}(t).$$

We calculate

$$v^*(2^{-n}) \geqslant \sum_{k=1}^{n} u_k^* \geqslant n u_n^*.$$

Therefore

$$\left\| \left\{ \frac{v^{**}(e^{-n})}{n} \right\} \right\|_G \geqslant \left\| \left\{ \frac{v^{**}(2^{-n})}{n} \right\} \right\|_G \geqslant \left\| \left\{ \frac{v^{*}(2^{-n})}{n} \right\} \right\|_G$$
$$\geqslant \|\{u_n^*\}\|_G = \|u\|_G.$$

CHAPTER III

# SCALES OF BANACH SPACES

## §1. Scales of Banach spaces. Related spaces

### 1. *Scales of Banach spaces.*

DEFINITION 1.1. A family $E_\alpha$, $\alpha_0 \leqslant \alpha \leqslant \beta_0$, of Banach spaces is called a *scale of Banach spaces* if the following conditions are met:

1) For $\beta > \alpha$ the space $E_\beta$ is densely imbedded in $E_\alpha$, and consequently

$$\|x\|_{E_\alpha} \leqslant C(\alpha, \beta)\|x\|_{E_\beta}:$$

2) There exists a function $C(\alpha, \beta, \gamma)$, finite at all points of the domain $\alpha_0 \leqslant \alpha < \beta < \gamma \leqslant \beta_0$, such that

$$\|x\|_{E_\beta} \leqslant C(\alpha, \beta, \gamma)\|x\|_{E_\alpha}^{(\gamma-\beta)/(\gamma-\alpha)}\|x\|_{E_\gamma}^{(\beta-\alpha)/(\gamma-\alpha)} \tag{1.1}$$

for any $x \in E_\gamma$.

A scale of spaces is said to be *compact* if the imbedding of $E_\beta$ in $E_\alpha$ is compact for $\beta > \alpha$.

If the family of $E_\alpha$ has property 1) and $E_\beta$ is an interpolation space between $E_\alpha$ and $E_\gamma$ of type $\mu = (\beta - \alpha)/(\gamma - \alpha)$ for $\alpha < \beta < \gamma$ (see Chapter I, §4.3), then these spaces form a scale. Indeed, let $f_0$ be a functional from $E'_{\alpha_0}$. Then (see Chapter I, §2.1) $f_0 \in E'_\alpha$ for all $\alpha \in [\alpha_0, \beta_0]$. Let $x \in E_\gamma$. Consider the operator $T_x u = f_0(u)x$ acting in $E_\gamma$ and all spaces $E_\theta$ for $\theta < \gamma$. We determine the norm of $T_x$ in the spaces $E_\theta$ for $\theta < \gamma$:

$$\|T_x\|_{E_\theta} = \sup_{u \in E_\theta} \frac{\|f_0(u)x\|_{E_\theta}}{\|u\|_{E_\theta}} = \|x\|_{E_\theta}\|f_0\|_{E'_\theta}. \tag{1.2}$$

Since $E_\beta$ is an interpolation space between $E_\alpha$ and $E_\beta$ of type $\mu$, we have

$$\|T_x\|_{E_\beta \to E_\beta} \leqslant C_1(\alpha, \beta, \gamma)\|T_x\|_{E_\alpha \to E_\alpha}^{(\gamma-\beta)/(\gamma-\alpha)}\|T_x\|_{E_\gamma \to E_\gamma}^{(\beta-\alpha)/(\gamma-\alpha)}.$$

Substituting expression (1.2) for the norm of the operator $T_x$ into this inequality, we obtain (1.1).

In what follows, for brevity we write

$$\mu = (\gamma - \beta)/(\gamma - \alpha) \quad \text{and} \quad \nu = (\beta - \alpha)/(\gamma - \alpha), \tag{1.3}$$

where $\mu + \nu = 1$ and $\mu\alpha + \nu\gamma = \beta$.

Inequality (1.1) can then be written in the form

$$\|x\|_{E_\beta} \leqslant C(\alpha, \beta, \gamma)\|x\|^{\mu}_{E_\alpha}\|x\|^{\nu}_{E_\gamma}.$$

**2. *Properties of scales of Banach spaces.*** 1°. If in a scale $E_\alpha$, $\alpha_0 \leqslant \alpha \leqslant \beta_0$, we replace the index $\alpha$ by an index $\alpha'$ according to the formula $\alpha = k\alpha' + \theta$, where $k > 0$, then the spaces $F_{\alpha'} = E_{k\alpha'+\theta}$ will form a scale on the interval $[(\alpha_0 - \theta)/k, (\beta_0 - \theta)/k]$ relative to the index $\alpha'$.

In connection with this, without loss of generality we may assume that $\alpha_0 = 0$ and $\beta_0 = 1$.

2°. If equivalent norms are introduced in the spaces $E_\alpha$ of a scale, then they will again form a scale.

3°. On the interval $[\alpha_0, \beta_1]$ let a family of Banach spaces $E_\alpha$ be given which form a scale on the intervals $[\alpha_0, \beta_0]$ and $[\beta_0, \beta_1]$ $(\alpha_0 < \beta_0 < \beta_1)$. For the spaces $E_\alpha$ to form a scale on the whole interval $[\alpha_0, \beta_1]$ it is necessary and sufficient that for all $\alpha < \beta_0$ and $\gamma > \beta_0$

$$\|x\|_{E_{\beta_0}} \leqslant C(\alpha, \beta_0, \gamma)\|x\|^{\mu}_{E_\alpha}\|x\|^{\nu}_{E_\gamma}.$$

4°. If a family of normed spaces is given which satisfies conditions 1) and 2) of Definition 1.1 and at least one of the spaces is not complete, then we shall say that an *incomplete scale* of spaces is given. Consider the completions $\overline{E}_\alpha$ of the spaces $E_\alpha$. Assume that the following condition is satisfied:

$\pi$) If $\{x_n\}$ is a Cauchy sequence in $E_\beta$ and $\|x_n\|_{E_{\alpha_0}} \to 0$ as $n \to \infty$, then also $\|x_n\|_{E_\beta} \to 0$.

Then the spaces $\{\overline{E}_\alpha\}$ form a scale under the natural imbeddings $\overline{E}_\beta \subset \overline{E}_\alpha$ $(\alpha < \beta)$.

We often encounter the special case of incomplete scales where all spaces $E_\alpha$ coincide as sets: $E_\alpha = M$ for $\alpha_0 < \alpha < \beta$, but differ in norm. Such an incomplete scale is called an *incomplete scale with base M*. Every scale $\{E_\alpha\}$, $\alpha_0 \leqslant \alpha \leqslant \beta$, can be obtained by completing the scale with base $E_{\beta_0}$ and the norms of the spaces $E_\alpha$.

**3. *Normal scales.***

Definition 1.2. A scale $E_\alpha$, $\alpha_0 \leqslant \alpha \leqslant \beta_0$, of Banach spaces is said to be *normal* if we can set $C(\alpha, \beta) = C(\alpha, \beta, \gamma) \equiv 1$, i.e. if the following conditions are satisfied:

1) $\|x\|_{E_\alpha} \leqslant \|x\|_{E_\beta}$ $(\alpha < \beta, x \in E_\beta)$,

2) $\|x\|_{E_\beta} \leqslant \|x\|^{\mu}_{E_\alpha}\|x\|^{\nu}_{E_\gamma}$.

The last inequality says that the function $\varphi_x(\beta) = \|x\|_{E_\beta}$ $(x \in E_\gamma)$ is logarithmically convex on $[\alpha_0, \gamma]$.

Let $\{E_\alpha\}$ be an incomplete normal scale with base $M$, i.e. a linear space $M$ on which a family $\|x\|_{E_\alpha}$ $(\alpha_0 \leqslant \alpha \leqslant \beta_0)$ of norms is given satisfying the preceding inequalities 1) and 2). The logarithmically convex function $\varphi_x(\alpha) = \|x\|_{E_\alpha}$ $(x \in M)$ is continuous on $[\alpha_0, \beta_0]$ except perhaps at the endpoint $\beta_0$. If the norm with index $\beta_0$ is replaced by a new norm by setting

$$\|x\|^*_{E_{\beta_0}} = \lim_{\beta \to \beta_0} \|x\|_{E_\beta}, \tag{1.4}$$

then a normal scale is obtained again.

Inequalities 1) and 2) imply that condition $\pi$) is satisfied for all $\beta < \beta_0$. The spaces obtained by completing the set $M$ in the norms $\|x\|_{E_\alpha}$ form a normal scale on every interval $[\alpha_0, \beta_1]$ with $\beta_1 < \beta_0$, and if the norm with index $\beta_0$ is defined by (1.4), the construction is valid for the whole interval $[\alpha_0, \beta_0]$.

DEFINITION 1.3. A normal scale $\{E_\alpha\}$ $(\alpha_0 \leqslant \alpha \leqslant \beta_0)$, for which the function $\varphi_x(\alpha) = \|x\|_{E_\alpha}$, $x \in E_{\beta_0}$, is continuous on the interval $[\alpha_0, \beta_0]$ is called a *continuous normal scale*. In this case we say that *the scale* $\{E_\alpha\}$ *connects the spaces* $E_{\alpha_0}$ *and* $E_{\beta_0}$.

**4. *Related spaces.***

DEFINITION 1.4. Two Banach spaces $E_0$ and $E_1$ are said to be *related* if $E_1$ is normally imbedded in $E_0$ and there exists a continuous normal scale on the interval $[0, 1]$ connecting $E_0$ and $E_1$.

LEMMA 1.1. *If* $\{E_\alpha\}$, $0 \leqslant \alpha \leqslant 1$, *is a normal scale, then*

$$\lim_{\alpha \to 1} \|x\|_{E_\alpha} \leqslant \|x\|_{E_{01}}, \tag{1.5}$$

*for* $x \in E_1$, *where* $E_{01}$ *is the completion of* $E_1$ *relative to* $E_0$.

PROOF. By the definition of the norm in $E_{01}$, there exists a sequence of elements $x_n \in E_1$ such that $\|x_n\|_{E_1} = \|x\|_{E_{01}}$ and $x_n \to x$ in $E_0$. Then

$$\|x - x_n\|_{E_\alpha} \leqslant \|x - x_n\|^{1-\alpha}_{E_0} \|x - x_n\|^{\alpha}_{E_1} \leqslant (\|x\|_{E_1} + \|x\|_{E_{01}})^\alpha \|x - x_n\|^{1-\alpha}_{E_0} \to 0$$

as $n \to \infty$. In particular, $\|x_n\|_{E_\alpha} \to \|x\|_{E_\alpha}$. We choose $\alpha$ so that $\|x\|_{E_\alpha} > \lim_{\alpha \to 1} \|x\|_{E_\alpha} - \varepsilon$, and then $n$ so large that

$$\|x\|_{E_{01}} = \|x_n\|_{E_1} \geqslant \|x_n\|_{E_\alpha} \geqslant \|x\|_{E_\alpha} - \varepsilon \geqslant \lim_{\alpha \to 1} \|x\|_{E_\alpha} - 2\varepsilon.$$

Since $\varepsilon$ is arbitrary, inequality (1.5) is proved.

COROLLARY. *For* $\beta < 1$ *the space* $E_\beta$ *is isometrically imbedded in its completion relative to* $E_0$.

Indeed, for $x \in E_1$ the function $\|x\|_E$ is continuous for $\alpha < 1$, and therefore

$$\|x\|_{E_{0\beta}} \geqslant \lim_{\alpha \to \beta - 0} \|x\|_{E_\alpha} = \|x\|_{E_\beta}.$$

The reverse inequality always holds, and so $\|x\|_{E_{0\beta}} = \|x\|_{E_\beta}$. Since $E_1$ is dense in $E_0$, this equality extends to all elements of $E_\beta$.

THEOREM 1.1. *Let the space $E_1$ be normally imbedded in the space $E_0$. For $E_0$ to be related with $E_1$ it is necessary and sufficient that $E_1$ be isometrically imbedded in its completion relative to $E_0$.*

PROOF. *Necessity.* If $E_0$ is related with $E_1$, then there exists a continuous normal scale $\{E_\alpha\}$ connecting these spaces. By Lemma 1.1 we then have $\|x\|_{E_1} = \lim_{\alpha\to 1}\|x\|_{E_\alpha} \leqslant \|x\|_{E_{01}}$, and since the reverse inequality always holds, the norms of the spaces $E_1$ and $E_{01}$ coincide on $E_1$.

*Sufficiency.* In the linear space $E_1$ we introduce a family of norms by the formula

$$\|x\|_{E_\alpha} = \sup_{f\in E_0'} \frac{|f(x)|}{\|f\|_{E_0'}^{1-\alpha}\|f\|_{E_1'}^{\alpha}}. \tag{1.6}$$

The quantity $\|x\|_{E_\alpha}$ obviously has all properties of a norm. Moreover, for every $x \in E_1$ and $f \in E_0'$ the function

$$\frac{|f(x)|}{\|f\|_{E_0'}^{1-\alpha}\|f\|_{E_1'}^{\alpha}} = \frac{|f(x)|}{\|f\|_{E_0'}}\left(\frac{\|f\|_{E_0'}}{\|f\|_{E_1'}}\right)^{\alpha} \tag{1.7}$$

is continuous and logarithmically convex in $\alpha$ on $[0, 1]$. Since $\|f\|_{E_1'} \leqslant \|f\|_{E_0'}$ $(f \in E_0')$, the function (1.7) is increasing in $\alpha$. Then the supremum of all functions (1.7) over $f \in E_0'$, i.e. $\|x\|_{E_\alpha}$, is a continuous increasing logarithmically convex function of $\alpha$. The set $E_1$ with the norms (1.6) is a continuous incomplete normal scale of spaces with base $E_1$. Moreover, for $\alpha = 0$ we have

$$\|x\|_{E_{\alpha-0}} = \sup_{f\in E_0'} \frac{|f(x)|}{\|f\|_{E_0'}} = \|x\|_{E_0},$$

and, by the completeness of $E_1$ in $E_0$ in the norm of $E_0$, the completion of $E_1$ in the norm $\|x\|_{E_0}$ can be identified with the space $E_0$.

From the hypotheses of Theorem 1.1 and Theorem 2.1 of Chapter I it follows that the set $E_0'$ is normative in the space $E_1'$. Therefore

$$\|x\|_{E_{\alpha-1}} = \sup_{f\in E_0'} \frac{|f(x)|}{\|f\|_{E_1'}} = \|x\|_{E_1}.$$

Thus, the completion of $E_1$ in the norms $\|x\|_{E_\alpha}$ gives a continuous normal scale $E_\alpha$ connecting $E_0$ with $E_1$.

The theorem is proved.

By Theorem 2.1 in Chapter I, the hypothesis of Theorem 1.1 admits various equivalent formulations. Combining them, we obtain the following assertion.

THEOREM 1.2. *For a space $E_1$ normally imbedded in $E_0$ to be related with $E_0$, it is necessary and sufficient that one of the following equivalent conditions be satisfied*:

1°. *The hypothesis of Theorem* 1.1 *holds.*

2°. *The space $E_0'$ is normative for $E_1$.*

3°. *The ball of $E_1$ is closed in the topology induced by the topology of $E_0$.*

Taking Lemma 1.3 of Chapter I into account, we obtain the following corollaries.

COROLLARY 1. *If the space $E_1$ is normally imbedded in $E_0$, then the completion $E_{01}$ of $E_1$ relative to $E_1$ is related with $E_0$.*

COROLLARY 2. *If the space $E_1$ is normally imbedded in $E_0$ and is complete relative to it, then $E_0$ and $E_1$ are related. In particular, a reflexive space is related with any space in which it is imbedded.*

Lemma 2.4 of Chapter I leads to the next assertion.

COROLLARY 3. *If the space $E_1$ is normally imbedded in $E_0$ and the dual space $E_0'$ is densely imbedded in the dual space $E_1'$, then $E_0$ and $E_1$ are related.*

Theorem 2.2 of Chapter I implies the following corollaries.

COROLLARY 4. *Under the hypotheses of Corollary* 3, *the spaces $E_0'$ and $E_1'$ are related.*

COROLLARY 5. *Let $E_2$ be normally imbedded in $E_1$ and $E_1$ normally imbedded in $E_0$. If $E_0$ and $E_2$ are related, so are $E_1$ and $E_2$.*

COROLLARY 6. *Let $E_2$ be related with $E_1$ and imbedded in $E_1$ compactly, and $E_1$ normally imbedded in $E_0$. Then $E_0$ and $E_2$ are related.*

**5. *Condensation of a normal scale by means of relative completion.*** We consider a normal scale of Banach spaces $E_\alpha$, $0 \leqslant \alpha \leqslant 1$. We construct a family of spaces $E_{0\alpha}$ which are completions of $E_\alpha$ relative to the space $E_0$ containing them, and establish a number of properties of this family.

1°. *For $\alpha < \gamma$ the space $E_{0\gamma}$ is imbedded in $E_{0\alpha}$ with imbedding constant* 1.

Indeed, if $x \in E_{0\gamma}$, then there exists a sequence $x_n \to x$ in $E_0$ such that $\|x_n\|_{E_\gamma} = \|x\|_{E_{0\gamma}}$. Then $\|x_n\|_{E_\alpha} \leqslant \|x_n\|_{E_\gamma} = \|x\|_{E_{0\gamma}}$, which implies that $x \in E_{0\alpha}$ and $\|x\|_{E_{0\alpha}} \leqslant \|x\|_{E_{0\gamma}}$.

2°. *The completion $E_{\alpha\gamma}$ of $E_\gamma$ relative to $E_\alpha$ $(\alpha < \gamma)$ coincides with $E_{0\gamma}$.*

Let $x \in E_{\alpha\gamma}$. Then there exists a sequence $x_n \to x$ in $E_\alpha$ such that $\|x_n\|_{E_\gamma} = \|x\|_{E_{\alpha\gamma}}$. By the imbedding $E_\alpha \subset E_0$ the sequence $x_n$ converges to $x$ in $E_0$, and therefore $x \in E_{0\gamma}$ and $\|x\|_{E_{0\gamma}} \leqslant \|x\|_{E_{\alpha\gamma}}$. Conversely, if $x \in E_{0\gamma}$, then one can construct a sequence $x_n \to x$ in $E_0$ so that $\|x_n\|_{E_\gamma} = \|x\|_{E_{0\gamma}}$. Now we use the fact that the spaces $E_\alpha$ form a normal scale. We have

$$\|x_n - x_m\|_{E_\alpha} \leqslant \|x_n - x_m\|_{E_0}^{1-\alpha/\gamma}\|x_n - x_m\|_{E_\gamma}^{\alpha/\gamma}$$
$$\leqslant \|x_n - x_m\|_{E_0}^{1-\alpha/\gamma}(2\|x\|_{E_{0\gamma}})^{\alpha/\gamma}.$$

Thus, the sequence $x_n$ is Cauchy in $E_\alpha$, and by virtue of the imbedding $E_\alpha \subset E_0$ it converges to the same element $x$ in $E_\alpha$. This means that $x \in E_{\alpha\gamma}$ and $\|x\|_{E_{\alpha\gamma}} \leqslant \|x\|_{E_{0\gamma}}$. The reverse inequality proved earlier implies the equality $\|x\|_{E_{\alpha\gamma}} = \|x\|_{E_{0\gamma}}$.

Property 2° and the properties of relative completion immediately imply

3°. *The space $E_{0\gamma}$ is imbedded in every space $E_\alpha$ for $\alpha < \gamma$ and does not coincide with it (if the spaces $E_\alpha$ and $E_\gamma$ did not coincide).*

From this we obtain

4°. *If the space $E_\alpha$ is not complete relative to $E_0$, i.e. $E_\alpha \neq E_{0\alpha}$, then the spaces $E_{0\gamma}$ are imbedded not densely in $E_{0\alpha}$ for $\alpha < \gamma$.*

Indeed, $E_\alpha$ is related with $E_0$, and therefore, if it is not complete, relative to $E_0$, then by Theorem 1.2 it is a proper subspace of $E_{0\alpha}$. On the other hand, $E_{0\gamma} \subset E_\alpha$ for $\alpha < \gamma$, and consequently $E_{0\gamma}$ is not dense in $E_{0\alpha}$.

5°. *For $\alpha < \beta < \gamma$ the following inequality holds:*

$$\|x\|_{E_{0\beta}} \leqslant \|x\|_{E_{0\alpha}}^{\mu}\|x\|_{E_{0\gamma}}^{\nu} \qquad (x \in E_{0\gamma}).$$

Let $x \in E_{0\gamma}$, $x_n \to x$ in $E_0$ and $\|x_n\|_{E_\gamma} = \|x\|_{E_{0\gamma}}$. Then

$$\|x_n\|_{E_\beta} \leqslant \|x_n\|_{E_\alpha}^{\mu}\|x_n\|_{E_\gamma}^{\nu} = \|x_n\|_{E_\alpha}^{\mu}\|x\|_{E_{0\gamma}}^{\nu}. \tag{1.8}$$

In the proof of property 2° we showed that $x_n \to x$ in $E_\alpha$. Theorem 1.1 implies that $E_\alpha$ is imbedded isometrically in $E_{0\alpha}$, and therefore $x_n \to x$ in $E_{0\alpha}$. Then for any $\varepsilon > 0$ and sufficiently large $n$ we have $\|x_n\|_{E_\alpha} \leqslant \|x\|_{E_{0\alpha}} + \varepsilon$. It follows from (1.8) that

$$\|x\|_{E_{0\beta}} \leqslant (\|x\|_{E_{0\alpha}} + \varepsilon)^{\mu}\|x\|_{E_{0\gamma}}^{\nu},$$

and, since $\varepsilon$ is arbitrary

$$\|x\|_{E_{0\beta}} \leqslant \|x\|_{E_{0\alpha}}^{\mu}\|x\|_{E_{0\gamma}}^{\nu}.$$

Thus, the family of spaces $E_{0\gamma}$ has all properties of a normal scale except that of dense imbedding of one space in another.

## §2. Maximal and minimal normal scales

**1.** ***Maximal normal scales of Banach spaces.*** Let a space $F_1$ be normally imbedded in a space $F_0$. On the interval $[0, 1]$ we consider all incomplete continuous normal scales $E_\alpha$ with base $F_1$ such that

$$\|x\|_{E_0} \leqslant \|x\|_{F_0}, \qquad x \in F_1, \tag{2.1}$$

$$\|x\|_{E_1} \leqslant \|x\|_{F_1}, \qquad x \in F_1. \tag{2.2}$$

We note that there always exists one such scale, the trivial scale for which $\|x\|_{E_\alpha} = \|x\|_{F_0}$ $(0 \leqslant \alpha \leqslant 1)$.

In $F_1$ we now introduce a family of norms $\|x\|_\alpha$, $0 \leqslant \alpha \leqslant 1$, by the formula

$$\|x\|_\alpha = \sup \|x\|_{E_\alpha} \qquad (x \in F_1), \tag{2.3}$$

where the supremum is taken over all scales with the above property. For a fixed scale $E_\alpha$ and fixed $x \in F_1$ the function $\|x\|_{E_\alpha}$ is increasing, logarithmically convex, and depends continuously on $\alpha$ in $[0, 1]$. The collection of such functions is uniformly bounded for fixed $x$:

$$\|x\|_{E_\alpha} \leqslant \|x\|_{E_1} \leqslant \|x\|_{F_1},$$

and therefore $\sup\|x\|_{E_\alpha}$ for this collection of functions is an increasing, logarithmically convex and continuous function on $[0, 1]$. The completions of the space $F_1$ in the norms $\|x\|_{E_\alpha}$ form a continuous normal scale of spaces $E_\alpha^{\max}$ on $[0, 1]$. From (2.1), (2.3) and the remark on the trivial scale $\|x\|_{E_\alpha} = \|x\|_{F_0}$ it follows that $\|x\|_0 = \|x\|_{F_0}$, and consequently the spaces $E_0^{\max}$ and $F_0$ coincide. The scale thus obtained is called the *maximal normal scale* constructed from the spaces $F_0$ and $F_1$ on the interval $[0, 1]$.

Thus, the maximal normal scale constructed from the normally imbedded spaces $F_1$ and $F_0$ has the following properties.

1) $E_0^{\max} = F_0$.

2) $F_1$ is normally imbedded in $E_1^{\max}$.

3) If for some continuous scale $E_\alpha$ on $[0, 1]$ the space $F_1$ is normally imbedded in $E_1$ and (2.1) is satisfied, then

$$\|x\|_{E_\alpha} \leqslant \|x\|_{E_\alpha^{\max}} \qquad (x \in F_1, 0 \leqslant \alpha \leqslant 1). \tag{2.4}$$

From the last property it is obvious that the maximal normal scale is defined in a unique way.

If $F_0$ and $F_1$ are related, then the maximal normal scale constructed from $F_0$ and $F_1$ connects these spaces. The following assertion is stronger.

LEMMA 2.1. *If a space $F_1$ is normally imbedded in a space $F_0$, then the space $E_1^{\max}$ coincides with the closure of $F_1$ in its completion $F_{01}$ relative to $F_0$.*

PROOF. By Corollary 1 of Theorem 1.2 the space $F_{01}$ is related with $F_0$. If $F_{0\alpha}$ is a continuous normal scale connecting $F_{01}$ with $F_0$ ($F_{00} = F_0$), then, since $\|x\|_{F_{01}} \leqslant \|x\|_{F_1}$ $(x \in F_1)$, properties (2.1) and (2.2) are satisfied for it. Therefore $\|x\|_{F_{01}} \leqslant \|x\|_{E_1^{\max}}$.

Now let $E_\alpha$ be an arbitrary continuous normal scale having property (2.2) and such that $E_0 = F_0$. For any element $x \in F_1$ there exists a sequence $x_n \to x$ in $F_0$ such that $\|x_n\|_{F_1} = \|x\|_{F_{01}}$.

Then $\|x_n\|_{E_1} \leqslant \|x_n\|_{F_1} = \|x\|_{F_{01}}$. This means that $\|x\|_{E_{01}} \leqslant \|x_n\|_{F_{01}}$. The space $E_1$ is related with $E_0$, and so by Theorem 1.2 $E_1$ is imbedded isometrically in $E_{01}$. From this it follows that $\|x\|_{E_1} \leqslant \|x\|_{F_{01}}$ $(x \in F_1)$. Now taking the scale $E_\alpha^{\max}$ as $E_\alpha$ and taking account of the inequality obtained earlier, we obtain the equality $\|x\|_{E_\alpha^{\max}} = \|x\|_{F_{01}}$ $(x \in F_1)$, which proves the lemma.

LEMMA 2.2. *Assume that the space $F_1$ is normally imbedded in the space $F_0$ and a functional $\Phi(x, \alpha)$, $0 \leqslant \alpha \leqslant 1$, is defined on $F_1$, which is a seminorm for fixed $\alpha$ and is a logarithmically convex function of $\alpha$ for fixed $x$ for which $\Phi(x, \alpha) \not\equiv 0$. If $\Phi(x, 0) \leqslant \|x\|_{F_0}$ and $\Phi(x, 1) \leqslant \|x\|_{F_1}$, then*

$$\Phi(x, \alpha) \leqslant \|x\|_{E_\alpha^{\max}} \qquad (x \in F_1, 0 \leqslant \alpha \leqslant 1). \tag{2.5}$$

PROOF. We define the functional $\Psi(x, \alpha) = \sup_{\alpha_1 < \alpha} \Phi(x, \alpha_1)$, $\Psi(x, 0) = \Phi(x, 0)$. This functional is also a seminorm for fixed $\alpha$ and is a continuous increasing logarithmically convex function of $\alpha$ for fixed $x$. Moreover, $\Psi(x, 0) \leqslant \|x\|_{F_0}$ and $\Psi(x, 1) \leqslant \|x\|_{F_1}$. We construct an incomplete continuous normal scale with norm $\max\{\Psi(x, \alpha), \|x\|_{E_\alpha^{\max}}\}$. This scale has properties (2.1) and (2.2), and consequently inequality (2.4) holds for it, from which (2.5) follows.

THEOREM 2.1. *The maximal normal scale $E_\alpha^{\max}$ constructed from the spaces $F_0$ and $F_1$ on the interval* [0, 1] *is the maximal scale constructed from the spaces $E_{\alpha_1}^{\max}$ and $E_1^{\max}$ on every interior interval* $[\alpha_1, \beta_1]$.

PROOF. Let $H_\alpha$ be the maximal scale constructed from the spaces $E_{\alpha_1}^{\max}$ and $E_{\beta_1}^{\max}$ on $[\alpha_1, \beta_1]$. Then

$$\|x\|_{E_\alpha^{\max}} \leqslant \|x\|_{H_\alpha} \quad \text{for } x \in H_{\beta_1}^{\max} \text{ and } \alpha \in (\alpha_1, \beta_1), \tag{2.6}$$

and $\|x\|_{E_\alpha^{\max}} = \|x\|_{H_\alpha}$ for $\alpha$ equal to $\alpha_1$ or $\beta_1$. We construct a family of spaces $E_\alpha$ on [0, 1] such that $E_\alpha = E_\alpha^{\max}$ for $\alpha \notin [\alpha_1, \beta_1]$ and $E_\alpha = H_\alpha$ for $\alpha \in [\alpha_1, \beta_1]$. By what was said in §1.2, the spaces $E_\alpha$ form a normal scale on [0, 1]. Besides, this scale is continuous and has properties (2.1) and (2.2), and so (2.4) holds.

In our case, for $\alpha \in [\alpha_1, \beta_1]$ inequality (2.4) becomes

$$\|x\|_{H_\alpha} \leqslant \|x\|_{E_\alpha^{\max}} \qquad (x \in F_1). \tag{2.7}$$

Since $F_1 \subset E_1^{\max} \subset E_{\beta_1}^{\max}$, from (2.6) and (2.7) it follows that $\|x\|_{H_\alpha} = \|x\|_{E_\alpha^{\max}}$ for $x \in F_1$ and $\alpha \in [\alpha_1, \beta_1]$. However, the space $F_1$ is densely imbedded in both $E_\alpha^{\max}$ and $H_\alpha$, and therefore $H_\alpha = E_\alpha^{\max}$.

The theorem is proved.

We consider two families $E_\alpha$ and $F_{\alpha'}$ of Banach spaces depending on the parameters $\alpha \in [\alpha_0, \beta_0]$ and $\alpha' \in [\alpha'_0, \beta'_0]$, respectively, and having the property that for $\alpha < \beta$ the space $E_\beta$ is imbedded densely in $E_\alpha$ (for $\alpha' < \beta'$ the space $F_{\beta'}$ is imbedded densely in $F_{\alpha'}$).

DEFINITION 2.1. We shall say that the family $E_\alpha$, $\alpha_0 \leqslant \alpha \leqslant \beta_0$, has

1) the *interpolation property* relative to the family $F_{\alpha'}$ of Banach spaces $F_{\alpha'}$ if the triple $(E_{\alpha_0}, E_{\beta_0}, E_\alpha)$ $(\alpha_0 \leqslant \alpha \leqslant \beta_0)$ of Banach spaces is an interpolating triple relative to the triple $(F_{\alpha'_0}, F_{\beta'_0}, F_{\alpha'})$, where $\alpha'$ satisfies the equality

$$\frac{\alpha - \alpha_0}{\beta_0 - \alpha_0} = \frac{\alpha' - \alpha'_0}{\beta'_0 - \alpha'_0}; \tag{2.8}$$

2) the *normalized interpolation property* relative to the family $F_{\alpha'}$ of Banach spaces if the triple $(E_{\alpha_0}, E_{\beta_0}, E_\alpha)$ of Banach spaces is a normalized interpolation triple of type $\theta = (\alpha - \alpha_0)/(\beta_0 - \alpha_0)$ relative to the triple $(F_{\alpha'_0}, F_{\beta'_0}, F_{\alpha'})$;

3) the *strong interpolation property* relative to the family $F_{\alpha'}$ $(\alpha'_0 \leqslant \alpha' \leqslant \beta'_0)$, if each part of it consisting of spaces whose index runs through any interval $[\alpha_1, \beta_1] \subset [\alpha_0, \beta_0]$ has the normalized interpolation property relative to the corresponding part of the family of the spaces $F_{\alpha'}$. The correspondence is given by (2.8).

THEOREM 2.2. *The maximal normal scale has the strong interpolation property relative to any normal scale.*

PROOF. Let $E_\alpha^{\max}$, $0 \leqslant \alpha \leqslant 1$, be the maximal scale, let $F_\alpha$, $0 \leqslant \alpha \leqslant 1$, be an arbitrary normal scale, and let $A$ be a linear operator mapping $E_1^{\max}$ into $F_1$ such that

$$\|Ax\|_{F_1} \leqslant C_1\|x\|_{E_1^{\max}} \quad (x \in E_1^{\max}), \qquad \|Ax\|_{F_0} \leqslant C_0\|x\|_{E_0^{\max}} \quad (x \in E_1^{\max}).$$

The function $C_0^{\alpha-1}C_1^{-\alpha}\|Ax\|_{F_\alpha} = \Phi(x, \alpha)$ has the properties indicated in Lemma 2.2. From (2.5) it follows that $\Phi(x, \alpha) \leqslant \|x\|_{E_\alpha^{\max}}$, or, in other words,

$$\|Ax\|_{F_\alpha} \leqslant C_0^{1-\alpha}C_1^{\alpha}\|x\|_{E_\alpha^{\max}}.$$

Thus we have proved that the scale $E_\alpha^{\max}$ has the normalized interpolation property relative to $F_\alpha$ on $[0, 1]$. Since $E_\alpha^{\max}$ is maximal on any interval $[\alpha_1, \beta_1] \subset [0, 1]$, this implies that it has the strong interpolation property.

**2.** ***Regular scales.*** Let $E_\alpha$, $0 \leqslant \alpha \leqslant 1$, be a family of Banach spaces such that $E_\beta$ is densely imbedded in $E_\alpha$ for $\alpha < \beta$. In the space $E'_0$ dual to $E_0$ we introduce a family of norms $\|f\|_{E'_\alpha}$. We denote the completion of $E'_0$ in the norm $\|f\|'_E$ by $\tilde{E}_\alpha$. We call the family $\{\tilde{E}_\alpha\}$ *dual* to the family $E_\alpha$. We emphasize that the spaces $\tilde{E}_\alpha$ of the dual family may not coincide with the spaces $E'_\alpha$ dual to $E_\alpha$ if $E'_0$ is not dense in $E'_\alpha$. The space $\tilde{E}_\alpha$ is a subspace of $E'_\alpha$.

We note that $\tilde{E}_\alpha \subset \tilde{E}_\beta$ for $\alpha < \beta$. This follows immediately from the construction of $\tilde{E}_\alpha$ and the fact that $E'_\alpha$ is imbedded in $E'_\beta$ for $\alpha < \beta$.

DEFINITION 2.2. A continuous normal scale $E_\alpha$, $0 \leqslant \alpha \leqslant 1$, is said to be *regular* if the norm in the dual spaces $E'_\alpha$ is a logarithmically convex function of $\alpha$, i.e.

$$\|f\|_{E'_\beta} \leqslant \|f\|^\mu_{E'_\alpha} \|f\|^\nu_{E'_\gamma}, \tag{2.9}$$

for $\alpha < \beta < \gamma$ and $f \in E'_\alpha$, where

$$\mu = (\gamma - \beta)(\gamma - \alpha)^{-1} \quad \text{and} \quad \nu = (\beta - \alpha)(\gamma - \alpha)^{-1}.$$

The family of spaces $L_p(0, 1)$ $(1 \leqslant p_0 \leqslant p \leqslant p_1 \leqslant \infty)$ is a regular scale. Indeed, the family $L_p(0, 1)$ $(p_0 \leqslant p \leqslant p_1)$ is a normal scale $E_\alpha$ $(\alpha_0 \leqslant \alpha \leqslant \alpha_1)$, if we set $\alpha = 1 - 1/p$. In view of the Hölder inequality, the norm in the dual spaces $E'_\alpha = L_{p'}$ $(1/p + 1/p' = 1)$ is a logarithmically convex function of $\alpha$.

If the spaces $E_\alpha$ form a regular scale, than the spaces $\tilde{E}_\alpha$ constitute a normal scale with respect to the index $\sigma = 1 - \alpha$ $(0 \leqslant \sigma \leqslant 1)$. This scale is continuous. Indeed, the function

$$\varphi(\sigma) = \sup_{x \in E_1} \frac{f(x)}{\|x\|_{E_{1-\sigma}}} = \|f\|_{\tilde{E}_{1-\sigma}}$$

is left continuous at $\sigma = 1$, as a supremum of continuous increasing functions of $\sigma$. We shall call the scale $\tilde{E}_{1-\sigma}$ $(\sigma = 1 - \alpha)$ the *dual scale*.

It is obvious that the scale $\tilde{E}_{1-\sigma}$ connects the spaces $E'_1$ and $E'_0$, provided that $E'_0$ is densely imbedded in $E'_1$.

THEOREM 2.3. *The maximal normal scale is regular.*

PROOF. Assume that $\alpha < \beta < \gamma$ and $f \in (E^{\max}_\alpha)'$, where $E^{\max}_\alpha$ is the maximal scale connecting $E_0$ and $E_1$. Since $(E^{\max}_\alpha)' \subset (E^{\max}_\gamma)'$, for every element $x \in E^{\max}_\gamma$ the functional

$$\Phi(x, \beta) = \frac{|f(x)|}{\|f\|^\mu_{(E^{\max}_\alpha)'} \|f\|^\nu_{(E^{\max}_\gamma)'}} \qquad (\alpha \leqslant \beta \leqslant \gamma),$$

where $\mu = (\gamma - \beta)(\gamma - \alpha)^{-1}$ and $\nu = (\beta - \alpha)(\gamma - \alpha)^{-1}$. This functional is a seminorm for fixed $\beta$ and an increasing, logarithmically convex function of $\beta$ for every fixed $x$ for which $\Phi(x, \beta) \not\equiv 0$. Besides,

$$\Phi(x, \alpha) \leqslant \|x\|_{E^{\max}_\alpha} \quad \text{and} \quad \Phi(x, \gamma) \leqslant \|x\|_{E^{\max}_\gamma}.$$

Since the scale $E_\beta^{\max}$ is the maximal scale connecting $E_\alpha^{\max}$ with $E_\gamma^{\max}$, on the basis of Lemma 2.2 we have

$$\Phi(x, \beta) = \frac{|f(x)|}{\|f\|^\mu_{(E_\alpha^{\max})'}\|f\|^\nu_{(E_\gamma^{\max})'}} \leqslant \|x\|_{E_\beta^{\max}} \qquad (x \in E_\gamma^{\max}).$$

The space $E_\gamma^{\max}$ is dense in $E_\beta^{\max}$, and so the last inequality implies that

$$\|f\|_{(E_\beta^{\max})'} \leqslant \|f\|^\mu_{(E_\alpha^{\max})'}\|f\|^\nu_{(E_\gamma^{\max})'}.$$

**3.** ***Minimal scale of spaces.*** In the proof of Theorem 1.1, as the criterion for the two spaces $E_0$ and $E_1$ to be related, we constructed a continuous normal scale of spaces by completing the space $E_1$ relative to the system of norms

$$\|x\|_{E_\alpha} = \sup_{f \in E_0'} \frac{|f(x)|}{\|f\|^{1-\alpha}_{E_0'}\|f\|^\alpha_{E_1'}}. \tag{2.10}$$

If the spaces $E_0$ and $E_1$ are related, then the scale connects them. We call the scale *minimal* and denote its spaces by $E_\alpha^{\min}$.

THEOREM 2.4. *Let $E_1 \subset E_0$ and $F_1 \subset F_0$ be two couples of related spaces. The minimal scales $E_\alpha^{\min}$ and $F_\alpha^{\min}$ have the normalized interpolation property relative to each other.*

PROOF. If the linear operator $A$ satisfies the condition

$$\|Ax\|_{F_0} \leqslant C_0\|x\|_{E_0}, \qquad \|Ax\|_{F_1} \leqslant C_1\|x\|_{E_1} \qquad (x \in E_1), \tag{2.11}$$

then the adjoint operator $A^*$ maps $F_0'$ into $E_0'$ and $F_1'$ into $E_1'$, and

$$\|A^*f\|_{E_0'} \leqslant C_0\|f\|_{F_0'}, \qquad \|A^*f\|_{E_1'} \leqslant C_1\|f\|_{F_1'}.$$

Let $x \in E_1$. Then

$$\begin{aligned}
\|Ax\|_{F_\alpha^{\min}} &= \sup_{f \in F_0'} \frac{|f(Ax)|}{\|f\|^{1-\alpha}_{F_0'}\|f\|^\alpha_{E_1'}} = \sup_{f \in F_0'} \frac{|A^*f(x)|}{\|f\|^{1-\alpha}_{F_0'}\|f\|^\alpha_{F_1'}} \\
&\leqslant C_0^{1-\alpha}C_1^\alpha \sup_{f \in F_0'} \frac{|A^*f(x)|}{\|A^*f\|^{1-\alpha}_{E_0'}\|A^*f\|^\alpha_{E_1'}} \leqslant C_0^{1-\alpha}C_1^\alpha \sup_{g \in E_0'} \frac{|g(x)|}{\|g\|^{1-\alpha}_{E_0'}\|g\|^\alpha_{E_1'}} \\
&= C_0^{1-\alpha}C_1^\alpha\|x\|_{E_\alpha^{\min}}.
\end{aligned} \tag{2.12}$$

DEFINITION 2.3. Let $E_\alpha$ and $\overline{E}_\alpha$, $\alpha_0 \leqslant \alpha \leqslant \alpha_1$, be two normal scales of Banach spaces connecting the space $E_{\alpha_0}$ with the space $E_{\alpha_1}$. We say that the scale $E_\alpha$ *majorizes* $\overline{E}_\alpha$ if

$$\|x\|_{\overline{E}_\alpha} \leqslant \|x\|_{E_\alpha} \qquad (x \in E_{\alpha_1}, \alpha_0 \leqslant \alpha \leqslant \alpha_1)$$

for every $x \in E_{\alpha_1}$.

THEOREM 2.5. *Let $E_0$ and $E_1$ be two related spaces. The minimal scale $E_\alpha^{\min}$ constructed for $E_0$ and $E_1$ is majorized by any regular scale $E_\alpha$ connecting $E_0$ and $E_1$.*

PROOF. Let $x \in E_1$. The spaces $E_\alpha$ and $E_0$ are related, and therefore

$$\|x\|_{E_\alpha} = \sup_{f \in E_0'} \frac{|f(x)|}{\|f\|_{E_\alpha'}}$$

(see Theorem 1.2). In view of the regularity of the scale $E_\alpha$ we have

$$\|f\|_{E_\alpha'} \leqslant \|f\|_{E_0'}^{1-\alpha} \|f\|_{E_1'}^{\alpha} \qquad (f \in E_0').$$

Then

$$\|x\|_{E_\alpha} \geqslant \sup_{f \in E_0'} \frac{|f(x)|}{\|f\|_{E_0'}^{1-\alpha} \|f\|_{E_1'}^{\alpha}} = \|x\|_{E_\alpha^{\min}}. \tag{2.13}$$

REMARK. In the proof of Theorem 2.5 the regularity property (2.9) was used only for $\alpha = 0$ and $\gamma = 1$.

Theorems 2.4 and 2.5 imply that every regular scale $E_\alpha$ has the normalized interpolation property relative to any minimal scale. Indeed, by means of (2.12) and (2.13) we obtain from (2.11) that

$$\|Ax\|_{F_\alpha^{\min}} \leqslant C_0^{1-\alpha} C_1^{\alpha} \|x\|_{E_\alpha} \qquad (x \in E_1), \tag{2.14}$$

which implies our assertion. The following more general theorem holds (see [186]).

THEOREM 2.6. *For a continuous normal scale to have the strong interpolation property relative to any minimal scale, it is necessary and sufficient that it be regular.*

The following assertion establishes a connection between the notions of minimal and maximal scales.

THEOREM 2.7. *Let $E_\alpha^{\min}$ be the minimal scale connecting the spaces $E_0$ and $E_1$ of a reflexive couple. If its dual family $\tilde{E}_{1-\sigma}^{\min}$ $(0 \leqslant \sigma \leqslant 1)$ forms a normal scale, then this scale is the maximal normal scale connecting $F_0 = \tilde{E}_1$ and $F_1 = \tilde{E}_0 = E_0'$.*

PROOF. The spaces $\tilde{E}_1$ and $E_0'$ are related. Let $F_\alpha$ be a normal scale connecting $F_0$ and $F_1$. From the dense imbedding $F_\sigma \subset F_0 \subset \tilde{E}_1$ we obtain the imbedding $(\tilde{E}_1)' \subset F_\sigma'$. We always have the natural imbedding $E_1 \subset (\tilde{E}_1)'$, and consequently $E_1 \subset F'_\sigma$. Using the density of $F_1 = E'_0$ in $F_\sigma$ for $x \in E_1$, we obtain

$$\|x\|_{F'_\sigma} = \sup_{f\in E'_0} \frac{|f(x)|}{\|f\|_{F_\sigma}} \geqslant \sup_{f\in E'_0} \frac{|f(x)|}{\|f\|_{F_0}^{1-\sigma}\|f\|_{F_1}^{\sigma}}$$

$$= \sup_{f\in E'_0} \frac{|f(x)|}{\|f\|_{E'_1}^{1-\sigma}\|f\|_{E'_0}^{\sigma}} = \|x\|_{E_{1-\sigma}^{\min}} \qquad (x \in E_1). \tag{2.15}$$

Next, the spaces $F_0$ and $F_\sigma$ are related, and therefore the set $F'_0 = (\tilde{E}_1)'$ is normative for $F_\sigma$. By the hypothesis of the theorem $(\tilde{E}_1)' = E_1$, and so

$$\|f\|_{F_\sigma} = \sup_{x\in(\tilde{E}_1)'} \frac{|f(x)|}{\|x\|_{F'_\sigma}} = \sup_{x\in E_1} \frac{|f(x)|}{\|x\|_{F'_\sigma}} \leqslant \sup_{x\in E_1} \frac{|f(x)|}{\|x\|_{E_{1-\sigma}^{\min}}}.$$

Now if $f \in E'_0$, then the quantity on the right is equal to $\|f\|_{\tilde{E}_{1-\sigma}}$. Since $E'_0$ is dense in $\tilde{E}_{1-\sigma}$, the preceding inequality implies the imbedding $\tilde{E}_{1-\sigma} \subset F_\sigma$ and the inequality

$$\|f\|_{F_\sigma} \leqslant \|f\|_{\tilde{E}_{1-\sigma}} \qquad (f \in \tilde{E}_{1-\sigma}).$$

Now if $F_\sigma$ is the maximal scale connecting $E'_0$ with $\tilde{E}_1$, then the reverse inequality is satisfied on $E'_0$, and therefore $F_\sigma$ and $\tilde{E}_{1-\sigma}$ coincide.

COROLLARY. *If the minimal scale connecting the spaces of a reflexive couple $E_0$ and $E_1$ is regular, then the scale dual to it is the maximal normal scale connecting $\tilde{E}_1$ and $E'_0$.*

The following assertion is dual to Theorem 2.7 in a certain sense.

THEOREM 2.8. *Let $E_\alpha$, $0 \leqslant \alpha \leqslant 1$, be a continuous normal scale, and let $\tilde{E}_\alpha$ be the family dual to it. If*

$$\|f\|_{\tilde{E}_\alpha} \leqslant \sup_{x\in E_1} \frac{|f(x)|}{\|x\|_{E_0}^{1-\alpha}\|x\|_{E_1}^{\alpha}}, \tag{2.16}$$

*for every $f \in E'_0$, then $E_\alpha$ is the maximal normal scale connecting $E_0$ and $E_1$.*

PROOF. If $E_\alpha^{\max}$ is the maximal normal scale constructed for $E_0$ and $E_1$, then $\|x\|_{E_\alpha} \leqslant \|x\|_{E_\alpha^{\max}}$. This implies that $\|f\|_{(E_\alpha^{\max})'} \leqslant \|f\|_{\tilde{E}_\alpha}$ for every $f \in E'_0$. On the other hand, by (2.16) we have

$$\|f\|_{(E_\alpha^{\max})'} = \sup_{x\in E_1} \frac{|f(x)|}{\|x\|_{E_\alpha^{\max}}} \geqslant \sup_{x\in E_1} \frac{|f(x)|}{\|x\|_{E_0}^{1-\alpha}\|x\|_{E_1}^{\alpha}} \geqslant \|f\|_{\tilde{E}_\alpha}.$$

The inequalities obtained above imply that $\|f\|_{(E_\alpha^{\max})'} = \|f\|_{\tilde{E}_\alpha}$ $(f \in E_0')$. Since the spaces $E_0$ and $E_\alpha^{\max}$ are related, we have

$$\|x\|_{E_\alpha^{\max}} = \sup_{f \in E_0'} \frac{|f(x)|}{\|f\|_{(E_\alpha^{\max})'}} = \sup_{f \in E_0'} \frac{|f(x)|}{\|f\|_{E_\alpha'}} = \|x\|_{E_\alpha} \qquad (x \in E_1).$$

## §3. The scale of Hölder spaces

The main notions of the theory of scales of spaces can be well illustrated by classical examples of function spaces, from which we only choose the Hölder spaces. For simplicity we consider functions on the interval [0, 1].

**1. *Hölder spaces.*** The Hölder spaces $C_{0,\alpha}(0, 1)$, $0 \leqslant \alpha \leqslant 1$, are intermediate between the space $C(0, 1)$ of functions continuous on [0, 1] and the space $C_1(0, 1)$ of functions continuously differentiable on [0, 1]. They consist of all continuous functions satisfying the Hölder condition

$$|x(t) - x(\tau)| \leqslant c|t - \tau|^\alpha$$

with index $\alpha$ for every $t, \tau \in [0, 1]$. The norm in $C_{0,\alpha}(0, 1)$ is defined by

$$\|x\|_{C_{0,\alpha}} = \max_{0 \leqslant t \leqslant 1} |x(t)| + \sup_{0 \leqslant t, \tau \leqslant 1} \frac{|x(t) - x(\tau)|}{|t - \tau|^\alpha}. \tag{3.1}$$

An equivalent norm can be introduced by the formula

$$\|x\|_{C_{0,\alpha}} = |x(0)| + \sup_{0 \leqslant t, \tau \leqslant 1} \frac{|x(t) - x(\tau)|}{|t - \tau|^\alpha}. \tag{3.2}$$

If in $C_1$ we introduce the norm

$$\|x\|_{C_1} = \max_{0 \leqslant t \leqslant 1} |x(t)| + \max_{0 \leqslant t \leqslant 1} |x'(t)|,$$

then the space $C_{0,\alpha}$ with norm (3.1) will be imbedded in $C$ and in it will be imbedded the space $C_1$ with imbedding constants equal to 1, i.e. $C_1 \overset{1}{\subset} C_{0,\alpha} \overset{1}{\subset} C$.

The presence of two terms in the norms (3.1) and (3.2) causes some inconvenience. Therefore, we shall consider the quotient space of $C_{0,\alpha}$ modulo the one-dimensional subspace consisting of the constant functions. We shall denote this quotient space by $H_\alpha$. If we introduce the norm (3.2) in $C_{0,\alpha}$, then this quotient space will be isometric to the subspace of all functions in $C_{0,\alpha}$ vanishing at zero. Consequently $H_\alpha$ consists of classes of functions differing by a constant and satisfying a Hölder condition of order $\alpha$, the norm being

$$\|x\|_{H_\alpha} = \sup_{0 \leqslant t, \tau \leqslant 1} \frac{|x(t) - x(\tau)|}{|t - \tau|^\alpha} \qquad (0 \leqslant \alpha \leqslant 1). \tag{3.3}$$

We note that $H_0$ is isometric to the quotient space of $C$ modulo the subspace of constant functions, if in $C$ we introduce the norm $\|x\|_C = |x(0)| + \max_{0 \leqslant t, \tau \leqslant 1} |x(t) - x(\tau)|$.

It is obvious that $H_\beta$ is imbedded in $H_\alpha$ if $\beta > \alpha$. This imbedding is not dense for $\alpha > 0$. Indeed, $x_0(t) = t^\alpha \in H_\alpha$, but $\|x_0 - x\|_{H_\alpha} > 1$ for all $x \in H_\beta$. Of course, all spaces $H_\alpha$ are densely imbedded in $H_0$.

According to (3.3) the norm of a given function $x \in H_1$ is a continuous logarithmically convex function of $\alpha$, being a supremum of continuous logarithmically convex functions. However, the family $H_\alpha$ does not form a scale of spaces in our sense, since the spaces $H_\alpha$ are not imbedded densely into each other. We may consider the incomplete normal scale with base $H_1$ and norms (3.3) and complete it.

**2. *The Hölder scale.*** We denote by $H_\alpha^0$ the completion of $H_1$ in the norm (3.3) of $H_\alpha$. In view of what has been said in §1.3, the spaces $H_\alpha^0$ form a continuous normal scale connecting $H_0$ and $H_1$. This scale is called the *Hölder scale.*

THEOREM 3.3. *The space $H_\alpha^0$, $0 \leqslant \alpha < 1$, consists of all functions (classes of functions) in $H_\alpha$ such that*

$$\lim_{|t-\tau|\to 0} \frac{|x(t) - x(\tau)|}{|t - \tau|^\alpha} = 0. \tag{3.4}$$

PROOF. All functions in $H_1$ obviously have property (3.4). Now let $x_n \in H_1$ and $\|x - x_n\|_{H_\alpha} \to 0$. We have

$$\frac{|x(t) - x(\tau)|}{|t - \tau|^\alpha} \leqslant \frac{|x(t) - x_n(t) - [x(\tau) - x_n(\tau)]|}{|t - \tau|^\alpha} + \frac{|x_n(t) - x_n(\tau)|}{|t - \tau|^\alpha}$$

$$\leqslant \|x - x_n\|_{H_\alpha} + \|x_n\|_{H_1}|t - \tau|^{1-\alpha}.$$

First choosing $n$ to make the first summand sufficiently small, and then choosing $|t - \tau|$ to make the second summand sufficiently small, we obtain that the left side is arbitrarily small if $|t - \tau|$ is sufficiently small. Consequently, every function $x \in H_\alpha$ has property (3.4).

Now let $x$ be an arbitrary function in $H_\alpha$ having property (3.4). We show that it belongs to $H_\alpha^0$. We consider the functions

$$x_n(t) = n\int_t^{t+1/n} x(\tau)\,d\tau - n\int_0^{1/n} x(\tau)\,d\tau \qquad (n = 1, 2),$$

setting $x(t) = x(1)$ if $t > 1$. The functions $x_n(t)$ are continuously differentiable, and consequently belong to $H_1$. Performing the change of variables $\tau = \theta + t$ in the first integral, we obtain

$$x_n(t) = n\int_0^{1/n} [x(t + \theta) - x(\theta)]\,d\theta.$$

Hence

$$x_n(t) - x(t) = n\int_0^{1/n} [x(t + \theta) - x(\theta) - x(t)]\,d\theta.$$

We write $[x_n(t+h) - x(t+h)] - [x_n(t) - x(t)] = \Psi(x_n - x, h)$. With this notation we have

$$\|x_n - x\|_{H_\alpha} = \sup_{0 < t+h < 1} \frac{|\Psi(x_n - x, h)|}{h^\alpha}.$$

In view of condition (3.4), for every $\varepsilon > 0$ we may choose $h_0 > 0$ such that $h^{-\alpha}|x(t+h) - x(t)| \leqslant \varepsilon/2$ for all $t$ and $h \leqslant h_0$. Then

$$\begin{aligned} &h^{-\alpha}|\Psi(x_n - x, h)| \\ &\quad = h^{-\alpha} n \left| \int_0^{1/n} \left[ x(t+\theta+h) - x(t+h) - x(t+\theta) + x(t) \right] d\theta \right| \\ &\quad \leqslant n \int_0^{1/n} \left[ \frac{|x(t+h+\theta) - x(t+\theta)|}{h^\alpha} + \frac{|x(t+h) - x(t)|}{h^\alpha} \right] d\theta \\ &\quad \leqslant n \cdot \frac{1}{n} \left( \frac{\varepsilon}{2} + \frac{\varepsilon}{2} \right) = \varepsilon \end{aligned}$$

for $h \leqslant h_0$ and $n \geqslant 1$. Now let $h_0 \leqslant h \leqslant 1$. Then

$$\begin{aligned} &h^{-\alpha}|\Psi(x_n - x, h)| \\ &\quad \leqslant n \int_0^{1/n} \frac{\theta^\alpha}{h^\alpha} \left[ \frac{|x(t+h+\theta) - x(t+h)|}{\theta^\alpha} + \frac{|x(t+\theta) - x(t)|}{\theta^\alpha} \right] d\theta \\ &\quad \leqslant 2n\|x\|_{H_\alpha} \int_0^{1/n} \frac{\theta^\alpha}{h^\alpha}\, d\theta \leqslant \frac{2\|x\|_{H_\alpha}}{(1+\alpha) h_0^\alpha n^\alpha}, \end{aligned}$$

and we again have $h^{-\alpha}|\Psi(x_n - x, h)| \leqslant \varepsilon$ for sufficiently large $n$. Thus, $\|x - x_n\|_{H_\alpha} < \varepsilon$ for sufficiently large $n$, i.e. $x \in H_\alpha^0$.

LEMMA 3.1. *The space $H_\beta^0$ is imbedded compactly in $H_\alpha^0$ for $\beta > \alpha$.*

PROOF. Let $x_n$ be a bounded sequence in $H_\beta^0$. We assume that $x_n(0) = 0$. Then the sequence $x_n(t)$ is uniformly bounded and equicontinuous, and therefore, compact in $C$. Let $x_{n'}(t)$ be a uniformly convergent subsequence. We show that it is a Cauchy sequence in $H_\alpha^0$. Indeed,

$$\begin{aligned} &\frac{|x_{n'}(t) - x_{m'}(t) - x_{n'}(\tau) + x_{m'}(\tau)|}{|t-\tau|^\alpha} \\ &\qquad = |t-\tau|^{\beta-\alpha} \left| \frac{x_{n'}(t) - x_{n'}(\tau)}{|t-\tau|^\beta} - \frac{x_{m'}(t) - x_{m'}(\tau)}{|t-\tau|^\beta} \right| \\ &\qquad \leqslant |t-\tau|^{\beta-\alpha} \left( \|x_{n'}\|_{H_\beta^0} + \|x_{m'}\|_{H_\beta^0} \right) < \varepsilon \end{aligned} \tag{3.5}$$

for

$$|t - \tau| < \left( \frac{\varepsilon}{2 \max_n \|x_n\|_{H_0^\beta}} \right)^{1/(\beta-\alpha)} = h.$$

For $|t - \tau| > h$ the expression on the left side of (3.5) can be made smaller than $\varepsilon$ for sufficiently large $m'$ and $n'$, in view of the uniform convergence of $x_{n'}(t)$. For such $m'$ and $n'$ we have $\|x_{n'} - x_{m'}\|_{H^0_\alpha} < \varepsilon$.

REMARK. The same proof shows that the assertion of the lemma also holds for the spaces $H_\alpha$.

We have obtained the Hölder scale $H^0_\alpha$ by completing an incomplete scale with norms of the spaces $H_\alpha$. The family of the spaces $H_\alpha$ can be recovered from the scale $H^0_\alpha$ by completing the spaces $H^0_\alpha$ relative to the space $H_0$ described in §1.5.

First we establish an auxiliary assertion. Let $0 = t_0 < t_1 < \cdots < t_{N-1} < t_N = 1$ be an arbitrary partition of $[0, 1]$ and let $x_N(t)$ be a continuous function which is linear on every interval $[t_{k-1}, t_k]$. This function obviously belongs to $H_1$.

LEMMA 3.2. *For a continuous piecewise linear function,*

$$\|x_N\|_{H_\alpha} = \max_{\substack{0 \leqslant i,j \leqslant N \\ i \neq j}} \frac{|x_N(t_i) - x_N(t_j)|}{|t_i - t_j|^\alpha}. \tag{3.6}$$

PROOF. Let us denote the right side of (3.6) by $\theta(x)$. It is obvious that

$$\|x_N\|_{H_\alpha} = \sup_{0 \leqslant t,\tau \leqslant 1} \frac{|x_N(t) - x_N(\tau)|}{|t - \tau|^\alpha} \geqslant \max_{0 \leqslant i,j \leqslant N} \frac{|x_N(t_i) - x_N(t_j)|}{|t_i - t_j|^\alpha} = \theta(x).$$

If the points $t$ and $\tau$ belong to $[t_{k-1}, t_k]$, where the function $x_N(t)$ is linear, then

$$\begin{aligned} \frac{|x_N(t) - x_N(\tau)|}{|t - \tau|^\alpha} &= \left| \frac{x_N(t_k) - x_N(t_{k-1})}{t_k - t_{k-1}} \right| |t - \tau|^{1-\alpha} \\ &\leqslant \frac{|x_N(t_k) - x_N(t_{k-1})|}{|t_k - t_{k-1}|^\alpha} \leqslant \theta(x). \end{aligned}$$

Assume that for the points $\bar{t}$, $t_{k-1}$, and $t_k$, where $\bar{t} \notin [t_{k-1}, t_k]$, we have proved that

$$\frac{|x_N(\bar{t}) - x_N(t_{k-1})|}{|\bar{t} - t_{k-1}|^\alpha} \leqslant \theta(x) \quad \text{and} \quad \frac{|x_N(\bar{t}) - x_N(t_k)|}{|\bar{t} - t_k|^\alpha} \leqslant \theta(x). \tag{3.7}$$

We show that

$$|x_N(\bar{t}) - x_N(t)|/|\bar{t} - t|^\alpha \leqslant \theta(x), \tag{3.8}$$

where $t = \lambda t_{k-1} + (1 - \lambda) t_k$ $(0 \leqslant \lambda \leqslant 1)$ is an arbitrary point of $[t_{k-1}, t_k]$.

Indeed, by virtue of the linearity of $x_N(t)$ on $[t_{k-1}, t_k]$ and (3.7) we have

$$\frac{|x_N(\bar{t}) - x_N(t)|}{|\bar{t} - t|^\alpha} = \frac{\left|\lambda\left[x_N(\bar{t}) - x_N(t_{k-1})\right] + (1-\lambda)\left[x_N(\bar{t}) - x_N(t_k)\right]\right|}{|\lambda(\bar{t} - t_{k-1}) + (1-\lambda)(\bar{t} - t_k)|^\alpha}$$

$$\leqslant \frac{\lambda|\bar{t} - t_{k-1}|^\alpha + (1-\lambda)|\bar{t} - t_k|^\alpha}{\left|\lambda(\bar{t} - t_{k-1}) + (1-\lambda)(\bar{t} - t_k)\right|^\alpha}\theta(x) \leqslant \theta(x).$$

The last inequality follows from the concavity of the power function with exponent $\alpha$, $0 < \alpha \leqslant 1$.

Now let $\tau \in [t_{i-1}, t_i]$ and $t \in [t_{k-1}, t_k]$. Inequalities (3.7) are obviously true for $\bar{t} = t_{i-1}$ and $\bar{t} = t_i$. Then it follows from the preceding that (3.8) holds for these values of $\bar{t}$. The $t$ inequality

$$|x_N(\tau) - x_N(t)|/|\tau - t|^\alpha \leqslant \theta(x)$$

follows from this by the same arguments.

Thus, $\|x_N\|_{H_\alpha} \leqslant \theta(x)$, and so (3.6) is satisfied.

THEOREM 3.4. *The completion of $H$ relative to $H_0$ coincides with $H_\alpha$.*

PROOF. Let $x \in H_\alpha$. We construct a sequence of piecewise linear functions $x_N(t)$ uniformly converging to a function $x(t)$ on $[0, 1]$ as $N \to \infty$ so that the graphs of these functions are inscribed in the graph of $x(t)$. Then by Lemma 3.2 we have

$$\|x_N\|_{H_\alpha^0} = \max_{\substack{1 \leqslant i,j \leqslant N \\ i \neq j}} \frac{|x_N(t_i) - x_N(t_j)|}{|t_i - t_j|^\alpha} = \max_{\substack{1 \leqslant i,j \leqslant N \\ i \neq j}} \frac{|x(t_i) - x(t_j)|}{|t_i - t_j|^\alpha} \leqslant \|x\|_{H_\alpha}.$$

The sequence $x_N$ belongs to $H_1 \subset H_\alpha^0$, converges to $x$ in $H_0$, and is bounded in $H_\alpha^0$. Therefore $x$ belongs to the completion of $H_\alpha^0$ relative to $H_0$. Conversely, if $x_n \to x$ in $H_0$ and $\|x_n\|_{H_\alpha^0} \leqslant M$, then

$$|x_n(t) - x_n(\tau)|/|t - \tau|^\alpha \leqslant M,$$

for every $t, \tau \in [0, 1]$, and, by the uniform convergence of $x_n(t)$ to $x(t)$,

$$|x(t) - x(\tau)|/|t - \tau|^\alpha \leqslant M,$$

i.e. $x \in H_\alpha$. The theorem is proved.

Thus, the spaces $H_\alpha^0$ are not complete relative to $H_0$, and in accordance with what has been said in §1.5 the spaces $H_\alpha$ are not imbedded densely into each other.

In order to elucidate the interpolation properties of the Hölder scale, we have to study some properties of the dual spaces of the $H_\alpha$. First we exhibit these properties in the case of finite-dimensional analogues of Hölder spaces.

For a given partition $0 = t_0 < t_1 < \cdots < t_{N-1} < t_N = 1$ of the interval $[0, 1]$ the collection of continuous piecewise linear functions that appears in Lemma 3.2 forms an $N$-dimensional subspace of $H_\alpha$, which we denote by $H_\alpha^N$. The norm in $H_\alpha^N$ is given by the right side of (3.6). We denote by $G_\alpha^N$ the space dual to $H_\alpha^N$. The functionals

$$g_{t_i t_j} = x(t_i) - x(t_j)$$

belong to this space. As follows from (3.6), we have

$$\| g_{t_i t_j} \|_{G_\alpha^N} \leqslant \| g_{t_i t_j} \|_{H_\alpha'} \leqslant |t_i - t_j|^\alpha.$$

Let $t_i < t_j$. We consider the piecewise linear function

$$x_0(t) = \begin{cases} 0 & \text{if } 0 \leqslant t \leqslant t_i, \\ (t - t_i)|t_i - t_j|^{\alpha - 1} & \text{if } t_i \leqslant t \leqslant t_j, \\ |t_i - t_j|^\alpha & \text{if } t_j \leqslant t \leqslant 1. \end{cases}$$

For it we have $\|x_0\|_{H_\alpha} = \|x_0\|_{H_\alpha^N} = 1$ and $|g_{t_i t_j}(x_0)| = |t_i - t_j|^\alpha$. Thus,

$$\| g_{t_i t_j} \|_{G_\alpha^N} = |t_i - t_j|^\alpha. \tag{3.9}$$

We write $f_{ij}(x) = |t_i - t_j|^{-\alpha} g_{t_i t_j}(x)$. Then $\|f_{ij}\|_{G_\alpha^N} = 1$, and formula (3.6) for the norm of the space $H_\alpha^N$ can be written in the form

$$\|x\|_{H_\alpha^N} = \max_{\substack{1 \leqslant i, j \leqslant N \\ i \neq j}} |f_{ij}(x)|. \tag{3.10}$$

This implies that the unit ball of $G_\alpha^N$ coincides with the convex hull of the functionals $f_{ij}$. Indeed, if $f$ does not belong to this convex hull, then there exists an element $\bar{x} \in H_\alpha^N$ such that $f(\bar{x}) > |f_{ij}(\bar{x})|$. Then by (3.10) we have $f(\bar{x}) > \|\bar{x}\|_{H_\alpha^N}$, i.e. $\|f\|_{G_\alpha^N} > 1$. Thus, all extreme points of the unit ball of $G_\alpha^N$ are among the functionals $f_{ij}$.

Let $S$ be a face of the unit sphere of $G_\alpha^N$. This means that there exists an element $x_S$ with $\|x_S\|_{H_\alpha^N} = 1$, such that $f(x_S) = 1$ for all $f \in S$. If the functional $f_{ij}$ belongs to $S$, then no functional of the form $f_{ki}$ belongs to $S$. Indeed, if $f_{ki} \in S$, then

$$\begin{aligned} f_{kj}(x_S) &= \frac{x_S(t_k) - x_S(t_j)}{|t_k - t_j|^\alpha} = \frac{x_S(t_k) - x_S(t_i)}{|t_k - t_j|^\alpha} + \frac{x_S(t_i) - x_S(t_j)}{|t_k - t_j|^\alpha} \\ &= f_{ki}(x_S)\frac{|t_k - t_i|^\alpha}{|t_k - t_j|^\alpha} + f_{ij}(x_S)\frac{|t_i - t_j|^\alpha}{|t_k - t_j|^\alpha} \\ &= \frac{|t_k - t_i|^\alpha + |t_j - t_i|^\alpha}{|t_k - t_j|^\alpha} > 1, \end{aligned}$$

which contradicts the equality $\|f_{kj}\|_{G_\alpha^N} = 1$.

Thus, for the system of functionals $f_{ij}$ lying on the face $S$, the set of first indices does not intersect with the set of second indices.

Now let $f$ be any functional with $\|f\|_{G_\alpha^N} = 1$. It belongs to some face $S$ of the unit sphere in $G_\alpha^N$, and consequently

$$f = \sum_{i,j} \alpha_{ij} f_{ij} \qquad (f_{ij} \in S),$$

where $\alpha_{ij} \geqslant 0$ and $\Sigma_{ij}\alpha_{ij} = 1$. We calculate the norm of $f$ in $G_0^N$. We have

$$\|f\|_{G_0^N} \leqslant \sum_{i,j} \alpha_{ij}\|f_{ij}\|_{G_0^N} = \sum_{i,j} \alpha_{ij}|t_i - t_j|^{-\alpha}.$$

Now we consider the continuous piecewise linear function $\bar{x} \in H_\alpha^N$ equal to 1 for all $t_i$ for which $\alpha_{ij} > 0$, and equal to zero for the remaining $t_i$. In view of the fact that the sets of first and second indices for which $\alpha_{ij} > 0$ do not intersect, the definition of $\bar{x}(t)$ is not contradictory. It is obvious that $\|\bar{x}\|_{H_0^N} = 1$ and

$$f(\bar{x}) = \sum_{i,j} \alpha_{ij} \frac{\bar{x}(t_i) - \bar{x}(t_j)}{|t_i - t_j|^\alpha} = \sum_{i,j} \alpha_{ij}|t_i - t_j|^{-\alpha}.$$

We have shown that

$$\|f\|_{G_0^N} = \sum_{i,j} \alpha_{ij}|t_i - t_j|^{-\alpha}. \tag{3.11}$$

Now let $0 \leqslant \beta \leqslant \alpha$. We calculate

$$\begin{aligned}\|f\|_{G_\beta^N} &\leqslant \sum_{i,j} \alpha_{ij}\|f_{ij}\|_{G_\beta^N} = \sum_{i,j} \alpha_{ij}|t_i - t_j|^{\beta-\alpha} \\ &= \sum_{i,j} \left(\alpha_{ij}^{-1/\alpha}|t_i - t_j|\right)^{-(\alpha-\beta)} \alpha_{ij}^{\beta/\alpha}.\end{aligned}$$

Applying the Hölder inequality with exponents $\alpha/(\alpha - \beta)$ and $\alpha/\beta$ and taking account of (3.11), we obtain

$$\|f\|_{G_\beta^N} \leqslant \left(\sum_{i,j} \alpha_{ij}|t_i - t_j|^{-\alpha}\right)^{(\alpha-\beta)/\alpha} \left(\sum_i \alpha_{ij}\right)^{\beta/\alpha} = \|f\|_{G_0^N}^{(\alpha-\beta)/\alpha}.$$

We recall that $\|f\|_{G_\alpha^N} = 1$. Then

$$\|f\|_{G_\beta^N} \leqslant \|f\|_{G_0^N}^{(\alpha-\beta)/\alpha} \|f\|_{G_\alpha^N}^{\beta/\alpha} \tag{3.12}$$

for any $f \in G_\alpha^N$.

We have prepared everything for the proof of the following assertion.

LEMMA 3.3. *For any functional $f \in H_0'$*

$$\|f\|_{(H_\beta^0)'} \leqslant \|f\|_{H_0'}^{(\alpha-\beta)/\alpha} \|f\|_{(H_\alpha^0)'}^{\beta/\alpha}. \tag{3.13}$$

PROOF. Let $f \in H_0'$. We denote by $x^0$ a function in $H_\beta^0$ such that $\|x^0\|_{H_\beta^0} = 1$, and $f(x^0) > \|f\|_{(H_\beta^0)'} - \varepsilon$. This function can be approximated with any accuracy in the norm of $H_0$ by a continuous piecewise linear function $x_N^0(t)$ whose graph is inscribed in the graph of $x^0(t)$. Since $f$ is continuous in the space $H_0$, we may choose $x_N^0$ so that $f(x_N^0) > \|f\|_{(H_\beta^0)'} - \varepsilon$; then $\|x_N^0\|_{H_\beta^0} \leqslant 1$, as we saw in the proof of Theorem 3.4. If we denote by $\tilde{f}$ the restriction of $f$ to the subspace $H_\beta^N$, then it follows from the preceding that $\|\tilde{f}\|_{G_\beta^N} > \|f\|_{(H_\beta^0)'} - \varepsilon$.

It follows from (3.13) that

$$\|f\|_{(H_\beta^0)'} - \varepsilon \leqslant \|\tilde{f}\|_{G_\beta^N} \leqslant \|\tilde{f}\|_{G_0^N}^{(\alpha-\beta)/\alpha}\|\tilde{f}\|_{G_\alpha^N}^{\beta/\alpha} \leqslant \|f\|_{(H_0)'}^{(\alpha-\beta)/\alpha}\|f\|_{(H_\alpha^0)'}^{\beta/\alpha}.$$

Since $\varepsilon > 0$ is arbitrary, the lemma is proved.

THEOREM 3.5. *The Hölder scale $H_\alpha^0$ is the minimal scale connecting the spaces $H_0$ and $H_1$.*

PROOF. It can be proved by means of Theorem 2.5 that inequality (3.13) for $\alpha = 1$ implies that the Hölder scale majorizes the minimal Scale constructed for $H_0$ and $H_1$, i.e. $\|x\|_{H_\alpha^{\min}} \leqslant \|x\|_{H_\alpha^0}$ $(x \in H_1,\ 0 \leqslant \alpha \leqslant 1)$. We prove the reverse inequality. We recall that

$$\|x\|_{H_\alpha^{\min}} = \sup_{f \in H_0'} \frac{|f(x)|}{\|f\|_{H_0'}^{1-\alpha}\|f\|_{H_1'}^{\alpha}}.$$

As above, we may show that for the functional $g_{\tau\tau'} = x(\tau) - x(\tau')$ we have $\|g_{\tau\tau'}\|_{H_\alpha} = |\tau - \tau'|^\alpha$. Therefore

$$\|x\|_{H_\alpha^{\min}} \geqslant \sup_{0 \leqslant \tau,\tau' \leqslant 1} \frac{|g_{\tau\tau'}(x)|}{\|g_{\tau\tau'}\|_{H_0'}^{1-\alpha}\|g_{\tau\tau'}\|_{H_1'}^{\alpha}}$$

$$= \sup_{0 \leqslant \tau,\tau' \leqslant 1} \frac{|x(\tau) - x(\tau')|}{|\tau - \tau'|^\alpha} = \|x\|_{H_\alpha^0}.$$

From Theorems 2.4 and 2.6 we obtain the following assertion.

THEOREM 3.6. *Every minimal scale has the normalized interpolation property relative to the Hölder scale. Every regular scale has the strong interpolation property relative to the Hölder scale.*

By Lemma 4.4 of Chapter I and Theorem 3.4, the fact that the Hölder scale is a normalized interpolation family implies that the family of spaces $H_\alpha$ is also a normalized interpolation family.

# CHAPTER IV

# INTERPOLATION METHODS

## Introduction

**1.** ***Spaces of strongly measurable functions with values in a Banach space.*** We recall (see [18]) that a function $x(t)$ defined on a space $\mathfrak{M}$ with measure $\mu$ and taking values in a Banach space $A$ is said to be *strongly measurable* if there exists a sequence of strictly simple functions on $\mathfrak{M}$ converging to $x(t)$ almost everywhere. The collection $S(\mathfrak{M}, \mu, A)$ of all strongly measurable functions (classes of functions) is obviously a linear space. This space is metrizable by means of the metric

$$\rho(x, y) = \int_{\mathfrak{M}} \frac{\|x(t) - y(t)\|_A}{1 + \|x(t) - y(t)\|_A} \nu(t)\, d\mu(t), \qquad \nu(t) > 0, \quad \nu \in L_1(\mathfrak{M}, \mu).$$

The space $S(\mathfrak{M}, \mu, A)$ is complete.

For functions belonging to $S(\mathfrak{M}, \mu, A)$ the analogue of Egorov's theorem holds. From it we can derive the following assertion.

LEMMA 1. *Every strongly measurable function is the limit of a uniformly convergent sequence of measurable elementary functions.*

For a strictly simple function $x(t)$ assuming the nonzero values $x_i$ on the sets $e_i$ the integral is defined by

$$\int x(s)\, d\mu = \sum x_i \mu(e_i).$$

A function $x(t)$ is said to be *(Bochner) integrable* if there exists a sequence of strictly simple functions $x_n(t)$ converging to $x(t)$ almost everywhere and such that

$$\lim_{n\to\infty} \int \|x(s) - x_n(s)\| d\mu = 0.$$

The *integral* of the function $x(s)$ is, by definition, the limit

$$\lim_{n\to\infty} \int x_n(s)\, d\mu = \int x(s)\, d\mu.$$

For a function $x(t)$ to be integrable it is necessary and sufficient that it be strongly measurable and

$$\|x\| \equiv \int \|x(s)\|\, d\mu < \infty. \tag{1}$$

The space of integrable functions with the norm (1) is a Banach space and is denoted by $L_1(A)$.

Let $E$ be an ideal lattice of functions of $\mathfrak{M}$. We denote by $E(A)$ the collection of strongly measurable functions $x(t)$ with values in $A$ for which $\|x(t)\|_A \in E$. It is easy to verify that the space $E(A)$ is linear.

LEMMA 2. *$E(A)$ is a Banach space relative to the norm*

$$\|x\|_{E(A)} = \|\|x(t)\|_A\|_E.$$

PROOF. Let $x_n \in E(A)$ and $\Sigma_1^\infty \|x_n\|_{E(A)} < \infty$. Since the space $E$ is complete, the series $\Sigma_1^\infty \|x_n(t)\|_A$ is convergent in $E$, and so its sum is finite almost everywhere. It follows from the completeness of $A$ that $\Sigma_1^\infty x_n(t)$ converges to a (strongly measurable) function $x(t)$ almost everywhere, and

$$\sum_{n=1}^\infty \|x_n(t)\|_A \geqslant \|x(t)\|_A. \tag{2}$$

Since $E$ is an ideal lattice, this inequality implies that

$$\|x\|_{E(A)} \leqslant \sum_{n=1}^\infty \|x_n\|_{E(A)} < \infty.$$

Thus, $x \in E(A)$. Applying (2) to the function $x(t) - \Sigma_{n=1}^N x_n(t) = \Sigma_{n=N+1}^\infty x_n(t)$, we obtain

$$\left\| x - \sum_{n=1}^N x_n \right\|_{E(A)} \leqslant \sum_{n=N+1}^\infty \|x_n\|_{E(A)} \to 0 \quad \text{as } N \to \infty.$$

The lemma is proved.

By arguments similar to the proof of Theorem 1 of Chapter II we can establish that $E(A)$ is imbedded in $S(\mathfrak{M}, \mu, A)$.

Lemma 1 admits the following sharpening.

LEMMA 3. *If $x \in S(\mathfrak{M}, \mu, A)$ and $\rho(t)$ is a measurable function, positive on the support of $x$, then there exists a measurable elementary function $\tilde{x}(t)$ such that $\|x(t) - \tilde{x}(t)\|_A \leqslant \rho(t)$ $(t \in \mathfrak{M})$.*

PROOF. We set $\mathfrak{M}_n = \{t \in \mathfrak{M}: 2^n < \rho(t) \leqslant 2^{n+1}\}$. By Lemma 1 there exists an elementary function $x_n(t)$ with support in $\mathfrak{M}_n$ such that $\|x(t) - x_n(t)\|_A < 2^n < \rho(t)$ for $t \in \mathfrak{M}_n$. Then the function $\tilde{x}(t)$ defined by $\tilde{x}(t) = x_n(t)$ for $t \in \mathfrak{M}_n$ and zero outside the support of $x$ has the required properties.

COROLLARY. *The set of elementary functions in $E(A)$ is dense in $E(A)$.*

Indeed, if $x \in E(A)$, then, choosing a function $\rho(t)$ positive on the support $E$ such that $\|\rho\|_E \leqslant \varepsilon$, according to Lemma 3 we construct an elementary function $\tilde{x}(t)$ such that $\|x(t) - \tilde{x}(t)\|_A \leqslant \rho(t)$, from which it follows that $\|x - \tilde{x}\|_{E(A)} \leqslant \|\rho\|_E \leqslant \varepsilon$ and

$$\tilde{x} = x + (\tilde{x} - x) \in E(A).$$

LEMMA 4. *If the lattice $E$ is regular, then the set of strictly simple functions from $E(A)$ is dense in $E(A)$.*

PROOF. It is sufficient to prove that any elementary function

$$x(t) = \sum_{k=1}^{\infty} x_k \chi_{e_k}(t) \qquad (x_k \in A, e_i \cap e_j = \varnothing, i \neq j)$$

in $E(A)$ can be approximated by a sequence of strictly simple functions. We set $x_n(t) = \sum_{k=1}^{n} x_k \chi_{e_k}(t)$. The intersection of the decreasing sets $\tilde{e}_n = \bigcup_{n+1}^{\infty} e_k$ is empty, and therefore the absolute continuity of the norm in $E$ implies that

$$\|x - x_n\|_{E(A)} = \|x\chi_{\tilde{e}_n}\|_{E(A)} = \|\chi_{\tilde{e}_n}(t)\|x(t)\|_A\|_E \to 0$$

as $n \to \infty$.

We need a modification of Lemma 4.

LEMMA 5. *If $\mathfrak{M} = \bigcup_1^{\infty} \mathfrak{M}_n$, where the measurable sets $\mathfrak{M}_n$ are disjoint and have finite measures, then the collection of those strictly simple functions whose supports can be covered by a finite number of sets $\mathfrak{M}_n$ is dense in $E(A)$.*

PROOF. Let $x \in E(A)$. The function $x_N(t) = x(t)\chi_N(t)$, where $\chi_N(t)$ is the characteristic function of $\bigcup_1^N \mathfrak{M}_n$, converges to $x$ in $E(A)$ because of the absolute continuity of the norm in $E$. According to the above, the function $x_N$ can be approximated by strictly simple functions with support in $\bigcup_1^N \mathfrak{M}_n$.

**2. *The dual spaces of the spaces*** $E(A)$. It is natural to assume that the dual space $(E(A))'$ has to coincide with $E'(A')$ by the duality

$$\int \langle x(t), x'(t) \rangle d\mu,$$

where $\langle\ ,\ \rangle$ is the value of a functional in $A'$ at an element of $A$. However, this is not true in the general case. For example, there is no such coincidence in the case where $A = l_1$ and $E = L_2(0, 1)$ (see [151a]). We indicate a class of ideal lattices for which the indicated coincidence takes place.

We assume that $\mathfrak{M} = N_+ = \{1, 2, \ldots\}$ and $\mu(\{n\}) = \mu_n > 0$.

THEOREM 1. *If $E$ is a regular lattice on $N_+$ (with support $= N_+$), then the space $(E(A))'$ is isometrically isomorphic to $E'(A')$.*

PROOF. Let $f \in [E(A)]'$. We consider a numerical sequence $\varphi = (\varphi_1, \varphi_2, \ldots) \in E$ and an element $x_0 \in A$. Then $\varphi x_0 = \{\varphi_1 x_0, \varphi_2 x_0, \ldots\} \in E(A)$ and

$$|f(\varphi x_0)| \leqslant \|f\|_{(E(A))'}\|\varphi x_0\|_{E(A)} = \|f\|_{(E(A))'}\|\varphi\|_E\|x_0\|_A. \tag{3}$$

It is obvious from this that $f(\varphi x_0)$ is a bounded linear functional on $E$, and so by the regularity of $E$ there exists a numerical sequence $\varphi'(x_0) = \{\varphi_1'(x_0), \varphi_2'(x_0), \ldots\} \in E'$, such that $f(\varphi x_0) = \sum_{n=1}^\infty \varphi_n\varphi_n'(x_0)\mu_n$. The uniqueness of the sequence $\varphi'(x_0)$ implies that it depends linearly on $x_0 \in A$. On $A$ we define the linear functionals $\langle x_0, f_n\rangle = \varphi_n'(x_0)$ $(x_0 \in A)$. Setting $\varphi = \psi_n = (0, \ldots, 0, 1, 0, \ldots)$ (1 in the $n$th place) in (3), we obtain

$$|\langle x_0, f_n\rangle| = |\varphi_n'(x_0)| = \mu_n^{-1}|f(\psi_n x_0)| \leqslant \mu_n^{-1}\|f\|_{E(A)'}\|\psi_n\|_E\|x_0\|_A,$$

which implies that $\varphi_n'$ is bounded on $A$. From the definition we obtain

$$f(\varphi x_0) = \sum_{n=1}^\infty \varphi_n\langle x_0, f_n\rangle\mu_n.$$

Any finitary sequence $x = (x_1, x_2, \ldots) \in E(A)$ (only finitely many terms are different from zero) can be represented in the form $\sum_1^N \psi_k x_k$. Therefore

$$f(x) = \sum_{k=1}^N f(\psi_k x_k) = \sum_{k=1}^N \langle x_k, f_k\rangle\mu_k. \tag{4}$$

Now we take $\varphi = (\varphi_1, \varphi_2, \ldots) \geqslant 0$ from $E$ and in place of $x_k$ we substitute the elements $\varphi_k y_k$ in (4), where $\|y_k\|_A \leqslant 1$. We have $x = \{\varphi_1 y_1, \ldots, \varphi_N y_N, 0, \ldots\} \in E(A)$, and the norm of this sequence in $E(A)$ does not exceed $\|\varphi\|_F$. Consequently $|f(x)| \leqslant \|f\|_{(E(A))'}\|\varphi\|_E$. Making $y_k$ run independently through the unit ball of $A$, we can make the right side of (4) arbitrarily close to $\sum_1^N \varphi_k\|f_k\|_{A'}\mu_k$. Hence

$$\sum_{k=1}^N \varphi_k\|f_k\|_{A'}\mu_k \leqslant \|f\|_{(E(A))'}\|\varphi\|_E.$$

Letting $N$ tend to $\infty$, we obtain

$$\sum_{k=1}^\infty \varphi_k\|f_k\|_{A'}\mu_k \leqslant \|f\|_{(E(A))'}\|\varphi\|_E.$$

From this it follows that the sequence $\tau f = \{f_1, f_2, \ldots\}$ belongs to $E'(A')$ and

$$\|\tau f\|_{E'(A')} \leqslant \|f\|_{(E(A))'}. \tag{5}$$

On $E(A)$ we consider the functional

$$\tilde{f}(x) = \sum_{k=1}^{\infty} \langle x_k, f_k \rangle \mu_k. \tag{6}$$

We have

$$|\tilde{f}(x)| = \left| \sum_{k=1}^{\infty} \langle x_k, f_k \rangle \mu_k \right| \leqslant \sum_{k=1}^{\infty} \|x_k\|_A \|f_k\|_{A'} \mu_k \leqslant \|x\|_{E(A)} \|\tau f\|_{E'(A')}.$$

Thus,

$$\|\tilde{f}\|_{(E(A))'} \leqslant \|\tau f\|_{E'(A')} \leqslant \|f\|_{(E(A))'}. \tag{7}$$

By (4) and (6), the bounded linear functionals $f$ and $\tilde{f}$ coincide on finitary sequences from $E(A)$, which are dense in $E(A)$ by Lemma 4. Consequently $f = \tilde{f}$.

Thus, to every functional $f \in [E(A)]'$ there corresponds the functional $\tau f = (f_1, f_2, \ldots) \in E'(A')$, and

$$f(x) = \sum_{k=1}^{\infty} \langle x_k, f_k \rangle \mu_k. \tag{8}$$

From (5) and (7) it follows that $\|f\|_{(E(A))'} = \|\tau f\|_{E'(A')}$. It is clear from the construction that the operator $\tau$ is linear.

We note that equality (8) determines the sequence $\{f_k\}$ uniquely, as can be verified by replacing $x$ by sequences having only one nonzero element.

If $\{f_k\}$ is an arbitrary sequence from $E'(A')$, then, as was shown earlier, (8) determines a linear functional on $E(A)$. From what we said above, it follows that $\tau f = \{f_k\}$. Thus, the mapping $f \to \tau f$ is surjective, i.e. the operator $\tau$ effects an isometric isomorphism between the spaces $(E(A))'$ and $E'(A')$.

The theorem is proved.

In the case of lattices on spaces with continuous measures the situation becomes more complicated. We mention one theorem, without proof, which will be used in the sequel (see [12], p. 807).

THEOREM 2. *Let $E = L_1$ on the axis $(-\infty, \infty)$ with Lebesgue measure. For any linear functional $F$ on the space $E(A) = L_1(A)$ there exists a function $\psi(t)$ with values in the space $A'$, bounded in the norm of $A'$ and such that the function $\langle x, \psi(t) \rangle$ is measurable for every $x \in E$ and*

$$F(x) = \int_{-\infty}^{\infty} \langle x(t), \psi(t) \rangle \, dt$$

*for $x(t) \in L_1(A)$. Moreover,* $\|F\|_{(L_1(A))'} = \operatorname{ess\,sup} \|\psi(t)\|_{A'}$.

## §1. The complex method of interpolation

1. ***Scalar harmonic and analytic functions in a strip.*** We recall that a complex-valued function $u(\zeta)$ ($|\zeta| \leqslant 1$) continuous and harmonic in the unit disk of the complex plane is representable in terms of its boundary values by means of a Poisson integral:

$$u(\zeta) = \frac{1}{2\pi}\int_{-\pi}^{\pi} u(e^{i\theta})\mathrm{Re}\frac{e^{i\theta}+\zeta}{e^{i\theta}-\zeta}\,d\theta. \tag{1.1}$$

If $\varphi(\theta)$ is an arbitrary continuous $2\pi$-periodic function, then the formula

$$u(\zeta) = \frac{1}{2\pi}\int_{-\pi}^{\pi} \varphi(\theta)\mathrm{Re}\frac{e^{i\theta}+\zeta}{e^{i\theta}-\zeta}\,d\theta \tag{1.2}$$

defines a function, harmonic inside the disk and continuous in the closed disk, that coincides with $\varphi(\theta)$ on the boundary of the disk (at the points $e^{i\theta}$).

Formula (1.2) retains its sense for any bounded measurable function $\varphi(\theta)$. In this case the function $u(\zeta)$ is also harmonic inside the disk and bounded in it. Moreover, at almost every point of the circle the boundary value of $u(\zeta)$ exists and is equal to $\varphi(\theta)$ on any path which is not tangent to the circle. If we denote this limit value by $u(e^{i\theta})$ (assigning it arbitrarily at all points where there is no limit value), then the Poisson formula (1.1) remains true for every bounded harmonic function (see, for example, [19]).

A special case of (1.2) is the assertion of the mean value theorem:

$$u(0) = \frac{1}{2\pi}\int_{-\pi}^{\pi} \varphi(\theta)\,d\theta. \tag{1.3}$$

There arises the question of when (1.2) generates a function analytic inside the disk. For this it is necessary and sufficient that all Fourier coefficients of $\varphi(\theta)$ with negative indices be equal to zero, i.e. that

$$\int_{-\pi}^{\pi} \varphi(\theta)e^{in\theta}\,d\theta = 0 \qquad (n = 1, 2, \dots). \tag{1.4}$$

These conditions can be formulated in a more concise form: For the function (1.2) to be analytic in the disk it is necessary and sufficient that, for any function analytic inside the disk, continuous in the closed disk and equal to zero at $\zeta = 0$, the equality

$$\int_{-\pi}^{\pi} \varphi(\theta)\psi(\theta)\,d\theta = 0 \tag{1.5}$$

hold.

The necessity of this condition follows from the fact that the product $u(\zeta)\psi(\zeta)$ is an analytic function, and consequently by (1.3) the integral is equal to $2\pi u(0)\psi(0) = 0$. This condition is sufficient, since it implies (1.4) (for $\psi_n(\zeta) = \zeta^n$).

Under a conformal mapping of the unit disk onto some domain, harmonic functions turn into harmonic functions and analytic functions into analytic functions. Therefore every such domain has its Poisson formula with properties similar to those described above. The Poisson formula for the half-plane is well known:

$$u(\zeta) = \frac{1}{\pi}\int_{-\infty}^{\infty} u(s)\operatorname{Im}\frac{1}{s-\zeta}\,ds. \tag{1.6}$$

We shall be interested in functions harmonic and analytic inside the strip $\Pi$: $0 \leqslant \operatorname{Re} z \leqslant 1$. The mapping $\zeta = e^{-\pi i z}$ maps this strip conformally onto the half-plane. By making the appropriate substitutions in the integral (1.6), we arrive at a Poisson formula of the form

$$\begin{aligned} u(z) &= \frac{1}{2}\int_{-\infty}^{\infty} u(i\tau)\frac{\sin \pi s}{\cosh \pi(\tau - t) - \cos \pi s}\,d\tau \\ &\quad + \frac{1}{2}\int_{-\infty}^{\infty} u(1 + i\tau)\frac{\sin \pi s}{\cosh \pi(\tau - t) + \cos \pi s}\,d\tau, \end{aligned}$$

where $z = s + it$. The kernels

$$\mu_0(z, \tau) = \frac{1}{2}\frac{\sin \pi s}{\cosh \pi(\tau - t) - \cos \pi s}$$

and

$$\mu_1(z, \tau) = \frac{1}{2}\frac{\sin \pi s}{\cosh \pi(\tau - t) + \cos \pi s}$$

are nonnegative, and

$$\int_{-\infty}^{\infty} \mu_0(\alpha, \tau)\,d\tau = 1 - \alpha, \qquad \int_{-\infty}^{\infty} \mu_1(\alpha, \tau)\,d\tau = \alpha. \tag{1.7}$$

A criterion of analyticity of a function representable in the form

$$u(z) = \int_{-\infty}^{\infty} g_0(\tau)\mu_0(z, \tau)\,d\tau + \int_{-\infty}^{\infty} g_1(\tau)\mu_1(z, \tau)\,d\tau, \tag{1.8}$$

where the functions $g_0(\tau)$ and $g_1(\tau)$ are continuous and bounded on $(-\infty, \infty)$, can be obtained from the analyticity criterion for the disk formulated above: For a function $u(z)$ of the form (1.8) to be analytic inside the strip it is necessary and sufficient that, for any function $h(z)$ continuous in the closed strip, analytic inside the strip, having limits as $\operatorname{Im} z \to \pm\infty$ and vanishing at $z = \alpha$,

$$\int_{-\infty}^{\infty} h(i\tau)g_0(\tau)\mu_0(\alpha, \tau)\,d\tau + \int_{-\infty}^{\infty} h(1 + i\tau)g_1(\tau)\mu_1(\alpha, \tau)\,d\tau = 0$$

hold.

We denote by $\mathfrak{F}$ the space of functions analytic inside $\Pi$, continuous and bounded in its closure, and equipped with the norm

$$\|u\|_{\mathfrak{F}} = \max\Big\{ \sup_{\tau} |u(i\tau)|,\ \sup_{\tau} |u(1 + i\tau)| \Big\}.$$

The completeness of this space follows easily from the maximum principle. It is obvious that the product of two functions from $\mathfrak{F}$ again belongs to $\mathfrak{F}$. Therefore $\mathfrak{F}$ is a Banach algebra.

**2.** ***Abstract analytic functions in a strip.*** Let $A$ be a complex Banach space. We consider the collection $\mathfrak{F}(A)$ of all functions with values in $A$, analytic inside $\Pi$, and continuous and bounded in the closed strip. If $f \in \mathfrak{F}(A)$, then we have the representation

$$f(z) = \int_{-\infty}^{\infty} f(i\tau)\mu_0(z, \tau)\, d\tau + \int_{-\infty}^{\infty} f(1 + i\tau)\mu_1(z, \tau)\, d\tau. \tag{1.9}$$

To prove this it is sufficient to apply (1.8) to scalar functions $\varphi(f(z))$, where $\varphi \in A'$, and use the fact that the integrals in (1.9) are convergent in $A$ by virtue of the boundedness of $f(z)$ and the integrability of the functions $\mu_0(z, \tau)$ and $\mu_1(z, \tau)$.

We introduce the norm

$$\|f\|_{\mathfrak{F}(A)} = \max\Big\{ \sup_{\tau} \|f(i\tau)\|_A,\ \sup_{\tau} \|f(1 + i\tau)\|_A \Big\}$$

in $\mathfrak{F}(A)$, which makes it a Banach space. (We recall that the maximum principle remains valid for abstract analytic functions (see [18], §3.13).) The space $\mathfrak{F}(A)$ is a module over the Banach algebra $\mathfrak{F}$.

**3.** ***The space*** $\mathfrak{F}(A_0, A_1)$. Now let $A_0$ and $A_1$ be a Banach couple of complex spaces. By $\mathfrak{F}(A_0, A_1)$ we denote the linear space consisting of all functions $f(z)$ defined in the strip $\Pi$: $0 \leqslant \operatorname{Re} z \leqslant 1$, with values in the space $A_0 + A_1$ and having the following properties:

1) $f(z)$ is continuous and bounded in the norm of $A_0 + A_1$ in the closed strip $\Pi$.

2) $f(z)$ is analytic relative to the norm of $A_0 + A_1$ inside the strip.

3) $f(i\tau)$ assumes values in the space $A_0$ and is continuous and bounded in the norm of this space, while $f(1 + i\tau)$ assumes values in $A_1$ and is continuous and bounded in the norm of $A_1$.

We equip the linear space $\mathfrak{F}(A_0, A_1)$ with the norm

$$\|f\|_{\mathfrak{F}(A_0, A_1)} = \max\Big\{ \sup_{\tau} \|f(i\tau)\|_{A_0},\ \sup_{\tau} \|f(1 + i\tau)\|_{A_1} \Big\}. \tag{1.10}$$

We note that, by the maximum principle, for an analytic function we have

$$\|f(z)\|_{A_0+A_1} \leqslant \max\left\{ \sup_\tau \|f(i\tau)\|_{A_0+A_1}, \sup_\tau \|f(1+i\tau)\|_{A_0+A_1}\right\}$$
$$\leqslant \max\left\{ \sup_\tau \|f(i\tau)\|_{A_0}, \sup_\tau \|f(1+i\tau)\|_{A_1}\right\} = \|f\|_{\mathfrak{F}(A_0,A_1)} \tag{1.11}$$

for $f \in \mathfrak{F}(A_0, A_1)$. From this it follows in particular that $f(z) = 0$, provided $\|f\|_{\mathfrak{F}(A_0,A_1)} = 0$.

We prove the completeness of $\mathfrak{F}(A_0, A_1)$. Let $f_n$ be a Cauchy sequence in $\mathfrak{F}(A_0, A_1)$. From (1.11) we obtain that $\|f_n(z) - f_m(z)\|_{A_0+A_1} \leqslant \|f_n - f_m\|_{\mathfrak{F}(A_0,A_1)}$, and consequently the sequence $f_n(z)$ is uniformly convergent in the norm of $A_0 + A_1$ in $\Pi$. Its limit $f(z)$ is a function having properties 1) and 2). The functions $f_n(j + i\tau)$ ($j = 0, 1$) converge uniformly in the norm of $A_j$, and by virtue of the imbedding $A_j \subset A_0 + A_1$ their limit is the function $f(j + i\tau)$, which is continuous and bounded in the norm of $A_j$. Thus, $f \in \mathfrak{F}(A_0, A_1)$ and $\|f - f_n\|_{\mathfrak{F}(A_0,A_1)} \to 0$ as $n \to \infty$.

We denote by $\mathfrak{F}_0(A_0, A_1)$ the subspace of $\mathfrak{F}(A_0, A_1)$ consisting of all functions $f(z)$ for which $f(j + i\tau)$ converges to zero in the norm of $A_j$ ($j = 0, 1$) as $|\tau| \to \infty$.

If $f \in \mathfrak{F}(A_0, A_1)$, then the function

$$f_\delta(z) = e^{\delta z^2} f(z) \tag{1.12}$$

obviously belongs to $\mathfrak{F}_0(A_0, A_1)$ for every $\delta > 0$. It is easy to verify that the functions $f_\delta(z)$ converge to $f(z)$ in the norm of $\mathfrak{F}(A_0, A_1)$ as $\delta \to 0$.

For the sequel we need several properties of the spaces $\mathfrak{F}(A_0, A_1)$ and $\mathfrak{F}_0(A_0, A_1)$.

LEMMA 1.1. *If $f \in \mathfrak{F}(A_0, A_1)$, and if the sequence of functions $g_n \in \mathfrak{F}_0(A_0, A_1)$ is bounded in $\mathfrak{F}(A_0, A_1)$ and has the property that the functions $g_n(j + i\tau)$ converge uniformly to $f(j + i\tau)$ in the norm of $A_j$ ($j = 0, 1$) on every bounded subdomain of $\tau$ values, then there exists a sequence of positive numbers $s_n$ such that the functions $e^{s_n z^2} g_n(z)$ converge to $f(z)$ in $\mathfrak{F}(A_0, A_1)$.*

PROOF. As we said above, the functions $e^{s_n z^2} f(z)$ converge to $f(z)$ in $\mathfrak{F}_0(A_0, A_1)$ as $s_n \to 0$. Therefore it is sufficient to choose a sequence of $s_n$ so that $s_n \to 0$ and $e^{s_n z^2}(g_n(z) - f(z)) \to 0$ in $\mathfrak{F}(A_0, A_1)$. Let us consider the functions

$$\varphi_k(t) = \max_{j=0,1}\left\{ \sup_{\substack{|\tau|<t \\ n>k}} \|f(j+i\tau) - g_n(j+i\tau)\|_{A_j}\right\}.$$

The functions $\varphi_k(t)$ are increasing in $t$, decreasing in $k$, and converge to zero as $k \to \infty$ for every $t \geqslant 0$ by the hypothesis of the lemma. Finally, the sequence $\varphi_k(t)$ is uniformly bounded: $\sup_{0 \leqslant t < \infty} \varphi_k(t) \leqslant C$.

We denote by $[0, a_n]$ the largest interval on which $t\varphi_n(t) \leqslant 1$. Obviously $a_n \leqslant a_{n+1}$. Moreover, $a_n \to \infty$, because otherwise $a_n < a < \infty$ and $\varphi_n(a) \geqslant 1/a$, which contradicts the fact that $\varphi_n(a) \to 0$ as $n \to \infty$. Now we choose $s_n = a_n^{-1}$. Then $s_n \to 0$ and

$$\begin{aligned}\|e^{s_n z^2}(g_n - f)\|_{\mathfrak{F}(A_0, A_1)} &= \max_{j=0,1}\left\{\sup_\tau e^{-s_n\tau^2 + js_n}\|f(j + i\tau) - g_n(j + i\tau)\|_{A_j}\right\}\\ &\leqslant e^{s_1}\varphi_n(a_n) + e^{s_1}e^{-a_n}\max_{j=0,1}\left\{\sup_{|\tau| > a_n}\|f(j + i\tau) - g_n(j + i\tau)\|_{A_j}\right\}\\ &\leqslant e^{s_1}a_n^{-1} + e^{s_1}e^{-a_n}\left(\|f\|_{\mathfrak{F}(A_0, A_1)} + \|g_n\|_{\mathfrak{F}(A_0, A_1)}\right) \to 0 \quad \text{as } n \to \infty.\end{aligned}$$

LEMMA 1.2. *For every function $f \in \mathfrak{F}_0(A_0, A_1)$ there exists a sequence of functions $g_n \in \mathfrak{F}(A_0, A_1)$ satisfying the hypotheses of Lemma 1.1 and consisting of functions periodic in $\tau$.*

PROOF. It is sufficient to prove the lemma for functions of the form (1.12), where $f \in \mathfrak{F}_0(A_0, A_1)$. We define the functions

$$g_T(z) = \sum_{k=-\infty}^{\infty} f_\delta(z + ikT) \qquad (T \geqslant 1). \tag{1.13}$$

To study the convergence of this series we consider the function series

$$\sum_{k=-\infty}^{\infty} e^{-\delta(\tau + kT)^2}.$$

The sum of this series depends on $\tau$ periodically with period $T$. Therefore it is sufficient to study the convergence of the series on the interval $0 \leqslant \tau \leqslant T$. For such $\tau$ we have

$$\begin{aligned}\sum_0^\infty e^{-\delta(\tau + kT)^2} + \sum_1^\infty e^{-\delta(\tau - kT)^2} &\leqslant \sum_0^\infty e^{-\delta k^2T^2} + \sum_1^\infty e^{-\delta(k-1)^2T^2}\\ &\leqslant 2\sum_0^\infty e^{-\delta k^2} = C(\delta) < \infty.\end{aligned}$$

The inequality

$$\|f_\delta(z + ikT)\| \leqslant e^{\delta\sigma^2 - \delta(\tau + kT)^2}\|f\|_{\mathfrak{F}(A_0, A_1)}, \qquad z = \sigma + i\tau,$$

which is true when we have on the left side the norm of $A_0 + A_1$ for $z$ inside $\Pi$ and the norm of $A_j$ for $z = j + i\tau$ $(j = 0, 1)$, implies the convergence in $\mathfrak{F}(A_0, A_1)$ of (1.13) to the function $g_T \in \mathfrak{F}(A_0, A_1)$. Moreover, $\|g_T\|_{\mathfrak{F}(A_0, A_1)} \leqslant e^\delta c(\delta)\|f\|_{\mathfrak{F}(A_0, A_1)}$. It is obvious that the functions $g_T(z)$ are periodic in $\tau$ with period $T$.

We show that the functions $g_T(j + i\tau)$ converge to $f_\delta(j + i\tau)$ in the norm of $A_j$ uniformly on every bounded subdomain of $\tau$ values as $T \to \infty$. Let $|\tau| \leqslant M$. We choose $T \geqslant 2M$. Then

$$\begin{aligned} \|g_T(i\tau) - f_\delta(i\tau)\|_{A_0} &\leqslant \sum e^{-\delta(\tau+kT)^2}\|f(i\tau + ikT)\|_{A_0} \\ &\leqslant 2\sum_1^\infty e^{-\delta(kT - T/2)^2}\|f\|_{\mathfrak{F}(A_0,A_1)} \\ &\leqslant 2e^{-\delta T^2/4}\sum_1^\infty e^{-\delta k(k-1)T^2}\|f\|_{\mathfrak{F}(A_0,A_1)} \to 0 \quad \text{as } T \to \infty. \end{aligned}$$

LEMMA 1.3. *If a function $h \in \mathfrak{F}(A_0, A_1)$ is periodic in $\tau$ with period $T$, then it can be represented as a limit in $\mathfrak{F}(A_0, A_1)$ of functions of the form*

$$h_N(z) = \sum_{k=-N}^{N} x_k e^{2k\pi z/T},$$

*where $x_k \in A_0 \cap A_1$.*

PROOF. For every $\sigma$ we expand $h(\sigma + i\tau)$ in Fourier series in $\tau$:

$$h(\sigma + i\tau) = \sum_{k=-\infty}^{\infty} b_k(\sigma)e^{i2k\pi\tau/T},$$

where

$$b_k(\sigma) = \frac{1}{T}\int_{-T/2}^{T/2} h(\sigma + i\tau)e^{-i2k\pi\tau/T}\,d\tau.$$

Writing $a_k(\sigma) = b_k(\sigma)e^{-2k\pi\sigma/T}$, we obtain

$$h(z) = \sum_{k=-\infty}^{\infty} a_k(\sigma)e^{2k\pi z/T}, \tag{1.14}$$

where

$$a_k(\sigma) = \frac{1}{T}\int_{-T/2}^{T/2} h(\sigma + i\tau)e^{-2k\pi(\sigma+i\tau)/T}\,d\tau. \tag{1.15}$$

Due to the periodicity of the integrand, the last formula can also be written as

$$a_k(\sigma) = \frac{1}{mT}\int_{-mT/2}^{mT/2} h(\sigma + i\tau)e^{-2k\pi(\sigma+i\tau)/T}\,d\tau. \tag{1.16}$$

We show that the coefficients $a_k(\sigma)$ are in fact independent of $\sigma$. Let $0 \leqslant \sigma_0 \leqslant \sigma_1 \leqslant 1$. Denote by $\Gamma_m$ the contour formed by the intervals $\{\sigma_0 \leqslant \operatorname{Re} z \leqslant \sigma_1,\ \operatorname{Im} z = \pm\, mT/2\}$ and $\{\operatorname{Re} z = \sigma_j\ (j = 0, 1),\ |\operatorname{Im} z| < mT/2\}$.

Since $h(z)e^{2k\pi z/T}$ is analytic and continuous within $\Pi$ in the norm of $A_0 + A_1$, its integral over this contour is equal to zero. Then

$$\begin{aligned}
&-\frac{i}{mT}\int_{-mT/2}^{mT/2} h(\sigma_0 + i\tau)e^{-2k\pi(\sigma_0+i\tau)/T}\,d\tau \\
&+\frac{i}{mT}\int_{-mT/2}^{mT/2} h(\sigma_1 + i\tau)e^{-2k\pi(\sigma_1+i\tau)/T}\,d\tau \\
&+\frac{1}{mT}\int_{\sigma_0}^{\sigma_1} h\left(\sigma - i\frac{mT}{2}\right)e^{-2k\pi(\sigma - imT/2)/T}\,d\sigma \\
&-\frac{1}{mT}\int_{\sigma_0}^{\sigma_1} h\left(\sigma + i\frac{mT}{2}\right)e^{-2k\pi(\sigma + imT/2)/T}\,d\sigma = 0.
\end{aligned}$$

The function $h(z)e^{2k\pi z/T}$ is bounded in the strip $0 \leqslant \operatorname{Re} z \leqslant 1$ in the norm of $A_0 + A_1$. Therefore the two last terms converge to zero as $m \to \infty$. The first two integrals do not depend on $m$ according to (1.15) and (1.16). Passing to the limit as $m \to \infty$, we obtain that $a_k(\sigma_0) = a_k(\sigma_1) = a_k$. In particular,

$$a_k = \frac{1}{T}\int_{-T/2}^{T/2} h(i\tau)e^{-2ik\pi\tau/T}\,d\tau = \frac{1}{T}\int_{-T/2}^{T/2} h(1 + i\tau)e^{-2k\pi(1+i\tau)/T}\,d\tau.$$

Since $h(i\tau)$ is continuous in the topology of $A_0$, the first integral belongs to $A_0$. Similarly, the second integral belongs to $A_1$, and therefore $a_k \in A_0 \cap A_1$.

Now we form the Cesàro sums of the series (1.14):

$$h_N(z) = \sum_{k=-N}^{N}\left(1 - \frac{|k|}{N+1}\right)a_k e^{2k\pi z/T} = \sum_{k=-N}^{N} x_k e^{2k\pi z/T}.$$

Since the functions $h(i\tau)$ and $h(1 + i\tau)$ are continuous in the topologies of $A_0$ and $A_1$, respectively, by Fejér's theorem the functions $h_N(i\tau)$ and $h_N(1 + i\tau)$ converge to the functions $h(i\tau)$ and $h(1 + i\tau)$ in the norms of $A_0$ and $A_1$ uniformly in $\tau$ as $N \to \infty$. Thus,

$$\|h_N - h\|_{\mathfrak{F}(A_0,A_1)} \to 0.$$

By combining the lemmas proved above, we obtain the following important assertion.

THEOREM 1.1. *The set of functions of the form*

$$g(z) = e^{\delta z^2}\sum_{n=1}^{N} x_n e^{\lambda_n z}, \tag{1.17}$$

*where $x_n \in A_0 \cap A_1$, the $\lambda_n$ are real, and $\delta > 0$, is dense in $\mathfrak{F}_0(A_0, A_1)$.*

REMARK. It is obvious that the elements $x_n$ may also be chosen from an arbitrary set $M$ dense in $A_0 \cap A_1$.

**4.** ***The interpolation functor*** $[A_0, A_1]_\alpha$. We denote by $[A_0, A_1]_\alpha$ $(0 \leqslant \alpha \leqslant 1)$ the collection of all elements $x \in A_0 + A_1$ representable in the form $x = f(\alpha)$ for some functions $f \in \mathfrak{F}(A_0, A_1)$. This collection is linear. In this collection we introduce the norm

$$\|x\|_{[A_0,A_1]_\alpha} = \|x\|_\alpha = \inf_{f(\alpha)=x} \|f\|_{\mathfrak{F}(A_0,A_1)}. \tag{1.18}$$

We denote by $N_\alpha$ the linear manifold of all functions from $\mathfrak{F}(A_0, A_1)$ vanishing at $z = \alpha$. It follows from (1.11) that $N_\alpha$ is a closed subspace of $\mathfrak{F}(A_0, A_1)$. The definition of the norm in $[A_0, A_1]_\alpha$ shows that this space is isometric to the quotient space of $\mathfrak{F}(A_0, A_1)$ modulo $N_\alpha$, and consequently $[A_0, A_1]_\alpha$ is a Banach space.

From (1.11) we obtain

$$\|x\|_{A_0+A_1} = \|f(\alpha)\|_{A_0+A_1} \leqslant \|f\|_{\mathfrak{F}(A_0,A_1)},$$

for a function $f(z)$ with $f(\alpha) = x$. Therefore for $x \in [A_0, A_1]_\alpha$ we have

$$\|x\|_{A_0+A_1} \leqslant \inf_{f(\alpha)=x} \|f\|_{\mathfrak{F}(A_0,A_1)} = \|x\|_{[A_0,A_1]_\alpha}.$$

Hence it follows that $[A_0, A_1]_\alpha$ is imbedded in $A_0 + A_1$ with imbedding constant 1.

The space $A_0 \cap A_1$ is imbedded in $[A_0, A_1]_\alpha$. To verify this, we consider an arbitrary element $x \in A_0 \cap A_1$ and function $f_x(z) = x$. Then $f_x \in \mathfrak{F}(A_0, A_1)$ and

$$\|x\|_\alpha \leqslant \|f_x\|_{\mathfrak{F}(A_0,A_1)} = \max\{\|x\|_{A_0}, \|x\|_{A_1}\} = \|x\|_{A_0 \cap A_1}.$$

Thus, $A_0 \cap A_1 \overset{1}{\subset} [A_0, A_1]_\alpha \overset{1}{\subset} A_0 + A_1$, and the space $[A_0, A_1]_\alpha$ turns out to be intermediate between $A_0$ and $A_1$.

We note that the mapping $f(z) \to f(1 - z)$ is an isometry of the space $\mathfrak{F}(A_0, A_1)$ onto $\mathfrak{F}(A_1, A_0)$, and therefore $[A_1, A_0]_\alpha = [A_0, A_1]_{1-\alpha}$.

THEOREM 1.2 (interpolation theorem). *Let $A_0$, $A_1$ and $B_0$, $B_1$ be two Banach couples. The triple $(A_0, A_1, [A_0, A_1]_\alpha)$ of Banach spaces is a normalized interpolation triple of type $\alpha$ relative to the triple $(B_0, B_1, [B_0, B_1]_\alpha)$.*

PROOF. Let $T \in L(A_0A_1, B_0B_1)$. We consider an element $x \in [A_0, A_1]_\alpha$. For any $\varepsilon > 0$ there exists an $f \in \mathfrak{F}(A_0, A_1)$ such that $f(\alpha) = x$ and $\|f\|_{\mathfrak{F}(A_0,A_1)} \leqslant \|x\|_\alpha + \varepsilon$.

The function $g(z) = \|T\|^{z-1}_{A_0\to B_0}\|T\|^{-z}_{A_1\to B_1}T(f(z))$ belongs to $\mathfrak{F}(B_0, B_1)$. Next,

$$\begin{aligned}\|g\|_{\mathfrak{F}(B_0,B_1)} &= \max\left\{\sup_\tau \|g(i\tau)\|_{B_0}, \sup_\tau \|g(1+i\tau)\|_{B_1}\right\}\\ &\leqslant \max\left\{\sup_\tau \|T\|^{-1}_{A_0\to B_0}\|T\|_{A_0\to B_0}\|f(i\tau)\|_{A_0},\right.\\ &\qquad\left.\sup_\tau \|T\|^{-1}_{A_1\to B_1}\|T\|_{A_1\to B_1}\|f(1+i\tau)\|_{A_1}\right\}\\ &= \|f\|_{\mathfrak{F}(A_0,A_1)} \leqslant \|x\|_\alpha + \varepsilon.\end{aligned}$$

This implies that

$$\begin{aligned}\|x\|_\alpha + \varepsilon \geqslant \|g\|_{\mathfrak{F}(B_0,B_1)} &\geqslant \|g(\alpha)\|_{[B_0,B_1]_\alpha}\\ &= \|\,\|T\|^{\alpha-1}_{A_0\to B_0}\|T\|^{-\alpha}_{A_1\to B_1}T(f(\alpha))\|_{[B_0,B_1]_\alpha}\\ &= \|T\|^{\alpha-1}_{A_0\to B_0}\|T\|^{-\alpha}_{A_1\to B_1}\|Tx\|_{[B_0,B_1]_\alpha}.\end{aligned}$$

Since $\varepsilon$ is arbitrary, we have

$$\|Tx\|_{[B_0,B_1]_\alpha} \leqslant \|T\|^{1-\alpha}_{A_0\to B_0}\|T\|^{\alpha}_{A_1\to B_1}\|x\|_{[A_0,A_1]_\alpha}.$$

REMARK 1. In calculating the norm $\|x\|_{[A_0,A_1]_\alpha}$ by (1.18), the infimum may be taken over functions from $\mathfrak{F}(A_0, A_1)$ only.

Indeed, if $f \in \mathfrak{F}(A_0, A_1)$ and $f(\alpha) = x$, then $f_1(z) = e^{\delta(z^2-\alpha^2)}f(z) \in \mathfrak{F}_0(A_0, A_1)$ and $f_1(\alpha) = x$ for any $\delta > 0$. Now let $\|f\|_{\mathfrak{F}(A_0,A_1)} \leqslant \|x\|_\alpha + \varepsilon$. For sufficiently small $\delta$ we have $\|f_1\|_{\mathfrak{F}(A_0,A_1)} \leqslant \|f\|_{\mathfrak{F}(A_0,A_1)} + \varepsilon$. Therefore

$$\|x\|_\alpha = \inf_{\substack{\varphi\in\mathfrak{F}(A_0,A_1)\\ \varphi(\alpha)=x}} \|\varphi\|_{\mathfrak{F}(A_0,A_1)} \leqslant \|f_1\|_{\mathfrak{F}(A_0,A_1)} \leqslant \|x\|_\alpha + 2\varepsilon,$$

which proves our assertion.

REMARK 2. If $x \in A_0 \cap A_1$ and $0 < \alpha < 1$, then in the calculation of the norm according to (1.18) the infimum may be taken over functions of the form $\sum_1^N a_k(z)x_k$ only, where $x_k \in A_0 \cap A_1$.

Indeed, according to the preceding remark we choose the function $f \in \mathfrak{F}_0(A_0, A_1)$ so that $f(\alpha) = x$ and $\|f\|_{\mathfrak{F}(A_0,A_1)} \leqslant \|x\|_\alpha + \varepsilon$. Let $r(z)$ be a function mapping the strip $0 < \operatorname{Re} z < 1$ conformally onto the unit disk so that $r(\alpha) = 0$. Then $|r(i\tau)| = |r(1 + i\tau)| = 1$, and therefore

$$\frac{f(z) - e^{\delta(z-\alpha)^2}x}{r(z)} \in \mathfrak{F}_0(A_0, A_1).$$

By Theorem 1.1 there exists a function $g(z)$ of the form (1.17) such that $\|[f(z) - e^{\delta(z-\alpha)^2}x]r^{-1}(z) - g(z)\|_{\mathfrak{F}(A_0,A_1)} < \varepsilon$. We set $f_1(z) = e^{\delta(z-\alpha)^2}x + r(z)g(z)$. This function has the required form, $f_1(\alpha) = x$, and

$$\|f_1\|_{\mathfrak{F}(A_0,A_1)} \leqslant \|f\|_{\mathfrak{F}(A_0,A_1)} + \|f - f_1\|_{\mathfrak{F}(A_0,A_1)} \leqslant \|x\|_\alpha + 2\varepsilon.$$

THEOREM 1.3. *The space $A_0 \cap A_1$ is densely imbedded in any space $[A_0, A_1]_\alpha$ for $0 \leqslant \alpha \leqslant 1$.*

PROOF. Let $x \in [A_0, A_1]_\alpha$ and $f \in \mathfrak{F}_0(A_0, A_1)$ with $f(\alpha) = x$. By Theorem 1.1, for any $\varepsilon > 0$ we can construct a function $g(z)$ of the form (1.17) so that $\|f - g\|_{\mathfrak{F}(A_0,A_1)} < \varepsilon$. Then

$$\|x - g(\alpha)\|_{[A_0,A_1]_\alpha} = \|f(\alpha) - g(\alpha)\|_{[A_0,A_1]_\alpha} \leqslant \|f - g\|_{\mathfrak{F}(A_0,A_1)} < \varepsilon$$

and the theorem is proved, since $g(z)$ takes values in $A_0 \cap A_1$.

Remark 2 and Theorem 1.3 show that the spaces $[A_0, A_1]_\alpha$ could be defined in the following way: We consider the space $\mathfrak{F}(A_0 \cap A_1)$, for the elements $x \in A_0 \cap A_1$ we introduce the norm

$$\|x\|_\alpha = \inf_{\substack{g(\alpha)=x \\ g\in\mathfrak{F}(A_0\cap A_1)}} \|g\|_{\mathfrak{F}(A_0,A_1)},$$

and we complete $A_0 \cap A_1$ in this norm.

We investigate what the extreme spaces $[A_0, A_1]_0$ and $[A_0, A_1]_1$ are. If $x \in [A_0, A_1]_0$, then $f(0) = x$ for some $f \in \mathfrak{F}(A_0, A_1)$. On the other hand, $f(i\tau) \in A_0$ for all $\tau$, and therefore $f(0) = x \in A_0$. Moreover,

$$\begin{aligned}\|f\|_{\mathfrak{F}(A_0,A_1)} &= \max\Big\{\sup_\tau \|f(i\tau)\|_{A_0}, \sup_\tau \|f(1 + i\tau)\|_{A_1}\Big\} \\ &\geqslant \|f(0)\|_{A_0} = \|x\|_{A_0}.\end{aligned}$$

Hence it follows that

$$\|x\|_{[A_0,A_1]_0} \geqslant \|x\|_{A_0}. \tag{1.19}$$

On the other hand, on the basis of Theorem 1.3 for $\varepsilon > 0$ there exists an element $x_1 \in A_0 \cap A_1$ such that $\|x - x_1\|_{[A_0,A_1]_0} < \varepsilon$. We construct the functions $f_n(z) = e^{z^2 - nz}x_1 \in \mathfrak{F}(A_0, A_1)$. We have $f_n(0) = x_1$ and

$$\|f_n\|_{\mathfrak{F}(A_0,A_1)} = \max\big\{\|x_1\|_{A_0}, e^{1-n}\|x_1\|_{A_1}\big\}.$$

Consequently

$$\|x_1\|_{[A_0,A_1]_0} \leqslant \max\big\{\|x_1\|_{A_0}, e^{1-n}\|x_1\|_{A_1}\big\}.$$

Letting $n$ tend to $\infty$, we obtain that $\|x_1\|_{[A_0,A_1]_0} \leqslant \|x_1\|_{A_0}$. By (1.19) we have $\|x - x_1\|_{A_0} \leqslant \|x - x_1\|_{[A_0,A_1]_0} < \varepsilon$, and therefore

$$\|x\|_{[A_0,A_1]_0} \leqslant \|x_1\|_{[A_0,A_1]_0} + \varepsilon \leqslant \|x_1\|_{A_0} + \varepsilon \leqslant \|x\|_{A_0} + 2\varepsilon.$$

Since $\varepsilon$ is arbitrary, this inequality and (1.19) imply

$$\|x\|_{[A_0,A_1]_0} = \|x\|_{A_0} \qquad \big(x \in [A_0, A_1]_0\big).$$

Thus, $[A_0, A_1]_0$ is isometric to some subspace $\overline{A}_0$ of $A_0$.

It can be proved similarly that $[A_0, A_1]_1$ is a subspace $\bar{A}_1$ of $A_1$. It follows from Theorem 1.3 that $[A_0, A_1]_j = \bar{A}_j$ ($j = 0, 1$) coincides with the closure of $A_0 \cap A_1$ in the norm of $A_j$. In particular, if $A_0 \cap A_1$ is dense in $A_j$, then $[A_0, A_1]_j = A_j$.

From the equivalent definition of the spaces $[A_0, A_1]_\alpha$ given above it follows that $[A_0, A_1]_\alpha = [\bar{A}_0, \bar{A}_1]_\alpha$, and so without loss of generality we may always assume that the intersection of the initial spaces is dense in each.

We establish some additional important properties of $[A_0, A_1]_\alpha$.

LEMMA 1.4. *Let $f \in \mathfrak{F}(A_0, A_1)$ and $\alpha \in (0, 1)$. Then*

$$\ln\|f(\alpha)\|_{[A_0,A_1]_\alpha} \leqslant \int_{-\infty}^{\infty} \left[\ln\|f(i\tau)\|_{A_0}\right] \mu_0(\alpha, \tau)\, d\tau + \int_{-\infty}^{\infty} \left[\ln\|f(1 + i\tau)\|_{A_1}\right] \mu_1(\alpha, \tau)\, d\tau, \tag{1.20}$$

$$\|f(\alpha)\|_{[A_0,A_1]_\alpha} \leqslant \left[\frac{1}{1-\alpha} \int_{-\infty}^{\infty} \|f(i\tau)\|_{A_0} \mu_0(\alpha, \tau)\, d\tau\right]^{1-\alpha} \times \left[\frac{1}{\alpha} \int_{-\infty}^{\infty} \|f(1 + i\tau)\|_{A_1} \mu_1(\alpha, \tau)\, d\tau\right]^{\alpha}, \tag{1.21}$$

$$\|f(\alpha)\|_{[A_0,A_1]_\alpha} \leqslant \int_{-\infty}^{\infty} \|f(i\tau)\|_{A_0} \mu_0(\alpha, \tau)\, d\tau + \int_{-\infty}^{\infty} \|f(1 + i\tau)\|_{A_1} \mu_1(\alpha, \tau)\, d\tau. \tag{1.22}$$

PROOF. We choose bounded functions $\varphi_0(\tau)$ and $\varphi_1(\tau)$, infinitely differentiable on $(-\infty, \infty)$, such that $\varphi_0(\tau) \geqslant \ln\|f(i\tau)\|_{A_0}$ and $\varphi_1(\tau) \geqslant \ln\|f(1 + i\tau)\|_{A_1}$. We construct a function $\Phi(z)$ analytic inside $\Pi$ whose real part is bounded and continuous in the closed strip, and with $\operatorname{Re} \Phi(i\tau) = \varphi_0(\tau)$ and $\operatorname{Re} \Phi(1 + i\tau) = \varphi_1(\tau)$. For $\operatorname{Re} \Phi(z)$ we have

$$\operatorname{Re} \Phi(z) = \int_{-\infty}^{\infty} \varphi_0(\tau)\mu_0(z, \tau)\, d\tau + \int_{-\infty}^{\infty} \varphi_1(\tau)\mu_1(z, \tau)\, d\tau.$$

It follows from the differentiability of the functions $\varphi_0(\tau)$ and $\varphi_1(\tau)$ that $\Phi(z)$ is continuous in the closed strip $\Pi$. Therefore $e^{-\Phi(z)}f(z) \in \mathfrak{F}(A_0, A_1)$, and since

$$\|e^{-\Phi(i\tau)}f(i\tau)\|_{A_0} = e^{-\varphi_0(\tau)}\|f(i\tau)\|_{A_0} \leqslant 1,$$

$$\|e^{-\Phi(1+i\tau)}f(1 + i\tau)\|_{A_1} = e^{-\varphi_1(\tau)}\|f(1 + i\tau)\|_{A_1} \leqslant 1,$$

we have

$$\|e^{-\Phi(z)}f(z)\|_{\mathfrak{F}(A_0,A_1)} \leqslant 1,$$

and consequently

$$\|e^{-\Phi(\alpha)}f(\alpha)\|_{[A_0,A_1]_\alpha} \leqslant 1.$$

This implies that

$$\begin{aligned}\ln\|f(\alpha)\|_{[A_0,A_1]_\alpha} &\leqslant \operatorname{Re}\Phi(\alpha)\\ &= \int_{-\infty}^{\infty}\varphi_0(\tau)\mu_0(\alpha,\tau)\,d\tau + \int_{-\infty}^{\infty}\varphi_1(\tau)\mu_1(\alpha,\tau)\,d\tau. \qquad (1.23)\end{aligned}$$

Now choosing decreasing sequences of functions $\varphi_0(\tau)$ and $\varphi_1(\tau)$ converging to $\ln\|f(i\tau)\|_{A_0}$ and $\ln\|f(1+i\tau)\|_{A_1}$, respectively, and passing to the limit, we obtain (1.20) from (1.23).

The convexity of the function $e^t$ and (1.7) imply that

$$\begin{aligned}&\exp\left[\frac{1}{1-\alpha}\int_{-\infty}^{\infty}\left[\ln\|f(i\tau)\|_{A_0}\right]\mu_0(\alpha,\tau)\,d\tau\right]\\ &\qquad\leqslant \frac{1}{1-\alpha}\int_{-\infty}^{\infty}\|f(i\tau)\|_{A_0}\mu_0(\alpha,\tau)\,d\tau, \qquad (1.24)\\ &\exp\left[\frac{1}{\alpha}\int_{-\infty}^{\infty}\left[\ln\|f(1+i\tau)\|_{A_1}\right]\mu_1(\alpha,\tau)\,d\tau\right]\\ &\qquad\leqslant \frac{1}{\alpha}\int_{-\infty}^{\infty}\|f(1+i\tau)\|_{A_1}\mu_1(\alpha,\tau)\,d\tau. \qquad (1.25)\end{aligned}$$

Multiplying and dividing the first and second terms on the right side of (1.20) by $1-\alpha$ and $\alpha$, respectively, taking powers, and using (1.24) and (1.25), we obtain (1.21). Finally, if in the inequality

$$\begin{aligned}\exp(a+b) &= \exp\left[(1-\alpha)a(1-\alpha)^{-1} + \alpha b\alpha^{-1}\right]\\ &\leqslant (1\quad\alpha)e^{a(1-\alpha)^{-1}} + \alpha e^{b\alpha^{-1}} \qquad (1.26)\end{aligned}$$

we replace $a$ and $b$ by the first and second terms on the right side of (1.20) and again use (1.24) and (1.25), we obtain (1.22).

LEMMA 1.5. *Let the sequence $f_n(z)$ be bounded in $\mathfrak{F}(A_0, A_1)$ and convergent for every $i\tau$ from some set $e$ of positive measure on the imaginary semiaxis in the norm of $A_0$. Then the sequence $f_n(\alpha)$ converges in $[A_0, A_1]_\alpha$ $(0 < \alpha < 1)$.*

PROOF. Applying (1.20), we obtain

$$\ln\|f_n(\alpha) - f_m(\alpha)\|_{[A_0,A_1]_\alpha}$$

$$\leqslant \int_e \left[\ln\|f_n(i\tau) - f_m(i\tau)\|_{A_0}\right] \mu_0(\alpha, \tau)\, d\tau$$

$$+ \int_{\mathbf{R}\setminus e} \left[\ln\|f_n(i\tau) - f_m(i\tau)\|_{A_0}\right] \mu_0(\alpha, \tau)\, d\tau$$

$$+ \int_{-\infty}^{\infty} \left[\ln\|f_n(1 + i\tau) - f_m(1 + i\tau)\|_{A_1}\right] \mu_1(\alpha, \tau)\, d\tau. \qquad (1.27)$$

By hypothesis, the first integral tends to $-\infty$ as $n, m \to \infty$ and the last two are bounded above since the sequence $f_n$ is bounded in $\mathfrak{F}(A_0, A_1)$ and the functions $\mu_0(\alpha, \tau)$ and $\mu_1(\alpha, \tau)$ are integrable. Therefore the left side of (1.27) tends to $-\infty$, and consequently $\|f_n(\alpha) - f_m(\alpha)\|_{[A_0,A_1]_\alpha} \to 0$ as $n, m \to \infty$.

**5. *Conditions for the reflexivity of* $[A_0, A_1]_\alpha$.**

THEOREM 1.4. *If the space $A_0$ is reflexive, then the spaces $[A_0, A_1]_\alpha$, $0 < \alpha < 1$, are also reflexive.*

PROOF. By Alaoglu's theorem it is sufficient to prove that the unit ball of $[A_0, A_1]_\alpha$ is compact in the weak topology, and by V. L. Šmul'jan's theorem the ball will have this property if every decreasing sequence of closed convex subsets $Q_n$ of the ball has a nonempty intersection (see [11], Chapter V, §6.2, Theorem 2). In every $Q_n$ we choose an arbitrary element $x_n$. Since $\|x_n\|_{[A_0,A_1]_\alpha} \leqslant 1$, there exist functions $f_n \in \mathfrak{F}(A_0, A_1)$ such that $f_n(\alpha) = x_n$ and $\|f_n\|_{\mathfrak{F}(A_0,A_1)} \leqslant 2$. We consider the restrictions of the functions $f_n(z)$ to some interval $I$ of the imaginary axis as elements of $L_2(A_0)$. By a theorem of Phillips (see [18]) this space is reflexive, and therefore every ball in it is compact in the weak topology. Since the indicated sequence of elements of $L_2(A_0)$ is bounded, it contains a weakly convergent subsequence with indices $n'$. Then there exists a sequence of convex combinations of these elements converging in the norm of $L_2(A_0)$. Thus, from the sequence $f_{n'}(z)$ we can construct another sequence

$$g_k(z) = \sum_{n'>k} a_k^{(n')} f_{n'}(z)$$

($a_k^{(n')} \geqslant 0$, $\Sigma_{n'} a_k^{(n')} = 1$, and only a finite number of coefficients $a_k^{(n')} \neq 0$ for a given $k$) having the property that the restrictions of the functions $g_k(z)$ to $I$ converge in $L_2(A_0)$, and consequently converge in measure on $I$. Then there exists a subsequence $g_{k_i}(z)$ which converges in the norm of $A_0$ almost everywhere on $I$. By construction, $\|g_{k_i}\|_{\mathfrak{F}(A_0,A_1)} \leqslant 2$, and so by Lemma 1.5 the sequence

$$g_{k_i}(\alpha) = \sum_{n'>k_i} a_{k_i}^{(n')} f_{n'}(\alpha) = \sum_{n'>k_i} a_{k_i}^{(n')} x_{n'}$$

converges in $[A_0, A_1]_\alpha$ to an element $y$. Since the sets $Q_n$ are convex, we have $g_{k_i}(\alpha) \in Q_{k_i}$, and since they are closed, the element $y$ belongs to all $Q_n$.

**6. *Completion of the spaces*** $[A_0, A_1]_\alpha$. The spaces $[A_0, A_1]_\alpha$ are imbedded in $A_0 + A_1$, and so we may consider their completions $\widetilde{[A_0, A_1]}_\alpha$ relative to $A_0 + A_1$. From the interpolation Theorem 1.2 and Lemma 4.5 of Chapter I we immediately obtain the following theorem.

THEOREM 1.5. *The triple* $(A_0, A_1, \widetilde{[A_0, A_1]}_\alpha)$ *of Banach spaces is a normalized interpolation triple of type* $\alpha$ *relative to the triple* $(B_0, B_1, \widetilde{[B_0, B_1]}_\alpha)$.

An example shows that the space $[A_0, A_1]_\alpha$ need not be complete relative to $A_0 + A_1$. We consider the space $A_0 = L_\infty(0, 1)$ with norm $\|x\|_0$ and the space $A_1$ consisting of all functions with finite norm $\|x\|_1 = \operatorname{ess\,sup} t^{-1}|x(t)|$. In subsection 9 we shall show that in this case $[A_0, A_1]_\alpha$ consists of all functions for which $\|x\|_\alpha = \operatorname{ess\,sup} t^{-\alpha}|x(t)|$ and $t^{-\alpha}|x(t)| \to 0$ as $t \to 0$. The completion of $[A_0, A_1]_\alpha$ relative to $A_0$ (here $A_0 \supset A_1$) consists, however, of all functions with finite norm $\|x\|_0$. Indeed, if for such a function we set $x_n(t) = \chi_{(1/n,1]}(t)x(t)$, then $x_n \in [A_0, A_1]_\alpha$ and

$$\|x_n\|_{[A_0,A_1]_\alpha} \leqslant \|x\|_\alpha,$$

$$\|x_n - x\|_{A_0} = \operatorname*{ess\,sup}_{0<t<1/n} |x(t)| \leqslant (1/n)^\alpha \|x\|_\alpha \to 0 \quad \text{as } n \to \infty.$$

**7. *Duality.*** We assume that $A_0 \cap A_1$ is dense in $A_0$ and $A_1$. Then the dual spaces $A_0'$ and $A_1'$ form a Banach couple imbedded in $(A_0 \cap A_1)'$, and on this couple the functor $[A_0', A_1']_\alpha$ is defined. By Theorem 1.3 the space $A_0 \cap A_1$ is dense in $[A_0, A_1]_\alpha$, and so we have the imbedding $([A_0, A_1]_\alpha)' \subset (A_0 \cap A_1)'$. Thus, both spaces $[A_0', A_1']_\alpha$ and $([A_0, A_1]_\alpha)'$ are imbedded in $(A_0 \cap A_1)'$, and we may pose the question of their mutual position in $(A_0 \cap A_1)'$.

LEMMA 1.6.

$$[A_0', A_1']_\alpha \overset{1}{\subset} ([A_0, A_1]_\alpha)'. \tag{1.28}$$

PROOF. Let $u \in [A_0', A_1']_\alpha$. This means that there exists a function $\psi \in \mathfrak{F}(A_0', A_1')$ such that $\psi(\alpha) = u$. The function $\psi(z)$ is analytic inside $\Pi$ and continuous in the norm of $A_0' + A_1'$ in the closed strip. By Theorem 3.1 of Chapter II the space $A_0' + A_1'$ is isometric to $(A_0 \cap A_1)'$, and so $\psi(z)$ is analytic in $\Pi$ and continuous as a function with values in $(A_0 \cap A_1)'$ in the closed strip. If now $f \in \mathfrak{F}(A_0 \cap A_1)$, then the scalar-valued function $\langle f(z), \psi(z)\rangle$ is analytic in $\Pi$, and by the maximum principle we have

$$|\langle f(z), \psi(z)\rangle| \leqslant \max\{\sup|\langle f(i\tau), \psi(i\tau)\rangle|, \sup|\langle f(1 + i\tau), \psi(1 + i\tau)\rangle|\}.$$

Let $x \in A_0 \cap A_1$. For the function $f_1 \in \mathfrak{F}(A_0 \cap A_1)$ with $f_1(\alpha) = x$ this implies that

$$|\langle x, u\rangle| = |\langle f_1(\alpha), \psi(\alpha)\rangle| \leqslant \|f_1\|_{\mathfrak{F}(A_0, A_1)} \|\psi\|_{\mathfrak{F}(A_0', A_1')}.$$

Taking the infimum over $f_1$ and $\psi$, we obtain (by virtue of Remark 2) that

$$|\langle x, u\rangle| \leqslant \|x\|_{[A_0, A_1]_\alpha} \|u\|_{[A_0', A_1']_\alpha} \qquad (x \in A_0 \cap A_1).$$

This inequality says that the functional $u$ as an element of $(A_0 \cap A_1)'$ belongs to the set $([A_0, A_1]_\alpha)'$ and

$$\|u\|_{[A_0, A_1]_\alpha'} \leqslant \|u\|_{[A_0', A_1']_\alpha'}.$$

COROLLARY. *By Theorem* 2.2 *of Chapter* I *the space* $([A_0, A_1]_\alpha)'$ *is complete relative to* $(A_0 \cap A_1)'$. *Therefore it follows from the imbedding* (1.28) *that*

$$\widetilde{[A_0', A_1']}_\alpha \overset{1}{\subset} ([A_0, A_1]_\alpha)', \tag{1.29}$$

*where* $\sim$ *is the symbol for completion relative to* $(A_0 \cap A_1)'$.

Our next goal is to prove that in the above inclusion we have the equality sign.

LEMMA 1.7. *Assume that the functions* $\psi(z)$ $(0 < \operatorname{Re} z < 1)$, $\psi_0(\tau)$ *and* $\psi(\tau)$ $(-\infty < \tau < \infty)$ *with values in* $A_0' + A_1'$ *have the following properties*:

1) $\psi(z)$ *is analytic and bounded inside* $\Pi$.

2) *For any* $x \in A_0 \cap A_1$ *the function* $\langle x, \psi(z)\rangle$ *has boundary values for* $z = i\tau$ *and* $z = 1 + i\tau$ *coinciding with the values of the functions* $\langle x, \psi_0(\tau)\rangle$ *and* $\langle x, \psi_1(\tau)\rangle$, *respectively*.

3)
$$\psi_j(\tau) \in A_j' \quad \text{and} \quad \|\psi_j(\tau)\|_{A_j'} \leqslant c \quad (j = 0, 1). \tag{1.30}$$

*Then*

$$\psi(\alpha) \in \widetilde{[A_0', A_1']}_\alpha \quad \text{and} \quad \|\psi(\alpha)\|_{\widetilde{[A_0', A_1']}_\alpha} \leqslant c. \tag{1.31}$$

PROOF. We consider the function

$$\varphi(z) = \int_{\frac{1}{2}}^{z} \psi(\xi)\, d\xi \qquad (0 < \operatorname{Re} z < 1).$$

Since $\psi(z)$ is bounded, it follows that $\varphi(z)$ can be extended continuously in the norm of $A_0' + A_1'$ (or, similarly, the norm of $(A_0 \cap A_1)'$) to the closed strip. Next, for $x \in A_0 \cap A_1$ we have

$$\begin{aligned}\langle x, \varphi(z + ih)\rangle - \langle x, \varphi(z)\rangle &= \int_z^{z+ih} \langle x, \psi(\xi)\rangle\, d\xi \\ &= i\int_0^h \langle x, \psi(z + i\tau)\rangle\, d\tau \quad (0 < \operatorname{Re} z < 1).\end{aligned}$$

Setting $z = s + it$ and letting $s$ tend to zero, we obtain

$$\langle x, \varphi(it + ih)\rangle - \langle x, \varphi(it)\rangle = i\int_0^h \langle x, \psi_0(t + \tau)\rangle \, d\tau.$$

By (1.30) the expression on the right gives a continuous linear functional on $A_0$ whose norm does not exceed $ch$. Consequently

$$\varphi(it + ih) - \varphi(it) \in A_0' \quad \text{and} \quad \left\| \tfrac{1}{h}[\varphi(it + ih) - \varphi(it)] \right\|_{A_0'} \leqslant c.$$

It can be shown similarly that

$$\left\| \tfrac{1}{h}[\varphi(1 + it + ih) - \varphi(1 + it)] \right\|_{A_1'} \leqslant c.$$

From the last inequalities it follows that the function

$$\Phi_h(z) = \tfrac{1}{h}[\varphi(z + ih) - \varphi(z)]$$

belongs to $\mathfrak{F}(A_0', A_1')$, and $\|\Phi_h\|_{\mathfrak{F}(A_0', A_1')} \leqslant c$. This implies that

$$\Phi_h(\alpha) \in [A_0', A_1']_\alpha \quad \text{and} \quad \|\Phi_h(\alpha)\|_{[A_0', A_1']_\alpha} \leqslant c.$$

On the other hand, $\Phi_h(\alpha) \to \psi(\alpha)$ as $h \to 0$ in $(A_0 \cap A_1)'$, and so (1.31) is satisfied.

THEOREM 1.6. *If $A_0 \cap A_1$ is dense in $A_0$ and $A_1$, then the dual space $([A_0, A_1]_\alpha)'$ is isometric to $\widetilde{[A_0', A_1']}_\alpha$.*

PROOF. Let $u \in ([A_0, A_1]_\alpha)'$. Since $[A_0, A_1]_\alpha$ has been defined as the quotient space of $\mathfrak{F}(A_0, A_1)$ modulo the subspace $N_\alpha$, the functional $u$ induces a linear functional $\bar{u}$ on $\mathfrak{F}(A_0, A_1)$ with the same norm and equal to zero on $N_\alpha$.

Now we consider the mapping $\theta$ of $\mathfrak{F}(A_0, A_1)$ into $L_1(A_0) \times L_1(A_1)$, defined by

$$\theta(f) = \{f(i\tau)\mu_0(\alpha, \tau), f(1 + i\tau)\mu_1(\alpha, \tau)\}.$$

This mapping is linear and one-to-one. On its range we define a linear functional $\lambda$ by the formula

$$\langle \theta(f), \lambda\rangle = \langle f, \bar{u}\rangle.$$

By (1.21) we have

$$\begin{aligned} |\langle \theta(f), \lambda\rangle| &= |\langle f, \bar{u}\rangle| = |\langle f(\alpha), u\rangle| \leqslant \|f(\alpha)\|_{[A_0, A_1]_\alpha} \|u\|_{([A_0, A_1]_\alpha)'} \\ &\leqslant \|u\|_{([A_0, A_1]_\alpha)'} \left[ \int_{-\infty}^{\infty} \|f(i\tau)\|_{A_0} \mu_0(\alpha, \tau)\, d\tau \right. \\ &\qquad \left. + \int_{-\infty}^{\infty} \|f(1 + i\tau)\|_{A_1} \mu_1(\alpha, \tau)\, d\tau \right] \\ &= \|u\|_{([A_0, A_1]_\alpha)'} \|\theta(f)\|_{L_1(A_0) \times L_1(A_1)}. \end{aligned} \tag{1.32}$$

Thus, on the range of $\theta$ the functional $\lambda$ is bounded in the norm of $L_1(A_0) \times L_1(A_1)$. We extend this functional with preservation of the norm to the whole space $L_1(A_0) \times L_1(A_1)$. By the theorem on the general form of a linear functional on $L_1(A)$ (see the Introduction to this chapter, Theorem 2) there exist two bounded functions $\psi_0(t)$ and $\psi_1(t)$ with values in $A_0'$ and $A_1'$, respectively, such that

$$\langle f, \bar{u}\rangle = \langle \theta(f), \lambda\rangle = \int_{-\infty}^{\infty} \langle f(i\tau)\mu_0(\alpha, \tau), \psi_0(\tau)\rangle \, d\tau + \int_{-\infty}^{\infty} \langle f(1 + i\tau)\mu_1(\alpha_0, \tau), \psi_1(\tau)\rangle \, d\tau. \tag{1.33}$$

Moreover, by (1.32) we have

$$\max_j \{\operatorname{ess\,sup} \|\psi_j(\tau)\|_{A_j}\} \leqslant \|u\|_{([A_0, A_1]_\alpha)'}. \tag{1.34}$$

Let $x \in A_0 \cap A_1$. We consider the function $\psi(x, z)$ defined by the following Poisson integral:

$$\psi(x, z) = \int_{-\infty}^{\infty} \langle x, \psi_0(\tau)\rangle \mu_0(z, \tau) \, d\tau + \int_{-\infty}^{\infty} \langle x, \psi_1(\tau)\rangle \mu_1(z, \tau) \, d\tau. \tag{1.35}$$

This function depends linearly on $x$ for given $z$, and by (1.34) we have

$$|\psi(x, z)| \leqslant \max_j \{\operatorname{ess\,sup} |\langle x, \psi_j(\tau)\rangle|\} \leqslant \|x\|_{A_0 \cap A_1} \|u\|_{([A_0, A_1]\alpha)'}. \tag{1.36}$$

Thus, there exists a functional $\psi(z) \in (A_0 \cap A_1)'$ such that $\psi(x, z) = \langle x, \psi(x)\rangle$. From (1.36) it follows that $\psi(z)$ is bounded in $\Pi$. We show that it is analytic inside $\Pi$. For this we take a scalar-valued function $h(z)$ analytic inside $\Pi$ and continuous in the closed strip, having limits at $\pm i\infty$, and such that $h(\alpha) = 0$. In (1.33) we put $f(z) = h(z)x$. Then

$$\begin{aligned} 0 = \langle h(\alpha)x, u\rangle = \langle f(\alpha), u\rangle &= \langle f, \bar{u}\rangle \\ &= \int_{-\infty}^{\infty} h(i\tau)\langle x, \psi_0(\tau)\rangle \mu_0(\alpha, \tau) \, d\tau \\ &\quad + \int_{-\infty}^{\infty} h(1 + i\tau)\langle x, \psi_1(\tau)\rangle \mu_1(\alpha, \tau) \, d\tau. \end{aligned}$$

By virtue of what we said in subsection 1 it follows that $\langle x, \psi(z)\rangle$ is analytic for any $x \in A_0 \cap A_1$, and so $\psi(z)$ is analytic in the norm of $(A_0 \cap A_1)'$.

The functions $\psi(z)$, $\psi_0(z)$, and $\psi_1(z)$ satisfy the hypotheses of Lemma 1.7, therefore $\psi(\alpha) \in \widetilde{[A_0', A_1']}_\alpha$ and

$$\|\psi(\alpha)\|_{\widetilde{[A_0', A_1']}_\alpha} \leqslant \|u\|_{([A_0, A_1]_\alpha)'}.$$

Setting $f(z) \equiv x \in A_0 \cap A_1$ in (1.33), by (1.35) we obtain

$$\langle x, u\rangle = \psi(x, \alpha) = \langle x, \psi(\alpha)\rangle,$$

i.e. $\psi(\alpha) = u$ as an element of $(A_0 \cap A_1)'$, and consequently $u \in \widetilde{[A_0', A_1']}_\alpha$ and

$$\|u\|_{\widetilde{[A_0', A_1']}_\alpha} \leqslant \|u\|_{([A_0, A_1]_\alpha)'}.$$

The theorem now follows by virtue of the imbedding (1.28).

REMARK. If $A_0$ is reflexive, then $([A_0, A_1]_\alpha)'$ and $[A_0', A_1']_\alpha$ are isometric.

Indeed, this follows immediately from Theorems 1.6 and 1.4, and Theorem 2.3 of Chapter I.

**8. *A reiteration theorem.*** From a Banach couple $A_0$, $A_1$ we construct the spaces $A_\alpha = [A_0, A_1]_\alpha$ and $A_\beta = [A_0, A_1]_\beta$ $(0 \leqslant \alpha < \beta \leqslant 1)$. These two spaces again form a Banach couple imbedded, for example, in $A_0 + A_1$. Therefore we may construct the space $[A_\alpha, A_\beta]_\sigma$ $(0 \leqslant \sigma \leqslant 1)$.

LEMMA 1.8. *The space* $[A_0, A_1]_s$, *where* $s = (1 - \sigma)\alpha + \sigma\beta$, *is imbedded in* $[A_\alpha, A_\beta]_\sigma$ *with imbedding constant not exceeding one.*

PROOF. Let $f \in \mathfrak{F}(A_0, A_1)$ and $f(s) = x \in [A_0, A_1]_s$. We introduce the function $g(z) = f((1 - z)\alpha + z\beta)$. Then $g(\sigma) = f(s) = x$,

$$g(i\tau) = f(\alpha + i(\beta - \alpha)\tau) \quad \text{and} \quad g(1 + i\tau) = f(\beta + i(\beta - \alpha)\tau).$$

The function $f_\tau(z) = f(z + i(\beta - \alpha)\tau)$ belongs to $\mathfrak{F}(A_0, A_1)$ for every fixed $\tau$, and therefore $g(i\tau) \in A_\alpha$ and $g(1 + i\tau) \in A_\beta$. Next,

$$\|g(i\tau)\|_{A_\alpha} \leqslant \|f_\tau\|_{\mathfrak{F}(A_0, A_1)} = \|f\|_{\mathfrak{F}(A_0, A_1)}.$$

Similarly, $\|g(1 + i\tau)\|_{A_\beta} \leqslant \|f\|_{\mathfrak{F}(A_0, A_1)}$. This implies that $x \in [A_\alpha, A_\beta]_\sigma$ and

$$\|x\|_{[A_\alpha, A_\beta]_\sigma} \leqslant \max\{\sup\|g(i\tau)\|_{A_\alpha}, \sup\|g(1 + i\tau)\|_{A_\beta}\} \leqslant \|f\|_{\mathfrak{F}(A_0, A_1)}.$$

Taking the infimum over $f$, we obtain

$$[A_0, A_1]_s \overset{1}{\subset} [A_\alpha, A_\beta]_\sigma. \tag{1.37}$$

We shall be interested in conditions under which the inclusion reverse to (1.37) holds. For this we study the problem of imbedding the dual spaces of the spaces considered in the lemma. However, if the space $[A_0, A_1]_s$ is not dense in $[A_\alpha, A_\beta]_\sigma$, then the dual spaces are not imbedded. In this connection, we make the additional assumption that $A_0 \cap A_1$ is dense in the spaces $A_0$, $A_1$, and $A_\alpha \cap A_\beta$. By Theorem 1.3 the space $A_\alpha \cap A_\beta$ is densely imbedded in $[A_\alpha, A_\beta]_\sigma$, and therefore $A_0 \cap A_1$, and so also $[A_0, A_1]_s$, is densely imbedded in $[A_\alpha, A_\beta]_\sigma$.

Due to the assumption made above, the dual spaces of all indicated spaces can be considered as sets in the space $(A_0 \cap A_1)' = A_0' + A_1'$.

LEMMA 1.9. *Under the above assumption, the spaces* $([A_0, A_1]_s)'$ *and* $([A_\alpha, A_\beta]_\sigma)'$ *coincide.*

PROOF. The imbedding (1.37) immediately implies the imbedding $([A_\alpha, A_\beta]_\sigma)' \overset{1}{\subset} ([A_0, A_1]_s)'$.

By Theorem 1.6 the space $([A_\alpha, A_\beta]_\sigma)'$ coincides isometrically with $\widetilde{[A'_\alpha, A'_\beta]}_\sigma^{\alpha\beta}$, where $\sim\alpha\beta$ means completion relative to $A'_\alpha + A'_\beta$. This last space is imbedded in $A'_0 + A'_1$, and $([A_\alpha, A_\beta]_\sigma)'$ is complete relative to $(A_0 \cap A_1)' = A'_0 + A'_1$ by Theorem 2.2 of Chapter I. We find ourselves in the situation considered in Corollary 2 to Lemma 1.6 of Chapter I, according to which $([A_\alpha, A_\beta]_\sigma)' = \widetilde{[A'_\alpha, A'_\beta]}_\sigma$, where $\sim$ denotes completion relative to $A'_0 + A'_1$ (a notation used without reference in the sequel).

According to Lemma 1.8, $[A'_0, A'_1]_s \overset{1}{\subset} [[A'_0, A'_1]_\alpha, [A'_0, A'_1]_\beta]_\sigma$. Further, from Theorem 1.6 we obtain

$$[A'_0, A'_1]_\alpha \overset{1}{\subset} \widetilde{[A'_0, A'_1]}_\alpha = A'_\alpha \quad \text{and} \quad [A'_0, A'_1]_\beta \overset{1}{\subset} \widetilde{[A'_0, A'_1]}_\beta = A'_\beta,$$

and therefore $[A'_0, A'_1]_s \overset{1}{\subset} [A'_\alpha, A'_\beta]_\sigma$.

Passing to completions relative to $A'_0 + A'_1$, we obtain $\widetilde{[A'_0, A'_1]}_s \overset{1}{\subset} \widetilde{[A'_\alpha, A'_\beta]}_\sigma$, or, similarly, $([A_0, A_1]_s)' \overset{1}{\subset} ([A_\alpha, A_\beta]_\sigma)'$.

We have obtained two mutually reverse imbeddings. The lemma is proved.

Lemmas 1.8 and 1.9 immediately imply the following theorem.

THEOREM 1.7. *If $A_0 \cap A_1$ is densely imbedded in $A_0$, $A_1$, and $A_\alpha \cap A_\beta$, then the spaces $[A_0, A_1]_s$, where $s = (1 - \sigma)\alpha + \sigma\beta$, and $[A_\alpha, A_\beta]_\sigma$ $(0 \leqslant \sigma \leqslant 1)$ coincide isometrically.*

**9.** ***Connection with the theory of scales.*** We consider the family of spaces $[E_0, E_1]_\alpha$, where $E_1$ is normally imbedded in $E_0$. In this case $E_0 \cap E_1$ coincides isometrically with $E_1$ and $E_0 + E_1$ with $E_0$. By the maximum principle for a function $f \in \mathfrak{F}(E_0, E_1)$ we obtain

$$\|f(z)\|_{E_0} \leqslant \|f\|_{\mathfrak{F}(E_0,E_1)}. \tag{1.38}$$

We show that for $\alpha < \beta$ the space $E_\beta = [E_0, E_1]_\beta$ is normally imbedded in $E_\alpha = [E_0, E_1]_\alpha$. Let the function $f \in \mathfrak{F}(E_0, E_1)$ be such that $\|f\|_{\mathfrak{F}(E_0,E_1)} \leqslant \|x\|_{E_\beta} + \varepsilon$ and $f(\beta) = x$. We consider the function

$$\Phi(z) = f\left(\frac{1-\beta}{1-\alpha} z + \frac{\beta-\alpha}{1-\alpha}\right).$$

It is analytic inside the strip $\Pi$ and continuous in the closed strip. Moreover, by (1.38) we have

$$\sup_\tau \|\Phi(i\tau)\|_{E_0} = \sup_\tau \left\| f\left(\frac{\beta-\alpha}{1-\alpha} + i\frac{1-\beta}{1-\alpha}\tau\right)\right\|_{E_0} \leqslant \|f\|_{\mathfrak{F}(E_0,E_1)}$$

and

$$\sup_\tau \|\Phi(1 + i\tau)\|_{E_1'} = \sup_\tau \left\|f\left(1 + i\frac{1-\beta}{1-\alpha}\tau\right)\right\|_{E_1} \leqslant \|f\|_{\mathfrak{F}(E_0, E_1)}.$$

Consequently $\Phi \in \mathfrak{F}(E_0, E_1)$ and $\|\Phi\|_{\mathfrak{F}(E_0,E_1)} \leqslant \|f\|_{\mathfrak{F}(E_0,E_1)}$. Finally, $\Phi(\alpha) = f(\beta) = x$. This means that $x \in E_\alpha$ and

$$\|x\|_{E_\alpha} \leqslant \|\Phi\|_{\mathfrak{F}(E_0, E_1)} \leqslant \|f\|_{\mathfrak{F}(E_0,E_1)} \leqslant \|x\|_{E_\beta} + \varepsilon.$$

Since $\varepsilon$ is arbitrary, we have $\|x\|_{E_\alpha} \leqslant \|x\|_{E_\beta}$ ($x \in E_\beta$), and since $E_1$ is dense in all spaces $E_\alpha$ (Theorem 1.3), the space $E_\beta$ is normally imbedded in $E_\alpha$.

The spaces $[E_0, E_1]_j$ ($j = 0, 1$) coincide with the spaces $E_j$.

We prove that the function $\|x\|_{E_\alpha}$ ($x \in E_1$) is logarithmically convex in $\alpha$. We consider the function

$$\varphi(z) = \left(\|x\|_{E_0}/\|x\|_{E_1}\right)^{z-\alpha}x.$$

Obviously $\varphi \in \mathfrak{F}(E_0, E_1)$ and

$$\sup_\tau \|\varphi(i\tau)\|_{E_0} = \|x\|_{E_0}^{1-\alpha}\|x\|_{E_1}^{\alpha}, \qquad \sup \|\varphi(1 + i\tau)\|_{E_1} = \|x\|_{E_0}^{1-\alpha}\|x\|_{E_1}^{\alpha}.$$

Next, $\varphi(\alpha) = x$, and therefore

$$\|x\|_{E_\alpha} \leqslant \|\varphi\|_{\mathfrak{F}(E_0,E_1)} = \|x\|_{E_0}^{1-\alpha}\|x\|_{E_1}^{\alpha}. \tag{1.39}$$

Now let $0 \leqslant \alpha < \beta < \gamma \leqslant 1$. By the reiteration Theorem 1.7 the space $E_\beta$ coincides isometrically with $[E_\alpha, E_\gamma]_\sigma$, where $\sigma = (\beta - \alpha)/(\gamma - \alpha)$. Applying (1.39) to $E_\alpha$ and $E_\gamma$, for $x \in E_\gamma$ we obtain

$$\|x\|_{E_\beta} \leqslant \|x\|_{E_\alpha}^{1-\sigma}\|x\|_{E_\gamma}^{\sigma} = \|x\|_{E_\alpha}^{(\gamma-\beta)/(\gamma-\alpha)}\|x\|_{E_\gamma}^{(\beta-\alpha)/(\gamma-\alpha)}.$$

Thus, the spaces $\{E_\alpha\}$ form a normal scale.

The spaces $[E_0', E_1']_\alpha$ also form a normal scale (we recall that $[E_0', E_1']_1$ is the closure of $E_0'$ in the norm of $E_1'$). By Theorem 1.6 the spaces $E_\alpha'$ coincide with the completions of the spaces $[E_0', E_1']_\alpha$ relative to $E_1'$, or, what is the same, relative to the subspace $[E_0', E_1']_1$ of it. Thus, the family $E_\alpha'$ coincides with the condensation of the normal scale $[E_0', E_1']_\alpha$ (see Chapter III, §1.5). From properties 1° and 5° of the condensation of a normal scale it follows that $\|f\|_{E_\alpha'}$ is a decreasing logarithmically convex function of $\alpha$. In particular, it is continuous at $\alpha = 1$. This circumstance permits us to obtain the properties of the scale $\{E_\alpha\}$ as $\alpha \to 1$. Indeed, let $x \in E_1$. Then by (2.1) of Chapter I we have

$$\|x\|_{E_{01}} = \sup_{f \in E_0'} \frac{f(x)}{\|f\|_{E_1'}}.$$

We choose $f_0 \in E_0'$ so that

$$\|x\|_{E_{01}} \leqslant |f_0(x)|/\|f_0\|_{E_1'} + \varepsilon.$$

Using the continuity of $\|f_0\|_{E_\alpha'}$ at $\alpha = 1$, we choose $\alpha$ so close to one that

$$\|x\|_{E_{01}} \leqslant \frac{|f_0(x)|}{\|f_0\|_{E'_\alpha}} + 2\varepsilon \leqslant \|x\|_{E_\alpha} + 2\varepsilon.$$

This implies that $\|x\|_{E_{01}} \leqslant \lim_{\alpha\to 1}\|x\|_{E_\alpha}$. By Lemma 1.1 of Chapter III the reverse inequality holds, and therefore

$$\lim_{\alpha\to 1} \|x\|_{[E_0,E_1]_\alpha} = \|x\|_{E_{01}} \qquad (x \in E_1).$$

The preceding considerations lead to the following assertion.

THEOREM 1.8. *If the spaces $E_0$ and $E_1$ are related, then the family $[E_0, E_1]_\alpha$ forms a regular normal scale.*

COROLLARY. *If the spaces $E_0$ and $E_1$ are related, then the scale $[E_0, E_1]_\alpha$ has the strong interpolation property relative to any minimal scale.*

We now introduce the notion of analytic scale of spaces. Let $M$ be a normed linear space in which a family of linear operators $T(z)$ acts satisfying the following conditions:

1°. For every $x \in M$ the function $T(z)x$ is an entire function of the complex variable $z$.

2°. The function $\|T(z)x\|_M$ is bouned on every straight line parallel to the imaginary axis.

3°. $T(0)x = x$.

4°. $\sup_{\mu,\nu}\|T(\alpha + i\mu)T(\beta + i\nu)\|_M \leqslant \sup_\tau\|T(\alpha + \beta + i\tau)x\|_M$.

5°.
$$T(i\mu)\frac{T(z + \Delta z)x - T(z)x}{\Delta z} \to T(i\mu)(T(z)x)'$$

uniformly in $\mu$ as $\Delta z \to 0$ (property 5° follows from property 1° if the operators $T(i\mu)$ are uniformly bounded in norm).

We introduce the family of norms

$$\|x\|_\alpha = \sup_{-\infty<\tau<\infty} \|T(\alpha + i\tau)x\|_M$$

in the space $M$ and complete $M$ to a Banach space $E_\alpha$ in each of these norms. The family $E_\alpha$, $-\infty < \alpha < \infty$, of Banach spaces will be called an *analytic scale of spaces*.

By the three lines theorem for the logarithmically subharmonic function $\|T(z)x\|_M$ (see [39]), the function $\|x\|_\alpha$, is a logarithmically convex function of $\alpha$. If $\|x\|_\alpha$ $(x \in M)$ is in addition an increasing function of $\alpha$, then the analytic scale will be a continuous normal scale on any interval $[\alpha_0, \beta_0]$.

Property 4° can be written in the following form:

$$\|T(\beta + i\nu)x\|_{E_\alpha} \leqslant \|x\|_{\alpha+\beta}. \tag{1.40}$$

Finally, from property 3° it follows that

$$\|x\|_M = \|T(0)x\|_M \leqslant \|x\|_{E_0}. \tag{1.41}$$

We give an example of an analytic scale. We consider the set of all continuous functions on [0, 1], equal to zero in a neighborhood of zero (the neighborhood may vary with the function). On this set we define the family of operators $T(z)x(t) = t^{-z}x(t)$. The operators $T(z)$ will be considered as linear operators in the space $M$ equipped with the norm of $L_p(0, 1)$ $(1 \leqslant p \leqslant \infty)$. We denote by $L_p^\alpha$ $(-\infty < \alpha < \infty)$ the scale of spaces constructed from these operators. The space $L_p^\alpha$ consists of measurable functions for which

$$\|x\|_{L_p^\alpha} = \left\{\int_0^1 |t^{-\alpha}x(t)|^p\,dt\right\}^{1/p} < \infty,$$

if $p < \infty$ and

$$\|x\|_{L_\infty^\alpha} = \operatorname{ess\,sup}|t^{-\alpha}x(t)| < \infty, \qquad \operatorname*{ess\,sup}_{0 \leqslant t \leqslant \tau}|t^{-\alpha}x(t)| \to 0 \quad (\tau \to 0)$$

if $p = \infty$.

THEOREM 1.9. *Let $E_\alpha$, $0 \leqslant \alpha \leqslant 1$, be an analytic scale of spaces, and let $E_1$ be normally imbedded in $E_0$. The space $E_\alpha$ coincides isometrically with $[E_0, E_1]_\alpha$.*

PROOF. Let $x$ belong to the set $M$ from which the scale $E_\alpha$ is constructed.

By property 5° the function $f(z) = T(\alpha - z)x$ is analytic in the norm of $E_0$. Next, by 4° we have

$$\|f(z)\|_{E_0} = \|T(\alpha - z)x\|_{E_0} = \sup_\tau \|T(i\tau)T(\alpha - z)x\|_M$$
$$\leqslant \sup_\mu \|T(\alpha - \operatorname{Re} z + i\mu)x\|_M \leqslant \max\{\|x\|_{E_\alpha}, \|x\|_{E_{\alpha-1}}\},$$

for $0 \leqslant \operatorname{Re} z \leqslant 1$, and so $\|f(z)\|_{E_0}$ is bounded in the strip $\Pi$.

According to (1.40), on the boundary of $\Pi$ we have

$$\|f(i\tau)\|_{E_0} = \|T(\alpha - i\tau)x\|_{E_0} \leqslant \|x\|_{E_\alpha},$$
$$\|f(1 + i\tau)\|_{E_0} = \|T(\alpha - 1 - i\tau)x\|_{E_1} \leqslant \|x\|_{E_\alpha}.$$

Finally, by property 3° we have $f(\alpha) = T(0)x = x$. Consequently, $x \in [E_0, E_1]_\alpha$ and $\|x\|_{[E_0,E_1]_\alpha} \leqslant \|x\|_{E_\alpha}$.

We prove the reverse inequality. By virtue of the remark to Theorem 1.1 and the proof of Remark 2 of subsection 4 for $x \in M$ we may construct a function $f(z) = \sum_1^N a_k(z)x_k$ such that $a_k \in \mathfrak{F}(\mathbf{C})$, $x_k \in M$, $f(\alpha) = x$ and

$$\|f\|_{\mathfrak{F}(E_0,E_1)} \leqslant \|x\|_{[E_0,E_1]_\alpha} + \varepsilon. \tag{1.42}$$

We consider the function $\Psi(z) = T(z + i\mu)f(z)$, where $\mu$ is a fixed real number. This function is analytic in $E_0$ inside $\Pi$ and continuous and bounded in the closed strip. Further,

$$\|\Psi(i\tau)\|_{E_0} \leqslant \sup_{t,\tau} \|T(it)f(i\tau)\|_{E_0} \leqslant \sup_{\tau} \|f(i\tau)\|_{E_0} \leqslant \|f\|_{\mathfrak{F}(E_0,E_1)},$$

$$\|\Psi(1 + i\tau)\|_{E_0} \leqslant \sup_{t,\tau} \|T(1 + it)f(1 + i\tau)\|_{E_0} \leqslant \sup_{\tau} \|f(1 + i\tau)\|_{E_1} \leqslant \|f\|_{\mathfrak{F}(E_0,E_1)}.$$

By the maximum principle, taking account of (1.42), we obtain

$$\|\Psi(\alpha)\|_{E_0} = \|T(\alpha + i\mu)f(\alpha)\|_{E_0} = \|T(\alpha + i\mu)x\|_{E_0} \leqslant \|x\|_{[E_0,E_1]_\alpha} + \varepsilon.$$

Finally, from (1.41) it follows that

$$\|x\|_{E_\alpha} = \sup_{\mu} \|T(\alpha + i\mu)x\|_M \leqslant \sup_{\mu} \|T(\alpha + i\mu)x\|_{E_0} \leqslant \|x\|_{[E_0,E_1]_\alpha} + \varepsilon.$$

Thus, $\|x\|_{E_\alpha} = \|x\|_{[E_0,E_1]_\alpha}$ on $M$. On the other hand, since $M$ is dense in all spaces $E_\alpha$ and also in $[E_0, E_1]_\alpha$ by Theorem 1.3, the theorem is proved.

From the interpolation Theorem 1.2 and the reiteration Theorem 1.7 we obtain the following theorem.

THEOREM 1.10. *Two analytic scales connecting two pairs of normally imbedded spaces have the strong interpolation property relative to each other.*

REMARK. It follows from the theorem that the spaces $L_p^\alpha(0, 1)$ considered above are obtained by the complex method from the spaces $L_p(0, 1)$ and $L_p^1(0, 1)$.

**10. *Hilbert scales.*** In applications an important role is played by Hilbert scales, which constitute a special class of analytic scales.

Let $H_0$ be a complex Hilbert space with inner product $(\cdot, \cdot)$. Let an unbounded positive definite selfadjoint operator $j$ be given in $H_0$ with domain $D(j)$, such that

$$\|x\|_{H_0} \leqslant \|jx\|_{H_0} \qquad (x \in D(j)). \tag{1.43}$$

We denote by $E_\lambda$ the spectral resolution of the identity corresponding to $j$, and consider the set $M$ of elements representable in the form $x = \int_1^N \lambda \, dE_\lambda x$ for some $N < \infty$. The set $M$ is dense in $H_0$. On this set the operators $T(z) = j^z$ are defined:

$$j^z x = \int_1^N \lambda^z \, dE_\lambda x.$$

It is easy to verify that these operators have properties 1°–5° necessary for the construction of an analytic scale from them. On $M$ we introduce the norms

$$\|x\|_\alpha = \sup_{-\infty<\tau<\infty} \|j^{\alpha+i\tau}x\|_{H_0} = \left\{\int_1^N \lambda^{2\alpha}(dE_\lambda x, x)\right\}^{1/2} = \|j^\alpha x\|_{H_0}.$$

The completions $H_\alpha$ of $M$ in these norms are Hilbert spaces; they form an analytic scale which is called a *Hilbert scale.* By virtue of (1.43) the norm $\|x\|_\alpha$ is an increasing function of $\alpha$, and therefore the scale $\{H_\alpha\}$, $-\infty < \alpha < \infty$, is a continuous normal scale on any interval.

For $\alpha > 0$, $H_\alpha \subset H_0$ and $H_\alpha$ constitutes the domain $D(j^\alpha)$ of $j^\alpha$. For $\alpha < 0$ the space $H_\alpha$ already contains generalized (ideal) elements relative to the initial space. Let $-\alpha = \beta > 0$. We clarify the structure of the space $H'_\alpha$ dual to $H_\alpha$. If $\varphi \in H'_\alpha$, then for elements of $M$ we have

$$|\varphi(x)| \leqslant \|\varphi\|_{H'_\alpha}\|x\|_\alpha = \|\varphi\|_{H'_\alpha}\|j^{-\beta}x\|_0.$$

We perform the change of variables $j^{-\beta}x = y$. Then $|\varphi(j^\beta y)| \leqslant \|\varphi\|_{H'_\alpha}\|y\|_0$, i.e. the functional $\varphi(j^\beta y)$ is bounded in the norm of $H_0$ on the set $j^{-\beta}M$. By the density of this set in $H_0$, this functional admits a unique representation of the form $\varphi(j^\beta y) = (y, z)$, where $z \in H_0$. Performing the inverse change of variables, we obtain

$$\varphi(x) = (j^{-\beta}x, z) = (x, j^{-\beta}z) = (x, u),$$

where $u = j^{-\beta}z \in D(j^\beta) = H_\beta$. Conversely, for $u \in H_\beta$ every functional $(x, u)$ is a linear functional on $M$, bounded in the norm of $H_\alpha$:

$$|(x, u)| = |(j^{-\beta}x, j^\beta u)| \leqslant \|u\|_{H_\beta}\|x\|_{H_\alpha},$$

and therefore it can be extended by continuity to a functional belonging to $H'_\alpha$. From the last inequality it follows that $\|\varphi\|_{H'_\alpha} \leqslant \|u\|_{H_\beta}$. We show that we have equality here. Let $u_n$ be a sequence of elements of $M$ such that $u_n \to u$ in $H_\beta$. We set $x_n = j^{2\beta}u_n$. We have

$$(x_n, u) = (j^{2\beta}u_n, u) = (j^\beta u_n, j^\beta u) \to \|u\|_\beta^2.$$

For any $\varepsilon > 0$ and sufficiently large $n$ we then have

$$\begin{aligned}\varphi(x_n) &= (x_n, u) \geqslant (1 - \varepsilon)\|u\|_\beta^2 \\ &\geqslant (1 - 2\varepsilon)\|u_n\|_\beta\|u\|_\beta = (1 - 2\varepsilon)\|x_n\|_\alpha\|u\|_\beta.\end{aligned}$$

From this and the above it follows that $\|\varphi\|_{H'_\alpha} = \|u\|_\beta$. We have proved that $H'_\alpha$ is isometric to $H_\beta = H_{-\alpha}$; an isometry is given by the equality $\varphi(x) = (x, u)$ ($\varphi \in H'_\alpha$, $u \in H_{-\alpha}$, $x \in M$). In what follows all values of the functional $\varphi(x)$ will be denoted by $(x, u)$ ($x \in H_\alpha$, $u \in H_{-\alpha}$).

Since Hilbert spaces are reflexive, $H_\alpha$ is isometric to $H'_{-\alpha}$.

Theorem 1.10 shows that two Hilbert scales have the strong interpolation property relative to each other. We consider a special case of this assertion.

THEOREM 1.11. *In a Hilbert space $H_0$ let two positive definite selfadjoint operators $j$ and $j_1$ having property* (1.43) *be given. If $D(j_1) \supset D(j)$ and*

$$\|j_1 x\|_{H_0} \leqslant \|jx\|_{H_0}, \tag{1.44}$$

*then for $0 \leqslant \alpha \leqslant 1$ we have $D(j_1^\alpha) \supset D(j^\alpha)$ and*

$$\|j_1^\alpha x\|_{H_0} \leqslant \|j^\alpha x\|_{H_0} \qquad (0 \leqslant \alpha \leqslant 1, x \in D(j^\alpha)).$$

PROOF. We construct the Hilbert scales $H_\alpha$ and $H_\alpha^1$ generated by the operators $j$ and $j_1$, respectively. Inequality (1.44) says that the identity operator $I$ (the imbedding operator) is a bounded operator from $H_1$ into $H_1^1$ with norm not exceeding 1. Obviously $I$ is bounded as an operator from $H_0$ into $H_0$. By the interpolation theorem the identity operator acts from $H_\alpha$ into $H_\alpha^1$ and has norm not exceeding 1. This gives the assertion of the theorem.

THEOREM 1.12. *If a Hilbert space $H_1$ is imbedded normally in $H_0$, then there exists a unique Hilbert scale connecting these spaces.*

PROOF. By virtue of the inequality

$$|(x, y)_{H_0}| \leqslant \|y\|_{H_0}\|x\|_{H_0} \leqslant \|y\|_{H_0}\|x\|_{H_1}$$

for a fixed $y \in H_0$ the functional $(y, x)$ is bounded on $H_1$, and so there is a uniquely determined element $z \in H_1$ such that $(x, y)_{H_0} = (x, z)_{H_1}$ $(x \in H_1)$. We write $z = Vy$. Then

$$(x, y)_{H_0} = (x, Vy)_{H_1} \qquad (x \in H_1). \tag{1.45}$$

Moreover, $\|z\|_{H_1} = \|Vy\|_{H_1} \leqslant \|y\|_{H_0}$. The operator $V$ is linear, defined on the whole of $H_0$, and bounded: $\|Vy\|_{H_0} \leqslant \|Vy\|_{H_1} \leqslant \|y\|_{H_0}$. The operator $V$ is selfadjoint and nonnegative. Indeed, by (1.45) for $x, y \in H_0$ we have

$$(Vx, y)_{H_0} = (Vx, Vy)_{H_1} = \overline{(Vy, Vx)}_{H_1} = \overline{(Vy, x)}_{H_0} = (x, Vy)_{H_0}$$

and $(Vy, y)_{H_0} = (Vy, Vy)_{H_1} \geqslant 0$. Next, if $Vy = 0$, then $(x, y)_{H_0} = 0$ for every $x \in H_1$, and by the density of $H_1$ we have $y = \theta$. Thus, $V$ is positive. We write $V^{-1} = U$. Then $D(U) = R(V) \subset H_1$, and (1.45) can be written in the form

$$(Uz, x)_{H_0} = (z, x)_{H_1}. \tag{1.46}$$

We introduce the operator $j = U^{1/2}$. The domain of $U^{1/2}$ can be obtained by closing $D(U)$ in the norm

$$\|U^{1/2}z\|_{H_0} = \sqrt{(Uz, z)_{H_0}} = \sqrt{(z, z)_{H_1}} = \|z\|_1. \tag{1.47}$$

Thus, $D(j)$ is contained in $H_1$ and is closed in it. We show that $D(j)$ coincides with $H_1$. Otherwise there would exist $x_0 \in H_1$ such that $(z, x_0)_{H_1} = 0$ for all $z \in D(j)$ and, in particular, for all $z \in D(U)$. Then it follows from (1.46) that

$(Uz, x_0) = 0$ ($z \in D(U)$), and consequently $x_0 = 0$, since $R(U) = H_0$. Finally, it follows from (1.47) that $\|jz\|_{H_0} = \|z\|_{H_1}$ ($z \in H_1$).

The Hilbert scale generated by the selfadjoint operator $j$ connects $H_0$ and $H_1$.

We show the uniqueness of the Hilbert scale connecting $H_0$ and $H_1$. If there were two such scales $H_\alpha$ and $H'_\alpha$, then for their generating operators $j$ and $j_1$ the relations $D(j) = H_1 = D(j_1)$ and $\|jx\|_{H_0} = \|j_1 x\|_{H_0}$ would be satisfied. By Theorem 1.11 we would then have $D(j^{1/2}) = D(j_1^{1/2})$ and $\|j^{1/2}x\|_{H_0} = \|j_1^{1/2}x\|_{H_0}$. In particular, for $x \in H_1$ we would have $(jx, x)_{H_0} = (j_1 x, x)_{H_0}$. From the equality of the quadratic forms we obtain the equality of the bilinear forms: $(jx, y)_{H_0} = (j_1 x, y)_{H_0}$ ($x, y \in H_1$). By the density of $H_1$ in $H_0$ this implies that $jx = j_1 x$ for $x \in H_1$, and consequently the scales $H_\alpha$ and $H^1_\alpha$ coincide. The theorem is proved.

Now, by means of interpolation we establish an inequality important in operator theory.

THEOREM 1.13. *In a Hilbert space $H_0$ let there be given two positive definite selfadjoint operators $j$ and $j_1$ normalized by condition* (1.43) *and a linear operator $T$ with $D(T) \supset D(j)$. If for any $x \in D(j)$ and $y \in D(j_1)$*

$$(Tx, y) \leqslant \|jx\|_{H_0}\|y\|_{H_0} \quad \textit{and} \quad (Tx, y) \leqslant \|x\|_{H_0}\|j_1 y\|_{H_0}, \tag{1.48}$$

*then*

$$(Tx, y) \leqslant \|j^\alpha x\|_{H_0}\|j_1^{1-\alpha} y\|_{H_0} \qquad (0 \leqslant \alpha \leqslant 1). \tag{1.49}$$

PROOF. We denote by $H_\alpha$ and $H^1_\alpha$ the Hilbert scales generated by the operators $j$ and $j_1$. From the first inequality in (1.48) it follows that $\|Tx\|_{H_0} \leqslant \|jx\|_{H_0}$, and so $T$ acts boundedly from $H_1$ into $H_0$ with norm not exceeding 1. The second inequality in (1.48) may be written in the form $(j_1^{-1}Tx, j_1 y) \leqslant \|x\|_{H_0}\|j_1 y\|_{H_0}$, which implies that $\|j_1^{-1}Tx\|_{H_0} \leqslant \|x\|_{H_0}$. In other words, we may say that $T$ is defined on the set $D(j)$ dense in $H_0$, and acts boundedly from $H_0$ into $H^1_{-1}$ with norm not exceeding 1. The operator $T$ can be extended by continuity to an operator $\tilde{T}$ defined on the whole space $H_0$ with the same norm. Applying the interpolation theorem to $\tilde{T}$, we obtain that it acts from any space $H_\alpha$ into $H^1_{\alpha-1}$ ($0 \leqslant \alpha \leqslant 1$) with norm not exceeding 1, i.e. $\|\tilde{T}x\|_{H^1_{\alpha-1}} \leqslant \|x\|_{H_\alpha}$.

If now $x \in D(j)$ and $y \in D(j_1)$, then

$$\begin{aligned}(Tx, y) &= (j_1^{\alpha-1}Tx, j_1^{1-\alpha}y) \\ &\leqslant \|j_1^{\alpha-1}Tx_1\|_{H_0}\|j_1^{1-\alpha}y\|_{H_0} \leqslant \|j^\alpha x\|_{H_0}\|j_1^{1-\alpha}y\|_{H_0}.\end{aligned}$$

REMARK. Condition (1.43) has been imposed on the operators $j$ and $j_1$ only to simplify the exposition. A multiplication of these operators by normalizing factors and a repetition of the arguments leads again to the conclusion that (1.48) implies (1.49). Moreover, by passing to the limit, this conclusion can be extended to the case where $j$ and $j_1$ are only positive.

**11.** ***The complex method of interpolation in ideal lattices.*** Let $E_0$, $E_1$ be a pair of ideal lattices of functions on a space $\mathfrak{M}$ with a $\sigma$-finite measure $\mu$. In the Introduction to this chapter we described how the Banach spaces $E_0(A)$ and $E_1(A)$ are constructed, where $A$ is an arbitrary Banach space. We consider the simple case where $A$ coincides with the field $\mathbf{C}$ of complex numbers, and write $E_0(\mathbf{C}) = X_0$ and $E_1(\mathbf{C}) = X_1$. We shall call these spaces *ideal lattices of complex-valued functions*.

We consider the space $[X_0, X_1]_\alpha$. It turns out that this space is also an ideal lattice. Indeed, let $x \in [X_0, X_1]_\alpha$ and $|y(t)| \leqslant |x(t)|$. We consider the function $\alpha(t) = y(t)/x(t)$, assuming that $\alpha(t) = 0$ if $x(t) = 0$. Clearly $|\alpha(t)| \leqslant 1$, and so the operator of multiplication by $\alpha(t)$ acts in the ideal lattices and has a norm not exceeding one there. By the interpolation Theorem 1.2 this operator also acts in the space $[X_0, X_1]_\alpha$ with norm not exceeding 1. Then $y \in [X_0, X_1]_\alpha$ and

$$\|y\|_{[X_0,X_1]_\alpha} = \|\alpha x\|_{[X_0,X_1]_\alpha} \leqslant \|x\|_{[X_0,X_1]_\alpha}.$$

We establish a series of properties of the ideal lattice $[X_0, X_1]_\alpha$.

If $x \in X_0 \cap X_1$, then by virtue of Remark 2 of subsection 4 there exists a function

$$g(t, z) = \sum_1^N a_k(z) x_k(t)$$

such that $x_k \in X_0 \cap X_1$, $a_k \in \mathfrak{F}(\mathbf{C})$, and

$$g(t, \alpha) = x(t), \qquad \|g\|_{\mathfrak{F}(X_0,X_1)} \leqslant (1 + \varepsilon)\|x\|_{[X_0,X_1]_\alpha}. \tag{1.50}$$

For fixed $t$ we apply (1.21) to the scalar-valued function $g(t, z)$. We obtain

$$|x(t)| = |g(t, \alpha)| \leqslant \left[\frac{1}{1-\alpha}\int_{-\infty}^{\infty} |g(t, i\tau)|\, \mu_0(\alpha, \tau)\, d\tau\right]^{1-\alpha}$$
$$\times \left[\frac{1}{\alpha}\int_{-\infty}^{\infty} |g(t, 1 + i\tau)|\, \mu_1(\alpha, \tau)\, d\tau\right]^{\alpha}. \tag{1.51}$$

We consider the functions

$$y_0(t) = \frac{1}{1-\alpha}\int_{-\infty}^{\infty} |g(t, i\tau)|\, \mu_0(\alpha, \tau)\, d\tau$$

and

$$y_1(t) = \frac{1}{\alpha}\int_{-\infty}^{\infty} |g(t, 1 + i\tau)|\, \mu_1(\alpha, \tau)\, d\tau.$$

The right sides of these equalities may be understood as Riemann integrals of functions with values in $X_0$ and $X_1$, respectively. Therefore

$$\|y_0\|_{X_0} \leqslant \frac{1}{1-\alpha}\int_{-\infty}^{\infty} \|g(t, i\tau)\|_{X_0}\mu_0(\alpha, \tau)\,d\tau$$
$$\leqslant \sup_\tau \|g(t, i\tau)\|_{X_0} \leqslant (1+\varepsilon)\|x\|_{[X_0,X_1]_\alpha} \tag{1.52}$$

(the last inequality holds in virtue of (1.50)) and

$$\|y_1\|_{X_1} \leqslant \frac{1}{\alpha}\int_{-\infty}^{\infty} \|g(t, 1+i\tau)\|_{X_1}\mu_1(\alpha, \tau)\,d\tau$$
$$\leqslant \sup \|g(t, 1+i\tau)\|_{X_1} \leqslant (1+\varepsilon)\|x\|_{[X_0,X_1]_\alpha}. \tag{1.53}$$

Inequality (1.51) takes the form

$$|x(t)| \leqslant [y_0(t)]^{1-\alpha}[y_1(t)]^\alpha.$$

If we now write $x_0(t) = [\|y_0\|_{X_0}]^{-1}y_0(t)$ and $x_1(t) = [\|y_1\|_{X_1}]^{-1}y_1(t)$, by (1.52) and (1.53) we obtain

$$|x(t)| \leqslant (1+\varepsilon)\|x\|_{[X_0,X_1]_\alpha}[x_0(t)]^{1-\alpha}[x_1(t)]^\alpha, \tag{1.54}$$

where $\|x_0\|_{X_0} = 1$ and $\|x_1\|_{X_1} = 1$.

It is interesting to consider the problem of when the functions of the form of the right side of (1.54) belong to $[X_0, X_1]_\alpha$.

LEMMA 1.10. *If $x_0 \in X_0$ and $x_1 \in X_1$ and the values of $|x_0(t)|$ and $|x_1(t)|$ have finite positive lower and upper bounds, then $|x_0(t)|^{1-\alpha}|x_1(t)|^\alpha \in [X_0, X_1]_\alpha$ and*

$$\|\,|x_0|^{1-\alpha}|x_1|^\alpha\|_{[X_0,X_1]_\alpha} \leqslant \|x_0\|_{X_0}^{1-\alpha}\|x_1\|_{X_1}^\alpha. \tag{1.55}$$

PROOF. First let $\|x_0\|_{X_0} = \|x_1\|_{X_1} = 1$. We write $x_\alpha(t) = |x_0(t)|^{1-\alpha}|x_1(t)|^\alpha$. Let $e$ be the support of $x_\alpha(t)$. The characteristic function $\chi_e$ of the set $e$ belongs to $X_0 \cap X_1$. Indeed, the functions $|x_0(t)|$ and $|x_1(t)|$ have positive lower bounds $m_0$ and $m_1$ on $e$, and therefore $\chi_e(t) \leqslant |x_0(t)|/m_0 \in X_0$ and $\chi_e(t) \leqslant |x_1(t)|/m_1 \in X_1$.

We consider the function

$$f(t, z) = |x_0(t)|^{1-z}|x_1(t)|^z.$$

Since $|x_0(t)|$ and $|x_1(t)|$ are bounded by numbers $M_0$ and $M_1$, for every $z$ we have

$$|f(t, z)| \leqslant M_0^{1-\operatorname{Re} z}M_1^{\operatorname{Re} z}\chi_e(t) \in X_0 \cap X_1.$$

For fixed $t \in e$ the function $f(t, z)$ is entire, and

$$\tfrac{1}{\Delta z}[f(t, z+\Delta z) - f(t, z)] \to f_z'(t, z)$$

uniformly in $t$. Since in the last relation all terms are equal to zero for $t \notin e$, we may write the inequality

$$\left|\tfrac{1}{\Delta z}\left[f(t, z+\Delta z)-f(t, z)\right]-f_z'(t, z)\right| \leqslant \eta(\Delta z)\chi_e(t),$$

where $\eta(\Delta z) \to 0$ as $\Delta z \to 0$. From this it follows that $f(t, z)$ is entire as a function with values in $X_0 \cap X_1$. By the above, it is bounded in the strip $0 \leqslant \operatorname{Re} z \leqslant 1$. Next,

$$\|f(t, i\tau)\|_{X_0} = \|x_0\|_{X_0} \leqslant 1 \quad \text{and} \quad \|f(t, 1+i\tau)\|_{X_1} = \|x_1\|_{X_1} \leqslant 1.$$

Finally, $f(t, \alpha) = x_\alpha(t)$. This implies that

$$x_\alpha \in [X_0, X_1]_\alpha \quad \text{and} \quad \|x_\alpha\|_{[X_0, X_1]_\alpha} \leqslant 1.$$

If now $x_0$ and $x_1$ are any functions from $X_0$ and $X_1$, then

$$\|\,|x_0(t)|^{1-\alpha}|x_1(t)|^\alpha\|_{[X_0,X_1]_\alpha}$$

$$= \|x_0\|_{X_0}^{1-\alpha}\|x_1\|_{X_1}^{\alpha}\left\|\left|\frac{x_0(t)}{\|x_1\|_{X_0}}\right|^{1-\alpha}\left|\frac{x_2(t)}{\|x_1\|_{X_1}}\right|^{\alpha}\right\|_{[X_0,X_1]_\alpha} \leqslant \|x_0\|_{X_0}^{1-\alpha}\|x_1\|_{X_1}^{\alpha}.$$

COROLLARY. *If $[X_0, X_1]_\alpha$ has the Fatou property (see the Introduction), then* (1.55) *holds for every $x_0 \in X_0$ and $x_1 \in X_1$.*

Indeed, denote by $e_m$ the set of all $t$ for which $1/m \leqslant |x_0(t)| \leqslant m$ and $1/m \leqslant |x_1(t)| \leqslant m$. The union of all sets $e_m$, $m = 1, 2, \ldots$, forms the support of $x_\alpha(t)$, and therefore $\chi_{e_m}(t)x_\alpha(t) \to x_\alpha(t)$ almost everywhere. By virtue of the lemma we have

$$\|\chi_{e_m}|x_0|^{1-\alpha}|x_1|^\alpha\|_{[X_0,X_1]_\alpha} = \|[\chi_{e_m}|x_0|]^{1-\alpha}[\chi_{e_m}|x_1|]^\alpha\|_{[X_0,X_1]_\alpha}$$
$$\leqslant \|\chi_{e_m}x_0\|_{X_0}^{1-\alpha}\|\chi_{e_m}x_1\|_{X_1}^{\alpha} \leqslant \|x_0\|_{X_0}^{1-\alpha}\|x_1\|_{X_1}^{\alpha}.$$

Due to the Fatou property, this inequality implies (1.55).

Inequality (1.55) is a further generalization of the Hölder inequality.

The above properties suggest the introduction of the following space: For fixed $\alpha \in (0, 1)$ denote by $X$ the class of complex-valued functions $x(t)$ on $\mathfrak{M}$ such that

$$|x(t)| \leqslant \lambda|x_0(t)|^{1-\alpha}|x_1(t)|^\alpha \tag{1.56}$$

for some $\lambda > 0$, $x_0 \in X_0$ and $x_1 \in X_1$ with $\|x_0\|_{X_0} \leqslant 1$ and $\|x_1\|_{X_1} \leqslant 1$. The space $X$ is linear. Indeed, if (1.56) is satisfied for $x$ and $|y(t)| \leqslant \mu|y_0(t)|^{1-\alpha}|y_1(t)|^\alpha$ with $\|y_0\|_{X_0} \leqslant 1$ and $\|y_1\|_{X_1} \leqslant 1$, then

$$|x(t)+y(t)| \leqslant \lambda|x_0(t)|^{1-\alpha}|x_1(t)|^\alpha + \mu|y_0(t)|^{1-\alpha}|y_1(t)|^\alpha,$$

and by Hölder's inequality with weights $\lambda$ and $\mu$ we have

$$|x(t)+y(t)| \leqslant (\lambda|x_0(t)|+\mu|y_0(t)|)^{1-\alpha}(\lambda|x_1(t)|+\mu|y_1(t)|)^{\alpha}$$
$$= (\lambda+\mu)\left(\frac{\lambda}{\lambda+\mu}|x_0(t)|+\frac{\mu}{\lambda+\mu}|y_0(t)|\right)^{1-\alpha}$$
$$\times\left(\frac{\lambda}{\lambda+\mu}|x_1(t)|+\frac{\mu}{\lambda+\mu}|y_1(t)|\right)^{\alpha}. \tag{1.57}$$

The functions in parentheses belong to the unit balls of $X_0$ and $X_1$, respectively, and therefore $x+y \in X$. It is obvious that $cx \in X$ if $x \in X$. In $X$ we introduce the norm

$$\|x\|_X = \inf \lambda, \tag{1.58}$$

where the infimum is taken over all $\lambda$ satisfying (1.56) for some $x_0$ and $x_1$. From (1.57) it follows that the quantity (1.58) satisfies the triangle inequality. It is obvious that $\|cx\|_X = |c|\ \|x\|_X$. Now let $\|x\|_X = 0$. This means that there exist sequences of numbers $\lambda_n \to +0$ and functions $x_{0n} \in X_0$ and $x_{1n} \in X_1$ with $\|x_{0n}\|_{X_0}, \|x_{1n}\|_{X_1} \leqslant 1$ such that

$$|x(t)| \leqslant \lambda_n|x_{0n}(t)|^{1-\alpha}|x_{1n}(t)|^{\alpha}.$$

The sequences $\lambda_n^{1/2(1-\alpha)}|x_{0n}|$ and $\lambda_n^{1/2\alpha}|x_{1n}|$ converge to zero in $X_0$ and $X_1$, respectively. Since $X_0$ and $X_1$ are imbedded in $S(\mathfrak{M}, \mu)$, these sequences converge to zero in measure on every set of finite measure. The sequences of functions $\sqrt{\lambda_n}\ |x_{0n}|^{1-\alpha}(t)$ and $\sqrt{\lambda_n}\ |x_{1n}|_{\alpha}(t)$ have the same property. Since $|x(t)|$ does not exceed the product of these functions, we have $|x(t)| = 0$ almost everywhere. Thus, (1.58) has all properties of a norm.

It is obvious that if $|u(t)| \leqslant |x(t)|$ and $x \in X$, then $u \in X$ and $\|u\|_X \leqslant \|x\|_X$.

We prove the completeness of $X$. Let $x_n \in X$ and $\Sigma\|x_n\|_X < \infty$. Given any $\varepsilon > 0$, for each function $x_n$ we choose a number $\lambda_n$ and functions $x_{0n} \in X_0$ and $x_{1n} \in X_1$, with $\|x_{0n}\|_{X_0}, \|x_{1n}\|_{X_1} \leqslant 1$, such that

$$\lambda_n \leqslant \|x_n\|_X + \varepsilon/2^n \quad \text{and} \quad |x_n(t)| \leqslant \lambda_n|x_{0n}(t)|^{1-\alpha}|x_{1n}(t)|^{\alpha}.$$

Then $\Sigma\,\lambda_n < \infty$, and the same arguments as in the proof of (1.57) show that

$$\sum_1^{\infty}|x_n(t)| \leqslant \Sigma\lambda_n\left[\sum_1^{\infty}\frac{\lambda_n}{\Sigma\lambda_n}|x_{0n}(t)|\right]^{1-\alpha}\left[\sum_1^{\infty}\frac{\lambda_n}{\Sigma\lambda_n}|x_{1n}(t)|\right]^{\alpha} \tag{1.59}$$

for almost all $t$.

By virtue of Corollary 2 to Theorem 1 in the Introduction to Chapter II the functions in square brackets are defined almost everywhere and belong to the unit balls of $X_0$ and $X_1$. This implies that $\Sigma_1^{\infty}|x_n(t)|$ belongs to $X$. We write

$x(t) = \Sigma_1^\infty x_n(t)$. Then $|x(t)| \leqslant \Sigma_1^\infty |x_n(t)|$, and therefore $x \in X$. Inequality (1.59) implies that $\|x\|_X \leqslant \|\Sigma|x_n|\,\|_X \leqslant \Sigma\,\lambda_n$, and by the arbitrariness of $\varepsilon$ we have

$$\|x\|_X \leqslant \sum_1^\infty \|x_n\|_X.$$

Applying this inequality to the function $x(t) - \Sigma_1^N x_n(t) = \Sigma_{N+1}^\infty x_n(t)$, we obtain

$$\left\| x - \sum_1^N x_n \right\|_X \leqslant \sum_{N+1}^\infty \|x_n\|_X \to 0 \quad \text{as } N \to \infty.$$

The completeness of $X$ is proved. Thus, $X$ is an ideal lattice. We denote this lattice by $X_0^{1-\alpha}X_1^\alpha$.

The space $X_0^{1-\alpha}X_1^\alpha$ is intermediate between $X_0$ and $X_1$. Indeed, if $x \in X_0 \cap X_1$, then

$$|x(t)| = |x(t)|^{1-\alpha}|x(t)|^\alpha = \|x\|_{X_0}^{1-\alpha}\|x\|_{X_1}^\alpha \left( \frac{|x(t)|}{\|x\|_{X_0}} \right)^{1-\alpha} \left( \frac{|x(t)|}{\|x\|_{X_1}} \right)^\alpha,$$

which implies that $x \in X_0^{1-\alpha}X_1^\alpha$ and

$$\|x\|_{X_0^{1-\alpha}X_1^\alpha} \leqslant \|x\|_{X_0}^{1-\alpha}\|x\|_{X_1}^\alpha \leqslant \max\{\|x\|_{X_0}, \|x\|_{X_1}\}.$$

Next, if $x \in X_0^{1-\alpha}X_1^\alpha$, then (1.56) implies that

$$|x(t)| \leqslant \lambda((1-\alpha)|x_0(t)| + \alpha|x_1(t)|)$$

and so $x \in X_0 + X_1$ and

$$\|x(t)\|_{X_0+X_1} \leqslant \lambda[(1-\alpha)\|x_0\|_{X_0} + \alpha\|x_1\|_{X_1}] \leqslant \lambda.$$

Thus, $\|x\|_{X_0+X_1} \leqslant \|x\|_{X_0^{1-\alpha}X_1^\alpha}$.

From inequality (1.54), true for any function $x \in X_0 \cap X_1$, Theorem 1.3, and the definition of the norm in $X_0^{1-\alpha}X_1^\alpha$ it follows that $[X_0, X_1]_\alpha$ is imbedded in $X_0^{1-\alpha}X_1^\alpha$ with imbedding constant not exceeding one.

The reverse imbedding holds under certain conditions.

THEOREM 1.14. *If the space $X = X_0^{1-\alpha}X_1^\alpha$ is regular (has absolutely continuous norm), then it coincides isometrically with $[X_0, X_1]_\alpha$.*

PROOF. We denote by $U_\varepsilon$ the collection of all functions $x$ in $X$ for which the inequality

$$|x(t)| \leqslant (1+\varepsilon)\|x\|_X x_0^{1-\alpha}(t)x_1^\alpha(t) \quad \text{with } \|x_0\|_{X_0} \leqslant 1 \text{ and } \|x_1\|_{X_1} \leqslant 1$$

is satisfied and the nonzero values of the functions $x_0(t)$ and $x_1(t)$ have finite positive lower and upper bounds. By Lemma 1.10 and the fact that $[X_0, X_1]_\alpha$

is an ideal lattice, we have

$$\|x\|_{[X_0,X_1]_\alpha} \leqslant (1+\varepsilon)\|x\|_X \qquad (x \in U_\varepsilon). \tag{1.60}$$

We show that the set $U_\varepsilon$ is dense in $X$. Let $y \in X$, let $e$ be the support of $y$, and let

$$|y(t)| \leqslant (1+\varepsilon/2)\|y\|_X |y_0(t)|^{1-\alpha}|y_1(t)|^\alpha,$$

where $\|y_0\|_{X_0}, \|y_1\|_{X_1} \leqslant 1$. We denote by $e_m$ the set of all $t$ at which $|y(t)| > 0$, $1/m \leqslant |y_0(t)| \leqslant m$ and $1/m \leqslant |y_1(t)| \leqslant m$. Then the functions $z_m(t) = \chi_{e_m}(t)y(t)$ converge to $y(t)$ almost everywhere, and

$$\|z_m - y\|_X = \|\chi_{e\setminus e_m} y\|_X \to 0$$

by virtue of the absolute continuity of the norm in $X$.

In particular, $\|z_m\|_X \to \|y\|_X$, and so

$$\begin{aligned}|z_m(t)| &\leqslant (1+\varepsilon/2)\|y\|_X|\chi_{e_m}(t)y_0(t)|^{1-\alpha}|\chi_{em}(t)y_1(t)|^\alpha \\ &\leqslant (1+\varepsilon)\|z_m\|_X|\chi_{e_m}(t)y_0(t)\|^{1-\alpha}|\chi_{e_m}(t)y_1(t)|^\alpha\end{aligned}$$

for sufficiently large $m$. We note that $\|\chi_m y_0\|_{X_0} \leqslant 1$ and $\|\chi_m y_1\|_{X_1} \leqslant 1$; thus $z_m \in U_\varepsilon$.

Now we may apply Lemma 3.4 of Chapter I, from which it follows by virtue of (1.60) that the space $X$ is imbedded in $[X_0, X_1]_\alpha$ with imbedding constant not exceeding $1+\varepsilon$.

Since $\varepsilon$ is arbitrary, the theorem is proved.

REMARK. The hypothesis of the theorem is satisfied for $0 < \alpha < 1$ if at least one of the spaces $X_0$ or $X_1$ has absolutely continuous norm.

Indeed, let $X_0$ have this property. If $e_n$ is a decreasing sequence of sets with empty intersection and $x \in X$, then from (1.56) we obtain

$$\begin{aligned}|\chi_{e_n}(t)x(t)| &\leqslant \lambda|\chi_{e_n}(t)x_0(t)|^{1-\alpha}|\chi_{e_n}(t)x_1(t)|^{1-\alpha} \\ &\leqslant \lambda\|\chi_{e_n}x_0\|_{X_0}^{1-\alpha}\left|\frac{\chi_{e_n}(t)x_0(t)}{\|\chi_{e_n}x\|_{X_0}}\right|^{1-\alpha}|x_1(t)|^\alpha.\end{aligned}$$

This implies that $\|\chi_{e_n}x\|_X \leqslant \lambda\|\chi_{e_n}x_0\|_{X_0}^{1-\alpha} \to 0$ as $n \to \infty$.

The family of spaces $X_0^{1-\alpha}X_1^\alpha$ is called the *Calderón family* connecting $X_0$ and $X_1$.*

We note that in the general case the space $X_0^{1-\alpha}X_1^\alpha$ may not coincide with $[X_0, X_1]_\alpha$. Moreover, $X_0^{1-\alpha}X_1^\alpha$ may not be an interpolation space between $X_0$

---

* *Editor's note*. In the Russian edition Calderón families were called Calderón scales. However, a Calderón family may not be a scale in the sense of Chapter III, since the spaces $X_0$ and $X_1$ need not be related by an imbedding.

and $X_1$ (see Notes on the Literature on p. 352). Nevertheless, if the linear operator $T$ is positive and acts boundedly in $X_0$ and $X_1$, then it acts in $X_0^{1-\alpha}X_1^\alpha$ as well. Indeed, if $x(t) \geqslant 0$ and $x \in X_0^{1-\alpha}X_1^\alpha$, then by (1.56)

$$Tx(t) \leqslant \lambda T(|x_0|^{1-\alpha}|x_1|^\alpha)(t).$$

Similarly to the way we did in subsection 1 of the Introduction to Chapter II (property 6°), from this we obtain that

$$Tx(t) \leqslant \lambda(T|x_0|)^{1-\alpha}(t)(T|x_1|)^\alpha(t),$$

and, since $T|x_0| \in X_0$ and $T|x_1| \in X_1$, we get that $Tx \in X_0^{1-\alpha}X_1^\alpha$. For any real functions the assertion follows from the equality $Tx = Tx_+ - Tx_-$, and for complex-valued functions from the equality $Tx = T(\operatorname{Re} x) + iT(\operatorname{Im} x)$.

If $X_0 = L_{p_0}$ and $X_1 = L_{p_1}$, $1 \leqslant p_0 \leqslant \infty$, then $X_0^{1-\alpha}X_1^\alpha$ coincides isometrically with $L_p$, where $1/p = (1-\alpha)/p_0 + \alpha/p_1$. Indeed, (1.56) and Hölder's inequality imply that for $x \in X_0^{1-\alpha}X_1^\alpha$ we have

$$\int |x(t)|^p\,d\mu \leqslant \lambda^p\left[\int |x_0(t)|^{p_0}\,d\mu\right]^{(1-\alpha)p/p_0}\left[\int |x_1(t)|^{p_1}\,d\mu\right]^{\alpha p/p_1} \leqslant \lambda^p,$$

from which it follows that $x \in L_p$ and $\|x\|_{L_p} \leqslant \|x\|_{L_{p_0}^{1-\alpha}L_{p_1}^\alpha}$. Conversely, for any function in $L_p$ the representation

$$|x(t)| = \|x\|_{L_p}\left(\frac{|x(t)|^{p/p_0}}{\|\,|x|^{p/p_0}\|_{L_{p_0}}}\right)^{1-\alpha}\left(\frac{|x(t)|^{p/p_1}}{\|\,|x|^{p/p_1}\|_{L_{p_1}}}\right)^{\alpha}$$

holds (we have used the fact that $1/p = (1-\alpha)/p_0 + \alpha/p_1$), and therefore

$$x \in L_{p_0}^{1-\alpha}L_{p_1}^\alpha \quad \text{and} \quad \|x\|_{L_{p_0}^{1-\alpha}L_{p_1}^\alpha} \leqslant \|x\|_{L_p}.$$

The norm is absolutely continuous in at least one of the spaces (if $p_0 \neq p_1$), and thus by Theorem 1.4 we have the isometric equality

$$L_{p_0}^{1-\alpha}L_{p_1}^\alpha = [L_{p_0}, L_{p_1}]_\alpha = L_p \qquad \left(\frac{1}{p} = \frac{1-\alpha}{p_0} + \frac{\alpha}{p_1}\right). \tag{1.61}$$

The interpolation Theorem 1.2 leads us in particular to the proof of the Riesz-Thorin theorem formulated in Chapter I, §4.2.

## §2. The methods of constants and means ($\mathcal{K}$- and $\mathcal{J}$-methods)

**1. *The method of constants ($\mathcal{K}$-method).*** We recall that if $A$ is a Banach space and $E$ is an ideal lattice on the space $\mathfrak{M}$ with measure $\mu$, then by $E(A)$ we denote the Banach space of functions (classes) $u(t)$ with values in $A$, strongly measurable in $A$, and such that $\|u(t)\|_A \in E$, with the norm $\|u\|_{E(A)} = \|\,\|u(t)\|_A\|_E$. This space is imbedded in the metric linear space $S(\mathfrak{M}, \mu, A)$ of all strongly measurable functions with values in $A$.

Now let $A_0$ and $A_1$ be a Banach couple, and $E_0$ and $E_1$ be two ideal lattices on the same measure space. The space $E_i(A_i)$ is imbedded in $E_i(A_0 + A_1)$, and, as we said above, these spaces are imbedded in $S(\mathfrak{M}, \mu, A_0 + A_1)$. Therefore $E_0(A_0)$ and $E_1(A_1)$ form a Banach couple. We denote by $(A_0, A_1)^{\mathcal{K}}_{E_0,E_1}$ the part of $E_0(A_0) + E_1(A_1)$, consisting of all functions which are constant almost everywhere. The imbedding

$$E_0(A_0) + E_1(A_1) \subset S(\mathfrak{M}, \mu, A_0 + A_1)$$

implies easily that $(A_0, A_1)^{\mathcal{K}}_{E_0,E_1}$ is a closed subspace of the sum $E_0(A_1) + E_1(A_1)$. It can be identified with the Banach space of all elements $x \in A_0 + A_1$ for which there exist functions $u_i = u_i(t) \in E_i(A_i)$ $(i = 0, 1)$, such that

$$x = u_0(t) + u_1(t) \tag{2.1}$$

with norm

$$\|x\|_{(A_0,A_1)^{\mathcal{K}}_{E_0,E_1}} = \inf_{x = u_0(t) + u_1(t)} \left(\|u_0\|_{E_0(A_0)} + \|u_1\|_{E_1(A_1)}\right). \tag{2.2}$$

The space $(A_0, A_1)^{\mathcal{K}}_{E_0,E_1}$ may contain nonzero elements only in the case when $e \in E_0 + E_1$, where $e = e(t) \equiv 1$. Indeed, if $x \in (A_0, A_1)^{\mathcal{K}}_{E_0,E_1}$ and $x \neq 0$, then there exist $u_i(t) \in E_i(A_i)$ such that $u_0(t) + u_1(t) = x$. By the definition of the norm in $A_0 + A_1$ we have $0 < \|x\|_{A_0+A_1} \leqslant \|u_0(t)\|_{A_0} + \|u_1(t)\|_{A_1}$. This implies that

$$e_i(t) = \frac{\|u_i(t)\|_{A_i}}{\|u_0(t)\|_{A_0} + \|u_1(t)\|_{A_1}} \leqslant \frac{\|u_i(t)\|_{A_i}}{\|x\|_{A_0+A_1}} \in E_i.$$

Then $e(t) = e_0(t) + e_1(t) \in E_0 + E_1$.

In what follows we always assume that $e \in E_0 + E_1$.

LEMMA 2.1. *The spaces $(A_0, A_1)^{\mathcal{K}}_{E_0,E_1}$ are intermediate between $A_0$ and $A_1$.*

PROOF. Let $x \in A_0 \cap A_1$. If $e(t) = e_0(t) + e_1(t)$, where $e_i(t) \in E_i$, we set $u_i(t) = e_i(t)x$. Then $x = u_0(t) + u_1(t)$, and $\|u_i(t)\|_{A_i} = |e_i(t)|\|x\|_{A_i} \in E_i$. Consequently, $x \in (A_0, A_1)^{\mathcal{K}}_{E_0,E_1}$ and

$$\begin{aligned}\|x\|_{(A_0,A_1)^{\mathcal{K}}_{E_0,E_1}} &\leqslant \|e_0\|_{E_0}\|x\|_{A_0} + \|e_1\|_{E_1}\|x\|_{A_1} \\ &\leqslant \left(\|e_0\|_{E_0} + \|e_1\|_{E_1}\right)\|x\|_{A_0\cap A_1}.\end{aligned}$$

Taking into account that $\inf_{e=e_0+e_1}(\|e_0\|_{E_0} + \|e_1\|_{E_1}) = \|e\|_{E_0+E_1}$, we obtain

$$\|x\|_{(A_0,A_1)^{\mathcal{K}}_{E_0,E_1}} \leqslant \|e\|_{E_0+E_1}\|x\|_{A_0\cap A_1},$$

i.e. $A_0 \cap A_1 \subset (A_0, A_1)^{\mathcal{K}}_{E_0,E_1}$.

Now let $x \in (A_0, A_1)^{\mathcal{K}}_{E_0,E_1}$. The equality $x = u_0(t) + u_1(t)$, $u_i(t) \in E_i(A_i)$, implies the inequality $\|x\|_{A_0+A_1} \leqslant \|u_0(t)\|_{A_0} + \|u_1(t)\|_{A_1}$. By calculating the norm in $E_0 + E_1$ of the left and right sides, we have

$$\|x\|_{A_0+A_1}\|e\|_{E_0+E_1} \leqslant \| \, \|u_0(t)\|_{A_0} + \|u_1(t)\|_{A_1}\|_{E_0+E_1} \leqslant \|u_0\|_{E_0(A_0)} + \|u_1\|_{E_1(A_1)}.$$

This implies that

$$\|x\|_{A_0+A_1} \leqslant \frac{1}{\|e\|_{E_0+E_1}} \|x\|_{(A_0,A_1)^{\mathcal{K}}_{E_0,E_1}},$$

i.e. $(A_0, A_1)^{\mathcal{K}}_{E_0,E_1} \subset A_0 + A_1$.

THEOREM 2.1 (interpolation theorem). *Let $A_0$, $A_1$ and $B_0$, $B_1$ be two Banach couples. The triple $A_0$, $A_1$, $(A_0, A_1)^{\mathcal{K}}_{E_0,E_1}$ of spaces is an interpolation triple relative to the triple $B_0$, $B_1$, $(B_0, B_1)^{\mathcal{K}}_{E_0,E_1}$.*

PROOF. If $x \in (A_0, A_1)^{\mathcal{K}}_{E_0,E_1}$ and $T \in \pi(A_0, A_1;\ B_0, B_1)$, then $x = u_0(t) + u_1(t)$, where $u_i \in E_i(A_i)$, and so $Tx = v_0(t) + v_1(t)$, where $v_i(t) = Tu_i(t)$. Moreover, $\|v_i(t)\|_{B_i} \leqslant \|T\|_{A_i\to B_i}\|u_i(t)\|_{A_i}$. Consequently $v_i \in E_i(B_i)$ and

$$\begin{aligned}\|Tx\|_{(B_0,B_1)^{\mathcal{K}}_{E_0,E_1}} &\leqslant \|v_0\|_{E_0(B_0)} + \|v_1\|_{E_1(B_1)} \\ &\leqslant \|T\|_{A_0\to B_0}\|u_0\|_{E_0(A_0)} + \|T\|_{A_1\to B_1}\|u_1\|_{E_1(A_1)} \\ &\leqslant \max\{\|T\|_{A_i\to B_i}\}(\|u_0\|_{E_0(A_0)} + \|u_1\|_{E_1(A_1)}).\end{aligned}$$

Taking the infimum over all representations (2.1), we obtain

$$\|T\|_{(A_0,A_1)^{\mathcal{K}}_{E_0,E_1}\to(B_0,B_1)^{\mathcal{K}}_{E_0,E_1}} \leqslant \|T\|_{\pi(A_0,A_1;B_0,B_1)}.$$

Theorem 2.1 shows that we have defined an interpolation functor $(\ ,\ )^{\mathcal{K}}_{E_0,E_1}$.

REMARK. If $e \in E_0 \cap E_1$, then $(A_0, A_1)^{\mathcal{K}}_{E_0,E_1}$ is isomorphic to $A_0 + A_1$. Indeed, if $x \in A_0 + A_1$, then $x = x_0 + x_1$, $x_i \in A_i$. We set $u_i(t) = e(t)x_i$. Then $\|u_i\|_{E_i(A_i)} = \|e\|_{E_i}\|x_i\|_{A_i}$. Thus, $x \in (A_0, A_1)^{\mathcal{K}}_{E_0,E_1}$ and

$$\|x\|_{(A_0,A_1)^{\mathcal{K}}_{E_0,E_1}} \leqslant \|e\|_{E_0}\|x_0\|_{A_0} + \|e\|_{E_1}\|x_1\|_{A_1}.$$

Taking the infimum on the right, we obtain

$$\|x\|_{(A_0,A_1)^{\mathcal{K}}_{E_0,E_1}} \leqslant \|e\|_{E_0\cap E_1}\|x\|_{A_0+A_1}.$$

**2. *The method of means ($\mathcal{J}$-method).*** In the notation of the preceding subsection we assume that $e \in E_0^1 + E_1^1$, where $E_i^1$ is the ideal lattice associated with $E_i$.

LEMMA 2.2. *For any function $u \in E_0(A_0) \cap E_1(A_1)$ the integral $\int u(t)\,d\mu$ exists in $A_0 + A_1$. The mapping $\mathcal{J}: u \to \int u(t)\,d\mu$ is a continuous linear operator from $E_0(A_0) \cap E_1(A_1)$ into $A_0 + A_1$.*

PROOF. Let $e(t) = e_0(t) + e_1(t)$, where $e_i \in E_i^1$. Without loss of generality we may assume that the functions $e_i(t)$ are nonnegative. We have

$$\begin{aligned}\int \|u(t)\|_{A_0+A_1}\,d\mu &\leqslant \int e_0(t)\|u(t)\|_{A_0}\,d\mu + \int e_1(t)\|u(t)\|_{A_1}\,d\mu\\ &\leqslant \|e_0\|_{E_0^1}\|u\|_{E_0(A_0)} + \|e_1\|_{E_1^1}\|u\|_{E_1(A_1)}\\ &\leqslant \left(\|e_0\|_{E_0^1} + \|e_1\|_{E_1^1}\right)\|u\|_{E_0(A_0)\cap E_1(A_1)}.\end{aligned}$$

Taking the infimum over all representations $e(t) = e_0(t) + e_1(t)$, we obtain

$$\left\|\int u(t)\,d\mu\right\|_{A_0+A_1} \leqslant \|e\|_{E_0^1+E_1^1}\|u\|_{E_0(A_0)\cap E_1(A_1)}.$$

The lemma is proved.

COROLLARY. *The null space $N(\mathfrak{J})$ of the operator $\mathfrak{J}$ is a closed subspace of $E_0(A_0) \cap E_1(A_1)$.*

We denote by $(A_0, A_1)^{\mathfrak{J}}_{E_0,E_1}$ the quotient space $E_0(A_0) \cap E_1(A_1)/N(\mathfrak{J})$, which can be identified with the Banach space of all elements $x \in A_0 + A_1$ representable in the form

$$x = \int u(t)\,d\mu, \tag{2.3}$$

where $u \in E_0(A_0) \cap E_1(A_1)$ with norm

$$\|x\|_{(A_0,A_1)^{\mathfrak{J}}_{E_0,E_1}} = \inf_{x=\int u(t)\,d\mu} \|u\|_{E_0(A_0)\cap E_1(A_1)}. \tag{2.4}$$

The space $(A_0, A_1)^{\mathfrak{J}}_{E_0,E_1}$ may contain nonzero elements only in the case where $E_0 \cap E_1 \neq \{0\}$. Indeed, if $x \neq 0$, $x = \int u(t)\,d\mu$, $\|u(t)\|_{A_i} \in E_i$, then $u \neq 0$ and the function $\min(\|u(t)\|_{A_0}, \|u(t)\|_{A_1})$ belongs to $E_0 \cap E_1$ and is different from zero, because the functions $\|u(t)\|_{A_0}$ and $\|u(t)\|_{A_1}$ vanish at the same time.

In what follows we assume that $E_0 \cap E_1 \neq \{0\}$.

LEMMA 2.3. *The spaces $(A_0, A_1)^{\mathfrak{J}}_{E_0,E_1}$ are intermediate between $A_0$ and $A_1$.*

PROOF. The continuity of the imbedding $(A_0, A_1)^{\mathfrak{J}}_{E_0,E_1} \subset A_0 + A_1$ follows from the preceding lemma and the definition of the norm in $(A_0, A_1)^{\mathfrak{J}}_{E_0,E_1}$.

Now let $x \in A_0 \cap A_1$. We take a function $\psi = \psi(t)$ such that $\psi \in E_0 \cap E_1$ and $\int \psi(t)\,d\mu = 1$. We set $u(t) = \psi(t)x$. Then $u = u(t) \in E_0(A_0) \cap E_1(A_1)$ and $x = \int u(t)\,d\mu$. Thus, $x \in (A_0, A_1)^{\mathfrak{J}}_{E_0,E_1}$ and $\|x\|^{\mathfrak{J}}_{E_0,E_1} \leqslant \|\psi\|_{E_0\cap E_1}\|x\|_{A_0\cap A_1}$, which means we have the imbedding $A_0 \cap A_1 \subset (A_0, A_1)^{\mathfrak{J}}_{E_0,E_1}$.

THEOREM 2.2 (interpolation theorem). *Let $A_0$, $A_1$ and $B_0$, $B_1$ be two Banach couples. The triple of spaces $A_0$, $A_1$, $(A_0, A_1)^{\mathfrak{J}}_{E_0,E_1}$ is an interpolation triple relative to the triple $B_0$, $B_1$, $(B_0, B_1)^{\mathfrak{J}}_{E_0,E_1}$.*

PROOF. If $x \in (A_0, A_1)^{\mathfrak{J}}_{E_0,E_1}$, then $x = \int u(t)\,d\mu$, $u \in E_0(A_0) \cap E_1(A_1)$. The operator $T \in \pi(A_0, A_1;\ B_0, B_1)$ acts from $A_0 + A_1$ to $B_0 + B_1$. Therefore $Tx = \int Tu(t)\,d\mu = \int v(t)\,d\mu$. Next, $\|v\|_{E_i(B_i)} \leqslant \|T\|_{A_i \to B_i}\|u\|_{E_i(A_i)}$. From this we obtain

$$\|v\|_{E_0(B_1)\cap E_1(B_1)} \leqslant \max\{\|T\|_{A_i\to B_i}\}\|u\|_{E_0(A_0)\cap E_1(A_1)}.$$

By the definition (2.4) we have

$$\|Tx\|_{(B_0,B_1)^{\mathfrak{J}}_{E_0,E_1}} \leqslant \|v\|_{E_0(B_0)\cap E_1(B_1)} \leqslant \|T\|_{\pi(A_0,A_1;B_0,B_1)}\|u\|_{E_0(A_0)\cap E_1(A_1)}.$$

Taking the infimum over all representations (2.3), we obtain

$$\|T\|_{(B_0,B_1)^{\mathfrak{J}}_{E_0,E_1}\to(A_0,A_1)^{\mathfrak{J}}_{E_0,E_1}} \leqslant \|T\|_{\pi(A_0,A_1;B_0,B_1)}.$$

Thus, the interpolation functor $(\ ,\ )^{\mathfrak{J}}_{E_0,E_1}$ is defined.

Sometimes in constructing the spaces $(A_0, A_1)^{\mathcal{K}}_{E_0,E_1}$ and $(A_0, A_1)^{\mathfrak{J}}_{E_0,E_1}$ it is convenient to effect the discretization of the lattices $E_0$ and $E_1$ beforehand (see Chapter II, the Introduction, subsection 2). Let $\mathfrak{M} = \bigcup_1^\infty \mathfrak{M}_n$ be a partition of $\mathfrak{M}$ into disjoint sets of finite positive measure $\mu_n$ and such that $\chi_{\mathfrak{M}_n} \in E_0 \cap E_1$ (such a partition exists if the supports of $E_0$ and $E_1$ coincide with $\mathfrak{M}$). We assume that the averaging operator $T$ constructed for this partition acts boundedly in the spaces $E_i$. Then it maps $E_i$ onto the subspace $E_i^c$ of functions constant on the sets $\mathfrak{M}_n$. In order to obtain a quantity equivalent to the norm of $(A_0, A_1)^{\mathcal{K}}_{E_0,E_1}$, we need consider only those representations $x = u_0(t) + u_1(t)$ in (2.2) in which the functions $u_i(t)$ are constant on the sets $\mathfrak{M}_n$. Indeed, if $x = v_0(t) + v_1(t)$, where $v_i \in E_i(A_i)$, we set $u_i(t) = Tv_i(t)$. Then $x = Tx = u_0(t) + u_1(t)$, and

$$\|u_i\|_{E_i(A_i)} = \|\,\|Tv_i\|_{A_i}\|_{E_i} \leqslant \|T(\|v_i\|_{A_i})\|_{E_i} \leqslant \|T\|_{E_i\to E_i}\|v\|_{E_i(A_i)}.$$

Thus, the infimum in (2.2) calculated only over functions constant on the sets $\mathfrak{M}_n$ does not exceed the norm of (2.2) by more than a factor of $\max\{\|T\|_{E_i\to E_i}\}$. A similar assertion holds for the spaces $(A_0, A_1)^{\mathfrak{J}}_{E_0,E_1}$.

In the terms introduced in subsection 2 of the Introduction of Chapter II, the assertions obtained so far can be formulated in the following way.

REMARK 2.1. If the averaging operator corresponding to the partition $\mathfrak{M} = \bigcup_1^\infty \mathfrak{M}_n$ is bounded in the spaces $E_i$, then

$$(A_0, A_1)^{\mathcal{K}(\mathfrak{J})}_{E_0,E_1} = (A_0, A_1)^{\mathcal{K}(\mathfrak{J})}_{dE_0,dE_1}$$

up to isomorphism.

**3.** ***The methods of constants and means for spaces with power weights.*** In this subsection the semiaxis $(0, \infty)$ with the measure defined by the differential form $dt/t$ will serve as the space $\mathfrak{M}$. The spaces $L_p$ constructed for this measure will be denoted by $L_{p,*}$. Let $\alpha_0$ and $\alpha_1$ be real parameters, and assume that $\alpha_0 < 0$ and $\alpha_1 > 0$. Let $E_0$ and $E_1$ be ideal lattices which are intermediate spaces between $L_{1,*}$ and $L_{\infty,*}$. We consider the spaces with weights $E_{i,t^{\alpha_i}} = E_i^{\alpha_i}$ $(i = 0, 1)$. We show that the function $e(t) \equiv 1$ belongs to $E_0^{\alpha_0} + E_1^{\alpha_1}$. Indeed, $e(t) = \chi_{(1,\infty)}(t) + \chi_{(0,1)}(t)$. Next, $t^{\alpha_0}\chi_{(1,\infty)}(t) \in L_{1,*} \cap L_{\infty,*} \subset E_0$ and $t^{\alpha_1}\chi_{(0,1)}(t) \in L_{1,*} \cap L_{\infty,*} \subset E_1$, as required. This implies that $(A_0, A_1)^{\mathcal{K}}_{E_0^{\alpha_0}, E_1^{\alpha_1}}$ is an intermediate space between $A_0$ and $A_1$.

We shall assume that both $E_0$ and $E_1$ contain functions that are positive almost everywhere. Then $E_0 \cap E_1 \neq \{0\}$. The associated spaces $E_0^1$ and $E_1^1$ are ideal lattices on $(0, \infty)$. It is easy to see that they are also intermediate spaces between $L_{1,*}$ and $L_{\infty,*}$. Next, the spaces $(E_i^{\alpha_i})^1 = E_i^{1(-\alpha_i)}$ are associated with the $E_i^{\alpha_i}$. The same reasoning as above shows that $e \in (E_0^{\alpha_0})^1 + (E_1^{\alpha_1})^1$. Therefore we may define the intermediate space $(A_0, A_1)^{\mathcal{J}}_{E_0^{\alpha_0}, E_1^{\alpha_1}}$.

Now we assume that the lattices $E_0$ and $E_1$ are interpolation spaces between $L_{1,*}$ and $L_{\infty,*}$. We recall that in §8.5 of Chapter II we considered certain operators acting in the spaces $E_i$ and obtained estimates for their norms.

For the dilation operator we have

$$\|\sigma_{1/\tau}\varphi\|_{E_i} = \|\varphi(\tau t)\|_{E_i} \leqslant k_i\|\varphi\|_{E_i}, \tag{2.5}$$

where the $k_i$ are the interpolation constants of the spaces $E_i$.

For a power transformation we have

$$\|T_\lambda\varphi\|_{E_i} = \|\varphi(t^\lambda)\|_{E_i} \leqslant k_i \max\{1, |\lambda|^{-1}\}\|\varphi\|_{E_i}. \tag{2.6}$$

For the convolution we have the following Young inequality:

$$\left\|\int_0^\infty \psi\left(\frac{t}{s}\right)\varphi(s)\frac{ds}{s}\right\|_{E_i} \leqslant k_i\|\psi\|_{L_{1,*}}\|\varphi\|_{E_i}. \tag{2.7}$$

Finally, we considered the special averaging operator

$$T_1\varphi(t) = \sum_{-\infty}^{\infty} \chi_{(e^n, e^{n+1}]}(t)\int_{e^n}^{e^{n+1}} \varphi(s)\frac{ds}{s}, \tag{2.8}$$

for the norm of which we have obtained the estimates

$$\|T_1\varphi\|_{E_i^{\alpha_i}} \leqslant k_i e^{|\alpha_i|}\|\varphi\|_{E_i}.$$

From Remark 2.1 and the last assertion we obtain

REMARK 2.2. To obtain a quantity equivalent to the norm of $(A_0, A_1)^{\mathcal{K}}_{E_1^{\alpha_0}, E_1^{\alpha_1}}$, in (2.2) we need consider only representations of the form $x = u_0(t) + u_1(t)$ in which the functions $u_i(t)$ are constant on the intervals $(e^n, e^{n+1}]$ $(n = 0, \pm 1, \pm 2, \ldots)$.

A similar remark holds for the space $(A_0, A_1)^{\mathcal{J}}_{E_1^{\alpha_0}, E_1^{\alpha_1}}$.

THEOREM 2.3. *If the ideal lattices $E_0$ and $E_1$ are interpolation spaces between $L_{1,*}$ and $L_{\infty,*}$, then the spaces $(A_0, A_1)^{\mathcal{K}}_{E_0^{\alpha_0},E_1^{\alpha_1}}$ and $(A_0, A_1)^{\mathcal{J}}_{E_0^{\alpha_0},E_1^{\alpha_1}}$ coincide.*

PROOF. Let $x \in (A_0, A_1)^{\mathcal{J}}_{E_0^{\alpha_0},E_1^{\alpha_1}}$. This means that there exists a representation

$$x = \int_0^\infty u(t)\frac{dt}{t}, \tag{2.9}$$

where $u \in E_0^{\alpha_0}(A_0) \cap E_1^{\alpha_1}(A_1)$. We consider the functions

$$v_0(t) = \int_0^t u(s)\frac{ds}{s} \quad \text{and} \quad v_1(t) = \int_t^\infty u(s)\frac{ds}{s}. \tag{2.10}$$

Then $x = v_0(t) + v_1(t)$. The function $v_0(t)$ assumes its values in $A_0$. Indeed,

$$\begin{aligned}\int_0^t \|u(s)\|_{A_0}\frac{ds}{s} &\leqslant \int_0^t s^{-\alpha_0}(s^{\alpha_0}\|u(s)\|_{A_0})\frac{ds}{s}\\ &\leqslant \|s^{-\alpha_0}\chi_{(0,t]}(s)\|_{E_0^1}\|s^{\alpha_0}\|u(s)\|_{A_0}\|_{E_0}.\end{aligned} \tag{2.11}$$

Since $\alpha_0 < 0$, we have $s^{-\alpha_0}\chi_{(0,t]}(s) \in L_{1,*} \cap L_{\infty,*} \subset E_0^1$. Therefore (2.11) implies that the first integral in (2.10) is convergent in the norm of $A_0$, and consequently $v_0(t) \in A_0$.

It can be verified in a similar way that $v_1(t) \in A_1$.

We show that $v_i \in E_i^{\alpha_i}(A_i)$. Using the Young inequality (2.7), we obtain

$$\begin{aligned}\|t^{\alpha_0}\|v_0(t)\|_{A_0}\|_{E_0} &\leqslant \left\|t^{\alpha_0}\int_0^t \|u(s)\|_{A_0}\frac{ds}{s}\right\|_{E_0}\\ &= \left\|\int_0^\infty \left(\frac{t}{s}\right)^{\alpha_0}\chi_{(1,\infty)}\left(\frac{t}{s}\right)s^{\alpha_0}\|u(s)\|_{A_0}\frac{ds}{s}\right\|_{E_0}\\ &\leqslant k_0\int_0^\infty t^{\alpha_0}\chi_{(1,\infty)}(t)\frac{dt}{t}\|s^{\alpha_0}\|u(s)\|_{A_0}\|_{E_0}\\ &= -\frac{k_0}{\alpha_0}\|u\|_{E_0^{\alpha_0}(A_0)}.\end{aligned} \tag{2.12}$$

A similar check shows that

$$\|t^{\alpha_1}\|v_1(t)\|_{A_1}\|_{E_1} \leqslant \frac{k_1}{\alpha_1}\|u\|_{E_1^{\alpha_1}(A_1)}. \tag{2.13}$$

From the representation $x = v_0(t) + v_1(t)$ and (2.12) and (2.13) it follows that $x \in (A_0, A_1)^{\mathcal{K}}_{E_0^{\alpha_0},E_1^{\alpha_1}}$, and so we have the imbedding

$$(A_0, A_1)^{\mathcal{J}}_{E_0^{\alpha_0},E_1^{\alpha_1}} \subset (A_0, A_1)^{\mathcal{K}}_{E_0^{\alpha_0},E_1^{\alpha_1}}.$$

For the proof of the reverse imbedding we consider $x \in (A_0, A_1)^{\mathcal{K}}_{E_0^{\alpha_0},E_1^{\alpha_1}}$. Then $x = v_0(t) + v_1(t)$, where $t^{\alpha_i}v_i \in E_i(A_i)$ $(i = 0, 1)$. We show that in this representation of $x$ the functions $v_0(t)$ and $v_1(t)$ may be assumed to be smooth. For this we consider an infinitely differentiable nonnegative function $\psi(t)$ on $(0, \infty)$ with compact support such that $\int_0^\infty \psi(t)t^{-1}\,dt = 1$. We set

$$\tilde{v}_i(t) = \int_0^\infty \psi\left(\frac{t}{s}\right) v_i(s)\frac{ds}{s}.$$

The functions $\tilde{v}_i(t)$ assume values in $A_i$, as the functions $v_i(t)$ do, and are infinitely differentiable, and $x = \tilde{v}_0(t) + \tilde{v}_1(t)$. Next,

$$t^{\alpha_i}\|\tilde{v}_i(t)\|_{A_i} \leqslant \int_0^\infty \left(\frac{t}{s}\right)^{\alpha_i} \psi\left(\frac{t}{s}\right) s^{\alpha_i}\|v_i(s)\|_{A_i}\frac{ds}{s}$$

and by the Young inequality (2.7) we have

$$\|\tilde{v}_i\|_{E_i^{\alpha_i}(A_i)} \leqslant \left(\int_0^\infty t^{\alpha_i}\psi(t)\frac{dt}{t}\right)\|v_i\|_{E_i^{\alpha_i}(A_i)}.$$

Now we set

$$u(t) = t\tilde{v}_0'(t) = -t\tilde{v}_1'(t).$$

Then we have

$$t^{\alpha_0}u(t) = t^{\alpha_0+1}\tilde{v}_0'(t) = \int_0^\infty \left(\frac{t}{s}\right)^{\alpha_0+1} \psi'\left(\frac{t}{s}\right) s^{\alpha_0}v_0(s)\frac{ds}{s},$$

and by the Young inequality (2.7)

$$\|u\|_{E_0^{\alpha_0}(A_0)} \leqslant \left(\int_0^\infty t^{\alpha_0+1}\psi'(t)\frac{dt}{t}\right)\|v_0\|_{E_0^{\alpha_0}(A_0)}.$$

Similarly,

$$\|u\|_{E_1^{\alpha_1}(A_1)} \leqslant \left(\int_0^\infty t^{\alpha_1+1}\psi'(t)\frac{dt}{t}\right)\|v_1\|_{E_1^{\alpha_1}(A_1)}.$$

Thus, $u \in E_0^{\alpha_0}(A_0) \cap E_1^{\alpha_1}(A_1)$. Finally,

$$\int_0^t u(s)\frac{ds}{s} = \int_0^t \tilde{v}_0'(s)\,ds = \tilde{v}_0(t) - \tilde{v}_0(+0) = \tilde{v}_0(t), \tag{2.14}$$

$$\int_t^\infty u(s)\frac{ds}{s} = -\int_t^\infty \tilde{v}_1'(s)\,ds = -[\tilde{v}_1(\infty) - \tilde{v}_1(t)] = \tilde{v}_1(t), \tag{2.15}$$

and thus $x = \int_0^\infty u(s)s^{-1}\,ds$. In the calculation of (2.14) and (2.15) we have used the fact that

$$\|\tilde{v}_i(t)\|_{A_i} \leqslant \int_0^\infty \psi\left(\frac{t}{s}\right)\|v_i(s)\|_{A_i}\frac{ds}{s} \leqslant \left\|s^{-\alpha_i}\psi\left(\frac{t}{s}\right)\right\|_{E_i'}\|v_i\|_{E_i^{\alpha_i}(A_i)}.$$

The function $(s/t)^{-\alpha_i}\psi(t/s)$ can be obtained from $s^{\alpha_i}\psi(s)$ by means of the power transformation $T_{-1}$ and the dilation $\sigma_t$. From (2.5) and (2.6) it follows that these operators act in the spaces $E_i'$ and have norm $k_i$ there. Therefore

$$\|\tilde{v}_i(t)\|_{A_i} \leqslant k_i t^{-\alpha_i} \|s^{\alpha_i}\psi(s)\|_{E_i'} \|v_i\|_{E_i^{\alpha_i}(A_i)} \to 0$$

for $i = 0$ as $t \to 0$ and for $i = 1$ as $t \to \infty$.

In what follows, under the hypotheses of Theorem 2.3 we shall omit the letters $\mathcal{K}$ and $\mathcal{J}$ in the notation of our spaces, and, defining them up to isomorphism, we shall write $(A_0, A_1)_{E_0^{\alpha_0}, E_1^{\alpha_1}}$.

It turns out that the number of real parameters describing these spaces may be reduced to one.

LEMMA 2.4. *Under the hypotheses of Theorem 2.3, for $\lambda \neq 0$*

$$(A_0, A_1)_{E_0^{\lambda\alpha_0}, E_1^{\lambda\alpha_1}} = (A_0, A_1)_{E_0^{\alpha_0}, E_1^{\alpha_1}}$$

*up to isomorphism.*

PROOF. If $x \in (A_0, A_1)_{E_0^{\alpha_0}, E_1^{\alpha_1}}$, then $x = v_0(t) + v_1(t)$, where $t^{\alpha_i}v(t) \in E_i(A_i)$. Then $x = v_0(t^\lambda) + v_1(t^\lambda)$, and by (2.6), $t^{\alpha_i\lambda}v_i(t^\lambda) \in E_i(A_i)$. This means that $x \in (A_0, A_1)_{E_0^{\lambda\alpha_0}, E_1^{\lambda\alpha_1}}$. Moreover,

$$\begin{aligned}\|x\|_{(A_0,A_1)_{E_0^{\lambda\alpha_0}, E_1^{\lambda\alpha_1}}} &\leqslant \|\|t^{\alpha_0\lambda}\|v_0(t^\lambda)\|_{A_0}\|_{E_0} + \|\|t^{\alpha_1\lambda}\|v_1(t^\lambda)\|_{A_1}\|_{E_1} \\ &\leqslant \max\left\{1, \frac{k_0}{|\lambda|}, \frac{k_1}{|\lambda|}\right\}\left(\|\|t^{\alpha_0}\|v_0(t)\|_{A_0}\|_{E_0} + \|\|t^{\alpha_1}\|v_1(t)\|_{A_1}\|_{E_1}\right).\end{aligned}$$

From this it follows that

$$\|x\|_{(A_0,A_1)_{E_0^{\lambda\alpha_0}, E_1^{\lambda\alpha_1}}} \leqslant \max\left\{1, \frac{k_0}{|\lambda|}, \frac{k_1}{|\lambda|}\right\}\|x\|_{(A_0,A_1)_{E_0^{\alpha_0}, E_1^{\alpha_1}}}.$$

The reverse inequality with $\lambda$ replaced by $1/\lambda$ can be established in a similar way.

COROLLARY. *If we choose $\lambda = 1/(\alpha_1 - \alpha_0)$ and write $\theta = -\alpha_0/(\alpha_1 - \alpha_0)$, then for $0 < \theta < 1$*

$$(A_0, A_1)_{E_0^{\alpha_0}, E_1^{\alpha_1}} = (A_0, A_1)_{E_0^{-\theta}, E_1^{1-\theta}} \equiv (A_0, A_1)_{\theta, E_0, E_1}$$

*up to isomorphism.*

An important special case is where $E_0$ and $E_1$ coincide. It turns out that the general case may always be reduced to this special case (see [118]). For the spaces $E_i = L_{p_i}$ this will be proved in subsection 7.

We mention yet another property of the spaces $(A_0, A_1)_{\theta, E_0, E_1}$.

LEMMA 2.5. *The inequality*

$$\|x\|_{(A_0,A_1)^{\mathcal{J}}_{E_0^{\alpha_0}, E_1^{\alpha_1}}} \leqslant k_0^{1-\theta}k_1^\theta \inf \|u\|^{1-\theta}_{E_0^{\alpha_0}(A_0)} \|u\|^\theta_{E_1^{\alpha_1}(A_1)}$$

*holds, where $\theta = -\alpha_0/(\alpha_1 - \alpha_0)$ and the infimum is taken over all $u \in E_0^{\alpha_0}(A_0) \cap E_1^{\alpha_1}(A_1)$ yielding the representation* (2.9).

PROOF. Let $u \in E_0^{\alpha_0}(A_0) \cap E_1^{\alpha_1}(A_1)$ and $x = \int_0^\infty u(t)t^{-1}\,dt$. If $\tau > 0$, then the function $u_\tau(t) = u(\tau t)$ belongs to $E_0^{\alpha_0}(A_0) \cap E_1^{\alpha_1}(A_1)$ and $x = \int_0^\infty u_\tau(t)t^{-1}\,dt$.

By virtue of (2.5) we have

$$\|x\|^{\mathfrak{J}}_{E_0^{\alpha_0}, E_1^{\alpha_1}} \leqslant \max\left(\|t^{\alpha_0}u_\tau\|_{E_0(A_0)}, \|t^{\alpha_1}u_\tau\|_{E_1(A_1)}\right)$$
$$\leqslant \max\left(k_0\tau^{-\alpha_0}\|t^{\alpha_0}u\|_{E_0(A_0)}, k_1\tau^{-\alpha_1}\|t^{\alpha_1}u\|_{E_1(A_1)}\right).$$

From this inequality, setting

$$\tau = \left(\frac{k_1\|u\|_{E_1^{\alpha_1}(A_1)}}{k_0\|u\|_{E_0^{\alpha_0}(A_1)}}\right)^{1/(\alpha_1-\alpha_0)},$$

we obtain

$$\|x\|^{\mathfrak{J}}_{E_0^{\alpha_0}, E_1^{\alpha_1}} \leqslant k_0^{1-\theta}k_1^{\theta}\|u\|^{1-\theta}_{E_0^{\alpha_0}(A_0)}\|u\|^{\theta}_{E_1^{\alpha_1}(A_1)}.$$

COROLLARY. *There exists* $C = C(\theta, E_0, E_1)$ *such that*

$$\|x\|_{(A_0,A_1)_{\theta,E_0,E_1}} \leqslant C\|x\|^{1-\theta}_{A_0}\|x\|^{\theta}_{A_1}$$

*for all* $x \in A_0 \cap A_1$.

Indeed, in the representation (2.9) of the element $x$ we may take $u(t) = \psi(t)x$. Then

$$\|x\|_{(A_0,A_1)_{\theta,E_0,E_1}} \leqslant C_1\|t^{\alpha_0}\psi(t)x\|^{1-\theta}_{E_0(A_0)}\|t^{\alpha_1}\psi(t)x\|^{\theta}_{E_1(A_1)}$$
$$\leqslant C_1\max\left(\|t^{\alpha_0}\psi\|^{1-\theta}_{E_0}, \|t^{\alpha_1}\psi\|^{\theta}_{E_1}\right)\|x\|^{1-\theta}_{A_0}\|x\|^{\theta}_{A_1}.$$

Lemma 2.5 implies the following sharpening of the interpolation theorem in our case.

THEOREM 2.4 (interpolation theorem). *Let* $A_0$, $A_1$ *and* $B_0$, $B_1$ *be two Banach couples, and let the lattices* $E_0$ *and* $E_1$ *satisfy the hypotheses of Theorem 2.3. Then the triple* $A_0, A_1, (A_0, A_1)_{\theta,E_0,E_1}$ *is an interpolation triple of type* $\theta$ *relative to the triple* $B_0, B_1, (B_0, B_1)_{\theta,E_0,E_1}$.

PROOF. Let $x \in (A_0, A_1)_{\theta,E_0,E_1}$ and $T \in \pi(A_0, A_1; B_0, B_1)$. From (2.9) we obtain

$$Tx = \int_0^\infty Tu(t)\frac{dt}{t}.$$

By Lemma 2.5 we have

$$\|Tx\|_{(B_0,B_1)_{\theta,E_0,E_1}} \leqslant C\|Tu\|^{1-\theta}_{E_0^{-\theta}(B_0)}\|Tu\|^{\theta}_{E_1^{1-\theta}(B_1)}$$
$$\leqslant C\|T\|^{1-\theta}_{A_0\to B_0}\|T\|^{\theta}_{A_1\to B_1}\|u\|^{1-\theta}_{E_0^{-\theta}(A_0)}\|u\|^{\theta}_{E_1^{1-\theta}(A_1)}$$
$$\leqslant C\|T\|^{1-\theta}_{A_0\to B_0}\|T\|^{\theta}_{A_1\to B_1}\|u\|_{E_0^{-\theta}(A_0)\cap E_1^{1-\theta}(A_1)}.$$

Taking the infimum over $u$ giving representation (2.9), we obtain

$$\|T\|_{(A_0,A_1)_{\theta,E_0,E_1}\to(B_0,B_1)_{\theta,E_0,E_1}} \leqslant C\|T\|_{A_0\to B_0}^{1-\theta}\|T\|_{A_1\to B_1}^{\theta}.$$

REMARK 2.3. As can be seen from the proofs of the assertions in this subsection, the interpolation property of the lattices $E_0$ and $E_1$ has been used only with respect to four concrete operators ((2.5)–(2.8)). The fact that these operators act in $E_0$ and $E_1$ can often be verified immediately, without establishing the general interpolation properties of $E_0$ and $E_1$. In this case all assertions of this subsection remain valid.

**4.** ***Imbedding theorems and the reiteration theorem.*** Let four ideal lattices $E_0$, $E_1$, $F_0$ and $F_1$ be given on the semiaxis $(0, \infty)$ with measure $dt/t$ which are interpolation spaces between $L_{1,*}$ and $L_{\infty,*}$. We assume that there exist two ideal lattices $G_0$ and $G_1$ containing all functions of compact support and such that for the convolution operator we have the inequality

$$\left\|\int_0^\infty \psi\left(\frac{t}{s}\right)\varphi(s)\frac{ds}{s}\right\|_{F_i} \leqslant k_i\|\psi\|_{G_i}\|\varphi\|_{E_i} \tag{2.16}$$

(see Chapter II, §6.9).

THEOREM 2.5 (imbedding theorem). *If condition* (2.16) *is satisfied, then* $(A_0, A_1)_{\theta,E_0,E_1}$ *is imbedded in* $(A_0, A_1)_{\theta,F_0,F_1}$.

PROOF. Let $x \in (A_0, A_1)_{\theta,E_0,E_1}$. By means of a nonnegative smooth function $\psi(t)$ of compact support with $\int_0^\infty \psi(t)\,dt = 1$ we pass from (2.9) to the representation

$$x = \int_0^\infty v(t)\frac{dt}{t}, \quad \text{where } v(t) = \int_0^\infty \psi\left(\frac{t}{s}\right)u(s)\frac{ds}{s}. \tag{2.17}$$

Then

$$\|t^{-\theta}v\|_{F_0(A_0)} \leqslant \left\|\int_0^\infty \left(\frac{t}{s}\right)^{-\theta}\psi\left(\frac{t}{s}\right)s^{-\theta}\|u(s)\|_{A_0}\frac{ds}{s}\right\|_{F_0}$$
$$\leqslant \|t^{-\theta}\psi\|_{G_0}\|s^{-\theta}\|u(s)\|_{A_0}\|_{E_0}.$$

Similarly,

$$\|t^{1-\theta}v\|_{F_1(A_1)} \leqslant \|t^{1-\theta}\psi\|_{G_1}\|s^{1-\theta}\|u(s)\|_{A_1}\|_{E_1}.$$

From these inequalities and (2.17) it follows that

$$\|x\|_{(A_0,A_1)_{\theta,F_0,F_1}} \leqslant \max\left\{\|t^{-\theta}\psi\|_{G_0}, \|t^{1-\theta}\psi\|_{G_1}\right\}\|u\|_{E_0^\theta(A_0)\cap E_1^{1-\theta}(A_1)}.$$

Taking the infimum over all $u(t)$ providing representation (2.9), we obtain the required imbedding:

$$(A_0, A_1)_{\theta,E_0,E_1} \subset (A_0, A_1)_{\theta,F_0,F_1}.$$

COROLLARY.

$$(A_0, A_1)_{\theta,L_{1,*},L_{1,*}} \subset (A_0, A_1)_{\theta,E_0,E_1} \subset (A_0, A_1)_{\theta,L_{\infty,*},L_{\infty,*}}. \tag{2.18}$$

Indeed, the first imbedding follows from the theorem and (2.7), and the second from the fact that inequalities (2.16) hold for $F_i = L_{\infty,*}$ and $G_i = E_i'$.

DEFINITION 2.1. The space $B$ intermediate between $A_0$ and $A_1$ *has type* $\theta$, $0 < \theta < 1$, if

$$(A_0, A_1)_{\theta,L_{1,*},L_{1,*}} \subset B \subset (A_0, A_1)_{\theta,L_{\infty,*},L_{\infty,*}}.$$

From the preceding it follows that the space $(A_0, A_1)_{\theta,E_0,E_1}$ $(0 < \theta < 1)$ has type $\theta$.

LEMMA 2.6. *Let $B$ be an intermediate space between $A_0$ and $A_1$. For $(A_0, A_1)_{\theta,L_{1,*},L_{1,*}}$ to be imbedded in $B$ it is necessary and sufficient that for any $x \in A_0 \cap A_1$*

$$\|x\|_B \leqslant C\|x\|_{A_0}^{1-\theta}\|x\|_{A_1}^{\theta}, \tag{2.19}$$

*where $C$ does not depend on $x$.*

PROOF. Let $B \supset (A_0, A_1)_{\theta,L_{1,*},L_{1,*}}$. Then there exists a constant $C_1$ such that $\|x\|_B \leqslant C_1\|x\|_{(A_0,A_1)_{\theta,L_{1,*},L_{1,*}}}$ for all $x \in (A_0, A_1)_{\theta,L_{1,*},L_{1,*}}$. By the corollary of Lemma 2.4 we have

$$\|x\|_{(A_0,A_1)_{\theta,L_{1,*},L_{1,*}}} \leqslant C_2\|x\|_{A_0}^{1-\theta}\|x\|_{A_1}^{\theta}.$$

Combining the inequalities, we obtain (2.19).

Conversely, let (2.19) be satisfied. If $x \in (A_0, A_1)_{\theta,L_{1,*},L_{1,*}}$, then representation (2.9) holds, where $u \in L_{1,*}^{-\theta}(A_0) \cap L_{1,*}^{1-\theta}(A_1)$. From (2.19) we obtain that $\|u(t)\|_B \leqslant C\|u(t)\|_{A_0}^{1-\theta}\|u(t)\|_{A_1}^{\theta}$. From this and Hölder's inequality it follows that

$$\|x\|_B \leqslant \int_0^\infty \|u(t)\|_B \frac{dt}{t} \leqslant C\int_0^\infty \left(t^{-\theta}\|u(t)\|_{A_0}\right)^{1-\theta}\left(t^{1-\theta}\|u(t)\|_{A_1}\right)^{\theta}\frac{dt}{t}$$
$$\leqslant C\|u\|_{L_{1,*}^{-\theta}(A_0)}^{1-\theta}\|u\|_{L_{1,*}^{1-\theta}(A_1)}^{\theta}.$$

Consequently, $x \in B$ and $\|x\|_B \leqslant C\|x\|_{\theta,L_{1,*},L_{1,*}}$.

LEMMA 2.7. *For a space $B$ intermediate between $A_0$ and $A_1$ to be imbedded in $(A_0, A_1)_{\theta,L_{\infty,*},L_{\infty,*}}$, it is necessary and sufficient that there exist a constant $C$ such that, for every $x \in B$ and $s$, $x_i = x_i(s) \in A_i$ may be chosen so that $x = x_0 + x_1$ and*

$$\|x_0\|_{A_0} \leqslant Cs^{\theta}\|x\|_B, \qquad \|x_1\|_{A_1} \leqslant Cs^{\theta-1}\|x\|_B. \tag{2.20}$$

PROOF. Let $B \subset (A_0, A_1)_{\theta, L_{\infty,*}, L_{\infty,*}}$; consequently there exists a constant $C_1$ such that $\|x\|^{\mathcal{K}}_{\theta, L_{\infty,*}, L_{\infty,*}} \leqslant C_1 \|x\|_B$ $(x \in B)$. As is shown in the proof of Theorem 2.3, there exist smooth functions $v_0(s)$ and $v_1(s)$ such that $x = v_0(s) + v_1(s)$ and

$$\|s^{-\theta} v_0\|_{L_{\infty,*}(A_0)} + \|s^{1-\theta} v_1\|_{L_{\infty,*}(A_1)} \leqslant C_2 \|x\|^{\mathcal{K}}_{\theta, L_{\infty,*}, L_{\infty,*}}.$$

Setting $x_0 = v_0(s)$ and $x_1 = v_1(s)$, from the last two inequalities we obtain that

$$\|x_0\|_{A_0} \leqslant C_1 C_2 s^{\theta} \|x\|_B \quad \text{and} \quad \|x_1\|_{A_1} \leqslant C_1 C_2 s^{\theta - 1} \|x\|_B.$$

Conversely, let conditions (2.20) be satisfied. Let $x \in B$ and $s_0 > 0$. There exist $x_0 \in A_0$ and $x_1 \in A_1$ such that

$$x = x_0 + x_1, \qquad \|x_0\|_{A_0} \leqslant C s_0^{\theta} \|x\|_B, \qquad \|x_1\|_{A_1} \leqslant C s_0^{\theta-1} \|x\|_B.$$

In some interval $I$ containing $s_0$ the inequalities

$$\|x_0\|_{A_0} \leqslant 2Cs^{\theta} \|x\|_B, \qquad \|x_1\|_{A_1} \leqslant 2Cs^{\theta-1} \|x\|_B$$

will be satisfied. It is now clear that $(0, \infty)$ can be partitioned into intervals $I_1, I_2, \ldots$ so that for every $n$ there exist $x_0^n \in A_0$ and $x_1^n \in A_1$ such that $x = x_0^n + x_1^n$ and

$$\|x_0^n\|_{A_0} \leqslant 2Cs^{\theta} \|x\|_B, \qquad \|x_1^n\|_{A_1} \leqslant 2Cs^{\theta-1} \|x\|_B$$

for $s \in I_n$. To complete the proof it remains to set $v_i(s) = x_i^n$ for $s \in I_n$. From this it will follow that

$$x \in (A_0, A_1)_{\theta, L_{\infty,*}, L_{\infty,*}} \quad \text{and} \quad \|x\|_{(A_0, A_1)_{\theta, L_{\infty,*}, L_{\infty,*}}} \leqslant 2C \|x\|_B.$$

REMARK. It is easy to see that in Lemma 2.6 we need only require that $B \supset A_0 \cap A_1$, and in Lemma 2.7 that $B \subset A_0 + A_1$.

THEOREM 2.6. *Let* $0 < \theta_0 < \theta < \theta_1 < 1$. *If the spaces* $(A_0, A_1)_{\theta_i, L_{1,*}, L_{1,*}}$ *are imbedded in the spaces* $B_i$, $i = 0, 1$, *intermediate between* $A_0$ *and* $A_1$, *then*

$$(A_0, A_1)_{\theta, E_0, E_1} \subset (B_0, B_1)_{\theta', F_0, F_1},$$

*where* $\theta' = (\theta - \theta_0)/(\theta_1 - \theta_0)$ *and the* $F_i = E_0^{1-\theta_i} E_1^{\theta_i}$ *are spaces of the Calderón family connecting the lattices* $E_0$ *and* $E_1$ (*see* §1.10).

PROOF. If $x \in (A_0, A_1)_{\theta, E_0, E_1}$, then there exists a representation (2.9) for $x$ with $u \in E_0^{-\theta}(A_0) \cap E_1^{1-\theta}(A_1)$. By Lemma 2.6 we have

$$\|u(t)\|_{B_i} \leqslant C_i \|u(t)\|_{A_0}^{1-\theta_i} \|u(t)\|_{A_1}^{\theta_i}.$$

Hence it follows that

$$t^{-(\theta - \theta_0)} \|u(t)\|_{B_0} \leqslant C_0 \big(t^{-\theta} \|u(t)\|_{A_0}\big)^{1-\theta_0} \big(t^{1-\theta} \|u(t)\|_{A_1}\big)^{\theta_0}$$

for $i = 0$. This inequality implies that the function on the left belongs to the space $F_0 = E_0^{1-\theta_0}E_1^{\theta_0}$ and

$$\|t^{-(\theta-\theta_0)}u(t)\|_{F_0(B_0)} \leqslant C_0\|u\|^{1-\theta_0}_{E_0^{-\theta}(A_0)}\|u\|^{\theta_0}_{E_1^{1-\theta}(A_1)}. \tag{2.21}$$

The inequality

$$\|t^{\theta_1-\theta}u(t)\|_{F_1(B_1)} \leqslant C_1\|u\|^{1-\theta_1}_{E_0^{-\theta}(A_0)}\|u\|^{\theta_1}_{E_1^{1-\theta}(A_1)} \tag{2.22}$$

may be obtained similarly.

From (2.21) and (2.22) it follows that $u \in F_0^{-(\theta-\theta_0)}(B_0) \cap F_1^{\theta_1-\theta}(B_1)$, and since $a^{1-\theta_i}b^{\theta_i} \leqslant \max\{a, b\}$, we have

$$\|u\|_{F_0^{-(\theta-\theta_0)}(B_0)\cap F_1^{\theta_1-\theta}(B_1)} \leqslant \max\{C_0, C_1\}\|u\|_{E_0^{-\theta}(A_0)\cap E_1^{1-\theta}(A_1)}.$$

Thus, $x \in (B_0, B_1)_{F_0^{-(\theta-\theta_0)}, F_1^{\theta_1-\theta}}$. On the other hand, by Lemma 2.4 the last space coincides with $(B_0, B_1)_{\theta', F_0, F_1}$ up to equivalent norms.

THEOREM 2.7. *Let* $0 < \theta_0 < \theta < \theta_1 < 1$. *If some spaces* $B_i$ *intermediate between* $A_0$ *and* $A_1$ *are imbedded in the spaces* $(A_0, A_1)_{\theta_i, L_{\infty,*}, L_{\infty,*}}$ $(i = 0, 1)$, *then*

$$(A_0, A_1)_{\theta, E_0, E_1} \supset (B_0, B_1)_{\theta', F_0, F_1},$$

*where* $\theta' = (\theta - \theta_0)/(\theta_1 - \theta_0)$ *and* $F_i = E_0^{1-\theta_i}E_1^{\theta_i}$ *are spaces of the Calderón family connecting the lattices* $E_0$ *and* $E_1$.

PROOF. Let $x \in (B_0, B_1)_{\theta', F_0, F_1}$. We again use the fact that by Lemma 2.3 we have $x \in (B_0, B_1)_{F_0^{-(\theta-\theta_0)}, F_1^{\theta_1-\theta}}$. Then there exists a representation $x = v_0(t) + v_1(t)$, where $t^{-(\theta-\theta_0)}\|v_0(t)\|_{B_0} \in F_0$ and $t^{\theta_1-\theta}\|v_1(t)\|_{B_1} \in F_1$. According to Remark 2.2 the functions $v_i(t)$ can be assumed to be step functions. Since $v_0(t) \in B_0$, by Lemma 2.7 for every $s > 0$ we have the decomposition $v_0(t) = v_{00}(t, s) + v_{01}(t, s)$, where

$$\|v_{00}(t, s)\|_{A_0} \leqslant C_0 s^{\theta_0}\|v_0(t)\|_{B_0}$$

and

$$\|v_{01}(t, s)\|_{A_1} \leqslant C_0 s^{\theta_0 - 1}\|v_0(t)\|_{B_0}.$$

As we can see from the proof of Lemma 2.7, $v_{00}(t, s)$ and $v_{01}(t, s)$ can be assumed to be step functions in $s$ for every $t$ and, obviously, not depending on $t$ on every interval of constancy of $v_0(t)$. The following estimates hold:

$$t^{-\theta}\|v_{00}(t, s)\|_{A_0} \leqslant C_0 t^{-\theta_0}s^{\theta_0}\left(t^{-(\theta-\theta_0)}\|v_0(t)\|_{B_0}\right), \tag{2.23}$$

$$t^{1-\theta}\|v_{01}(t, s)\|_{A_1} \leqslant C_0 t^{1-\theta_0}s^{\theta_0-1}\left(t^{-(\theta-\theta_0)}\|v_0(t)\|_{B_0}\right). \tag{2.24}$$

The function in parentheses belongs to $F_0 = E_0^{1-\theta_0}E_1^{\theta_0}$, and so there exist functions $\varphi_{00} \in E_0$ and $\varphi_{01} \in E_1$ with $\|\varphi_{0i}\|_{E_i} \leqslant 1$ such that

$$t^{-(\theta-\theta_0)}\|v_0(t)\|_{B_0} \leqslant 2\|v_0\|_{F_0^{-(\theta-\theta_0)}(B_0)}|\varphi_{00}(t)|^{1-\theta_0}|\varphi_{01}(t)|^{\theta_0}.$$

Combining this inequality with (2.23) and (2.24), we obtain

$$t^{-\theta}\|v_{00}(t,s)\|_{A_0} \leqslant 2C_0\|v_0\|_{F_0^{-(\theta-\theta_0)}(B_0)} t^{-\theta_0} s^{\theta_0} |\varphi_{00}(t)|^{1-\theta_0} |\varphi_{01}(t)|^{\theta_0},$$
$$t^{1-\theta}\|v_{01}(t,s)\|_{A_1} \leqslant 2C_0\|v_0\|_{F_0^{-(\theta-\theta_0)}(B_0)} t^{1-\theta_0} s^{\theta_0-1} |\varphi_{00}(t)|^{1-\theta_0} |\varphi_{01}(t)|^{\theta_0}.$$

Now we choose $s = s_0(t)$ from the equality

$$s_0(t) = t|\varphi_{00}(t)|/|\varphi_{01}(t)|.$$

Then

$$t^{-\theta}\|v_{00}(t,s_0(t))\|_{A_0} \leqslant 2C_0\|v_0\|_{F_0^{-(\theta-\theta_0)}(B_0)}|\varphi_{00}(t)|,$$
$$t^{1-\theta}\|v_{01}(t,s_0(t))\|_{A_1} \leqslant 2C_0\|v_0\|_{F_0^{-(\theta-\theta_0)}(B_0)}|\varphi_{01}(t)|.$$

Similarly, $v_1(t)$ can be decomposed into two parts:

$$v_1(t) = v_{10}(t, s_1(t)) + v_{11}(t, s_1(t)),$$

so that

$$t^{-\theta}\|v_{10}(t,s_1(t))\|_{A_0} \leqslant 2C_1\|v_1\|_{F_1^{\theta_1-\theta}(B_1)}|\varphi_{10}(t)|,$$
$$t^{1-\theta}\|v_{11}(t,s_1(t))\|_{A_1} \leqslant 2C_1\|v_1\|_{F_1^{\theta_1-\theta}(B_1)}|\varphi_{11}(t)|,$$

where $\varphi_{10} \in E_0$, $\varphi_{11} \in E_1$ and $\|\varphi_{1i}\|_{E_i} \leqslant 1$.

We set

$$\bar{v}_0(t) = v_{00}(t, s_0(t)) + v_{10}(t, s_1(t)), \qquad \bar{v}_1(t) = v_{01}(t, s_0(t)) + v_{11}(t, s_1(t)).$$

The functions $\bar{v}_0(t)$ and $\bar{v}_1(t)$ are strongly measurable in $A_0$ and $A_1$, respectively, due to the fact that the $v_{ij}(t,s)$ are step functions. By construction, $x = \bar{v}_0(t) + \bar{v}_1(t)$. Finally, by the preceding inequalities we have

$$t^{-\theta}\|\bar{v}_0(t)\|_{A_0} \leqslant 2\max\{C_0, C_1\}\big(\|v_0\|_{F_0^{-(\theta-\theta_0)}(B_0)} + \|v_1\|_{F_1^{\theta_1-\theta}(B_1)}\big) \times \max(|\varphi_{00}(t)|, |\varphi_{10}(t)|),$$
$$t^{1-\theta}\|\bar{v}_1(t)\|_{A_1} \leqslant 2\max\{C_0, C_1\}\big(\|v_0\|_{F_0^{-(\theta-\theta_0)}(B_0)} + \|v_1\|_{F_1^{\theta_1-\theta}(B_1)}\big) \times \max(|\varphi_{01}(t)|, |\varphi_{11}(t)|).$$

The right sides of these inequalities belong to $E_0$ and $E_1$, respectively, and therefore $x \in (A_0, A_1)_{\theta, E_0, E_1}$ and

$$\|x\|_{(A_0,A_1)_{\theta,E_0,E_1}} \leqslant 8\max\{C_0, C_1\}\|x\|_{(B_0,B_1)_{F_0^{-(\theta-\theta_0)},F_1^{\theta_1-\theta}}}.$$

Combining Theorems 2.6 and 2.7, we arrive at the following very important assertion.

THEOREM 2.8 (reiteration theorem). *If the spaces $B_0$ and $B_1$, intermediate between $A_0$ and $A_1$, have types $\theta_0$ and $\theta_1$, respectively* $(0 < \theta_0 < \theta_1 < 1)$, *then* $(A_0, A_1)_{\theta, E_0, E_1}$ *is isomorphic to* $(B_0, B_1)_{\theta', F_0, F_1}$ *for* $\theta_0 < \theta < \theta_1$, *where* $\theta' = (\theta - \theta_0)/(\theta_1 - \theta_0)$ *and* $F_i = E_0^{1-\theta_i} E_1^{\theta_i}$ *are the spaces of the Calderón family connecting* $E_0$ *and* $E_1$.

REMARK 1. The operators (2.5), (2.6), and (2.8) are positive, and the convolution operator (2.7) in the proof of Theorem 2.3 was used in the case where its kernel was positive. $E_0$ and $E_1$ are interpolation spaces relative to $L_{1,*}$ and $L_{\infty,*}$, and therefore the operators (2.5)–(2.8) act in them and also in the spaces $F_0$ and $F_1$, in view of their positivity, according to what we said on p. 246 and in §1.10. By Remark 2.3, Theorem 2.3 holds for these spaces, i.e. the spaces $(B_0, B_1)^{\mathcal{K}}_{\theta', F_0, F_1}$ and $(B_0, B_1)^{\mathcal{J}}_{\theta', F_0, F_1}$ coincide, which fact we used in the proof of Theorems 2.6–2.8.

REMARK 2. The proof of the reiteration theorem also works in the case where $B_0 = A_0$ $(B_1 = A_1)$ if we set $\theta_0 = 0$ $(\theta_1 = 1)$.

The reiteration theorem allows us to describe intermediate interpolation spaces between the spaces $B_0$ and $B_1$ "of complex nature" by means of spaces constructed from the "simpler" spaces $A_0$ and $A_1$. Moreover, it is interesting that for this description we only need to know the type of the spaces $B_0$ and $B_1$.

**5. *Construction of spaces by means of the functionals* $\mathcal{K}(t, x)$ *and* $\mathcal{J}(t, x)$.** In some cases it is more convenient to calculate the norm in intermediate spaces constructed by the $\mathcal{K}$-method or $\mathcal{J}$-method in terms of the functional $\mathcal{K}(t, x)$ defined for $x \in A_0 + A_1$ by

$$\mathcal{K}(t, x) = \inf_{x = x_0 + x_1} \left(\|x_0\|_{A_0} + t\|x_1\|_{A_1}\right), \tag{2.25}$$

or $\mathcal{J}(t, x)$ defined for $x \in A_0 \cap A_1$ by

$$\mathcal{J}(t, x) = \max\left\{\|x\|_{A_0}, t\|x\|_{A_1}\right\}. \tag{2.26}$$

It is obvious that for given $t$ the functionals $\mathcal{K}(t, x)$ and $\mathcal{J}(t, x)$ are equivalent to the norms of $A_0 + A_1$ and $A_0 \cap A_1$, respectively.

For fixed $x$, $\mathcal{K}(t, x)$, as the infimum of functions linear in $t$, will be an increasing continuous concave function of $t$.

Now let $x \in (A_0, A_1)^{\mathcal{K}}_{E, E_{\rho(t)}}$, where $\rho(t)$ is a continuous almost everywhere positive weight. This means by definition that there exists a representation $x = u_0(t) + u_1(t)$, where $u_0 \in E(A_0)$ and $u_1 \in E_{\rho(t)}(A_1)$. Then

$$\begin{aligned}\mathcal{K}(\rho(t), x) &= \inf_{x = x_0 + x_1} \left(\|x_0\|_{A_0} + \rho(t)\|x_1\|_{A_1}\right)\\ &\leqslant \|u_0(t)\|_{A_0} + \rho(t)\|u_1(t)\|_{A_1}.\end{aligned}$$

Both terms on the right belong to $E$, and therefore $\mathcal{K}(\rho(t), x) \in E$ and

$$\|\mathcal{K}(\rho(t), x)\|_E \leqslant \| \, \|u_0(t)\|_{A_0}\|_E + \|\rho(t)\|u_1(t)\|_{A_1}\|_E.$$

Taking the infimum over all representations (2.1) on the right side, we obtain

$$\|\mathcal{K}(\rho(t), x)\|_E \leqslant \|x\|_{(A_0,A_1)^{\mathcal{K}}_{E,E_\rho}}.$$

Conversely, if $\mathcal{K}(\rho(t), x) \in E$, then for a given $t_0$ there are elements $x_0 \in A_0$ and $x_1 \in A_1$ such that $x = x_0 + x_1$ and

$$\|x_0\|_{A_0} + \rho(t_0)\|x_1\|_{A_1} < 2\mathcal{K}(\rho(t_0), x).$$

Since the functions on the right and left are continuous in $t$, the inequality

$$\|x_0\|_{A_0} + \rho(t)\|x_1\|_{A_1} < 2\mathcal{K}(\rho(t), x)$$

holds in a neighborhood $I$ of the point $t_0$. Hence it is clear that $(0, \infty)$ can be partitioned into intervals $I_n$ such that for $t \in I_n$ there exists a decomposition $x = x_0^n + x_1^n$ such that

$$\|x_0^n\|_{A_0} + \rho(t)\|x_1^n\|_{A_1} < 2\mathcal{K}(\rho(t), x).$$

We set $u_0(t) = x_0^n$ and $u_1(t) = x_1^n$ for $t \in I_n$. Then $x = u_0(t) + u_1(t)$, and

$$\|u_0(t)\|_{A_0} + \rho(t)\|u_1(t)\|_{A_1} < 2\mathcal{K}(\rho(t), x).$$

From this we obtain $\|u_0\|_{E(A_0)} \leqslant 2\|\mathcal{K}(\rho(t), x)\|_E$ and $\|u_1\|_{E_\rho(A_1)} < 2\|\mathcal{K}(\rho(t), x)\|_E$, and therefore

$$\|x\|_{(A_0,A_1)_{E,E_\rho}} \leqslant 4\|\mathcal{K}(\rho(t), x)\|_E.$$

We have proved the assertion.

LEMMA 2.8. *The space $(A_0, A_1)^{\mathcal{K}}_{E,E_\rho}$ consists of all elements $x$ for which $\mathcal{K}(\rho(t), x) \in E$, and the norm of the space is equivalent to $\|\mathcal{K}(\rho(t), x)\|_E$.*

We may establish the following lemma similarly.

LEMMA 2.9. *The space $(A_0, A_1)^{\mathcal{J}}_{E,E_\rho}$ consists of all elements $x$ representable in the form* (2.9) *for which $\mathcal{J}(\rho(t), u(t)) \in E$, and the norm of the space is equivalent to $\inf\|\mathcal{J}(\rho(t), u(t))\|_E$, where the infimum is taken over all representations* (2.9).

We leave the verification of this simple fact to the reader.

We note that the space $(A_0, A_1)^{\mathcal{K}}_{E,E_\rho}$ may contain nonzero elements only under the condition that $e(t) \equiv 1 \in E + E_\rho$. This condition is satisfied if $\varphi(t) = \min\{1, \rho(t)\} \in E$. Indeed, $e(t) \equiv \chi_1(t) + \chi_2(t)$, where $\chi_1(t)$ is the characteristic function of the set where $\rho(t) \leqslant 1$. Then $\rho(t)\chi_1(t) \leqslant \varphi(t) \in E$, and $\chi_2(t) \leqslant \varphi(t) \in E$, and therefore $e \in E + E_\rho$. Conversely, if $\varphi \notin E$, then one of the functions $\rho(t)\chi_1(t)$ or $\chi_2(t)$ does not belong to $E$, since $\varphi(t) = \rho(t)\chi_1(t) + \chi_2(t)$. We claim that then one of the functions $\chi_1(t)$ or $\chi_2(t)$ does not belong to $E + E_\rho$. Indeed, if $\chi_2(t) = \mu_1(t) + \mu_2(t)$, $\mu_1 \in E$ and $\mu_2 \in E_\rho$, then we may assume that $\mu_1(t) \geqslant 0$ and $\mu_2(t) \geqslant 0$ and the functions $\mu_1$ and $\mu_2$ have the same support as $\chi_2$. Therefore $\rho\mu_2 \in E$ implies that $\mu_2 \in E$. Then $\chi_2 = \mu_1 + \mu_2 \in E$. If $\chi_1(t) = \nu_1(t) + \nu_2(t)$, $\nu_1 \in E$, $\nu_2 \in E_\rho$, then, under the same assumptions, $\rho\nu_1 \in E$, and so $\rho\chi_1 = \rho\nu_1 + \rho\nu_2 \in E$. Thus, either $\chi_1 \notin E + E_\rho$ or $\chi_2 \notin E + E_\rho$, and therefore $e = \chi_1 + \chi_2 \notin E + E_\rho$.

Thus, *for the space* $(A_0, A_1)^{\mathcal{K}}_{E,E_\rho}$ *to be nontrivial it is necessary that* $\min\{1, \rho(t)\} \in E$.

In the construction of the spaces $(A_0, A_1)^{\mathcal{J}}_{E_0,E_1}$ we have used the fact that $e \in E_0^1 + E_1^1$. Therefore $(A_0, A_1)^{\mathcal{J}}_{E,E_\rho}$ is defined if $e \in E^1 + (E_\rho)^1 = E^1 + E^1_{\rho^{-1}}$, which is equivalent, by virtue of the preceding, to the relation $\min(1, \rho^{-1}(t)) \in E^1$.

Under the assumption that the $\mathcal{K}$- and $\mathcal{J}$-methods coincide, we obtain the following theorem.

THEOREM 2.9. *If $E$ is an ideal lattice which is an interpolation space between $L_{1,*}$ and $L_{\infty,*}$, then $(A_0, A_1)_{\theta,E,E}$ consists of all elements $x$ for which $\mathcal{K}(t, x) \in E^{-\theta}$ and also all elements $x$ representable in the form* (2.9) *with $\mathcal{J}(t, u(t)) \in E^{-\theta}$. The norm of $(A_0, A_1)_{\theta,E}$ is equivalent to*

$$\|\mathcal{K}(t, x)\|_{E^{-\theta}} \quad \text{and} \quad \inf\|\mathcal{J}(t, u(t))\|_{E^{-\theta}},$$

*where the infimum is taken over all representations* (2.9).

We apply this theorem to obtain another version of the reiteration theorem.

THEOREM 2.10. *Let $F_0$, $F_1$, $G_0$ and $G_1$ be ideal lattices which are interpolation spaces between $L_{1,*}$ and $L_{\infty,*}$, and let $B_0 = (A_0, A_1)_{\theta,G_0,G_0}$ and $B_1 = (A_0, A_1)_{\theta,G_1,G_1}$. Then*

$$(B_0, B_1)_{\theta',F_0,F_1} = (A_0, A_1)_{\theta,H,H},$$

*up to isomorphism, where*

$$H = (G_0, G_1)_{\theta',F_0,F_1} \qquad (0 < \theta, \theta' < 1).$$

PROOF. Let $x \in (B_0, B_1)_{\theta',F_0,F_1}$. Then

$$x = \int_0^\infty u(t)\frac{dt}{t}, \qquad t^{-\theta'}\|u(t)\|_{B_0} \in F_0, \qquad t^{1-\theta'}\|u(t)\|_{B_1} \in F_1. \quad (2.27)$$

To calculate the norms in $B_0$ and $B_1$, we apply Theorem 2.9. We obtain

$$t^{-\theta'}\|s^{-\theta}\mathcal{K}(s, u(t))\|_{G_0} \in F_0 \quad \text{and} \quad t^{1-\theta'}\|s^{-\theta}\mathcal{K}(s, u(t))\|_{G_1} \in F_1. \quad (2.28)$$

From (2.9) it follows that

$$s^{-\theta}\mathcal{K}(s, x) \leqslant \int_0^\infty s^{-\theta}\mathcal{K}(s, u(t))\frac{dt}{t}.$$

By virtue of (2.28), the function on the right belongs to $H = (G_0, G_1)_{\theta', F_0, F_1}$, and therefore so does $s^{-\theta}\mathcal{K}(s, x)$. This implies by Theorem 2.9 that $x \in (A_0, A_1)_{\theta, H, H}$.

Conversely, if $x \in (A_0, A_1)_{\theta, H, H}$, then by Theorem 2.9 we have the representation

$$x = \int_0^\infty u(s)\frac{ds}{s}, \qquad s^{-\theta}\mathcal{J}(s, u(s)) \in H = (G_0, G_1)_{\theta', F_0, F_1}. \tag{2.29}$$

Then by the definition of $H$ we may construct the decomposition

$$\begin{gathered} s^{-\theta}\mathcal{J}(s, u(s)) = s^{-\theta}\varphi_0(s, t) + s^{-\theta}\varphi_1(s, t), \\ t^{-\theta'}\|s^{-\theta}\varphi_0(s, t)\|_{G_0} \in F_0, \qquad t^{1-\theta'}\|s^{-\theta}\varphi_1(s, t)\|_{G_1} \in F_1. \end{gathered} \tag{2.30}$$

We set

$$v_0(t) = \int_0^\infty \frac{\varphi_0(s, t)}{\mathcal{J}(s, u(s))}u(s)\frac{ds}{s}, \qquad v_1(t) = \int_0^\infty \frac{\varphi_1(s, t)}{\mathcal{J}(s, u(s))}u(s)\frac{ds}{s}. \tag{2.31}$$

Using Theorem 2.9 again, we calculate the norm of $v_0(t)$ and $v_1(t)$ in $B_0$ and $B_1$. We have

$$\|v_0(t)\|_{B_0} \leqslant C_0\left\|s^{-\theta}\mathcal{J}\left(\frac{\varphi_0(s, t)}{\mathcal{J}(s, u(s))}u(s)\right)\right\|_{G_0} = C_0\|s^{-\theta}\varphi_0(s, t)\|_{G_0},$$

$$\|v_1(t)\|_{B_1} \leqslant C_1\left\|s^{-\theta}\mathcal{J}\left(\frac{\varphi_1(s, t)}{\mathcal{J}(s, u(s))}u(s)\right)\right\|_{G_1} = C_1\|s^{-\theta}\varphi_1(s, t)\|_{G_1}.$$

Then (2.30) implies that $t^{-\theta'}\|v_0(t)\|_{B_0} \in F_0$, $t^{1-\theta'}\|v_1(t)\|_{B_1} \in F_1$. Finally, by (2.29)–(2.31) we have $x = v_0(t) + v_1(t)$, i.e. $x \in (B_0, B_1)_{\theta', F_0, F_1}$.

**6. *Intermediate spaces between*** $L_1$ ***and*** $L_\infty$. As in the main part of Chapter II, we shall consider the spaces $L_1(0, \infty)$ and $L_\infty(0, \infty)$ with Lebesgue measure. We construct the space $(L_1, L_\infty)_{\theta, L_{p,*}, L_{p,*}}$. According to Theorem 2.9 we may calculate an equivalent norm in this space using the functional $\mathcal{K}(t, x)$. For this functional in Chapter II, §3.1 we saw that

$$\mathcal{K}(t, x) = \int_0^t x^*(\tau)\,d\tau = tx^{**}(t).$$

Then for the norm we obtain

$$\|x\|_{(L_1, L_\infty)_{\theta, L_{p,*}, L_{p,*}}} = \left\{\int_0^\infty (t^{1-\theta}x^{**}(t))^p\frac{dt}{t}\right\}^{1/p}.$$

Up to a constant we obtain the norm of the spaces considered in Chapter II, §6.8, with $1/r = 1 - \theta$. Thus, up to isomorphism we have

$$(L_1, L_\infty)_{1/r', L_{p,*}, L_{p,*}} = L_{r,p} \qquad (1 < r < \infty,\ 1 \leqslant p \leqslant \infty). \tag{2.32}$$

In particular cases we obtain

$$(L_1, L_\infty)_{1/p', L_{p,*}, L_{p,*}} = L_{p,p} = L_p,$$

$$(L_1, L_\infty)_{1/r', L_{\infty,*}, L_{\infty,*}} = M_{t^{1/r}} \quad \text{(Marcinkiewicz spaces)},$$

$$(L_1, L_\infty)_{1/r', L_{1,*}, L_{1,*}} = \Lambda_{t^{1/r}} \quad \text{(Lorentz spaces)}.$$

From (2.32) it follows that $L_{r,p}$ has type $1/r'$ with respect to $L_1$ and $L_\infty$. We consider two numbers $r_0$ and $r_1$, $1 < r_0 < r_1 < \infty$, and apply the reiteration Theorem 2.8 to $L_{r_0,p_0}$ and $L_{r_1,p_1}$. Then for $1/r_0' < \theta < 1/r_1'$ the space $(L_1, L_\infty)_{\theta, L_{p,*}, L_{p,*}}$ is isomorphic to $(L_{r_0,p_0}, L_{r_1,p_1})_{\theta', L_{p,*}, L_{p,*}}$, where $\theta' = (\theta - 1/r_0')/(1/r_1' - 1/r_0')$. Taking account of (2.32), up to isomorphism we obtain

$$(L_{r_0,p_0}, L_{r_1,p_1})_{\theta', L_{p,*}, L_{p,*}} = L_{r,p}, \quad \text{where } \frac{1}{r} = \frac{1-\theta'}{r_0} + \frac{\theta'}{r_1}. \tag{2.33}$$

Using the interpolation Theorem 2.4, by means of the relations obtained so far we may derive a number of concrete interpolation theorems. For example, setting $r_0 = p_0 = 1/\alpha_0$, $r_1 = p_1 = 1/\alpha_1$ and $r = p = 1/\alpha$, where $\alpha = (1-\theta')\alpha_0 + \theta'\alpha_1$, and then $r_0 = p_0 = 1/\beta_0$, $r_1 = p_1 = 1/\beta_1$ and $r = 1/\beta$, where $\beta = (1-\theta')\beta_0 + \theta'\beta_1$, we obtain the triples $(L_{1/\alpha_0}, L_{1/\alpha_1}, L_{1/\alpha})$ and $(L_{1/\beta_0}, L_{1/\beta_1}, L_{1/\beta, 1/\alpha})$ mentioned in Theorem 6.14 of Chapter II, which are interpolation triples relative to each other.

Setting $r_0 = p_0 = 1/\alpha_0$, $r_1 = p_1 = 1/\alpha_1$ and $r = p = 1/\alpha$ again, where $\alpha = (1-\theta')\alpha_0 + \theta'\alpha_1$, and then $r_0 = 1/\beta_0$, $p_0 = \infty$, $r_1 = 1/\beta_1$, $p_1 = \infty$ and $r = 1/\beta$, where $\beta = (1-\theta')\beta_0 + \theta'\beta_1$, we obtain that $(L_{1/\alpha_0}, L_{1/\alpha_1}, L_{1/\alpha})$ is an interpolation triple of type $\theta'$ relative to the triple $(M_{t^{1-1/\beta_0}}, M_{t^{1-1/\beta_1}}, L_{1/\beta,1/\alpha})$. If we now assume that $\beta_i \leqslant \alpha_i$, then $1/\beta > 1/\alpha$, and $L_{1/\beta,1/\alpha}$ is imbedded in $L_{1/\beta,1/\beta} = L_{1/\beta}$. We have arrived at the following assertion.

MARCINKIEWICZ'S THEOREM. *If* $0 < \beta_i \leqslant \alpha_i < 1$, *then the triple* $(L_{1/\alpha_0}, L_{1/\alpha_1}, L_{1/\alpha})$ *of spaces is an interpolation triple of type* $\theta$ *relative to the triple* $(M_{t^{1-1/\beta_0}}, M_{t^{1-1/\beta_1}}, L_{1/\beta})$, *where*

$$\alpha = (1-\theta)\alpha_0 + \theta\alpha_1, \qquad \beta = (1-\theta)\beta_0 + \theta\beta_1.$$

As we have seen above, a sharper assertion holds in which $L_{1/\beta}$ must be replaced by $L_{1/\beta,1/\alpha}$.

7. ***Intermediate spaces between abstract $L^p$ spaces with weight.*** In this subsection we consider a more general situation than in the preceding one. Let $A_0$ and $A_1$ be a Banach couple. We introduce the spaces $L_{p_0}^{w_0}(A_0)$ and $L_{p_1}^{w_1}(A_1)$ of functions defined on the semiaxis $(0, \infty)$ with values in $A_0$ and $A_1$, where $w_1(t)$ and $w_2(t)$ are weights—in other words, positive measurable functions on $(0, \infty)$.

THEOREM 2.11. *If at least one of the indices $p_0$ or $p_1$ is finite, then*

$$(L_{p_0}^{w_0}(A_0), L_{p_1}^{w_1}(A_1))_{\theta, L_{p_0*}, L_{p_1,*}} = L_p^{w_0^{1-\theta} w_1^\theta}((A_0, A_1)_{\theta, L_{p_0,*}, L_{p_1,*}})$$

*up to isomorphism, where* $1/p = (1-\theta)/p_0 + \theta/p_1$.

PROOF. Let $x(s)$ belong to $(L_{p_0}^{w_0}(A_0), L_{p_1}^{w_1}(A_1))_{\theta, L_{p_0,*}, L_{p_1,*}}$. Then the representation

$$x = \int_0^\infty u(t)\frac{dt}{t}$$

holds, where $u \in L_{p_0,*}^{-\theta}(L_{p_0}^{w_0}(A_0)) \cap L_{p_1,*}^{1-\theta}(L_{p_1}^{w_1}(A_1))$, $u(t)$ is constant in each of the intervals $(e^n, e^{n+1})$, $n = 0, \pm 1, \pm 2, \ldots$, and the integral is taken in $L_{p_0}^{w_0}(A_0) + L_{p_1}^{w_1}(A_1)$. For every $t > 0$ the abstract function $u(t)$ is a function from $L_{p_0}^{w_0}(A_0) \cap L_{p_1}^{w_1}(A_1)$; we denote it by $u(s, t)$ and assume that $u(s, t) \equiv u(s, t')$ for $e^n < t, t' \leqslant e^{n+1}$ $(n = 0, \pm 1, \ldots)$. For a given $s$, $u(s, t)$ is a step function in $t$ (and consequently, strongly measurable in $A_0 \cap A_1$) and, in addition, belongs to $L_{p_0,*}^{-\theta}(A_0) \cap L_{p_1,*}^{1-\theta}(A_1)$ (as we shall see below when applying Fubini's theorem). Hence in $A_0 + A_1$ the integral $\int_0^\infty u(s, t)t^{-1}\,dt$ exists and is an element of $(A_0, A_1)_{\theta, L_{p_0,*}, L_{p_1,*}}$. Next, since $\int_{e^{-n}}^{e^n} u(t)t^{-1}\,dt$ and $\int_{e^{-n}}^{e^n} u(s, t)t^{-1}\,dt$ obviously coincide as functions of $s$, the sequence of the functions

$$x_n(s) = \int_{e^{-n}}^{e^n} u(s, t)\frac{dt}{t} \qquad (n = 1, 2, \ldots)$$

converges to $x(s)$ in $L_{p_0}^{w_0}(A_0) + L_{p_1}^{w_1}(A_1)$. This implies the equality

$$x(s) = \int_0^\infty u(s, t)\frac{dt}{t}.$$

Now from Lemma 2.5 we have

$$\|x(s)\|_{(A_0,A_1)_{\theta,L_{p_0,*},L_{p_1,*}}} \leqslant C\left\{\int_0^\infty \left(t^{-\theta}\|u(s, t)\|_{A_0}\right)^{p_0}\frac{dt}{t}\right\}^{(1-\theta)/p_0} \times \left\{\int_0^\infty \left(t^{1-\theta}\|u(s, t)\|_{A_1}\right)^{p_1}\frac{dt}{t}\right\}^{\theta/p_1}.$$

We note that $x(s)$ is strongly measurable in $(A_0, A_1)_{\theta, L_{p_0,*}, L_{p_1,*}}$. This is a consequence of the fact that the sequence of the functions $x_n(s)$, strongly measurable in $A_0 \cap A_1$, converges to $x(s)$ almost everywhere in $(A_0, A_1)_{\theta, L_{p_0,*}, L_{p_1,*}}$ in view of the inequality

$$\|x(s) - x_n(s)\|_{(A_0,A_1)_{\theta,L_{p_0,*},L_{p_1,*}}}$$

$$\leqslant C\|(1 - \chi_{(e^{-n},e^n)})u(s, \cdot)\|_{L^{-\theta}_{p_0,*}(A_0)}\|(1 - \chi_{(e^{-n},e^n)})u(s, \cdot)\|_{L^{1-\theta}_{p_1,*}(A_1)}$$

and the fact that one of the indices $p_0$ or $p_1$ is finite. Next,

$$(w_0^{1-\theta}(s)w_1^{\theta}(s)\|x(s)\|_{(A_0,A_1)_{\theta,L_{p_0,*},L_{p_1,*}}})^p$$

$$\leqslant C^p\left\{\int_0^\infty \left(t^{-\theta}w_0(s)\|u(s, t)\|_{A_0}\right)^{p_0}\frac{dt}{t}\right\}^{(1-\theta)p/p_0}$$

$$\times \left\{\int_0^\infty \left(t^{1-\theta}w_1(s)\|u(s, t)\|_{A_1}\right)^{p_1}\frac{dt}{t}\right\}^{\theta p/p_1}.$$

Integrating both sides of this inequality with respect to $s$ and applying Hölder's inequality and Fubini's theorem on the right side (the function $\|u(s, t)\|_{A_i}$ is obviously jointly measurable in all variables), we obtain

$$\|x\|_{L_p^{w_0^{1-\theta}w_1^{\theta}}((A_0,A_1)_{\theta,L_{p_0,*},L_{p_1,*}})}$$

$$\leqslant C\left\{\int_0^\infty \int \left(t^{-\theta}w_0(s)\|u(s, t)\|_{A_1}\right)^{p_0} ds\frac{dt}{t}\right\}^{(1-\theta)/p_0}$$

$$\times \left\{\int_0^\infty \int \left(t^{1-\theta}w_1(s)\|u(s, t)\|_{A_1}\right)^{p_1} ds\frac{dt}{t}\right\}^{\theta/p_1}$$

$$= C\|u\|^{1-\theta}_{L^{-\theta}_{p_0,*}(L_{p_0}^{w_0}(A_0))}\|u\|^{\theta}_{L^{1-\theta}_{p_1,*}(L_{p_1}^{w_1}(A_1))}.$$

Conversely, let

$$x \in L_p^{w_0^{1-\theta}w_1^{\theta}}((A_0, A_1)_{\theta,L_{p_0,*},L_{p_1,*}}).$$

Let the elementary functions $\bar{w}_0(s)$ and $\bar{w}_1(s)$ be such that $\bar{w}_i(s) \leqslant w_i(s) \leqslant 2\bar{w}_i(s)$. We shall assume that $x(s)$ is strictly simple and has support contained in the union of a finite number of sets on which both functions $\bar{w}_0(s)$ and $\bar{w}_1(s)$ are constant.

For every $s$ we have the representation

$$x(s) = \int_0^\infty u(s, t)\frac{dt}{t},$$

where $u(s, \cdot) \in L_{p_0,*}^{-\theta}(A_0) \cap L_{p_1,*}^{1-\theta}(A_1)$ and $u(s, \cdot)$ is constant for $e^n < t \leqslant e^{n+1}$, $n = 0, \pm 1, \ldots$. We may assume, first, that if $x(s) = x(s')$, then $u(s, t) \equiv u(s', t)$, and second, that

$$\max\left( \|t^{-\theta} \|u(s, t)\|_{A_0}\|_{L_{p_0*}}, \|t^{1-\theta}\|u(s, t)\|_{A_1}\|_{L_{p_1*}} \right) \leqslant 2\|x(s)\|_{(A_0,A_1)_{\theta,L_{p_0,*},L_{p_1,*}}}.$$

We consider the function

$$v(s, t) = u\left(s, \overline{w}_0^{\lambda}(s)\overline{w}_1^{\mu}(s)\|x(s)\|^{\nu}_{(A_0,A_1)_{\theta,L_{p_0*},L_{p_1*}}} t\right),$$

where the exponents $\lambda$, $\mu$, and $\nu$ will be chosen later ($v(s, t) = 0$ when $x(s) = 0$). For each fixed $t$ the function $v(s, t)$ is a strictly simple function of $s$ with values in $A_0 \cap A_1$. It is also easy to see that $v$ as an abstract function of $t$ with range in the space of strongly measurable functions of $s$ assuming values in $A_0 \cap A_1$ is a step fu. ction. Indeed, the support of $x(s)$ can be written in the form $\bigcup_1^N M_n$, where the measurable sets $M_n$ are disjoint and the functions $\overline{w}_0(s)$ and $\overline{w}_1(s)$ are constant on every $M_n$. Setting

$$r_n = \overline{w}_0^{\lambda}(s)\overline{w}_1^{\mu}(s)\|x(s)\|^{\nu}_{(A_0,A_1)_{\theta,L_{p_0*},L_{p_1*}}} \quad \text{and} \quad u_n(t) = u(s, r_n t),$$

for $s \in M_n$ we have the equality

$$v(s, t) = \sum_{n=1}^{N} u_n(t)\chi_{M_n}(s).$$

Now the required assertion follows from the fact that $u_n(t)$, $n = 1, \ldots, N$, is constant in the intervals $e^m r_n^{-1} < t \leqslant e^{m+1} r_n^{-1}$, $m = 0, \pm 1, \ldots$. Next,

$$\begin{aligned}
\|v\|^{p_0}_{L_{p_0,*}(L_{p_0}^{w_0}(A_0))} &= \int_0^{\infty} \int \left(t^{-\theta} w_0(s)\|v(s, t)\|_{A_0}\right)^{p_0} ds \frac{dt}{t} \\
&= \int \left( w_0(s)\overline{w}_0^{\lambda\theta}(s)\overline{w}_1^{\mu\theta}(s)\|x(s)\|^{\nu\theta}_{(A_0,A_1)_{\theta,L_{p_0*},L_{p_1*}}} \right)^{p_0} \\
&\quad \times \int_0^{\infty} \left(t^{-\theta}\|u(s, t)\|_{A_0}\right)^{p_0} \frac{dt}{t}\, ds \\
&\leqslant 2^{(|\lambda|+|\mu|)\theta p_0 + p_0} \int w_0^{(1+\lambda\theta)p_0}(s) w_1^{\mu\theta p_0}(s)\|x(s)\|^{(\nu\theta+1)p_0}_{(A_0,A_1)_{\theta,L_{p_0*},L_{p_1*}}} ds.
\end{aligned}$$

Similarly,

$$\begin{aligned}
\|v\|^{p_1}_{L_{p_1,*}^{1-\theta}(L_{p_0}^{w_1}(A_0))} &\leqslant 2^{(|\lambda|+|\mu|)(1-\theta)p_1+p_1} \int w_0^{-\lambda(1-\theta)p_1}(s) \\
&\quad \times \|x(s)\|^{-\nu(1-\theta)p_1+p_1}_{(A_0,A_1)_{\theta,L_{p_0*},L_{p_1*}}} ds.
\end{aligned}$$

It is natural to try to choose $\lambda$, $\mu$, and $\nu$ so that

$$(1+\lambda\theta)p_0 = (1-\theta)p, \qquad \mu\theta p_0 = \theta p, \qquad (\nu\theta + 1)p_0 = p$$

and

$$-\lambda(1-\theta)p_1 = (1-\theta)p, \qquad -\mu(1-\theta)p_1 + p_1 = \theta p,$$
$$-\nu(1-\theta)p_1 + p_1 = p.$$

It turns out that this can be done: $\lambda = -pp_1^{-1}$, $\mu = pp_0^{-1}$ and $\nu = p(p_0^{-1} - p_1^{-1})$.

Thus, in $L_{p_0}^{w_0}(A_0) + L_{p_1}^{w_1}(A_1)$ the integral of $v$ exists and is an element $x$ of the space $(L_{p_0}^{w_0}(A_0), L_{p_1}^{w_1}(A_1))_{\theta, L_{p_0,*}, L_{p_1,*}}$. On the other hand, since $\int_0^\infty v(s,t)t^{-1}\,dt = x(s)$ in $A_0 + A_1$, we have $\bar{x} = x$. Besides,

$$\begin{aligned}
\|x\|_{(L_{p_0}^{w_0}(A_0), L_{p_1}^{w_1}(A_1))_{\theta, L_{p_0,*}, L_{p_1,*}}} &\le C'\|v\|^{1-\theta}_{L^{1-\theta}_{p_0,*}(L_{p_0}^{w_0}(A_0))}\|v\|^{\theta}_{L^{1-\theta}_{p_1,*}(L_{p_1}^{w_1}(A_1))} \\
&\le C'\|x\|^{(1-\theta)p/p_0}_{L_p^{w_0^{1-\theta}w_1^{\theta}}((A_0,A_1)_{\theta,L_{p_0,*},L_{p_1,*}})} \\
&\quad\times \|x\|_{L_p^{w_0^{1-\theta}w_1^{\theta}}((A_0,A_1)_{\theta,L_{p_0,*},L_{p_1,*}})} \\
&= C''\|x\|_{L_p^{w_0^{1-\theta}w_1^{\theta}}((A_0,A_1)_{\theta,L_{p_0,*},L_{p_1,*}})}.
\end{aligned}$$

Finally, it can be seen easily that the strictly simple functions $x$ of the form considered are dense in the space

$$L_p^{w_0^{1-\theta}w_1^{\theta}}\big((A_0, A_1)_{\theta, L_{p_0,*}, L_{p_1,*}}\big).$$

This follows from the regularity of $L_p^{w_0^{1-\theta}w_1^{\theta}}$ ($p < \infty$) and Lemma 5 in the Introduction.

Consequently, we have the continuous imbedding

$$L_p^{w_0^{1-\theta}w_1^{\theta}}\big((A_0, A_1)_{\theta, L_{p_0,*}, L_{p_1,*}}\big) \subset (L_{p_0}^{w_0}(A_0), L_{p_1}^{w_1}(A_1))_{\theta, L_{p_0,*}, L_{p_1,*}}.$$

COROLLARY. *Up to isomorphism,*

$$\big(L_{p_0}(A_0), L_{p_1}(A_1)\big)_{\theta, L_{p_0,*}, L_{p_1,*}} = L_p\big((A_0, A_1)_{\theta, L_{p_0,*}, L_{p_1,*}}\big),$$

*where* $1/p = (1-\theta)/p_0 + \theta/p_1$.

This corollary implies, for example, the following interpolation theorem.

THEOREM 2.12. *For* $1/p = (1-\theta)/p_0 + \theta/p_1$ *the triple*

$$\Big(L_{p_0}(A_0), L_{p_1}(A_1), L_p\big((A_0, A_1)_{\theta, L_{p_0,*}, L_{p_1,*}}\big)\Big)$$

*of spaces is an interpolation triple of type* $\theta$ *relative to the triple* $(L_{p_0}(B_0), L_{p_1}(B_1), L_p((B_0, B_1)_{\theta, L_{p_0,*}, L_{p_1,*}}))$, *where* $A_0$, $A_1$ *and* $B_0$, $B_1$ *are arbitrary Banach couples.*

Theorem 2.11 permits us to simplify the construction of the spaces $(A_0, A_1)_{\theta, L_{p_0,*}, L_{p_1,*}}$.

THEOREM 2.13. *Up to isomorphism,*

$$(A_0, A_1)_{\theta, L_{p_0,*}, L_{p_1,*}} = (A_0, A_1)_{\theta, L_{p,*}, L_{p,*}},$$

*where* $1/p = (1-\theta)/p_0 + \theta/p_1$.

PROOF. We choose $\theta_0$ and $\theta_1$ so that $0 < \theta_0 < \theta < \theta_1 < 1$. We write $(1-\theta_i)/p_0 + \theta_i/p_1 = 1/\rho_i$ and introduce the spaces $(A_0, A_1)_{\theta_i, L_{\rho_i,*}, L_{\rho_i,*}}$ having types $\theta_i$ according to the corollary to Theorem 2.5.

By the reiteration Theorem 2.8 we have

$$(A_0, A_1)_{\theta, L_{p_0,*}, L_{p_1,*}} = \left((A_0, A_1)_{\theta_0, L_{\rho_0,*}, L_{\rho_0,*}}, (A_0, A_1)_{\theta_1, L_{\rho_1,*}, L_{\rho_1,*}}\right)_{\theta', F_0, F_1}, \tag{2.37}$$

where $\theta' = (\theta - \theta_0)/(\theta_1 - \theta_0)$, $F_i = (L_{p_0,*})^{1-\theta_i}(L_{p_i,*})^{\theta_i} = L_{\rho_i,*}$.

Now let $x \in (A_0, A_1)_{\theta, L_{p_0,*}, L_{p_1,*}}$. Then by (2.37) and Theorem 2.9 we have

$$x = \int_0^\infty u(t)\frac{dt}{t}, \qquad t^{-\theta'}s^{-\theta_0}\mathcal{K}(s, u(t)) \in L_{\rho_0,*}(L_{\rho_0,*})$$

$$t^{1-\theta'}s^{-\theta_1}\mathcal{K}(s, u(t)) \in L_{\rho_1,*}(L_{\rho_1,*}). \tag{2.38}$$

Hence

$$\mathcal{K}(s, x) \leqslant \int_0^\infty \mathcal{K}(s, u(t))\frac{dt}{t}.$$

By virtue of (2.38), the function on the right, and so also the function $\mathcal{K}(s, x)$, belongs to $(L_{\rho_0,*}^{-\theta_0}, L_{\rho_1,*}^{-\theta_1})_{\theta', L_{\rho_0,*}, L_{\rho_0,*}}$, which is isomorphic to $L_\rho^{-\theta''}$ by Theorem 2.11, where $\theta'' = (1-\theta')\theta_0 + \theta'\theta_1 = \theta$ and

$$\frac{1}{\rho} = \frac{1-\theta'}{\rho_0} + \frac{\theta'}{\rho_1} = (1-\theta')\left(\frac{1-\theta_0}{p_0} + \frac{\theta_0}{p_1}\right) + \theta'\left(\frac{1-\theta_1}{p_0} + \frac{\theta_1}{p_1}\right)$$

$$= \frac{1-\theta}{p_0} + \frac{\theta}{p_1} = \frac{1}{p}.$$

Thus, $\mathcal{K}(s, x) \in L_p^{-\theta}$, and hence $x \in (A_0, A_1)_{\theta, L_{p,*}, L_{p,*}}$.

Conversely, from the last membership relation it follows that

$$x = \int_0^\infty u(s)\frac{ds}{s}, \quad \text{where } s^{-\theta}\mathcal{J}(s, u(s)) \in L_{p,*}.$$

Taking into account that $L_{p,*}^{-\theta} = (L_{\rho_0,*}^{-\theta_0}, L_{\rho_1,*}^{-\theta_1})_{\theta', L_{p_0,*}, L_{p_1,*}}$, we obtain the existence of a decomposition

$$s^{-\theta}\mathcal{J}(s, u(s)) = s^{-\theta}\varphi_0(s, t) + s^{-\theta}\varphi_1(s, t),$$

where $t^{-\theta'}s^{-\theta_0}\varphi_0(s,t) \in L_{p_0,*}(L_{p_0,*})$ and $t^{1-\theta'}s^{-\theta_1}\varphi_1(s,t) \in L_{p_1,*}(L_{p_1,*})$. The functions $\varphi_0(s,t)$ and $\varphi_1(s,t)$ may be assumed nonnegative. We set

$$v_i(t) = \int_0^\infty \frac{\varphi_i(s,t)}{\mathcal{J}(s,u(s))} u(s)\frac{ds}{s}.$$

Then

$$v_0(t) + v_1(t) = \int_0^\infty u(s)\frac{ds}{s} = x. \tag{2.39}$$

Next,

$$t^{-\theta'}s^{-\theta_0}\mathcal{J}\left(s, \frac{\varphi_0(s,t)}{\mathcal{J}(s,u(s))}u(s)\right) = t^{-\theta'}s^{-\theta_0}\varphi_0(s,t) \in L_{p_0,*}(L_{p_0,*}),$$

$$t^{1-\theta'}s^{-\theta_1}\mathcal{J}\left(s, \frac{\varphi_1(s,t)}{\mathcal{J}(s,u(s))}u(s)\right) = t^{1-\theta'}s\varphi_1(s,t) \in L_{p_1,*}(L_{p_1,*}),$$

and so by Theorem 2.9 we have

$$\|v_0(t)\|_{(A_0,A_1)_{\theta_0,L_{p_0,*},L_{p_0,*}}} \in L^{-\theta'}_{p_0,*}$$

and

$$\|v_1(t)\|_{(A_0,A_1)_{\theta_1,L_{p_1,*},L_{p_1,*}}} \in L^{1-\theta'}_{p_1,*}.$$

By (2.39) and (2.37) this means that $x \in (A_0, A_1)_{\theta,L_{p_0,*},L_{p_1,*}}$.

**8. *The interpolation functor*** $(A_0, A_1)_{\theta,p}$. In applications we most often encounter the interpolation functor $(A_0, A_1)_{\theta,L_{p,*},L_{p,*}}$ $(0 < \theta < 1)$, $1 \leqslant p \leqslant \infty$, which is denoted briefly by $(A_0, A_1)_{\theta,p}$. For the convenience of the reader we give the main properties of this functor proved in more general situations earlier.

1°. Up to isomorphism,

$$(A_0, A_1)^{\mathcal{K}}_{\theta,L_{p_0,*},L_{p_1,*}} = (A_0, A_1)^{\mathcal{J}}_{\theta,L_{p_0,*},L_{p_1,*}} = (A_0, A_1)_{\theta,p},$$

where $1/p = (1 - \theta)/p_0 + \theta/p_0$ (Theorems 2.3 and 2.13).

2°. The space $A_0 \cap A_1$ is densely imbedded in $(A_0, A_1)_{\theta,q}$ for $1 \leqslant p < \infty$ (see Lemma 2.14, below).

3°. For $p < q$ the imbedding $(A_0, A_1)_{\theta,p} \subset (A_0, A_1)_{\theta,q}$ holds (Theorem 2.5 and Young's inequality).

4°. If the spaces $B_i$ have types $\theta_i$ $(\theta_0 < \theta_1)$, then for $\theta_0 < \theta < \theta_1$

$$(B_0, B_1)_{\theta',p} = (A_0, A_1)_{\theta,p},$$

up to isomorphism, where $\theta' = (\theta - \theta_0)/(\theta_1 - \theta_0)$ (Theorem 2.8).

Moreover, $B$ has type $\theta$ if

$$(A_0, A_1)_{\theta,1} \subset B \subset (A_0, A_1)_{\theta,\infty}.$$

The space $(A_0, A_1)_{\theta,p}$ has type $\theta$ by virtue of 3°.

5°. The equivalent norms in the space $(A_0, A_1)_{\theta,p}$ may be calculated as $\|t^{-\theta}\mathcal{K}(t, x)\|_{L_{p,*}}$ or $\inf\|\mathcal{J}(t, u(t))\|_{L_{p,*}}$, where the infimum is taken over all representations (2.9) (Theorem 2.9).

6°. Up to isomorphism,

$$((A_0, A_1)_{\theta,p_0}, (A_0, A_1)_{\theta,p_1})_{\theta',p} = (A_0, A_1)_{\theta,p}, \tag{2.40}$$

where $1/p = (1 - \theta')/p_0 + \theta'/p_1$.

Indeed, by 1° we have

$$((A_0, A_1)_{\theta,p_0}, (A_0, A_1)_{\theta,p_1})_{\theta',p} = ((A_0, A_1)_{\theta,p_0}, (A_0, A_1)_{\theta,p_1})_{\theta',L_{p_0,*},L_{p_1,*}}.$$

By Theorem 2.10 these spaces coincide with $(A_0, A_1)_{\theta,H,H}$, where $H = (L_{p_0,*}, L_{p_1,*})_{\theta',L_{p_0,*},L_{p_1,*}}$. By Theorem 2.11, taking into account that $L_{p,*} = L_p^{-1}$, we obtain that $H = L_{p,*}$, where $1/p = (1 - \theta')/p_0 + \theta'/p_1$. This proves equality (2.40).

**9. *The study of extreme spaces.*** We consider the space $(A_0, A_1)^{\mathcal{K}}_{0,L_\infty,L_\infty}$. Since $e(t) \equiv 1 \in L_\infty$, this space is intermediate between $A_0$ and $A_1$. Moreover, $A_0$ is imbedded in this space, since for $x \in A_0$ the equality $x = u_0(t) + u_1(t)$, where $u_0(t) \equiv x$ and $u_1(t) \equiv 0$, shows that $x \in (A_0, A_1)^{\mathcal{K}}_{0,L_\infty,L_\infty}$ and

$$\|x\|_{(A_0,A_1)^{\mathcal{K}}_{\theta,L_\infty,L_\infty}} \leqslant \sup\|u_0(t)\|_{A_0} + \sup t\|u_1(t)\|_{A_1} = \|x\|_{A_0}.$$

Now let $x$ be an arbitrary element of $(A_0, A_1)^{\mathcal{K}}_{0,L_\infty,L_\infty}$. Then $x = u_0(t) + u_1(t)$, where

$$\sup\|u_0(t)\|_{A_0} + \sup t\|u_1(t)\|_{A_1} \leqslant (1 + \varepsilon)\|x\|^{\mathcal{K}}_{(A_0,A_1)_{0,L_\infty,L_\infty}}.$$

This implies that $\|u_1(t)\|_{A_1} \to 0$ as $t \to \infty$, and therefore $u_0(t) \to x$ as $t \to \infty$ in the norm of $A_1$, and consequently in the norm of $A_0 + A_1$ as well. Recalling the definition of completion of a space relative to another one containing it (Chapter I, §1.4), we conclude that $x$ belongs to the completion $\tilde{A}_0$ of $A_0$ with respect to $A_0 + A_1$, and

$$\|x\|_{\tilde{A}_0} \leqslant \sup\|u_0(t)\|_{A_0} \leqslant (1 + \varepsilon)\|x\|^{\mathcal{K}}_{(A_0,A_1)_{0,L_\infty,L_\infty}}.$$

Conversely, let $x \in \tilde{A}_0$. For any $\varepsilon > 0$ there exists a sequence $x_n \in A_0$ such that $\|x_n\|_{A_0} = \|x\|_{\tilde{A}_0}$ and $\|x - x_n\|_{A_0+A_1} < \varepsilon/n$. The latter means that $x - x_n = y_n + z_n$, where $\|y_n\|_{A_0} + \|z_n\|_{A_1} < \varepsilon/n$. We set $v_0(t) = x_n + y_n$ for $n - 1 < t \leqslant n$, and $v_1(t) = x - v_0(t)$. Then

$$\|v_0(t)\|_{A_0} \leqslant \|x_n\|_{A_0} + \|y_n\|_{A_0} \leqslant \|x\|_{\tilde{A}_0} + \varepsilon/n \leqslant \|x\|_{\tilde{A}_0} + \varepsilon$$

for $n - 1 < t \leqslant n$. Next, $t\|v_1(t)\|_{A_1} \leqslant t\|z_n\|_{A_1} \leqslant t\varepsilon/n \leqslant \varepsilon$. From the inequalities obtained so far it follows that

$$x \in (A_0, A_1)_{0,L_\infty,L_\infty} \quad \text{and} \quad \|x\|_{(A_0,A_1)_{0,L_\infty,L_\infty}} \leqslant \|x\|_{\tilde{A}_0} + 2\varepsilon.$$

We have proved the following assertion for $i = 0$.

LEMMA 2.10. *The space* $(A_0, A_1)^{\mathcal{K}}_{i,L_\infty,L_\infty}$ $(i = 0, 1)$ *coincides isometrically with the completion of* $A_i$ *relative to* $A_0 + A_1$.

We note that the spaces $(A_0, A_1)^{\mathcal{K}}_{i,L_{1,*},L_{1,*}}$ $(i = 0, 1)$ consists of only zero elements, since the function $\min\{1, t\}$ belongs to neither $L_{1,*}$ nor $L^{-1}_{1,*}$. For the same reason we do not consider the spaces $(A_0, A_1)^{\mathcal{J}}_{i,L_\infty,L_\infty}$.

We consider the space $(A_0, A_1)^{\mathcal{J}}_{0,L_{1,*},L_{1,*}}$. It is intermediate between $A_0$ and $A_1$. Moreover, it is contained in $A_0$. Indeed, if

$$x = \int_0^\infty u(t)\frac{dt}{t},$$

where $\|u(t)\|_{A_0} \in L_{1,*}$ and $t\|u(t)\|_{A_1} \in L_{1,*}$, then the integral is convergent in $A_0$, and so $x \in A_0$ and

$$\|x\|_{A_0} \leqslant \|u\|_{L_{1,*}(A_0)} \leqslant \|x\|_{(A_0,A_1)^{\mathcal{J}}_{0,L_{1,*},L_{1,*}}}.$$

If now $x \in A_0 \cap A_1$, we consider the function

$$u(t) = [\ln(1 + 1/\varepsilon)]^{-1}\chi_{(\varepsilon,1+\varepsilon)}(t)x.$$

Obviously $x = \int_0^\infty u(t)t^{-1}\,dt$. Next,

$$\|u(t)\|_{L_{1,*}(A_0)} = \|x\|_{A_0} \qquad \text{and} \qquad \|tu(t)\|_{L_{1,*}(A_1)} = [\ln(1 + 1/\varepsilon)]^{-1}\|x\|_{A_1}.$$

We choose $\varepsilon$ so small that $[\ln(1 + 1/\varepsilon)]^{-1}\|x\|_{A_1} \leqslant \|x\|_{A_0}$. Then

$$\|x\|_{(A_0,A_1)^{\mathcal{J}}_{0,L_{1,\infty},L_{1,\infty}}} \leqslant \|x\|_{A_0}.$$

The two inequalities just obtained show that the norms of the spaces $A_0$ and $(A_0, A_1)^{\mathcal{J}}_{0,L_{1,*},L_{1,*}}$ coincide on $A_0 \cap A_1$. Below we prove Lemma 2.14 from which it follows that $A_0 \cap A_1$ is dense in $(A_0, A_1)^{\mathcal{J}}_{0,L_{1,*},L_{1,*}}$. We arrive at the following assertion.

LEMMA 2.11. *The space* $(A_0, A_1)^{\mathcal{J}}_{i,L_{1,*},L_{1,*}}$ $(i = 0, 1)$ *coincides isometrically with the closure of* $A_0 \cap A_1$ *in* $A_i$.

**10.** ***The duality of the*** $\mathcal{K}$***- and*** $\mathcal{J}$***-methods.*** Assuming the hypotheses of Remark 2.2 to be satisfied, we suppose that $E_0$ and $E_1$ are two ideal lattices of functions (sequences) on $\mathbf{N}_+ = \{1, 2, \dots\}$ with measure $\mu(\{n\}) = \mu_n > 0$.

LEMMA 2.12. *If the lattices* $E_0$ *and* $E_1$ *are regular, and for a Banach couple* $A_0, A_1$ *the intersection* $A_0 \cap A_1$ *is dense in both* $A_0$ *and* $A_1$, *then the set of* $A_0 \cap A_1$*-valued finitary functions is dense in* $E_i(A_i)$, $i = 0, 1$.

PROOF. The norm in $E_i(A_i)$ is absolutely continuous, and so in this space the finitary sequences with values in $A_i$ are dense. Every such sequence can be approximated arbitrarily well by an $A_0 \cap A_1$-valued finitary sequence in $E_i(A_i)$.

COROLLARY. *The space $E_0(A_0) \cap E_1(A_1)$ is dense in $E_0(A_0)$ and $E_1(A_1)$, and consequently the spaces $[E_0(A_0)]'$ and $[E_1(A_1)]'$ form a Banach couple.*

By virtue of Theorem 1 we have the isometric isomorphisms $[E_i(A_i)]' = E_i'(A_i')$ for the regular lattices $E_0$ and $E_1$. These isomorphisms $\tau_i f^i = \{f_1^i, f_2^i, \dots\}$ $(f^i \in (E_i(A_i))', f_k^i \in A_i')$ are such that relation (8) is satisfied:

$$f^i(x) = \sum_{k=1}^{\infty} \langle x_k, f_k^i \rangle \mu_k \qquad (x = \{x_1, x_2, \dots\} \in E_i(A_i)).$$

We consider $(E_0(A_0))' \cap (E_1(A_1))'$. Let $f$ belong to this intersection, and let $x \in E_0(A_0) \cap E_1(A_1)$. We have

$$f(x) = \sum_{k=1}^{\infty} \langle x_k, f_k^0 \rangle \mu_k = \sum_{k=1}^{\infty} \langle x_k, f_k^1 \rangle \mu_k.$$

This equality implies that the functionals $f_k^0$ and $f_k^1$ coincide on $A_0 \cap A_1$, which is dense in $A_i$ by assumption. This implies in turn that the functionals $f_k^0$ and $f_k^1$ as elements of $(A_0 \cap A_1)' = A_0' + A_0'$ are the same: $f_k^0 = f_k^1 = f_k$. Since $\{\|f_1^i\|_{A_i'}, \|f_2^i\|_{A_i'}, \dots\} \in E_i'$, we have $\{\|f_1\|_{A_i'}, \|f_2\|_{A_i'}, \dots\} \in E_i'$, and consequently $\{f_1, f_2, \dots\} \in E_0'(A_0') \cap E_1'(A_1')$. We have obtained a mapping $\tau f = \{f_k\}$ of $(E_0(A_0))' \cap (E_1(A_1))'$ into $E_0'(A_0') \cap E_1'(A_1')$. This mapping is isometric:

$$\begin{aligned}\|\tau f\|_{E_0'(A_0') \cap E_1'(A_1')} &= \max \|\tau f\|_{E_i'(A_i')} \\ &= \max \|f\|_{(E_i(A_i))'} = \|f\|_{(E_0(A_0))' \cap (E_i(A_1))'}.\end{aligned}$$

The surjectivity of $\tau$ can be proved in the same way as in the proof of Theorem 1.

We may prove similarly that $(E_0(A_0))' + (E_1(A_1))'$ is isometrically isomorphic to $E_0'(A_1') + E_1'(A_1')$.

Using Theorem 3.1 of Chapter I, we obtain the following asssertion.

LEMMA 2.13. *If the lattices $E_0$ and $E_1$ are regular and $A_0 \cap A_1$ is dense in the spaces of the Banach couple $A_0, A_1$, then the isometric isomorphisms*

$$[E_0(A_0) + E_1(A_1)]' = E_0'(A_0') \cap E_1'(A_1')$$

*and*

$$[E_0(A_0) \cap E_1(A_1)]' = E_0(A_0') + E_1'(A_1')$$

*hold.*

DEFINITION. We say that the lattices $E_0$ and $E_1$ *have property* D) relative to the couple $A_0, A_1$ if $A_0 \cap A_1$ is dense in $(A_0, A_1)^{\mathcal{K}}_{E_0, E_1}$.

THEOREM 2.14. *If* $A_0 \cap A_1$ *is dense in the spaces of the Banach couple* $A_0, A_1$ *and the regular lattices* $E_0$ *and* $E_1$ *on* $\mathbf{N}_+$ *have property* D) *relative to* $A_0, A_1$, *and* $E_0 + E_1 \ni e$,* *then the space* $[(A_0, A_1)^{\mathcal{K}}_{E_0,E_1}]'$ *is isometrically isomorphic to* $(A'_0, A'_1)^{\mathcal{J}}_{E'_0,E'_1}$.

PROOF. We note that by virtue of the inclusion $(E'_0)^1 + (E'_1)^1 \supset E_0 + E_1 \ni e$ the space $(A'_0, A'_1)^{\mathcal{J}}_{E'_0,E'_1}$ is defined.

The space $(A_0, A_1)^{\mathcal{K}}_{E_0,E_1}$ is a subspace of $E_0(A_0) + E_1(A_1)$ consisting of sequences of the form $\{x_0, x_0, \dots\}$, where $x_0 \in A_0 + A_1$. By Lemma 2.13, $E'_0(A'_0) \cap E'_1(A'_1)$ is the dual of $E_0(A_0) + E_1(A_1)$. By the Hahn-Banach theorem, every functional $f \in [(A_0, A_1)^{\mathcal{K}}_{E_0,E_1}]'$ is the restriction $S\tilde{f}$ of a functional $\tilde{f}$ belonging to $(E_0(A_0) + E_1(A_1))'$ having the form $\tilde{f} = \{f_1, f_2, \dots\}$ ($f_k \in A'_0 \cap A'_1$), and

$$\|f\|_{[(A_0,A_1)^{\mathcal{K}}_{E_0,E_1}]'} = \inf_{S\tilde{f}=f} \|\tilde{f}\|_{E'_0(A'_0)\cap E'_1(A'_1)}.$$

By virtue of the definition of this space, every functional $\varphi$ belonging to $(A'_0, A'_1)^{\mathcal{J}}_{E'_0,E'_1}$ can be obtained from a functional $\tilde{f} = \{f_1, f_2, \dots\} \in E'_0(A'_0) \cap E'_1(A'_1)$ by means of the operator $\mathcal{J}\tilde{f} = \sum_1^\infty f_k\mu_k = \varphi$, and

$$\|\varphi\|_{(A'_0,A'_1)^{\mathcal{J}}_{E'_0,E'_1}} = \inf_{\mathcal{J}\tilde{f}=\varphi} \|\tilde{f}\|_{E'_0(A'_0)\cap E'(A'_1)}.$$

If we now introduce the correspondence $S\tilde{f} = f \leftrightarrow \mathcal{J}\tilde{f} = \varphi$, to prove that it is well defined we have to verify that the kernels of the mappings $S$ and $\mathcal{J}$ coincide. Let $S\tilde{f} = 0$ and $x = \{x_0, x_0, \dots\}$, where $x_0 \in A_0 \cap A_1$. Then by (8) we have

$$\tilde{f}(x) = \sum f_k(x_0)\mu_k = 0 \qquad (x_0 \in A_0 \cap A_1). \tag{2.40'}$$

Since $\mathcal{J}\tilde{f} = \sum f_k\mu_k$ is convergent in $A'_0 + A'_1 = (A_1 \cap A_1)'$ by Lemma 2.2, this implies that $\mathcal{J}\tilde{f} = 0$.

Conversely, if $\mathcal{J}\tilde{f} = 0$, then (2.40′) holds, which means that $S\tilde{f}$ is equal to zero on the set $A_0 \cap A_1$ dense in $(A_0, A_1)^{\mathcal{K}}_{E_0,E_1}$. From this it follows that $Sf = 0$. The isometricity of the correspondence $f \leftrightarrow \varphi$ follows from the formulas for the calculation of norms.

COROLLARY. *If the hypotheses of the theorem are satisfied for two lattices* $E_0$ *and* $E_1$ *defined on an arbitrary measure space, and such that a common averaging operator acts in them, then the space* $[(A_0, A_1)^{\mathcal{K}}_{E_0,E_1}]'$ *is isomorphic to* $(A'_0, A'_1)^{\mathcal{J}}_{E'_0,E'_1}$.

---

* *Editor's note.* Concerning $e$ see subsection 1 of this section.

The corollary follows immediately from the theorem and Remark 2.1.

We may establish the next results by similar arguments.

THEOREM 2.15. *If $A_0 \cap A_1$ is dense in the spaces of the Banach couple $A_0, A_1$ and the regular lattices $E_0$ and $E_1$ on $\mathbf{N}_+$ are such that $E_0'$ and $E_1'$ have property D) and $E_0' + E_1' \ni e$, then the space $[(A_0, A_1)^{\mathcal{J}}_{E_0,E_1}]'$ is isometrically isomorphic to $(A_0', A_1')^{\mathcal{K}}_{E_0',E_1'}$.*

COROLLARY. *If the hypotheses of the theorem are satisfied for lattices $E_0$ and $E_1$ defined on an arbitrary measure space and such that an averaging operator acts in them, then $[(A_0, A_1)^{\mathcal{J}}_{E_0,E_1}]'$ is isomorphic to $(A_0', A_1')^{\mathcal{K}}_{E_0',E_1'}$.*

The verification of condition D) becomes simpler in cases where the spaces constructed by the $\mathcal{K}$- and $\mathcal{J}$-methods are isomorphic. Indeed, the following lemma holds.

LEMMA 2.14. *If $E_0$ and $E_1$ are regular lattices consisting of functions summable on every set of finite measure, then $A_0 \cap A_1$ is dense in $(A_0, A_1)^{\mathcal{J}}_{E_0,E_1}$.*

PROOF. We consider an increasing sequence of measurable sets $\mathfrak{M}_n$ such that $\mu(\mathfrak{M}_n) < \infty$ and $\mathfrak{M}_n \uparrow \mathfrak{M}$. Let $x \in (A_0, A_1)^{\mathcal{J}}_{E_0,E_1}$. Then $x = \int_{\mathfrak{M}} u(t)\, d\mu$, where $u \in E_0(A_0) \cap E_1(A_1)$. We write $x_n = \int_{\mathfrak{M}_n} u(t)\, d\mu$. It can be shown that $u(t)$ is strongly measurable as a function with values in $A_0 \cap A_1$ (we leave this to the reader). Besides,

$$\int_{\mathfrak{M}_n} \|u(t)\|_{A_0 \cap A_1}\, d\mu = \int_{\mathfrak{M}_n} \max\{\|u(t)\|_{A_0}, \|u(t)\|_{A_1}\}\, d\mu < \infty,$$

by the hypothesis of the theorem. Therefore $u(t)$ as a function with values in $A_0 \cap A_1$ is integrable on $\mathfrak{M}_n$, and consequently $x_n \in A_0 \cap A_1$. Next,

$$x - x_n = \int u(t)\left[1 - \chi_{\mathfrak{M}_n}(t)\right] d\mu,$$

and therefore

$$\|x - x_n\|_{(A_0,A_1)^{\mathcal{J}}_{E_0,E_1}} \leqslant \max \|u(1 - \chi_{\mathfrak{M}_n})\|_{E_i(A_i)} \to 0$$

as $n \to \infty$, by virtue of the regularity of the $E_i$.

We recall that by Theorem 2.3 the spaces $(A_0, A_1)^{\mathcal{K}}_{E_0^{\alpha_0},E_1^{\alpha_1}}$ and $(A_0, A_1)^{\mathcal{J}}_{E_0^{\alpha_0},E_1^{\alpha_1}}$ coincide if $E_0$ and $E_1$ are ideal lattices that are interpolation spaces between $L_{1,*}$ and $L_{\infty,*}$. Besides, if $E_0$ and $E_1$ are regular, so are $E_0^{\alpha_0}$ and $E_1^{\alpha_1}$. Then the dual spaces of $E_0^{\alpha_0}$ and $E_1^{\alpha_1}$ are $(E_0')^{-\alpha_0}$ and $(E_1')^{-\alpha_1}$.

Combining the above facts with the preceding results, we arrive at the following assertion.

THEOREM 2.16. *If $E_0$ and $E_1$ are regular lattices that are interpolation spaces between $L_{1,*}$ and $L_{\infty,*}$ and the intersection $A_0 \cap A_1$ of the spaces of the Banach couple $A_0$, $A_1$ is dense in each of these spaces, then the dual space of $(A_0, A_1)_{E_0^{\alpha_0}, E_1^{\alpha_1}}$ is isomorphic to $(A_0', A_1')_{(E_0')^{-\alpha_0},(E_1')^{-\alpha_1}}$, or, in other words,*

$$[(A_0, A_1)_{\theta, E_0, E_1}]' = (A_0', A_1')_{\theta, E_0', E_1'} \qquad (0 < \theta < 1),$$

*up to isomorphism.*

REMARK. As has been shown recently in [116], property D) is always satisfied if the lattices $E_0$ and $E_1$ are regular and do not contain the function identically equal to one.

**11.** ***Almost interpolation properties of scales of spaces.*** The theory of spaces constructed by the $\mathcal{K}$- and $\mathcal{J}$-methods is connected with the theory of scales of spaces which are not normal scales.

In this subsection we always assume that the space $A_1$ is imbedded in $A_0$.

LEMMA 2.15. *If $0 < \alpha < \beta < 1$, then every space $B_1$ of type $\beta$ intermediate between $A_0$ and $A_1$ is imbedded in every intermediate space $B_2$ of type $\alpha$.*

PROOF. By assumption, $(A_0, A_1)_{\alpha, L_{1,*}, L_{1,*}} \subset B_2$. By the remark to the reiteration of Theorem 2.8, the space $(A_0, A_1)_{\alpha, L_{1,*}, L_{1,*}}$ is isomorphic to $(A_0, B_1)_{\alpha/\beta, L_{1,*}, L_{1,*}}$ and consists of the same elements. Therefore

$$B_1 \subset (A_0, B_1)_{\alpha/\beta, L_{1,*}, L_{1,*}} \subset B_2.$$

LEMMA 2.16. *If $0 < \alpha < \beta < \gamma < 1$ and the spaces $B_\alpha$, $B_\beta$ and $B_\gamma$ intermediate between $A_0$ and $A_1$ have types $\alpha$, $\beta$, and $\gamma$, respectively, then*

$$\|x\|_{B_\beta} \leqslant C_{\alpha,\beta,\gamma} \|x\|_{B_\alpha}^{(\gamma-\beta)/(\gamma-\alpha)} \|x\|_{B_\gamma}^{(\beta-\alpha)/(\gamma-\alpha)}. \tag{2.41}$$

PROOF. By assumption, $(A_0, A_1)_{\beta, L_{1,*}, L_{1,*}} \subset B_\beta$. By the reiteration Theorem 2.8 we have $(A_0, A_1)_{\beta, L_{1,*}, L_{1,*}} = (B_\alpha, B_\gamma)_{\theta, L_{1,*}, L_{1,*}}$, where $\theta = (\beta - \alpha)/(\gamma - \alpha)$. Then $(B_\alpha, B_\gamma)_{\theta, L_{1,*}, L_{1,*}} \subset A_\beta$, and by Lemma 2.6 we have (2.41).

In accordance with Definition 1.2 of Chapter III, §1, Lemmas 2.15 and 2.16 lead to the following assertion.

THEOREM 2.17. *Let $\{A_\alpha\}$ be a family of Banach spaces having the property that the spaces $A_\alpha$, $0 < \alpha < 1$, are intermediate of type $\alpha$ between $A_0$ and $A_1$. Then the family $\{\hat{A}_\alpha\}$ consisting of the closures $\hat{A}_\alpha$ of $A_1$ in the norms of $A_\alpha$ forms a scale of spaces.*

The scales of spaces occurring in the theorem will be called *scales of means*.

The family of spaces $(A_0, A_1)_{\alpha, L_{1,*}, L_{1,*}}$ provides an example of a scale of means. According to Lemma 2.11 we have

$$(A_0, A_1)_{1, L_{1,*}, L_{1,*}} = (A_0, A)^{\oint}_{1, L_{1,*}, L_{1,*}} = A_1$$

and the space $(A_0, A_1)_{0, L_{1,*}, L_{1,*}} = (A_0, A_1)^{\oint}_{0, L_{1,*}, L_{1,*}} = \hat{A}_0$ coincides with the closure of $A_1$ in $A_0$. By Lemma 2.14 the space $A_1$ is dense in the spaces $(A_0, A_1)_{\alpha, L_{1,*}, L_{1,*}}$ for $0 < \alpha < 1$. Due to the property expressed in the next lemma it is natural to call the scale obtained in this way the *maximal scale of means*.

LEMMA 2.17. *The maximal scale of means is imbedded in every scale of spaces* $\{B_\alpha\}$, $0 \leqslant \alpha \leqslant 1$, *where* $B_1 = A_1$ *and* $B_0 = \hat{A}_0$.

PROOF. By the properties of the scale for $x \in A_1$ we have $\|x\|_{B_\alpha} \leqslant C\|x\|^{1-\alpha}_{A_0}\|x\|^{\alpha}_{A_1}$, from which by virtue of Lemma 2.6 we obtain the imbedding $(A_0, A_1)_{\alpha, L_{1,*}, L_{1,*}} \subset B_\alpha$.

The following lemma gives an important property of scales of means.

LEMMA 2.18. *Any scale of means is imbedded in the minimal scale constructed from the spaces* $A_0$ *and* $A_1$.

PROOF. For any scale of means $\{\hat{A}_\alpha\}$ we have the imbeddings $\hat{A}_\alpha \subset A_\alpha \subset (A_0, A_1)_{\alpha, L_\infty, L_{\infty,*}}$. For $x \in A_1$ we consider the representation

$$x = \int_0^\infty u(t)\frac{dt}{t},$$

where $u \in L^{-\alpha}_{\infty,*}(A_0) \cap L^{1-\alpha}_{\infty,*}(A_1)$. Next, for $f \in A_0'$ we have

$$|f(u(t))| \leqslant \|f\|_{A_0'} t^\alpha \|t^{-\alpha}u(t)\|_{A_0} \leqslant t^\alpha \|f\|_{A_0'}\|u\|_{L^{-\alpha}_{\infty,*}(A_0)\cap L^{1-\alpha}_{\infty,*}(A_1)}$$

and

$$|f(u(t))| \leqslant \|f\|_{A_1'} t^{\alpha-1}\|t^{1-\alpha}u(t)\|_{A_1} \leqslant t^{\alpha-1}\|f\|_{A_1'}\|u\|_{L^{-\alpha}_{\infty,*}(A_0)\cap L^{1-\alpha}_{\infty,*}(A_1)}.$$

Then

$$\begin{aligned} |f(x)| &\leqslant \int_0^\infty |f(u(t))|\frac{dt}{t} \\ &\leqslant \left(\int_0^N t^\alpha \|f\|_{A_0'}\frac{dt}{t} + \int_N^\infty t^{\alpha-1}\|f\|_{A_1'}\frac{dt}{t}\right)\|u\|_{L^{-\alpha}_{\infty,*}(A_0)\cap L^{1-\alpha}_{\infty,*}(A_1)} \\ &\leqslant \left(\frac{1}{\alpha}N^\alpha\|f\|_{A_0'} + \frac{1}{1-\alpha}N^{\alpha-1}\|f\|_{A_1'}\right)\|u\|_{L^{-\alpha}_{\infty,*}(A_0)\cap L^{1-\alpha}_{\infty,*}(A_1)}. \end{aligned}$$

By minimizing the right side with respect to $N$, we obtain

$$|f(x)| \leqslant \frac{1}{\alpha(1-\alpha)} \|f\|_{A_0'}^{1-\alpha} \|f\|_{A_1'}^{\alpha} \|u\|_{L_{\infty,*}^{-\alpha}(A_0) \cap L_{\infty,*}^{1-\alpha}(A_1)}.$$

Hence

$$|f(x)| \leqslant \frac{1}{\alpha(1-\alpha)} \|f\|_{A_0'}^{1-\alpha} \|f\|_{A_1'}^{\alpha} \|x\|_{(A_0, A_1)_{(\alpha, L_{\infty,*}, L_{\infty,*})}}.$$

Then

$$\|x\|_{A_\alpha^{\min}} = \sup \frac{|f(x)|}{\|f\|_{A_0'}^{1-\alpha} \|f\|_{A_1'}^{\alpha}}$$

$$\leqslant \frac{1}{\alpha(1-\alpha)} \|x\|_{(A_0, A_1)_{(\alpha, L_{\infty,*}, L_{\infty,*})}} \leqslant \frac{C}{\alpha(1-\alpha)} \|x\|_{\hat{A}_\alpha};$$

by virtue of the density of $A_1$ in $\hat{A}_\alpha$ this implies the imbedding $\hat{A}_\alpha \subset A_\alpha^{\min}$.

In Lemmas 2.17 and 2.18 we have used the first part of the following definition.

DEFINITION 2.2. If $\{A_\alpha\}$ and $\{B_\alpha\}$, $0 \leqslant \alpha \leqslant 1$, are two families of Banach spaces such that $A_\beta \subset A_\alpha$ and $B_\beta \subset B_\alpha$ for $\alpha < \beta$, then we say that the first family is *imbedded* in the second if $A_\alpha \subset B_\alpha$ for every $\alpha$, and *almost imbedded* in the second if $A_\beta \subset B_\alpha$ for every $\alpha$ and $\beta > \alpha$.

We recall that in §2.2 of Chapter III for a family $\{A_\alpha\}$, $0 \leqslant \alpha \leqslant 1$, of spaces for which $A_\beta$ is densely imbedded in $A_\alpha$ for $\alpha < \beta$ we introduced the dual family $\tilde{A}_\alpha$ which consists of the closures of $A_0'$ in the spaces $A_\alpha'$. Now let $\{A_\alpha\}$ be an arbitrary scale. Since $A_1$ is dense in $A_\alpha$, the space $(A_0, A_1)_{\alpha, L_{1,*}, L_{1,*}}$ is densely imbedded in $A_\alpha$, according to Lemma 2.17. Then by Theorem 2.16 we have $A_\alpha' \subset (A_0', A_1')_{\alpha, L_{\infty,*}, L_{\infty,*}}$. If, in addition,

$$\|f\|_{A_\alpha'} \leqslant C \|f\|_{A_0'}^{1-\alpha} \|f\|_{A_1'}^{\alpha} \qquad (f \in A_0'), \tag{2.42}$$

then $(A_0', A_1')_{\alpha, L_{1,*}, L_{1,*}} \subset A_\alpha'$, by Lemma 2.6. Consequently $A_\alpha'$ has type $\alpha$ relative to the couple $A_0'$, $A_1'$ and type $1-\alpha$ relative to the couple $A_1'$, $A_0'$. Theorem 2.17 leads to the following assertion.

THEOREM 2.18. *Let $\{A_\alpha\}$, $0 \leqslant \alpha \leqslant 1$, be an arbitrary scale of Banach spaces. For the dual family $\{\tilde{A}_{1-\sigma}\}$, $0 \leqslant \sigma \leqslant 1$, to form a scale it is necessary and sufficient that condition* (2.42) *be satisfied.*

REMARK. Under condition (2.42) the scale $\{\tilde{A}_{1-\sigma}\}$ will be a scale of means relative to $A_1'$ and $A_0'$.

COROLLARY. *The dual family of a minimal scale forms a scale.*

Indeed, for a minimal scale inequality (2.42) is established in Lemma 2.3 of Chapter III.

We note that condition (2.42) implies that the scale $\{A_\alpha\}$ is imbedded in the minimal scale. Indeed, for $x \in A_1$ we have

$$\|x\|_{A_\alpha} = \sup_{f \in A'_\alpha} \frac{|f(x)|}{\|f\|_{A'_\alpha}}$$

$$\geqslant \sup_{f \in A'_0} \frac{|f(x)|}{\|f\|_{A'_\alpha}} \geqslant \frac{1}{C} \sup_{f \in A'_0} \frac{|f(x)|}{\|f\|_{A'_0}^{1-\alpha} \|f\|_{A'_1}^{\alpha}} = \frac{1}{C} \|x\|_{A_\alpha^{\min}},$$

which implies the imbedding $A_\alpha \subset A_\alpha^{\min}$. Conversely, if the scale $\{A_\alpha\}$ is imbedded in $\{A_\alpha^{\min}\}$, then $(A_\alpha^{\min})' \subset A'_\alpha$, and therefore

$$\|f\|_{A'_\alpha} \leqslant C\|f\|_{(A_\alpha^{\min})'} \leqslant C_1 \|f\|_{A'_0}^{1-\alpha} \|f\|_{A'_1}^{\alpha}$$

according to Lemma 2.3 of Chapter III.

Thus, the following theorem holds.

THEOREM 2.19. *Condition* (2.42) *is necessary and sufficient for the scale* $\{A_\alpha\}$ *to be imbedded in the minimal scale constructed from* $A_0$ *and* $A_1$.

LEMMA 2.19. *The minimal scale* $\{A_\alpha^{\min}\}$ *is almost imbedded in every scale* $\{A_\alpha\}$ *imbedded in it.*

PROOF. By the above, the spaces $A'_\alpha$ and $(A_\alpha^{\min})'$ have type $1 - \alpha$ relative to $A'_1$ and $A'_0$. Then by Lemma 2.15 for $\alpha < \beta$ we have $(A_\alpha)' \subset (A_\beta^{\min})'$. For $x \in A_1$ we have

$$\|x\|_{A_\alpha} = \sup_{f \in A'_\alpha} \frac{|f(x)|}{\|f\|_{A'_\alpha}} \leqslant \sup_{f \in (E_\beta^{\min})'} \frac{|f(x)|}{\|f\|_{A'_\alpha}}$$

$$\leqslant \frac{1}{C} \sup_{f \in (E_\beta^{\min})'} \frac{|f(x)|}{\|f\|_{(A_\beta^{\min})'}} = \frac{1}{C} \|x\|_{A_\beta^{\min}},$$

where $C$ is the imbedding constant of $A'_\alpha$ in $(A_\beta^{\min})'$. The inequality thus obtained implies the imbedding $A_\beta^{\min} \subset A_\alpha$.

THEOREM 2.20. *The minimal scale* $\{A_\alpha^{\min}\}$ *is almost imbedded in every scale* $\{A_\alpha\}$.

PROOF. It is sufficient to consider the scale $\{\bar{A}_\alpha\}$ consisting of the closures of $E_1$ in $A_\alpha^{\min} \cap A_\alpha$. It is imbedded in $A_\alpha^{\min}$. Therefore $A_\beta^{\min} \subset \bar{A}_\alpha \subset A_\alpha$ for $\alpha < \beta$ according to Lemma 2.19.

DEFINITION 2.3. A family $\{A_\alpha\}$, $0 \leqslant \alpha \leqslant 1$, of Banach spaces has the *almost interpolation property* relative to a family $\{B_\alpha\}$ of Banach spaces if the condition that a linear operator acts boundedly from $A_0$ into $B_0$ and from $A_1$ into $B_1$ implies that it acts from every space $A_\beta$ into the space $B_\alpha$ for $\alpha < \beta$.

THEOREM 2.21. *Let $\{B_\alpha\}$ be an arbitrary scale and let $\{A_\alpha\}$ be a scale satisfying condition* (2.42). *Then $\{A_\alpha\}$ has the almost interpolation property relative to $\{B_\beta\}$.*

PROOF. Let the operator $T$ act boundedly from $A_0$ into $B_0$ and from $A_1$ into $B_1$. By Theorem 2.3 of Chapter III the operator $T$ acts from $A_\beta^{\min}$ into $B_\beta^{\min}$ (in the proof of Theorem 2.3 the relativity of spaces was not used). By Theorem 2.19 it acts from $A_\beta$ into $B_\beta^{\min}$, and by Theorem 2.20 it acts from $A_\beta$ into $B_\alpha$ for $\alpha < \beta$.

COROLLARY 1. *A minimal scale has the almost interpolation property relative to any scale.*

COROLLARY 2. *Any scale of means has the almost interpolation property relative to any scale.*

Indeed, by Theorem 2.18 any scale of means is imbedded in a minimal one, and so it satisfies (2.42) by virtue of Theorem 2.19.

COROLLARY 3. *Any regular scale has the almost interpolation property relative to any scale.*

**12. *Connection with the complex method.*** Let $A_0$ and $A_1$ be a Banach couple of complex spaces. We consider the spaces $[A_0, A_1]_\alpha$ constructed by the complex method of interpolation.

LEMMA 2.20. *The space $[A_0, A_1]_\alpha$ has type $\alpha$ relative to $A_0$ and $A_1$.*

PROOF. The fact that the space $(A_0, A_1)_{\alpha, L_{1,*}, L_{1,*}}$ is imbedded in $[A_0, A_1]_\alpha$ follows from (1.39) and Lemma 2.6. We show that

$$[A_0, A_1] \subset (A_0, A_1)_{\alpha, L_{\infty,*}, L_{\infty,*}}.$$

Let $x \in [A_0, A_1]_\alpha$ and estimate the functional $K(t, x)$ (see subsection 5). Let the function $f(z)$ from $\mathfrak{F}(A_0, A_1)$ be such that $f(\alpha) = x$. The function $f_1(z) = t^{\alpha - z} f(z)$ also belongs to $\mathfrak{F}(A_0, A_1)$, and $f_1(\alpha) = x$. We use the representation (1.9) for the function $f_1(z)$:

$$x = f_1(\alpha) = \int_{-\infty}^{\infty} f_1(i\tau)\mu_0(\alpha, \tau)\, d\tau + \int_{-\infty}^{\infty} f_1(1 + i\tau)\mu_s(\alpha, \tau)\, d\tau = x_0 + x_1,$$

where $x_0 \in A_0$ and $x_1 \in A_1$. Then

$$\begin{aligned} K(t, x) &\leq \|x_0\|_{A_0} + t\|x_1\|_{A_1} \\ &\leq (1 - \alpha) \sup_\tau \|f_1(i\tau)\|_{A_0} + \alpha t \sup \|f_1(1 + i\tau)\|_{A_1} \\ &= (1 - \alpha) t^\alpha \sup_\tau \|f(i\tau)\|_{A_0} + \alpha t^\alpha \sup \|f(1 + i\tau)\|_{A_1} \leq t^\alpha \|f\|_{\mathfrak{F}(A_0, A_1)}. \end{aligned}$$

On the right side we take the infimum over all $f \in \mathfrak{F}(A_0, A_1)$ for which $f(\alpha) = x$. This gives us

$$t^{-\alpha}K(t, x) \leqslant \|x\|_{[A_0,A_1]_\alpha}.$$

By Theorem 2.9 we have $x \in (A_0, A_1)_{\alpha,L_{1,*},L_{1,*}}$. The lemma is proved.

Now let $A_1$ be imbedded in $A_0$. Since $A_1$ is dense in $[A_0, A_1]_\alpha$, the spaces $[A_0, A_1]_\alpha$ form a scale of means. Therefore the following theorem holds.

THEOREM 2.22. *The family* $[A_0, A_1]_\alpha$ *of spaces has the almost interpolation property relative to any scale of spaces.*

Lemma 2.20 and reiteration Theorem 2.8 immediately imply one further corollary, important in applications.

THEOREM 2.23. *The space* $([A_0, A_1]_\alpha, [A_0, A_1]_\beta)_{\gamma,E,E}$, *where* $0 < \alpha < \beta < 1$ *and* $0 < \gamma < 1$, *is isomorphic to* $(A_0, A_1)_{\gamma',E,E}$, *where* $\gamma' = (1 - \gamma)\alpha + \gamma\beta$.

For completeness of the exposition we note that for $A_1 \subset A_0$ the scale $[A_0, A_1]_\alpha$ is not only imbedded in the minimal scale but also majorizes the minimal scale (see Definition 2.3 in Chapter III). Indeed, let $f \in \mathfrak{F}(A_0, A_1)$ and let $\varphi$ be a functional from $A_0'$. The function $\langle f(z), \varphi\rangle$ is analytic inside the strip $0 < \operatorname{Re} z < 1$, continuous in the closed strip, and bounded. By the three lines theorem we have

$$\begin{aligned}|\langle f(\alpha), \varphi\rangle| &\leqslant \left[\sup|\langle f(i\tau), \varphi\rangle|\right]^{1-\alpha}\left[\sup|\langle f(1+i\tau), \varphi\rangle|\right]^{\alpha}\\ &\leqslant \|\varphi\|_{A_0'}^{1-\alpha}\|\varphi\|_{A_1'}^{\alpha}\left[\sup\|f(i\tau)\|_{A_0}\right]^{1-\alpha}\left[\sup\|f(1+i\tau)\|_{A_1}\right]^{\alpha}\\ &\leqslant \|\varphi\|_{A_0'}^{1-\alpha}\|\varphi\|_{A_1'}^{\alpha}\|f\|_{\mathfrak{F}(A_0,A_1)}.\end{aligned}$$

Setting $f(\alpha) = x \in [A_0, A_1]_\alpha$, we obtain

$$\|x\|_{A_\alpha^{\min}} = \sup_{\varphi \in A_0'} \frac{|\langle x, \varphi\rangle|}{\|\varphi\|_{A_0'}^{1-\alpha}\|\varphi\|_{A_1'}^{\alpha}} \leqslant \|f\|_{\mathfrak{F}(A_0,A_1)}.$$

This implies that $x \in A_\alpha^{\min}$ and $\|x\|_{A_\alpha^{\min}} \leqslant \|x\|_{[A_0,A_1]_\alpha}$.

# CHAPTER V

# INTERPOLATION IN SPACES OF SMOOTH FUNCTIONS

BY S. G. KREĬN

## §1. Interpolation spaces constructed from an unbounded operator and a smoothing approximation process

**1.** ***Banach couples associated with unbounded operators.*** In a Banach space $E$ we consider a closed unbounded linear operator $A$ with domain $\mathfrak{D}(A)$ dense in $E$. If in $\mathfrak{D}(A)$ we introduce the graph norm $\|x\|_{E_A} = \|x\|_E + \|Ax\|_E$, then $\mathfrak{D}(A)$ becomes a Banach space, which will be denoted by $E_A$. Since $\|x\|_E \leqslant \|x\|_{E_A}$, the space $E_A$ is normally imbedded in $E$.

We denote by $N_A$ the kernel of $A$, i.e. the collection of all elements $x \in E$ for which $Ax = \theta$. The set $N_A$ is a subspace of $E$. Indeed, if $x_n \to x$ in $E$ and $Ax_n = 0$, then $Ax = 0$ since $A$ is closed, and consequently $x \in N_A$. On $N_A$ the norms of $E$ and $E_A$ coincide.

In applications $A$ is often a differential operator; therefore the membership relation $x \in \mathfrak{D}(A)$ expresses a certain smoothness. The problem of constructing intermediate spaces between $E$ and $E_A$ is connected with finding spaces of functions having intermediate smoothness. This construction is often facilitated by the presence of a *smoothing approximation process* in $E$. By such a process we understand a commutative family of bounded linear operators $Q_t$ $(0 < t < \infty)$ having the following properties:

1) The operators $Q_t$ are strongly continuous in $t$, bounded uniformly in $t$ on $(0, \infty)$, i.e.

$$\|Q_t x\|_E \leqslant M\|x\|_E, \tag{1.1}$$

and strongly converges to $I$ as $t \to 0$.

2) For every $t$ the operator $Q_t$ maps $E$ into $\mathfrak{D}(A)$ (smooths) and commutes with $A$ on $\mathfrak{D}(A)$.

We consider the operators $S_t = AQ_t$. From 1) and 2) it follows that

$$\|S_t x\|_E = \|Q_t Ax\|_E \leqslant M\|Ax\|_E \leqslant M\|x\|_{E_A} \tag{1.2}$$

for $x \in \mathfrak{D}(A)$.

Besides, $S_t$ is bounded for every $t > 0$. Indeed, it is closed as the product of a closed operator and a bounded operator, and is defined on the whole space. We note that if $A$ is not bounded, then the function $\|S_t\|_{E\to E}$ cannot be bounded near zero. Indeed, the inequality $\|S_t\|_{E\to E} \leqslant K$ for $0 < t \leqslant a$ would imply $\|AQ_t x\|_E \leqslant K\|x\|_E$. Next, $AQ_t x = Q_t Ax \to Ax$ on $\mathfrak{D}(A)$, and therefore the operators $AQ_t$ would be uniformly bounded and would converge to $A$ on the dense set $\mathfrak{D}(A)$. By the Banach-Steinhaus theorem $A$ would then be bounded.

We denote by $\rho(t)$ a positive increasing continuous function having the property that

$$\|\rho(t)S_t\|_{E\to E} \leqslant 1 \qquad (0 < t < \infty). \tag{1.3}$$

From what has been said above, it follows that $\rho(t) \to 0$ as $t \to 0$. Moreover, $\rho(t)S_t x \to 0$ for every $x \in E$. Indeed, for $x \in \mathfrak{D}(A)$ this follows from the preceding: $\rho(t)S_t x = \rho(t)Q_t Ax \to 0$, and since $\mathfrak{D}(A)$ is dense in $E$, and the operators $\rho(t)S_t$ are uniformly bounded, this is also true for any $x \in E$.

By (1.2), for the elements of $\mathfrak{D}(A)$ we have

$$\sup\|S_t x\|_E < \infty. \tag{1.4}$$

We pose the problem of describing all elements $x$ for which (1.4) holds.

LEMMA 1.1. *The set of elements $x \in E$ having property* (1.4) *coincides with the completion $E_{0A}$ of $E_A$ relative to $E$.*

PROOF. If $x \in E_{0A}$, then there exists a sequence $x_n \in E_A$ such that $x_n \to x$ in $E$ and $\|x_n\|_{E_A} = \|x\|_{E_{0A}}$. Then (1.2) leads to the inequality $\|S_t x_n\|_E \leqslant M\|x_n\|_{E_A} = M\|x\|_{E_{0A}}$. Using the boundedness of the operator $S_t$ in $E$, we obtain that $\|S_t x\|_E \leqslant M\|x\|_{E_{0A}}$. This implies that $\sup\|S_t x\|_E \leqslant M\|x\|_{E_{0A}}$.

Conversely, if (1.4) is satisfied, then $Q_t x \to x$ in $E$ and

$$\|Q_t x\|_{E_A} = \|Q_t x\|_E + \|AQ_t x\|_E \leqslant M\|x\|_E + \sup\|S_t x\|_E.$$

Hence $x \in E_{0A}$ and $\|x\|_{E_{0A}} \leqslant M\|x\|_E + \sup\|S_t x\|_E$.

The lemma is proved.

Since we always have $\|x\|_E \leqslant \|x\|_{0A}$ $(x \in E_{0A})$, from the proof of the lemma we obtain the inequality

$$C_1(\|x\|_E + \sup\|S_t x\|_E) \leqslant \|x\|_{E_{0A}} \leqslant C_2(\|x\|_E + \sup\|S_t x\|_E). \tag{1.5}$$

LEMMA 1.2. *The set $E_A$ is closed in $E_{0A}$.*

PROOF. Let $x_n \in E_A$ and $x_n \to x$ in $E_{0A}$. By (1.5) we have $x_n \to x$ in $E$ and $S_t x_n \to S_t x$ in $E$ uniformly in $t$. For any $\varepsilon > 0$ and sufficiently large $n$ and $m$ we have $\|S_t x_n - S_t x_m\|_E < \varepsilon$ for all $t > 0$. From this we obtain $\|Q_t(Ax_n - Ax_m)\|_E < \varepsilon$. Passing to the limit as $t \to 0$, from property 1) of the operators $Q_t$ we obtain that $\|Ax_n - Ax_m\|_E \leqslant \varepsilon$. Since $A$ is closed, from the last inequality and the fact that $x_n \to x$ it follows that $x \in \mathfrak{D}(A) = E_A$. The lemma is proved.

COROLLARY 1. *The norms* $\|x\|_E + \|Ax\|_E$ *and* $\|x\|_E + \sup\|S_t x\|_E$ *are equivalent on* $\mathfrak{D}(A)$.

COROLLARY 2. *The space* $E_A$ *is almost related with* $E$ (*see Chapter* III, *Theorem* 3.1).

**2. *Construction of intermediate spaces.*** A smoothing approximation process assigns, to every element $x$ of a Banach space $E$, a function $S_t x$ on $(0, \infty)$ with values in $E$. Lemma 1.1 says that the space $E_{0A}$ consists of all elements $x$ for which $\|S_t x\|_E \in L_\infty(0, \infty)$, or, in other words, $S_t x \in L_\infty(E; (0, \infty))$. The consideration of other function spaces on the semiaxis leads to the possibility of constructing spaces intermediate between $E_A$ and $E$. Let $F(0, \infty)$ be an ideal lattice on the semiaxis $(0, \infty)$ with respect to the measure $dt/t$. By definition, we set $x \in E_{A,F}$ if $\|S_t x\|_E \in F(0, \infty)$. The set $E_{A,F}$ is linear. In it we introduce the norm

$$\|x\|_{E_{A,F}} = \|x\|_E + \|\|S_t x\|_E\|_F. \tag{1.6}$$

Then $E_{0A}$ is isomorphic to $E_{A,L_\infty}$ in view of (1.5).

LEMMA 1.3. *The space* $E_{A,F}$ *is complete.*

PROOF. We consider a sequence of elements $x_n \in E_{A,F}$ for which $\sum_1^\infty \|x_n\|_{E_{A,F}} < \infty$. We show that $\sum x_n$ converges to some element of $E_{A,F}$. From (1.6) and the completeness of $E$ it follows that $\sum x_n$ converges in $E$ to an element $x$. Then $\|S_t x\|_E = \|S_t \sum x_n\|_E \leqslant \sum\|S_t x_n\|_E$. Since $F$ is complete and $\sum\| \|S_t x_n\|_E\|_F$ is convergent, the function $\sum\|S_t x_n\|_E$ belongs to $F$, and since $F$ is an ideal lattice, we have $\|S_t x\|_E \in F$. Consequently $x \in E_{A,F}$. Moreover,

$$\begin{aligned}\|x\|_{E_{A,F}} &= \Big\|\sum x_n\Big\|_E + \Big\| \Big\|S_t \sum x_n\Big\|_E\Big\|_F \\ &\leqslant \sum \|x_n\|_E + \sum \| \|S_t x_n\|_E\|_F = \sum \|x_n\|_{E_{A,F}}.\end{aligned} \tag{1.7}$$

It remains to prove that $\Sigma_1^N x_n \to x$ in $E_{A,F}$ as $N \to \infty$. This follows from (1.7) applied to the element $x - \Sigma_1^N x_n = \Sigma_{N+1}^\infty x_n$. We have

$$\left\| x - \sum_1^N x_n \right\|_{E_{A,F}} = \left\| \sum_{N+1}^\infty x_n \right\|_{E_{A,F}} \leqslant \sum_{N+1}^\infty \|x_n\|_{E_{A,F}} \to 0 \quad \text{as } N \to \infty.$$

The lemma is proved.

It is obvious that $\|x\|_{A,F} \geqslant \|x\|_E$, and so $E_{A,F}$ is imbedded in $E$.

We assume that the ideal lattice $F$ has the following property:

a) The function $\min\{1, \rho^{-1}(t)\}$ belongs to $F$.

By the obvious inequality $\min\{M, \rho^{-1}(t)\} \leqslant \max\{1, M\}\min\{1, \rho^{-1}(t)\}$ we have $\min\{M, \rho^{-1}(t)\} \in F$ for every $M > 0$. Next, for every $a > 0$ we have

$$\chi_{[0,a]} \leqslant \rho(a)\min\{\rho^{-1}(a), \rho^{-1}(t)\} \in F$$

and

$$\rho^{-1}(t)\chi_{(a,\infty)} \leqslant \min(\rho^{-1}(a), \rho^{-1}(t)) \in F.$$

LEMMA 1.4. *Under condition* a) *the space $E_A$ is imbedded in $E_{A,F}$, and so $E_{A,F}$ is intermediate between $E_A$ and $E$.*

PROOF. If $x \in E_A$, then from (1.2) and (1.3) it follows that $\|S_t x\|_E \leqslant \min\{M, \rho^{-1}(t)\}\|x\|_{E_A}$. Hence

$$\| \, \|S_t x\|_E \|_F \leqslant \|\min\{M, \rho^{-1}(t)\}\|_F \|x\|_{E_A}.$$

Condition a) implies that the first factor on the right is finite; that is, $E_A$ is imbedded in $E_{A,F}$. The lemma is proved.

The operators $Q_s$ map $E$ into $E_A$, and therefore they act in $E_{A,F}$. Moreover, they are uniformly bounded in $E_{A,F}$. Indeed,

$$\begin{aligned} \|Q_s x\|_{E_{A,F}} &= \|Q_s x\|_E + \| \, \|S_t Q_s x\|_E \|_F = \|Q_s x\|_E + \| \, \|Q_s S_t x\|_E \|_F \\ &\leqslant M(\|x\|_E + \| \, \|S_t x\|_E \|_F) \leqslant M \|x\|_{E_{A,F}}. \end{aligned} \tag{1.8}$$

LEMMA 1.5. *If an ideal lattice $F$ has the Fatou property, then $E_{A,F}$ is complete relative to $E$ and coincides with the completion of $E_A$ in the norm* (1.6) *relative to $E$.*

PROOF. If an element $x$ belongs to the completion of $E_{A,F}$ relative to $E$, then there exists a sequence $x_n \to x$ in $E$ such that $\|x_n\|_{E_{A,F}} \leqslant C$. Since the operators $S_t$ are bounded, $S_t x_n \to S_t x$ in $E$, and hence $\|S_t x_n\|_E \to \|S_t x\|_E$ for all $t$. Then from the Fatou property it follows that

$$\| \, \|S_t x\|_E \|_F \leqslant \lim \| \, \|S_t x_n\|_E \|_F \leqslant \lim \|x_n\|_{E_{A,F}} \leqslant C,$$

and so $x \in E_{A,F}$. The completeness of $E_{A,F}$ relative to $E$ is proved.

Now let $x$ be any element in $E_{A,F}$. Then $Q_s x \in E_A$, $Q_s x \to x$ in $E$, and by (1.1) we have $\|Q_s x\|_{E_{A,F}} \leqslant M\|x\|_{E_{A,F}}$. This means that $E_{A,F}$ is contained in the completion of $E_A$ in the norm (1.6) relative to $E$, and since $E_{A,F}$ is complete in $E$, it coincides with this completion.

The lemma is proved.

LEMMA 1.6. *If the ideal lattice $F$ has property* a) *and has absolutely continuous norm, the set $E_A$ is dense in $E_{A,F}$.*

PROOF. Let $x$ be an arbitrary element in $E_{A,F}$. Using (1.3) and (1.7), we have

$$\begin{aligned}\|x - Q_s x\|_{E_{A,F}} &\leqslant \|\chi_{[0,\delta]}(t)\|S_t(x - Q_s x)\|_E\|_F \\ &\quad + \|\chi_{[\delta,\infty)}\|S_t(x - Q_s x)\|_E\|_F + \|x - Q_s x\|_E \\ &\leqslant (M+1)\|\chi_{[0,\delta]}(t)\|S_t x\|_E\|_F \\ &\quad + \big(\|\chi_{[\delta,\infty)}(t)\rho^{-1}(t)\|_F + 1\big)\|x - Q_s x\|_E.\end{aligned}$$

Due to the absolute continuity of the norm in $F$, the first term on the right may be made arbitrarily small if we choose $\delta$ sufficiently small, and then so can the second term, if we choose $s$ sufficiently small. Thus, $Q_s x \in E_A$ and $Q_s x \to x$ in $E_{A,F}$.

The lemma is proved.

3. ***Interpolation intermediate spaces.*** There arises the problem, important in applications, of the relations between the spaces $E_{A,F}$ constructed by means of a smoothing approximation process and the spaces constructed from the Banach couple $E$, $E_A$ by means of certain interpolation constructions.

THEOREM 1.1. *The space $(E, E_A)^{\mathcal{K}}_{F_{\rho^{-1}(t)},F}$ constructed from the Banach couple $E$, $E_A$ by the $\mathcal{K}$-method by means of the ideal lattices $F_{\rho^{-1}(t)}$, $F$ is imbedded in $E_{A,F}$.*

PROOF. If $x \in (E, E_A)^{\mathcal{K}}_{F_{\rho^{-1}},F}$, then there exists a representation $x = u_0(t) + u_1(t)$, where $u_0 \in F_{\rho^{-1}}(E)$ and $u_1 \in F(E_A)$. Then $S_t x = S_t u_0(t) + S_t u_1(t)$. By (1.2) and (1.3) we have

$$\|S_t u_0(t)\|_E \leqslant \rho^{-1}(t)\|u_0(t)\|_E \in F$$

and

$$\|S_t u_1(t)\|_E \leqslant M\|u_1(t)\|_{E_A} \in F.$$

Hence $\|S_t x\|_E \leqslant \|S_t u_0(t)\|_E + \|S_t u_1(t)\|_E \in F$. From the inequalities obtained so far it also follows that

$$\|x\|_{E_{A,F}} \leqslant M\|x\|_{(E,E_A)^{\mathcal{K}}_{F_{\rho^{-1}},F}}.$$

The theorem is proved.

From the point of view of approximation theory, Theorem 1.1 belongs to assertions of the type of the classical Jackson theorem: a space is described on which the process $S_t$ behaves in a certain way ($\|S_t x\|_E \in F$) and the construction of the space is not connected with the process itself. The converse assertions in which from a certain behavior of the process $S_t$ we conclude that the element $x$ belongs to some space are analogues of the classical Bernstein theorem. To obtain such theorems we need a more delicate use of the connection between the operator $A$ and the process $Q_t$, or, what is the same, between the processes $S_t$ and $Q_t$.

In the examples given below the operators $S_t$ and $Q_t$ satisfy a relation of the following form:

$$I = \int_0^1 \rho(\lambda|t)L(\lambda, t)S_{\lambda t}\,d\alpha(\lambda) + \int_\delta^1 Q_{\lambda t}N(\lambda, t)\,d\beta(\lambda), \tag{1.9}$$

where $L(\lambda, t)$ and $N(\lambda, t)$ are operator-valued functions uniformly bounded in $\lambda$ and $t$ and commuting with $S_\tau$ and $Q_\tau$ for all $\tau > 0$, and $\alpha(\lambda)$ and $\beta(\lambda)$ are bounded increasing scalar-valued functions on $(0, 1]$ and $[\delta, 1]$, respectively.

THEOREM 1.2. *Assume that representation* (1.9) *is valid, and that condition* a) *and the inequality*

$$\int_0^1 M_\rho(\lambda)\|\sigma_{1/\lambda}\|_{F\to F}\,d\alpha(\lambda) + \int_\delta^1 \|\sigma_{1/\lambda}\|_{F\to F}\,d\beta(\lambda) < \infty, \tag{1.10}$$

*hold, where $\sigma_{1/\lambda}$ is a dilation operator and*

$$M_\rho(\lambda) = \sup_{t>0} \rho(\lambda t)\rho^{-1}(t).$$

*Then $E_{A,F}$ is isomorphic to $(E, E_A)_{F_{\rho^{-1}},F}$.*

PROOF. By virtue of Theorem 1.1 it is sufficient to obtain the imbedding $E_{A,F} \subset (E, E_A)^{\mathcal{K}}_{F_{\rho^{-1}},F}$. According to Lemma 2.8 of Chapter IV, for this it is sufficient to show that for any $x \in E_{A,F}$ we have $\rho^{-1}(t)\mathcal{K}(\rho(t), x) \in F$, where the function $\mathcal{K}(t, x)$ is constructed for the spaces $E$ and $E_A$. Using (1.9), we have

$$\begin{aligned}\mathcal{K}(\rho(t), x) &\leqslant \left\|\int_0^1 \rho(\lambda t)L(\lambda, t)S_{\lambda t}x\,d\alpha(\lambda)\right\|_E \\ &\quad + \rho(t)\left\|\int_\varepsilon^1 Q_{\lambda t}N(\lambda, t)x\,d\beta(\lambda)\right\|_{E_A} \\ &\leqslant C\Big(\int_0^1 \rho(\lambda t)\|S_{\lambda t}x\|_E\,d\alpha(\lambda) + \rho(t)\int_\delta^1 \|s_{\lambda t}x\|_E\,d\beta(\lambda) + \rho(t)\|x\|_E\Big).\end{aligned}$$

From this we obtain

$$\rho^{-1}(t)\mathcal{K}(\rho(t), x) \leqslant C\Big(\int_0^1 M_\rho(\lambda)\|S_{\lambda t}x\|_E\, d\alpha(\lambda) + \int_\delta^1 \|S_{\lambda t}x\|_E\, d\beta(\lambda) + \|x\|_E\Big).$$

By definition, $\mathcal{K}(t, x) \leqslant \|x\|_E$ for any $x \in E$, and therefore

$$\rho^{-1}(t)\mathcal{K}(t, x) \leqslant C\int_0^1 M_\rho(\lambda)\|S_{\lambda t}x\|_E\, d\alpha(\lambda) + C\int_\delta^1 \|S_{\lambda t}x\|_E\, d\beta(\lambda) + \min\{C, \rho^{-1}(t)\}\|x\|_E.$$

The last term on the right belongs to $F$ by condition a). For the second term we have

$$\begin{aligned}\Big\|\int_\delta^1 \|S_{\lambda t}x\|\, d\beta(\lambda)\Big\|_F &\leqslant \int_\delta^1 \|\, \|S_{\lambda t}x\|_E\|_F\, d\beta(\lambda)\\ &\leqslant \int_\delta^1 \|\sigma_{1/\lambda}\|_{F\to F} d\beta(\lambda)\|\; \|S_t x\|_E\|_F\\ &\leqslant \int_\delta^1 \|\sigma_{1/\lambda}\|_{F\to F} d\beta(\lambda)\|x\|_{E_{A,F}}.\end{aligned}$$

It can be shown similarly that under condition (1.10) the first term also belongs to $F$.

The theorem is proved.

To calculate the norm in $E_{A,F}$ the quantity

$$\omega(t, x) = \sup_{0<\tau<t} \|\rho(\tau)S_\tau x\|_E$$

is sometimes introduced. By (1.2) and (1.3) we have

$$\begin{aligned}\omega(t, x) \leqslant \|x\|_E \quad \text{and} \quad \omega(t, x) &\leqslant \sup_{0<\tau<t} \rho(\tau)M\|Ax\|_E\\ &\leqslant M\rho(t)\|Ax\|_E \qquad (x \in \mathfrak{D}(A)). \quad (1.11)\end{aligned}$$

LEMMA 1.7. *Under condition* (1.10) *the norm in* $E_{A,F}$ *is equivalent to* $\|\omega(t, x)\|_{F_{\rho^{-1}}}$.

PROOF. By means of (1.11) we obtain from representation (1.9) that

$$\begin{aligned}\omega(t, x) &\leqslant \int_0^1 \omega(t, \rho(\lambda t)L(\lambda, t)S_{\lambda t}x)\, d\alpha(\lambda) + \int_\delta^1 \omega(t, Q_{\lambda t}N(\lambda, t)x)\, d\beta(\lambda)\\ &\leqslant \int_0^1 \|\rho(\lambda t)L(\lambda, t)S_{\lambda t}x\|_E\, d\alpha(\lambda)\\ &\quad + M\int_\delta^1 \rho(t)\|AQ_{\lambda t}N(\lambda, t)x\|_E\, d\beta(\lambda).\end{aligned}$$

Hence

$$\rho^{-1}(t)\omega(t, x) \leqslant C\int_0^1 M_\rho(\lambda)\|S_{\lambda t}x\|_E d\alpha(\lambda) + C\int_\delta^1 \|S_{\lambda t}x\|_E\, d\beta(\lambda).$$

In the proof of Theorem 1.2 we proved that the function on the right belongs to $F$ and its norm does not exceed $C\|x\|_{E_{A,F}}$.

By definition, $\rho^{-1}(t)\omega(t, x) \geqslant \|S_t x\|_E$, and so the reverse inequality $\|x\|_{E_{A,F}} \leqslant \|\rho^{-1}\omega\|_F$ is obvious.

The lemma is proved.

As a consequence, Theorem 1.2 implies the following theorem.

THEOREM 1.3 (interpolation theorem). *Under conditions* a), (1.9), *and* (1.10) *$E_{A,F}$ is an interpolation space between $E$ and $E_A$.*

In what follows the function $\rho(t)$ will be proportional to $t^r$. Let $G$ be an interpolation space between $L_{1,*}$ and $L_{\infty,*}$. We set $F = G^{r-s}$, where $0 < s < r$. Condition a) turns into the condition $\min\{1, t^{-r}\} \in G^{r-s}$ or, what is the same, $\min\{t^{r-s}, t^{-s}\} \in G$. The last function belongs to $L_{1,*} \cap L_{\infty,*}$, and therefore also to $G$, i.e. condition a) is satisfied.

It is easy to verify that for the dilation operator we have

$$\|\sigma_{1/\lambda}\|_{G^{r-s}\to G^{r-s}} = \lambda^{-(r-s)}\|\sigma_{1/\lambda}\|_{G\to G}.$$

Since $G$ is an interpolation space between $L_{1,*}$ and $L_{\infty,*}$, $\|\sigma_{1/\lambda}\|_{G\to G}$ is bounded by a constant (see (2.5) in Chapter IV). The second term in (1.10) is then finite. Since in our case the function $M_\rho(\lambda)$ is equal to $\lambda^r$, the first term in (1.10) takes the form $\int_0^1 \lambda^s d\alpha(\lambda)$, and so it is also finite. Thus, (1.10) is satisfied.

The space $E_{A,F} = (E, E_A)^{\mathcal{K}}_{F_\rho^{-1},F}$ turns into the space $(E, E_A)^{\mathcal{K}}_{G^{-s},G^{r-s}}$. By Theorem 2.3 and Lemma 2.4 of Chapter IV this space is isomorphic to $(E, E_A)_{\theta,G,G}$, where $\theta = s/r$.

**4. *Subordinate operators.***

DEFINITION 1.1. A closed linear operator $B$ is *subordinate to the operator $A$ with order $\theta$* $(0 \leqslant \theta \leqslant 1)$ if $\mathfrak{D}(B) \supset \mathfrak{D}(A)$ and

$$\mathfrak{D}(B) \supset \mathfrak{D}(A) \quad \text{and} \quad \|Bx\|_E \leqslant C\|x\|_E^{1-\theta}\|x\|_{E_A}^{\theta} \qquad (x \in \mathfrak{D}(A)). \tag{1.12}$$

An operator $B$ *has order $\theta$ with respect to the operator $A$ and the smoothing approximation process $Q_t$* if (1.12) is satisfied and

$$\|\rho(t)S_t x\|_E \leqslant C\rho^\theta(t)\|x\|_{E_B} \tag{1.13}$$

for $x \in \mathfrak{D}(B)$.

From (1.12) it follows that

$$\|x\|_{E_B} \leqslant \max\{1, C\}\|x\|_E^{1-\theta}\|x\|_{E_A}^{\theta}$$

and by Lemma 2.6 of Chapter IV we have

$$(E, E_A)_{\theta,L_{1,*},L_{1,*}} \subset E_B.$$

If $\rho(t)$ is proportional to $t^r$, then (1.13) implies that

$$t^{r-\theta r}\|S_t x\| \leqslant C\|x\|_{E_B},$$

and consequently $E_B$ is imbedded in $E_{A,L^{r-\theta r}_{\infty,*}}$, which is isomorphic to $(E, E_A)_{\theta,L_{\infty,*},L_{\infty,*}}$ under conditions (1.9) and (1.2).

We have arrived at the following assertion.

LEMMA 1.8. *If condition* (1.9) *is satisfied and the operator $B$ has order $\theta$ with respect to the operator $A$ and the smoothing approximation process $Q_t$, then $E_B$ is an intermediate space of type $\theta$ between $E$ and $E_A$.*

The reiteration Theorem 2.8 of Chapter IV and the remarks to it lead us to the following important assertion.

THEOREM 1.4. *If condition* (1.9) *is satisfied and the operators $B_i$ $(i = 0, 1)$ have order $\theta_i$ with respect to the operator $A$ and the smoothing approximation process $S_t$, then for $0 < \theta_0 < \theta < \theta_1 < 1$*

$$(E_{B_0}, E_{B_1})_{\theta',G,G} = (E, E_A)_{\theta,G,G} = E_{A,G^{r-s}}, \tag{1.14}$$

*up to isomorphism, where $\theta' = (\theta - \theta_0)/(\theta_1 - \theta_0)$ and $s = \theta r$. Moreover, for $0 < \theta < \theta_1$*

$$(E, E_{B_1})_{\theta'',G,G} = (E, E_A)_{\theta,G,G},$$

*up to isomorphism, where $\theta'' = \theta/\theta_1$, and for $\theta_0 < \theta < 1$*

$$(E_{B_0}, E_A)_{\theta''',G,G} = (E, E_A)_{\theta,G,G},$$

*where $\theta''' = (\theta - \theta_0)/(1 - \theta_0)$.*

**5. *Smoothing by means of the resolvent.*** In a Banach space $E$ let an unbounded closed linear operator $B$ be given with domain $\mathfrak{D}(B)$ dense in $E$, having the property that all points of the negative semiaxis $(-\infty, 0)$ are regular for it and

$$\sup_{\lambda>0}\|\lambda(B + \lambda)^{-1}\|_E < \infty. \tag{1.15}$$

From this property it follows immediately that

$$\sup_{\lambda>0}\|B(B + \lambda)^{-1}\|_E = \sup_{\lambda>0}\|I - \lambda(B + \lambda)^{-1}\|_E < \infty.$$

We write

$$a = \sup_{\lambda>0}\|\lambda(B + \lambda)^{-1}\|_E < \infty \quad \text{and} \quad b = \sup_{\lambda>0}\|B(B + \lambda)^{-1}\|_E < \infty. \tag{1.16}$$

If $x \in \mathfrak{D}(B)$, then $\lambda(B + \lambda)^{-1}x = x - (B + \lambda)^{-1}Bx$, and $\lambda(B + \lambda)^{-1}x \to x$ as $\lambda \to \infty$ by (1.15). Since, again by (1.15), the operators $\lambda(B + \lambda)^{-1}$ are uniformly bounded, we have

$$\lambda(B+\lambda)^{-1}x \to x \quad \text{as } \lambda \to \infty \quad (x \in E).$$

We write $t = 1/\lambda$. Then (1.16) and the last relation can be written in the following form:

$$a = \sup_{t>0} \|(tB+I)^{-1}\|_E < \infty, \quad b = \sup_{t>0} \|tB(tB+I)^{-1}\|_E < \infty, \tag{1.17}$$

$$(tB+I)^{-1}x \to x \quad \text{as } t \to 0 \quad (x \in E). \tag{1.18}$$

We introduce the operator $A = B^r$. This operator also has dense domain $\mathfrak{D}(A) = \mathfrak{D}(B^r)$ in $E$ (see [18], §2.16). The space $E_A$ consists of all elements $x \in \mathfrak{D}(B^r)$ with norm

$$\|x\|_{E_{B^r}} = \|x\|_E + \|B^r x\|_E.$$

It is sometimes convenient to introduce another equivalent norm in $E_{B^r}$. This possibility is based on the following well-known assertion.

LEMMA 1.9. *If $A$ is a closed linear operator with domain $\mathfrak{D}(A)$ and $D$ is a linear operator that admits closure, whose domain $\mathfrak{D}(D)$ contains $\mathfrak{D}(A)$, then*

$$\|Dx\|_E \leqslant C(\|x\|_E + \|Ax\|_E), \tag{1.19}$$

*where $C$ does not depend on $x \in \mathfrak{D}(A)$ (the operator $D$ is subordinate to $A$ with order $\theta = 0$).*

PROOF. We consider the restriction of $D$ to $\mathfrak{D}(A)$ as an operator from $E_A$ to $E$. It is closed. Indeed, if $x_n \to x$ in $E_A$ and $Dx_n \to y$ in $E$, then $x_n \to x$ in $E$ and $Ax_n \to x$ in $E$. The operator $D$ admits closure $\bar{D}$ in $E$; therefore $y = \bar{D}x$, and then $y = Dx$, since $x \in \mathfrak{D}(A)$. The operator $D$ is defined on the whole space $E_A$, and is therefore bounded. This assertion is equivalent to (1.19).

The lemma is proved.

By the definition of powers of operators, we have $\mathfrak{D}(B^k) \supset \mathfrak{D}(B^r)$ for $k < r$, and therefore Lemma 1.9 implies the inequality

$$\|B^k x\|_E \leqslant C_{kr}(\|x\|_E + \|B^r x\|_E). \tag{1.20}$$

From this it follows that in $E_{B^r}$ an equivalent norm can be introduced by the formula

$$\|x\|_{E_{B^r}} = \sum_{k=0}^{r} \|B^k x\|_E. \tag{1.21}$$

Now for the operator $B^r$ we construct a smoothing approximation process in $E$ by the formula

$$Q_t x = (tB+I)^{-r}x. \tag{1.22}$$

The properties of the resolvent and relations (1.17) and (1.18) enable us to verify easily that this process has the required properties 1) and 2). We may set $M = a^r$.

Next,

$$S_t = AQ_t = B^r(tB + I)^{-r} = \left[B(tB + I)^{-1}\right]^r. \tag{1.23}$$

From (1.17) we obtain that

$$\|S_t x\|_E \leqslant b^r/t^r,$$

and so we may assume that

$$\rho(t) = b^{-r}t^r. \tag{1.24}$$

In the corresponding space $E_{B^r,G^{r-s}}$ the norm is defined by

$$\|x\|_{E_{B^r,G^{r-s}}} = \|x\|_E + \left\| t^{r-s} \left\| \left[B(tB + I)^{-1}\right]^r x \right\|_E \right\|_G.$$

Now we pass to the derivation of a representation of the type (1.9). For this we observe that the polynomials $(1 - u)^r$ and $u^r$ in $u$ are relatively prime, and thus there exist polynomials $p(u)$ and $q(u)$ such that $(1 - u)^r p(u) + u^r q(u) = I$. In this identity we substitute the bounded operator $(tB + I)^{-1}$ for $u$. Then we obtain

$$I = \left[tB(tB + I)^{-1}\right]^r p\left((tB + I)^{-1}\right) + (tB + I)^{-r} q\left((tB + I)^{-1}\right). \tag{1.25}$$

Taking account of (1.22)–(1.24), we see that (1.25) is precisely a representation of type (1.9) in which the functions $\alpha(\lambda)$ and $\beta(\lambda)$ are functions of unit jump at the point

$$\lambda = 1 \qquad (d\alpha(\lambda) = d\beta(\lambda) = \delta(1 - \lambda)d\lambda),$$

$$L(\lambda, t) = b^r p\left((\lambda tB + I)^{-1}\right) \quad \text{and} \quad N(\lambda, t) = q\left((\lambda tB + I)^{-1}\right).$$

We shall not reformulate all assertions proved earlier for the special form of the operator and the smoothing approximation process considered here, and restrict ourselves to the formulation of an assertion which follows from Theorem 1.2.

THEOREM 1.5. *Let the space $E_{B^r,G^{r-s}}$ be constructed from the operator $B^r$, the smoothing approximation process $Q_t = (tB + I)^{-r}$ and the space $G$ that is an interpolation space between $L_{1,*}$ and $L_{\infty,*}$. If the operator $B$ satisfies condition* (1.15), *then the space $E_{B^r,G^{r-s}}$ is isomorphic to the space $(E, E_{B^r})_{\theta,G,G}$, where $\theta = s/r$.*

Now we pass to subordinate operators.

LEMMA 1.10. *If condition* (1.15) *is satisfied, then for $m < r$ the operator $B^m$ is subordinate to the operator $B^r$ with order $m/r$.*

PROOF. We consider the polynomials $(1-u)^{r-m}$ and $u^m$ in $u$. Since they are relatively prime, there are polynomials $p_m(u)$ and $q_m(u)$ such that $(1-u)^{r-m}p_m(u) + u^m q_m(u) = 1$. Hence $(1-u)^m = (1-u)^r p_m(u) + u^m(1-u)^m q_m(u)$. In this identity we substitute the bounded operator $(tB + I)^{-1}$ for $u$. Then we obtain

$$[tB(tB+I)^{-1}]^m = [tB(tB+I)^{-1}]^r p_m((tB+I)^{-1})$$
$$+ (tB+I)^{-m}(tB(tB+I)^{-1})^m q_m((tB+I)^{-1}).$$

If now $x \in \mathfrak{D}(B^r)$, then applying the operators to this element on both sides of the above identity, we can write the result in the form

$$(tB+I)^{-m}B^m x = t^{r-m}(tB+I)^{-r}p_m((tB+I)^{-1})B^r x$$
$$+ t^{-m}(tB+I)^{-m}(tB(tB+I)^{-1})^m q_m((tB+I)^{-1})x.$$

Since the operator $(tB+I)^{-m}$ is invertible, we have

$$B^m x = t^{r-m}(tB+I)^{-r+m}p_m((tB+1)^{-1})B^r x$$
$$+ t^{-m}(tB(tB+I)^{-1})^m q_m((tB+I)^{-1})x.$$

Taking account of (1.17), we obtain

$$\|B^m x\|_E \leqslant C_1(a,b)[t^{r-m}\|B^r x\|_E + t^{-m}\|x\|_E].$$

Minimizing the right side over $t$, we finally get

$$\|B^m x\|_E \leqslant C(a,b)\|x\|_E^{1-m/r}\|B^r x\|_E^{m/r}. \tag{1.26}$$

The lemma is proved.

Now we show that (1.13) is satisfied. We have

$$\|\rho(t)S_t x\|_E = b^{-r}\|[tB(tB+I)^{-1}]^r x\|_E \leqslant b^{-m}\|[tB(tB+I)^{-1}]^m x\|_E$$
$$\leqslant b^{-m}t^m a^m\|B^m x\|_E = a^m \rho^{m/r}(t)\|B^m x\|_E.$$

Thus, the operator $B^m$ has order $m/r$ with respect to the operator $B^r$ and the process $(tB+I)^{-r}$.

Then Theorem 1.4 implies the following assertion.

THEOREM 1.6. *If the operator $B$ satisfies condition* (1.15), *then, up to isomorphism, for* $0 \leqslant m_0/r < \theta < m_1/r \leqslant 1$ *we have*

$$(E_{B^{m_0}}, E_{B^{m_1}})_{\theta',G,G} = (E, E_{B^r})_{\theta,G,G} = E_{B^r,G^{r-s}}, \tag{1.27}$$

*where* $\theta' = (s-m_0)/(m_1-m_0)$ *and* $s = \theta r$. (*We assume that* $E_{B^0} = E$.)

Hence we obtain the following important result.

COROLLARY 1. *The space* $E_{B^r,G^{r-s}}$ *is independent of* $r > s$ *up to isomorphism.*

Indeed, from Theorem 1.6 it follows that for $s < r_1 < r$ ($r$ and $r_1$ are integers) we have

$$E_{B^{r_1},G^{r_1-s}} = (E, E_{B^{r_1}})_{s/r_1,G,G} = (E, E_{B^r})_{s/r,G,G} = E_{B^r,G^{r-s}}.$$

COROLLARY 2. *If the integer $m$ is smaller than $s$, then $E_{B^r,G^{r-s}} \subset \mathfrak{D}(B^m)$.*

Indeed, in this case we have

$$E_{B^r,G^{r-s}} = (E_{B^m}, E_{B^r})_{(s-m)/(1-m),G,G} \subset E_{B^m} = \mathfrak{D}(B^m).$$

We attempt to describe the space on the left of (1.27) more effectively. On the space $E_{B^{m_0}} = \mathfrak{D}(B^{m_0})$ we may consider the operator $B$, which will satisfy condition (1.15) in the norm of this space too. The domain of $B^{m_1-m_0}$ on $\mathfrak{D}(B^{m_0})$ is exactly the space $\mathfrak{D}(B^{m_1})$. Finally, the operators $(tB+I)^{-(m_1-m_0)}$ form a smoothing approximation process for $B^{m_1-m_0}$. By Theorem 1.5, up to isomorphism, we then have

$$\begin{aligned}(E_{B^{m_0}}, E_{B^{m_1}})_{\theta',G,G} &= (\mathfrak{D}(B^{m_0}))_{B^{m_1-m_0},G^{(1-\theta')(m_1-m_0)}} \\ &= (\mathfrak{D}(B^{m_0}))_{B^{m_1-m_0},G^{m_1-s}}.\end{aligned}$$

This means that our space consists of all elements of $\mathfrak{D}(B^{m_0})$ for which

$$\begin{aligned}&t^{m_1-s}\left\|\left[B(tB+I)^{-1}\right]^{m_1-m_0}x\right\|_{\mathfrak{D}(B^{m_0})} \\ &\quad = t^{m_1-s}\left\|\left[B(tB+I)^{-1}\right]^{m_1-m_0}B^{m_0}x\right\|_E \\ &\qquad + t^{m_1-s}\left\|\left[B(tB+I)^{-1}\right]^{m_1-m_0}x\right\|_E \in G.\end{aligned}$$

We show that if the first term belongs to $G$, then $x$ belongs to $E_{B^r,G^{r-s}}$. By virtue of Corollary 1 it is sufficient to show that $x \in E_{B^{2m_1},G^{2m_1-s}}$. We consider the expression

$$t^{2m_1-s}\left[B(tB+I)^{-1}\right]^{2m_1} = t^{2m_1-s}B^{m_1}B^{m_1}(tB+I)^{-2m_1}$$

and to estimate it we apply Lemma 1.10 with exponents $r = m_1 + m_0$ and $m = m_1$. Then

$$\begin{aligned}&t^{2m_1-s}\left\|\left[B(tB+I)^{-1}\right]^{2m_1}x\right\|_E \\ &\quad \leqslant t^{2m_1-s}\left\{\|B^{m_1}(tB+I)^{-2m_1}x\|_E\right\}^{m_0/(m_0+m_1)} \\ &\qquad \times \left\{\|B^{2m_1+m_0}(tB+I)^{-2m_1}x\|_E\right\}^{m_1/(m_0+m_1)} \\ &\quad \leqslant a^{m_0}b^{m_1}t^{m_1-s}\|B^{m_1-m_0}(tB+I)^{-(m_1-m_0)}B^{m_0}x\|_E \in G.\end{aligned}$$

Thus we have proved the following theorem.

THEOREM 1.7. *The space* $E_{B^r, G^{r-s}}$ *consists of all elements* $x \in \mathfrak{D}(B^{m_0})$ *with* $m_0 < s$ *for which*

$$B^{m_0}x \in E_{B^\rho, G^{\rho-(s-m_0)}} \qquad (\rho > s - m).$$

We have proved the theorem for $\rho = m_1 - m_0$; nevertheless, by Corollary 1 it also holds in the above formulation.

From general considerations it follows that

$$\|x\|_E + \left\| t^{\rho-(s-m_0)} \left\| \left[ B(tB + I)^{-1} \right]^\rho B^{m_0}x \right\|_E \right\|_G \qquad (\rho > s - m_0)$$

is equivalent to the norm of $E_{B^r, G^{r-s}}$.

We mention two special cases of the last formula: if $s$ is not an integer, then by setting $m_0 = [s]$ and $\rho = 1$, we obtain that

$$\|x\|_E + \left\| t^{1-(s-[s])} \left\| \left[ B(tB + I)^{-1} \right] B^{[s]}x \right\|_E \right\|_G \tag{1.28}$$

is equivalent to the norm of $E_{B^r, G^{r-s}}$. If, on the other hand, $s$ is an integer, then, taking $m_0 = s - 1$ and $\rho = 2$, we obtain an equivalent norm of the form

$$\|x\|_E + \left\| t \left\| \left[ B(tB + I)^{-1} \right]^2 B^{s-1}x \right\|_E \right\|_G. \tag{1.29}$$

**6. *Smoothing by means of a semigroup.***(¹) A *strongly continuous semigroup of operators* is, by definition, a family of bounded linear operators $U(t)$ in a Banach space $E$ which depends on the parameter $t \in (0, \infty)$ continuously in the strong operator topology and has the property that $U(t + \tau) = U(t)U(\tau)$ $(0 < t, \tau < \infty)$. If, moreover, $U(t)$ strongly converges to the identity operator as $t \to 0$, we say that the semigroup *satisfies the* $C_0$*-condition*. For every semigroup with the $C_0$-condition we have

$$\| U(t) \|_{E \to E} \leqslant M e^{\omega t}.$$

We shall consider *bounded* semigroups, i.e. semigroups with the $C_0$-condition and the estimate

$$\| U(t) \|_{E \to E} \leqslant M. \tag{1.30}$$

For a semigroup with the $C_0$-condition the function $U(t)x$ is continuous at zero, and $U(0)x = x$. We consider the collection of all elements $x$ for which $U(t)x$ is differentiable at zero (from the right). For these elements the linear operator

$$U'(0)x = \lim_{h \to +0} \frac{U(h)x - x}{h} = Dx$$

---

(¹) The basic information on semigroups of linear operators can be found in the books [18], [11], [48], and [24].

is defined, and therefore we denote the indicated collection by $\mathfrak{D}(D)$. It turns out that the operator $D$ is closed and has dense domain $\mathfrak{D}(D)$ in $E$. The operator $D$ is called the *infinitesimal generator* of the semigroup $U(t)$.

The infinitesimal generator of a bounded semigroup has a resolvent $R(\lambda) = (D - \lambda)^{-1}$ for $\lambda > 0$, and this resolvent is related to the semigroup by the Laplace transformation

$$R(\lambda) = -\int_0^\infty e^{-\lambda t} U(t)\, dt. \tag{1.31}$$

For a closed operator $D$ with dense domain $\mathfrak{D}(A)$ in $E$ to be the infinitesimal generator of a bounded semigroup it is necessary and sufficient that it have a resolvent $R(\lambda)$ for $\lambda > 0$ satisfying the conditions

$$\|\lambda^k R^k(\lambda)\|_{E\to E} \leqslant M \qquad (k = 1, 2, \dots). \tag{1.32}$$

From this it is clear (for $k = 1$) that the operator $B = -D$ belongs to the class of operators satisfying condition (1.15) considered in subsection 5. We would like to mention that this class of operators is larger than the class of infinitesimal generators of bounded semigroups.

On $\mathfrak{D}(D)$ the semigroup $U(t)$ commutes with $D$, and

$$\frac{dU(t)x}{dt} = DU(t)x = U(t)Dx \qquad (x \in \mathfrak{D}(A)). \tag{1.33}$$

From this we obtain

$$B\int_0^t U(\tau)x\,d\tau = -D\int_0^t U(\tau)x\,d\tau = (I - U(t))x. \tag{1.34}$$

We consider the operator $B^r = (-D)^r$, and for it we construct a smoothing interpolation process by the formula

$$Q_t x = \left[\frac{1}{t}\int_0^t U(\tau)\, d\tau\right]^r x = \frac{1}{t^r}\int_0^t \cdots \int_0^t U(\tau_1 + \cdots + \tau_r)x\,d\tau_1 \dots d\tau_r. \tag{1.35}$$

The operators $Q_t$ form a commuting family of bounded operators which is strongly continuous in $t$ and strongly converges to $I$ as $t \to 0$ (by virtue of the strong continuity of the semigroup at zero, the $C_0$-condition). Next,

$$\|Q_t x\| \leqslant \frac{1}{t^r}\int_0^t \cdots \int_0^t \|U(\tau_1 + \cdots + \tau_r)x\|\,d\tau_1 \dots d\tau_r \leqslant M\|x\|.$$

Finally, by (1.33) and (1.34) the operator $Q_t$ maps $E$ into $\mathfrak{D}(B^r)$ and commutes with $B^r$ on $\mathfrak{D}(B^r)$. Thus, all conditions concerning a smoothing approximation process are satisfied.

From (1.35) we obtain

$$S_t x = \left[\frac{1}{t} B \int_0^t U(\tau)\, d\tau\right]^r x = \frac{1}{t^r}[I - U(t)]^r x. \tag{1.36}$$

From the estimate

$$\|S_t x\|_E \leqslant \frac{(M+1)^r}{t^r} \|x\|_E$$

it follows that as $\rho(t)$ we may take the function

$$\rho(t) = (M+1)^{-r} t^r. \tag{1.37}$$

Now we attempt to obtain a relation of the type (1.9) for our operators $S_t$ and $Q_t$. We have

$$\begin{aligned} x &= \left(\frac{r}{t}\right)^r \int_0^{t/r} \cdots \int_0^{t/r} [I - U(\tau_1 + \cdots + \tau_r)]^r x \, d\tau_1 \ldots d\tau_r \\ &\quad + \left(\frac{r}{t}\right)^r \int_0^{t/r} \cdots \int_0^{t/r} \left[I - (I - U(\tau_1 + \cdots + \tau_r))^r\right] x \, d\tau_1 \ldots d\tau_r \\ &= \int_0^1 \cdots \int_0^1 \left[I - U\left(\frac{t}{r}(s_1 + \cdots + s_r)\right)\right]^r x \, ds_1 \ldots ds_r \\ &\quad + \sum_{k=1}^r (-1)^{k-1} C_r^k \left(\frac{r}{t}\right)^r \int_0^{t/r} \cdots \int_0^{t/r} U(k(\tau_1 + \cdots + \tau_r)) x \, d\tau_1 \ldots d\tau_r. \end{aligned}$$

We use the elementary formula

$$\int_0^1 \cdots \int_0^1 \varphi\left(\frac{\tau_1 + \cdots + \tau_r}{r}\right) d\tau_1 \ldots d\tau_r = \int_0^1 \varphi(\lambda)\, a_r(\lambda) d\lambda. \tag{1.38}$$

Here the functions $a_r(\lambda)$ can be determined from the recurrence relation

$$a_n(\lambda) = \begin{cases} n \int_0^{n\lambda/(n-1)} a_{n-1}(\mu)\, d\mu, & 0 \leqslant \lambda \leqslant 1/n, \\ n \int_{(n\lambda-1)/(n-1)}^{n\lambda/(n-1)} a_{n-1}(\mu)\, d\mu, & 1/n \leqslant \lambda \leqslant (n-1)/n, \\ n \int_{(n\lambda-1)/(n-1)}^{1} a_{n-1}(\mu)\, d\mu, & (n-1)/n \leqslant \lambda \leqslant 1. \end{cases}$$

We note that $a_r(\lambda) = r^r \lambda^{r-1}/(r-1)!$ for $0 \leqslant \lambda \leqslant 1/r$. Applying (1.38), we obtain

$$\begin{aligned} x &= \int_0^1 [I - U(\lambda t)]^r x a_r(\lambda)\, d\lambda \\ &\quad + \sum_{k=1}^r (-1)^{k-1} C_r^k \frac{1}{(kt/r)^r} \int_0^{kt/r} \cdots \int_0^{kt/r} U(s_1 + \cdots + s_r) x \, ds_1 \ldots ds_r. \end{aligned} \tag{1.39}$$

If we use the notation of (1.35)–(1.37), then this identity can be written in the form (1.9), and the operator-valued functions $L(\lambda, t)$ and $N(\lambda, t)$ can be assumed to be identically equal to $I$, $d\alpha(\lambda) = (M + 1)^r a_r(\lambda)\, d\lambda$, and the function $\beta(\lambda)$ is the sum of jump functions with jumps $(-1)^{k-1} C_r^k$ at the points $\lambda = k/r$, $k = 1, \ldots, r$. Thus, (1.39) has all properties of (1.9).

Theorem 1.2 yields the following theorem.

THEOREM 1.8. *Let the space $E_{B^r, G^{r-s}}$ be constructed from the operator $B^r$, the smoothing approximation process $[t^{-1}\int_0^t U(\tau)\, d\tau]^r$, and an interpolation space $G$ between $L_{1,*}$ and $L_{\infty,*}$. If the operator $-B$ is the infinitesimal generator of a bounded semigroup $U(t)$, then the space $E_{B^r, G^{r-s}}$ is isomorphic to $(E, E_{B^r})_{\theta, G, G}$, where $\theta = s/r$.*

As has already been mentioned, the operator $B$ satisfies (1.15), and therefore Theorem 1.6 remains valid if on the right side of (1.27) we understand the space $E_{B^r, G^{r-s}}$ described in Theorem 1.8. The corollaries of Theorem 1.6 also remain true.

To make things more definite, we mention that in our case the space $E_{B^r, G^{r-s}}$ consists of those elements $x \in E$ for which

$$\|x\|_{E_{B^r, G^{r-s}}} = \|x\|_E + \left\| t^{-s} \left\| [I - U(t)]^r x \right\|_E \right\|_G < \infty, \tag{1.40}$$

according to the general construction and (1.6).

Repeating the arguments preceding the formulation of Theorem 1.6, we may show that for $m_0 < s < m_1$ and $x \in E_{B^r, G^{r-s}}$ we have

$$t^{m_0 - s} \left\| [I - U(t)]^{m_1 - m_0} B^{m_0} x \right\|_E \subset G.$$

Next we show by means of Lemma 1.10 that, conversely, from the last formula it follows that $x$ belongs to $E_{B^r, G^{r-s}}$. (In the proof $B(tB + I)$ must be replaced by $t^{-1} B \int_0^t U(\tau)\, d\tau$.) Thus Theorem 1.7 remains true for spaces with the norm (1.40).

We mention two formulas, analogous to (1.28) and (1.29), for equivalent norms in $E_{B^r, G^{r-s}}$. For nonintegral $s$ the norm

$$\|x\|_E + \left\| t^{[s]-s} \left\| [I - U(t)] B^{[s]} x \right\|_E \right\|_G, \tag{1.41}$$

and for integral $s$ the norm

$$\|x\|_E + \left\| \left\| t^{-1} [I - U(t)]^2 B^{s-1} x \right\|_E \right\|_G \tag{1.42}$$

is equivalent to (1.40).

**7. *A system of operators with commuting resolvents.*** Let a system of operators $B_j$ ($j = 1, \ldots, n$) act in a Banach space $E$, the operators satisfying condition (1.15) and having the property that the resolvents $(B_j + \lambda)^{-1}$ commute:

$$(B_i + \lambda)^{-1}(B_j + \mu)^{-1} = (B_j + \mu)^{-1}(B_j + \lambda)^{-1}.$$

From each of the operators $B_j$ we may construct the space $E_{B_j^r, G^{r-s}}$ ($j = 1, \ldots, n$) by the method described in subsection 5 and consider the intersection $\bigcap E_{B_j^r, G^{r-s}}$ of these spaces. It is convenient to define a norm in this space by the formula

$$\|x\|_{\cap E_{B^r_j, G^{r-s}}} = \|x\|_E + \sum_{j=1}^{n} \left\| t^{r-s} \left\| \left[ B_j(tB_j + I)^{-1} \right]^r x \right\|_E \right\|_G. \tag{1.43}$$

On the other hand, we may consider the intersection of the domains $\mathfrak{D}(B_j^r)$ of the operators $B_j$ and introduce the norm

$$\|x\|_{\cap E_{B_j^r}} = \|x\|_E + \sum \|B_j^r x\|_E$$

in it. With this norm it becomes a Banach space.

From the Banach couple $E$, $\bigcap E_{B_j^r}$ interpolation intermediate spaces may be constructed by the $\mathcal{K}$- and $\mathcal{J}$-methods. There arises the question of the relation of these spaces to the spaces $\bigcap E_{B_j^r, G^{r-s}}$.

THEOREM 1.9. *Up to isomorphism,*

$$\left(E, \bigcap E_{B_j^r}\right)_{\theta,G,G} = \bigcap E_{B_j^r, G^{r-s}} = \bigcap \left(E, E_{B_j^r}\right)_{\theta,G,G},$$

*where* $\theta = s/r$.

PROOF. The space $\bigcap E_{B_j^r}$ is imbedded in every space $E_{B_j^r}$, and therefore, by applying the interpolation Theorem 2.4 of Chapter IV to the imbedding operator, we conclude that

$$\left(E, \bigcap E_{B_j^r}\right)_{\theta,G,G} \subset \bigcap \left(E, E_{B_j^r}\right)_{\theta,G,G}.$$

We show that the reverse imbedding also holds. Let $x \in \bigcap E_{B_j^r, G^{r-s}}$. This means that the quantity (1.43) is finite for $x$. Using (1.25) and assuming that $t \leqslant 1$, we represent $x$ in the form

$$x = \left[ I - \prod_{j=1}^{n} (tB_j + I)^{-r} q\left((tB_j + I)^{-1}\right) \right] x + \prod_{j=1}^{n} (tB_j + I)^{-r} q\left((tB_j + I)^{-1}\right) x$$

$$= \sum_{j=1}^{n} \left( \left[ tB_j(tB_j + I)^{-1} \right]^r p\left((tB_j + I)^{-1}\right) \prod_{i=1}^{j-1} (tB_j + I)^{-r} q\left((tB_j + I)^{-1}\right) x \right.$$

$$\left. + \prod_{j=1}^{n} (tB_j + I)^{-r} q\left((tB_j + I)^{-1}\right) x \right).$$

If we denote the first term by $v_0(t)$ and the second by $v_1(t)$, then, taking into account that all factors commute, we obtain

$$t^{-s}\|v_0(t)\|_E \leqslant C\sum_{j=1}^{n} t^{r-s}\left\|\left[B_j(tB_j+I)^{-1}\right]^r x\right\|_E$$

and

$$\begin{aligned} t^{r-s}\|v_1(t)\|_{\cap E_{B_k^r}} &= t^{r-s}\left\|\prod_{j=1}^{n}(tB_j+I)^{-r}q\left((tB_j+I)^{-1}\right)x\right\|_E \\ &\quad + t^{r-s}\sum_{k=1}^{n}\left\|B_k^r\prod_{j=1}^{n}(tB_j+I)^{-r}q\left((tB_j+I)^{-1}\right)x\right\|_E \\ &\leqslant Ct^{r-s}\|x\|_E + C\sum_{k=1}^{n}t^{r-s}\left\|\left[B_k(tB_k+I)^{-1}\right]^r x\right\|_E. \end{aligned}$$

In the first term we have used the assumption that $t \leqslant 1$. If $t \geqslant 1$, we represent $x$ in the form

$$x = \left[I - t^{-r}v_1(1)\right]x + t^{-r}v_1(1)x.$$

Then $t^{-s}\|[I - t^{-r}v_1(1)]x\|_E \leqslant Ct^{-s}\|x\|_E$ and

$$t^{r-s}\|t^{-r}v_1(1)\|_{\cap E_{B_k^r}} \leqslant Ct^{-s}\|x\|_E.$$

Thus, we have obtained a representation $x = v_0(t) + v_1(t)$ in which the functions $t^{-s}\|v_0(t)\|_{E_0}$ and $t^{r-s}\|v_1(t)\|_{E_{\cap B_j^r}}$ are majorized by the function

$$C\sum_{j=1}^{n} t^{r-s}\left\|\left[B_j(tB_j+I)^{-1}\right]^r x\right\|_E + C\min\{t^{r-s}, t^{-s}\}\|x\|_E.$$

Both summands belong to the space $G$; the first one by assumption, and the second one by virtue of the relation $\min\{t^{r-s}, t^{-s}\} \in L_{1,*} \cap L_{\infty,*} \subset G$. Thus

$$\|x\|_{(E,\cap E_{B_j^r})_{\theta,G,G}} \leqslant \|t^{-s}\|v_0(t)\|_E\|_G + \|t^{r-s}\|v_1(t)\|_{E_{\cap B_j^r}}\|_G \leqslant C\|x\|_{\cap E_{B_j^r,G^{r-s}}}.$$

The theorem is proved.

The following theorem can be proved similarly.

THEOREM 1.10. *Up to isomorphism,*

$$\left(\cap E_{B_j^{m_0}}, \cap E_{B_j^{m_1}}\right)_{\theta',G,G} = \cap\left(E, E_{B_j^r}\right)_{\theta,G,G},$$

*where* $\theta' = (s - m_0)/(m_1 - m_0)$.

It is possible to introduce a space $E_{\bar{B}^r}$ smaller than $\cap E_{B_j^r}$, on which all possible products of the powers of the operators $B_j$ with total degree $r$ are defined. It is natural to introduce the norm

$$\|x\|_{E_{\vec{B}^r}} = \|x\|_E + \sum_{r_1+\cdots+r_n=r} \|B_1^{r_1}\cdots B_n^{r_n}x\|_E$$

in this space.

It turns out that the following theorem holds.

THEOREM 1.11. *Up to isomorphism,*

$$(E, E_{\vec{B}^r})_{\theta,G,G} = \bigcap E_{B_j^r,G^{r-s}} = \bigcap (E, E_{B_j^r})_{\theta,G,G}.$$

The proof of this theorem is somewhat more complicated than that of Theorem 1.9, and we refer the reader to the original publications [153 and 246].

**8.** ***A commuting system of semigroups.*** We assume that a system $U_1(t), \ldots, U_n(t)$ of bounded semigroups of linear operators commuting with one another: $U_i(t)U_j(\tau) = U_j(\tau)U_i(t)$ acts in the space $E$. If the resolvents of the infinitesimal generators $D_j$ of the semigroups $U_j(t)$ commute with each other, then so do the semigroups by virtue of (1.31). We may also verify the converse of this assertion by means of the inversion formula of the Laplace transformation. Thus, the operators $B_j = -D_j$ satisfy the conditions of subsection 7.

From the system of the semigroups $U_j(t)$ an $n$-parameter semigroup can be formed by setting

$$U(y) = U_1(y_1)\cdots U_n(y_n) \tag{1.44}$$

for $y = (y_1, \ldots, y_n) \in \mathbf{R}_+^n$ (+ means that $y_j \geqslant 0$).

We pose the problem of expressing the norm in the space $\bigcap E_{B_j^r,G^{r-s}} = (E, D(\vec{B}^r))_{s/r,G,G}$ in terms of the semigroup $U(y)$.

We assume that the space $G$ is constructed from the spaces $L_{1,*}(0, \infty)$ and $L_{\infty,*}(0, \infty)$ by means of some interpolation functor. We consider the spaces $L_{1,*}(\mathbf{R}_+^n)$ and $L_{\infty,*}(\mathbf{R}_+^n)$ of functions on the half-space $\mathbf{R}_+^n$ with measure

$$\frac{dy}{|y|^n} = \frac{dy_1\cdots dy_n}{|y|^n} \qquad (|y| = y_1^2 + \cdots + y_n^2).$$

We denote by $G(\mathbf{R}_+^n)$ the space constructed from $L_{1,*}(\mathbf{R}_+^n)$ and $L_{\infty,*}(\mathbf{R}_+^n)$ by means of the same interpolation functor. We introduce the quantity

$$\|\, |y|^{-s}\|(I - U(y))^r x\|_E\|_{G(\mathbf{R}_+^n)}.$$

LEMMA 1.11.

$$\|t^{-s}\|(I - U_j(t))^r x\|_E\|_G \leqslant C\|\, |y|^{-s}\|(I - U(y))^r x\|_E\|_{G(\mathbf{R}_+^n)}. \tag{1.45}$$

PROOF. Consider the identity

$$\sum_{k=0}^{r}(-1)^k C_r^k (z^k - w)^r = \sum_{k=0}^{r}(-1)^k C_r^k \sum_{l=0}^{r}(-1)^{r-l} z^{kl} w^{r-l} l_z^l$$

$$= \sum_{l=0}^{r}(-1)^{r-l} C_r^l w^{r-l} \sum_{k=0}^{r} C_r^k z^{kl}$$

$$= \sum_{l=0}^{r}(-1)^{r-l} C_r^l w^{r-l}(1 - z^l)^r.$$

Leaving only the term with $k = 0$ on the left, we obtain

$$(1 - w)^r = \sum_{k=1}^{r}(-1)^k C_r^k \{(z^k - 1)^r w^{r-k} - (z^k - w)^r\},$$

in which we substitute the commuting operators $U_1(t)U(y)$ and $U_1(t)$ for $z$ and $w$. Then we have

$$(I - U_1(t))^r = \sum_{k=1}^{r}(-1)^k C_r^k \{(U_1(kt)U(ky) - I)^r U_1((r-k)t)$$
$$- (U_1((k-1)t)U(ky) - I)^r U_1(rt)\}.$$

Hence

$$\|(I - U_1(t))^r x\|_E$$
$$\leqslant M_1 \sum_{k=1}^{r} C_r^k \{\|(U(ky + kte_1) - I)^r x\|_E + \|(U(ky + (k-1)te_1) - I)^r x\|_E\},$$

where $e_1 = (1, 0, \dots, 0)$ is a basis vector in $\mathbf{R}^n$. We integrate both sides of the inequality over the cube $0 \leqslant y_j \leqslant t$. Then we obtain

$$\|(I - U_1(t))^r x\|_E$$
$$\leqslant \frac{M_j}{t^n} \sum_{k=1}^{r} C_r^k \int_0^t \cdots \int_0^t \{\|(U(ky + kte_1) - I)^r x\|_E$$
$$+ \|(U(ky + (k-1)te_1) - I)^r x\|_E\} \, dy_1 \dots dy_n.$$

Replacing the domain of integration by a larger one in every summand, we arrive at

$$t^{-s}\|(I - U_1(t))^r x\|_E \leqslant \frac{C}{t^n}\int_0^{2rt}\int_0^{rt}\cdots\int_0^{rt}\|(U(z) - I)^r x\|_E \, dz_1 \dots dz_n$$
$$= C\int_0^{2rt}\int_0^{rt}\cdots\int_0^{rt}\frac{|z|^{n+s}}{t^{n+s}}|z|^{-s}\|(U(z) - I)^r x\|_E \frac{dz}{|z|^n}. \tag{1.46}$$

Consider the integral operator

$$\mathcal{K}\varphi = \int_0^{2rt}\int_0^{rt}\cdots\int_0^{rt}\frac{|z|^{n+s}}{t^{n+s}}\varphi(z)\frac{dz}{|z|^n}.$$

If $\varphi \in L_{\infty,*}(\mathbf{R}^n_+)$, then

$$\|\mathcal{K}\varphi\|_{L_{\infty,*}(0,\infty)} \leqslant \|\varphi\|_{L_{\infty,*}(\mathbf{R}^n_+)}\int_0^{2r}\int_0^{r}\cdots\int_0^{r}|u|^s\,du.$$

If $\varphi \in L_{1,*}(\mathbf{R}^n_+)$, then

$$\|\mathcal{K}\varphi\|_{L_{1,*}(0,\infty)} \leqslant \int_0^\infty \mathcal{K}|\varphi|\frac{dt}{t} = \int_0^\infty\int_0^\infty\cdots\int_0^\infty \frac{|z|^{n+s}\varphi(z)\int_1^\infty \frac{dt}{t^{n+s+1}}\frac{dz}{|z^n|}}{\frac{1}{r}\max\{z_1/2, z_2, \ldots, z_n\}}$$

$$= \frac{1}{n+s}\int_0^\infty\cdots\int_0^\infty\left(\frac{|z|}{\frac{1}{r}\max\{z_1/2, z_2, \ldots, z_n\}}\right)^{n+s}\varphi(z)\frac{dz}{|z|^n}$$

$$\leqslant C\|\varphi\|_{L_{1,*}(\mathbf{R}^n_+)}.$$

Thus, the operator $\mathcal{K}$ acts from $L_{1,*}(\mathbf{R}^n_+)$ to $L_{1,*}(0,\infty)$ and from $L_{\infty,*}(\mathbf{R}^n_+)$ to $L_{\infty,*}(0,\infty)$, and therefore it also acts from $G(\mathbf{R}^n_+)$ to $G$.

Inequality (1.46) can be written in the form

$$t^{-s}\|(I - U_1(t))^r x\|_E \leqslant C\mathcal{K}(|y|^{-s}\|(I - U(y))^r x\|_E).$$

From the indicated property of $\mathcal{K}$ we obtain (1.45) for $j = 1$. The case of an arbitrary $j$ can be treated similarly. The lemma is proved.

LEMMA 1.12.

$$\|\,|y|^{-s}\|(I - U(y))^r x\|_E\|_{G(\mathbf{R}^n_+)} \leqslant \sum_{j=1}^n \|t^{-s}\|(I - U_j(t))^r x\|_E\|_G. \tag{1.47}$$

PROOF. By virtue of the definition (1.44) we have

$$(I - U(y))^r = \left[\sum_{j=1}^n (I - U_j(y_j))U_1(y_1)\cdots U_{j-1}(y_{j-1})\right]^r.$$

Therefore

$$\|(I - U(y))^r x\|_E \leqslant C\sum_{r_1+\cdots+r_n=r}\left\|\prod_{j=1}^n (I - U_j(y_j))^{r_j}x\right\|_E. \tag{1.48}$$

Now we consider another identity:

$$1 = 2^{-\rho}(z+1)^\rho + (z-1)q(z),$$

where $q(z)$ is a polynomial of degree $\rho - 1$. From this identity it follows that

$$(z-1)^\rho = 2^{-\rho}(z^2-1)^\rho + (z-1)^{\rho+1}q(z).$$

Substituting the bounded operator $U_j(y_j)$ for the variable $z$ in the case $\rho = r_j$, we obtain

$$(U_j(y_j) - I)^{r_j} = 2^{-r_j}(U_j(2y_j) - I)^{r_j} + (U_j(y_j) - I)^{r_j+1}q(U_j(y_j)).$$

Multiplying these identities for $j = 1, \ldots, n$, we arrive at a relation of the form

$$\prod_{j=1}^{n} (U_j(y_j) - I)^{r_j} = 2^{-r} \prod_{j=1}^{n} (U_j(2y_j) - I)^{r_j} + \sum_{j=1}^{n} \prod_{k=1}^{n} (U_k(y_k) - I)^{r_k}(U_j(y_j) - I)\mathcal{P}_j,$$

where the $\mathcal{P}_j$ are polynomials in the operators $U_1(y_1), \ldots, U_n(y_n)$. In this equality we replace $y_j$ by $\lambda y_j$ and then carry out an estimate:

$$\sup_{\lambda>0} \sum_{r_1+\cdots+r_n=r} \left\| |y|^{-s} \left\| \prod_{j=1}^{n} (U_j(\lambda y_j) - I)^{r_j} x \right\|_E \right\|_{G(\mathbf{R}^n_+)}$$

$$\leqslant 2^{s-r} \sup_{\lambda>0} \sum_{r_1+\cdots+r_n=r} \left\| |2\lambda y|^{-s} \left\| \prod_{j=1}^{n} (U_j(2\lambda y_j) - I)^{r_j} x \right\|_E \right\|_{G(\mathbf{R}^n_+)}$$

$$+ C \sup_{\lambda>0} \sum_{r_1+\cdots+r_n=r} \sum_{j=1}^{n} \left\| |\lambda y|^{-s} \left\| \prod_{k=1}^{n} (U_k(\lambda y_k) - I)^{r_k}(U_j(\lambda y_j) - I) x \right\|_E \right\|_{G(\mathbf{R}^n_+)}$$

As is easy to see, the first summand on the right differs from the left side only by a factor $2^{s-r} < 1$. We bring it to the right side and observe that the dilation operators $\sigma_{1/\lambda}$ are bounded both above and below in the interpolation space $G(\mathbf{R}^n_+)$ between $L_{1,*}(\mathbf{R}^n_+)$ and $L_{\infty,*}(\mathbf{R}^n_+)$. This yields

$$\sum_{r_1+\cdots+r_n=r} \left\| |y|^{-s} \left\| \prod_{j=1}^{n} (U_j(y_j) - I)^{r_j} x \right\|_E \right\|_{G(\mathbf{R}^n_+)}$$

$$\leqslant C \sum_{r_1+\cdots+r_n=r} \sum_{j=1}^{n} \left\| |y|^{-s} \left\| \prod_{k=1}^{n} (U_k(y_k) - I)^{r_k}(U_j(y_j) - I) x \right\|_E \right\|_{G(\mathbf{R}^n_+)}.$$

The total degree of the operators $U_j(y_j) - I$ is equal to $r$ in all terms on the left side, and to $r + 1$ in all terms on the right side. Besides, the structure of the right side shows that the inequality can be integrated, and thus the total degree can be brought to any number, for example, to $rn$. Then at least one of the factors $U_j(y_j) - I$ in every term will occur to a power not smaller than $r$. Retaining this power, majorizing the remaining factors by constants, and taking account of (1.48), we get

$$\| |y|^{-s} \|(I - U(y))^r x\|_E \|_{G(\mathbf{R}^n_+)} \leqslant C \sum_{j=1}^{n} \| |y|^{-s} \|(I - U_j(y_j))^r x\|_E \|_{G(\mathbf{R}^n_+)}. \quad (1.49)$$

Now we consider the operator assigning to a function $\varphi(t)$ on $(0, \infty)$ the function $y_j^s|y|^{-s}\varphi(y_j)$ in $\mathbf{R}_+^n$. It is obvious that

$$y_j^s|y|^{-s}|\varphi(y_j)| \leqslant |\varphi(y_j)|,$$

and so this operator acts from $L_{\infty,*}(0, \infty)$ into $L_{\infty,*}(\mathbf{R}_+^n)$. Next,

$$\int_0^\infty \cdots \int_0^\infty y_j^s|y|^{-s}|\varphi(y_j)|\frac{dy_1 \cdots dy_n}{|y|^n}$$

$$= \int_0^\infty y_j^s|\varphi(y_j)|\int_0^\infty \cdots \int_0^\infty \frac{dy_1 \cdots dy_{j-1}dy_{j+1} \cdots dy_n}{|y|^{n+s}}\, dy_j$$

$$= \int_0^\infty |\varphi(y_j)|\frac{dy}{y_j}\int_0^\infty \cdots \int_0^\infty \frac{du_1 \cdots du_{j-1}}{\left(\sqrt{1+\sum u_i^2}\right)^{n+s}} = C\int_0^\infty |\varphi(t)|\frac{dt}{t}.$$

This means that the indicated operator acts from $L_{1,*}(0, \infty)$ into $L_{1,*}(\mathbf{R}_+^n)$, and so it also acts from $G$ into $G(\mathbf{R}_+^n)$. It is exactly this operator that occurs on the right side of (1.49) applied to the function $t^{-s}\|(I - U(t))^r x\|_E$. Consequently (1.49) implies the required inequality (1.47).

The lemma is proved.

The quantity on the right side of (1.49) is the main summand of the norm in the intersection of the spaces $E_{B_j^r, G^{r-s}}$. A combination of Lemmas 1.11 and 1.12 yields the following theorem.

THEOREM 1.12. *If $U_j(t)$, $j = 1, \ldots, n$, is a system of commuting bounded semigroups with infinitesimal generators $D_j = -B_j$, then the norm in $\bigcap E_{B_j^r, G^{r-s}}$ is equivalent to the quantity*

$$\|x\|_E + \| \, |y|^{-s}\|(I - U(y))^r x\|_E\|_{G(\mathbf{R}_+^n)}, \tag{1.50}$$

*where $G(\mathbf{R}_+^n)$ is constructed from $L_{1,*}(\mathbf{R}_+^n)$ and $L_{\infty,*}(\mathbf{R}_+^n)$ by means of the same interpolation functor used to construct $G$ from $L_{1,*}(0, \infty)$ and $L_{\infty,*}(0, \infty)$.*

**9. *Unbounded operators in a couple of Banach spaces. Imbedding theorem.*** Let $\mathfrak{S}$ be a separated topological linear space and let $E$ be a Banach space imbedded in $\mathfrak{S}$. If $R$ is a linear operator acting in $\mathfrak{S}$, then by $R_E$ we denote the operator defined by the equality $R_E x = Rx$ for elements $x \in E$ for which $Rx \in E$. It is easy to verify that, since the operator $R$ is continuous (closed, if $\mathfrak{S}$ is a Banach space), $R_E$ is closed. If in this case $R_E$ turns out to be defined on the whole of $E$, then it is bounded. Its powers also enjoy this property, and $(R_E)^m = (R^m)_E$.

Now in addition let $R$ have an inverse $R^{-1}$. Consider the operator $(R^{-1})_E$. It is defined for elements $x \in E$ for which $R^{-1}x \in E$ and $(R^{-1})_E x = R^{-1}x$. This implies that $R^{-1}x \in \mathfrak{D}(R_E)$ and $R_E(R^{-1}x) = R(R^{-1}x) = x$. Thus

$R_E(R^{-1})_E x = x$. On the other hand, if $x$ belongs to the range of $R_E$, then $x = R_E y = Ry$ $(x, y \in E)$, and consequently $R^{-1}x = y \in E$, i.e.

$$x \in \mathfrak{D}((R^{-1})_E) \quad \text{and} \quad y = (R^{-1})_E x = (R^{-1})_E R_E y.$$

We have proved the equality $(R^{-1})_E = (R_E)^{-1}$. The operator $R^m$ has the same properties as $R$, and so

$$(R^{-m})_E = ((R^m)_E)^{-1} = ((R_E)^m)^{-1} = (R_E)^{-m}.$$

Now let $E_0$ and $E_1$ be a couple of Banach spaces imbedded continuously in $\mathfrak{S}$, and let $B$ be a continuous linear operator in $\mathfrak{S}$ such that for $\lambda > 0$ the inverse $(B + \lambda I)^{-1}$ exists and is continuous on the whole space $\mathfrak{S}$. We briefly write $B_{E_i} = B_i$. Assume that the operators $((B + \lambda I)^{-1})_{E_i}$ are defined on the whole space $E_i$. According to the preceding (the role of $R$ is played by the operators $(B + \lambda I)^{-1}$) we then have

$$\begin{aligned} ((B + \lambda I)^{-1})_{E_i} = (B_i + \lambda I)^{-1} \quad \text{and} \quad \mathfrak{D}(B_i^m) &= \mathfrak{D}((B_i + \lambda I)^m) \\ &= \mathfrak{D}((B + \lambda I)^m_{E_i}). \end{aligned} \tag{1.51}$$

Here the operator $(B_i + \lambda I)^{-1}$ is bounded in $E_i$.

Consider an interpolation functor associating a Banach space $(A_0, A_1)_f$ to a Banach couple $A_0, A_1$. Since $(E_0, E_1, (E_0, E_1)_f)$ is an interpolation triple, the operator $(B + \lambda I)^{-1}$ acts in the space $(E_0, E_1)_f$, and so $Bf = (B)_{(E_0,E_1)_f}$ satisfies (1.51).

The operators $(B_i + \lambda_0 I)^{-m}$ $(\lambda_0 > 0)$, which are restrictions of $(B + \lambda_0 I)^{-m}$ to $E_i$, effect isomorphisms between the spaces $E_i$ and $\mathfrak{D}(B_i^m)$ (with the graph norm). Therefore $(\mathfrak{D}(B_0^m), \mathfrak{D}(B_1^m))_f$ is isomorphic to the image of $(E_0, E_1)_f$ under $(B + \lambda I)^{-m}$, and for $x = (B + \lambda I)^{-m} y$ $(y \in (E_0, E_1)_f)$ it may be assumed that

$$\|x\|_{(\mathfrak{D}(B_0^m), \mathfrak{D}(B_1^m))_f} = \|y\|_{(E_0,E_1)_f} = \|(B + \lambda_0 I)^m x\|_{(E_0,E_1)_f} \tag{1.52}$$

(see the Introduction to Chapter IV).

On the other hand, the indicated image is precisely the domain of $B_f^m$, and the norm (1.52) is equivalent to the graph norm on $\mathfrak{D}(B_f^m)$. Thus, up to isomorphism we have

$$(\mathfrak{D}(B_0^m), \mathfrak{D}(B_1^m))_f = \mathfrak{D}(B_f^m). \tag{1.53}$$

In particular,

$$[\mathfrak{D}(B_0^m), \mathfrak{D}(B_1^m)]_\theta = \mathfrak{D}(B_\theta^m), \quad \text{where } B_\theta = (B)_{[E_0,E_1]_\theta}, \tag{1.54}$$

$$(\mathfrak{D}(B_0^m), \mathfrak{D}(B_1^m))_{\theta,G,G} = \mathfrak{D}(B^m_{\theta,G,G}), \quad \text{where } B_{\theta,G,G} = (B)_{(E_0,E_1)_{\theta,G,G}}. \tag{1.55}$$

THEOREM 1.13 (imbedding theorem). *Let $E_0$ and $E_1$ be Banach spaces imbedded algebraically and topologically into a separated topological linear space $\mathfrak{E}$, $(E_0, E_1)$ is a Banach couple and let $B$ be a linear operator on $\mathfrak{E}$ having the following properties:*

1. *The operator $B$ is continuous, and the operators $B + \lambda I$ are continuously invertible on $\mathfrak{E}$ for any $\lambda > 0$.*

2. *The restrictions $(B_i + \lambda I)^{-1}$ of the operators $(B + \lambda I)^{-1}$ to the spaces $E_i$ act in $E_i$ and satisfy condition* (1.15):

$$\|\lambda(B_i + \lambda I)^{-1}\|_{E_i \to E_i} \leqslant a \qquad (\lambda > 0). \tag{1.56}$$

3. *The operator $(B_0 + \lambda I)^{-1}$ acts from $E_0$ into $E_1$, and*

$$\|(B_0 + \lambda I)^{-1}\|_{E_0 \to E_1} \leqslant c\lambda^{\nu - 1}, \tag{1.57}$$

*where $c$ and $\nu$ are positive constants independent of $\lambda$.*

4. *The domains of the $B_i$ are dense in $E_i$.*

*Then for $0 < \theta < \theta + \nu/r < 1$*

$$(E_0, E_{B_0^r})_{\theta + \nu/r, G, G} \subset (E, E_{B_1^r})_{\theta, G, G}. \tag{1.58}$$

PROOF. First we assume that $s = \theta r > 1$. If $x \in (E_0, E_{B_0^r})_{\theta + \nu/r, G, G}$ then by virtue of Theorem 1.5 this is equivalent to the relation

$$t^{r-(s+\nu)}\left\|\left[B_0(tB_0 + I)^{-1}\right]^r x\right\|_{E_0} \in G. \tag{1.59}$$

Since $s > 1$, Theorem 1.7 implies that $x \in \mathfrak{D}(B_0)$. It follows from property 3 that $\mathfrak{D}(B_0) \subset E_1$. Indeed, for $x \in \mathfrak{D}(B_0)$ we have

$$x = \lambda(B_0 + \lambda I)^{-1}x + (B_0 + \lambda I)^{-1}B_0 x \in E_1.$$

We consider the function

$$\begin{aligned}\Phi(t) &= t^{r+1-s}\left\|\left[B_1(tB_1 + I)^{-1}\right]^{r+1}x\right\|_{E_1}\\ &= t^{r+1-s}\left\|\left[B_1(tB_1 + I)^{-b}1\right]^r B_1(tB_1 + I)^{-1}x\right\|_{E_1}.\end{aligned}$$

Since $x \in E_0 \cap E_1$, the element $B_1(tB_1 + I)^{-1}x = B_0(tB_0 + I)^{-1}x = (tB_0 + I)^{-1}B_0 x$ also belongs to $E_0 \cap E_1$. By (1.57) we have

$$\|(tB_0 + I)^{-1}\|_{E_0 - E_1} \leqslant ct^{-\nu}$$

and therefore

$$\begin{aligned}\Phi(t) &= t^{r+1-s}\left\|\left[B_0(tB_0 + I)^{-1}\right]^r (tB_0 + I)^{-1}B_0 x\right\|_{E_1}\\ &\leqslant t^{r-(s+\nu-1)}\left\|\left[B_0(tB_0 + I)^{-1}\right]^r B_0 x\right\|_{E_0}.\end{aligned}$$

By Theorem 1.7 it follows from (1.59) that the function on the right belongs to $G$, and so we also have $\Phi(t) \in G$. Corollary 1 to Theorem 1.6 then says that $x \in E_{B_1^r, G^{r-s}} = (E_1, E_{B_1^r})_{\theta,G,G}$. Hence the imbedding (1.58) holds.

Now let $s = \theta r < 1$. The operator $(B + \lambda I)^{-1}$ implements an isomorphism between the couples $E_i, E_{B_i^r}$ and $E_{B_i}, E_{B_i^{r+1}}$, and therefore it also determines an isomorphism between $(E_1, E_{B_1^r})_{\theta,G,G}$ and $(E_{B_1}, E_{B_1^{r+1}})_{\theta,G,G}$, and also between $(E_0, E_{B_0^r})_{\theta+\nu/r,G,G}$ and $(E_{B_0}, E_{B_0^{r+1}})_{\theta+\nu/r,G,G}$.

By Theorem 1.6 we have

$$(E_{B_1}, E_{B_1^{r+1}})_{\theta,G,G} = (E_1, E_{B_1^r})_{\theta+1/r,G,G}$$

and

$$(E_{B_0}, E_{B_0^{r+1}})_{\theta+\nu/r,G,G} = (E_0, E_{B_0^r})_{\theta+(\nu+1)/r,G,G},$$

up to isomorphism. According to what has been proved, we have

$$(E_0, E_{B_0^r})_{\theta+(\nu+1)/r,G,G} \subset (E_1, E_{B_1^r})_{\theta+1/r,G,G}$$

and consequently

$$(E_{B_0}, E_{B_0^{r+1}})_{\theta+\nu/r,G,G} \subset (E_{B_1}, E_{B_1^{r+1}})_{\theta,G,G}.$$

This implies that

$$\begin{aligned}(E_0, E_{B_0^r})_{\theta+\nu/r,G,G} &= (B + \lambda I)(E_{B_0}, E_{B_0^{r+1}})_{\theta+\nu/r,G,G} \\ &\subset (B + \lambda I)(E_{B_1}, E_{B_1^{r+1}})_{\theta,G,G} = (E_1, E_{B_1^r})_{\theta,G,G}.\end{aligned}$$

The theorem is proved.

## §2. Trace theory

**1.** ***Generalized functions with locally summable derivatives, and traces.*** We recall that a *generalized function* on the semiaxis $(0, \infty)$ with values in a Banach space $A$ is, by definition, a continuous linear mapping of the space $D(0, \infty)$ of infinitely differentiable scalar-valued functions with compact support in $(0, \infty)$ into the space $A$. A generalized function is said to be *regular* if it is induced by a locally summable function $u(t)$ by the formula

$$u(\varphi) = \int_0^\infty u(t)\varphi(t)\,dt \qquad (\varphi \in \mathfrak{D}(0, \infty)).$$

The *derivative of order $m$ of a generalized function $u$* is, by definition, the generalized function $u^{(m)}$ defined by

$$u^{(m)}(\varphi) = (-1)^m u(\varphi^{(m)}).$$

We note that $(u^{(m-1)})'(\varphi) = -u^{(m-1)}(\varphi') = (-1)^m u(\varphi^{(m)}) = u^{(m)}(\varphi)$.

Now we assume that $u^{(m)}$ is a regular generalized function, i.e., it is induced by the locally summable function $u^{(m)}(t)$. We consider the function

$$v(t) = \int_a^t u^{(m)}(t)\,dt \qquad (a > 0) \tag{2.1}$$

and for any $\varphi \in \mathfrak{D}(0, \infty)$ we calculate

$$\begin{aligned}\int_0^\infty v(t)\varphi'(t)\,dt &= \int_0^\infty \int_a^t u^{(m)}(s)\,ds\,\varphi'(t)\,dt\\ &= -\int_0^a u^{(m)}(s)\int_0^s \varphi'(t)\,dt\,ds + \int_a^\infty u^{(m)}(s)\int_s^\infty \varphi'(t)\,dt\,ds\\ &= -\int_0^\infty u^{(m)}(s)\varphi(s)\,ds.\end{aligned}$$

From this it follows that $v(\varphi') = -u^{(m)}(\varphi) = u^{(m-1)}(\varphi')$ or $[v - u^{(m-1)}]'(\varphi) = 0$. This means that $v(t) = u^{(m-1)}(t) + \alpha_0$ almost everywhere, where $\alpha_0$ is a fixed element of $A$. Thus, the derivative $u^{(m-1)}$ is determined by an absolutely continuous function. It then follows from (2.1) that $v(t) = u^{(m-1)}(t) - u^{(m-1)}(a)$. In other words, we obtain the Newton-Leibniz formula

$$\int_a^t u^{(m)}(t)\,dt = u^{(m-1)}(t) - u^{(m-s)}(a). \tag{2.2}$$

It is obvious that all lower order derivatives $u^{(j)}(t)$, $j = 0, \ldots, m-2$, are continuously differentiable in the strong sense.

If $\psi(t)$ is an infinitely differentiable scalar-valued function, then the following formula holds for the derivative of a product:

$$\begin{aligned}(\psi u)'(\varphi) &= -u(\psi\varphi') = -u((\psi\varphi)') + u(\psi'\varphi) = u'(\psi\varphi) + u(\psi'\varphi)\\ &= (\psi u' + \psi' u)(\varphi).\end{aligned}$$

If $u'$ is regular, then, using (2.2), we obtain the ordinary formula for integration by parts:

$$\int_a^b \psi(t)u'(t)\,dt + \int_a^b \psi'(t)u(t)\,dt = \int_a^b (\psi u)'(t)\,dt = \psi(b)u(b) - \psi(a)u(a).$$

By passage to the limit this formula generalizes to functions $\psi \in C^1(a, b)$.

Now we choose two ideal lattices $F_0$ and $F_1$ which are interpolation spaces between $L_{1,*}(0, \infty)$ and $L_{\infty,*}(0, \infty)$, and two real numbers $\eta_0$ and $\eta_1$. As we saw in Chapter II, §8.5, these spaces consist of locally summable functions. We assume $u(t)$ is such that $u \in F_0^{\eta_0}(A)$ and $u^{(m)} \in F_1^{\eta_1}(A)$.

By integration by parts we obtain

$$u^{(j)}(b) = (-1)^{m+1}\int_a^b p_{m,j}(\tau)u^{(m)}(\tau)\,d\tau + \int_a^b p_{m,j}^{(m)}(\tau)u(\tau)\,d\tau$$
$$(0 < a < b < \infty),$$

where $p_{m,j}(\tau)$ is the polynomial of degree $2m-1$ satisfying the conditions $p(a) = p'(a) = \cdots = p^{(m-1)}(a) = 0$, $p(b) = p'(b) = \cdots = p^{(m+j-2)}(b) = p^{(m+j)}(b) = \cdots = p^{(m-1)}(b) = 0$ and $p^{(m-j-1)}(b) = (-1)^j$. The coefficients of this polynomial as functions of $b$ are uniformly bounded provided $b$ remains bounded. Therefore

$$\|u^{(j)}(b)\|_A \leqslant c(a,b)\left[\int_a^b \|u^{(m)}(\tau)\|_A d\tau + \int_a^b \|u(\tau)\|_A\, d\tau\right]$$
$$\leqslant c(a,b)\left[\|u\|_{F_0^{\eta_0}(A)} + \|u^{(m)}\|_{F_1^{\eta_1}(A)}\right]. \tag{2.3}$$

Now we study the behavior of the derivatives $u^{(j)}(t)$ as $t \to 0$. For this we use the formula

$$u^{(j)}(t) = (-1)^{m-j}\int_t^b \frac{(s-t)^{m-j-1}}{(m-j-1)!} u^{(m)}(s)\, ds$$
$$+ u^{(m-1)}(b)\frac{(b-t)^{m-j-1}}{(m-j-1)!} + \cdots + u^{(j)}(b). \tag{2.4}$$

Hence we obtain

$$\|u^{(j)}(t)\|_A \leqslant c\left[\int_t^b s^{m-j-1}\|u^{(m)}(s)\|_A\, ds + \sum_{i=j}^{m-1} \|u^{(i)}(b)\|_A\right].$$

By (8.43) of Chapter II we have

$$\int_0^b s^{m-j-1}\|u^{(m)}(s)\|_A\, ds \leqslant c\|u^{(m)}\|_{F_1^{\eta_1}(A)},$$

if $m - j > \eta_1$. In this case $u^{(j)}(t)$ is bounded down to zero. Passing to the limit in (2.4) according to the Lebesgue theorem, we obtain

$$\lim_{t\to+0} u^{(j)}(t) = \frac{(-1)^{m-j}}{(m-j-1)!}\int_0^b s^{m-j-1}u^{(m)}(s)\, ds$$
$$+ \sum_{i=j}^{m-1} \frac{1}{(i-j)!} b^{i-j}u^{(i)}(b). \tag{2.6}$$

DEFINITION 2.1. The limit $\lim_{t\to 0} u^{(j)}(t) = u^{(j)}(0)$, if it exists, is called the *trace of order $j$ of the function $u(t)$.*

Relation (2.6) and inequalities (2.3) and (2.5) lead to the following assertion.

THEOREM 2.1. *If $F_0$ and $F_1$ are two interpolation spaces between $L_{1,*}$ and $L_{\infty,*}$, and $u(t) \in F_0^{\eta_0}(A)$ and $u^{(m)}(t) \in F_1^{\eta_1}(A)$, then for $j < m - \eta_1$ the trace of order $j$ of the function $u(t)$ exists and*

$$\|u^{(j)}(0)\|_A \leqslant c\left(\|u\|_{F_0^{\eta_0}(A)} + \|u^{(m)}\|_{F_1^{\eta_1}(A)}\right) \tag{2.7}$$

REMARK. We have used the interpolation property of $F_0$ only in the estimation of $\int_a^b \|u(t)\|_A \, d\tau$ in terms of $\|u\|_{F_0^{\eta_0}(A)}$, which of course holds for larger classes of spaces.

2. ***Some simple operators in spaces of differentiable functions.*** We consider how the convolution and Hardy-Littlewood operators act in spaces of differentiable functions.

Let $\varphi$ be an infinitely differentiable scalar-valued function with compact support, and let $u(t)$ be a locally summable function with values in $A$. The function

$$\varphi * u(t) = \int_0^\infty \varphi\left(\frac{t}{s}\right) u(s) \frac{ds}{s} = \int_0^\infty \varphi(\tau) u\left(\frac{t}{\tau}\right) \frac{d\tau}{\tau} \tag{2.8}$$

is also infinitely differentiable, and its derivatives may be calculated according to the formulas

$$(\varphi * u)^{(j)}(t) = \int_0^\infty s^{-j} \varphi\left(\frac{t}{s}\right) u(s) \frac{ds}{s}. \tag{2.9}$$

If $u(t)$ itself is $j$ times continuously differentiable, then the derivative of the convolution can be written in the form

$$(\varphi * u)^{(j)}(t) = \int_0^\infty \varphi(\tau) \tau^{-j} u^{(j)}\left(\frac{t}{\tau}\right) d\tau = \int_0^\infty \left(\frac{s}{t}\right)^j \varphi\left(\frac{t}{s}\right) u^{(j)}(s) \frac{ds}{s}. \tag{2.10}$$

If the trace of order $j$ of $u(t)$ exists, then the first equality in (2.10) implies that

$$(\varphi * u)^{(j)}(0) = u^{(j)}(0) \int_0^\infty \tau^{-j} \varphi(\tau) \frac{d\tau}{\tau}. \tag{2.11}$$

We have already applied the convolution operator in order to smoothout functions, which we shall also do in what follows.

Now we study the Hardy-Littlewood operators

$$H_l u(t) = \frac{1}{t^{1+l}} \int_0^t s^l u(s) \, ds. \tag{2.12}$$

According to (8.33) of Chapter II, this operator can also be expressed in terms of convolution, but not with a function of compact support. If $l > \eta - 1$, this operator acts in the space $F^\eta(A)$ (see (8.40) in Chapter II).

If $u(t)$ is continuous, then

$$(H_l u)'(t) = \frac{1}{t} u(t) - \frac{1 + l}{t} H_l u(t). \tag{2.13}$$

If, in addition, $u \in F^\eta(A)$, then $(H_l u)' \in F^{\eta+1}(A)$. By using the equality

$$H_l u(t) = \int_1^\infty \tau^{-l-2} u\left(\frac{t}{\tau}\right) d\tau$$

for a $j$-fold continuously differentiable function we obtain

$$(H_l u)^{(j)}(t) = \int_1^\infty \tau^{-l-j-2} u^{(j)}\left(\frac{t}{\tau}\right) d\tau = H_{l+j} u^{(j)}(t). \tag{2.14}$$

If $u^{(j)} \in F^\eta(A)$, then $(H_l u)^{(j)} \in F^\eta(A)$ for $l + j > \eta - 1$.

If the trace $w^{(j)}(0)$ exists, then

$$(H_l u)^{(j)}(0) = u^{(j)}(0) \int_1^\infty \tau^{-l-j-2} d\tau = \frac{1}{l+j+1} u^{(j)}(0). \tag{2.15}$$

Thus, *the Hardy-Littlewood operator $H_l$ preserves the membership of $u^{(j)}(t)$ in $F^\eta(A)$ for $l + j > \eta - 1$, increases the smoothness of the function by one, and ensures that the derivatives of order $j$ of the function $t(H_l u)'(t)$ belong to the same classes $F^\eta(A)$.*

3. ***Spaces of traces.*** Now let $A_0$, $A_1$ be a Banach couple, let $F_0$ and $F_1$ be two interpolation spaces between $L_{1,*}$ and $L_{\infty,*}$, and let $\eta_0$, $\eta_1 \in \mathbf{R}$. We denote by $W_m(\eta_0, F_0, A_0; \eta_1, F_1, A_1)$ the space of functions with values in $A_0 + A_1$ for which $u \in F_0^{\eta_0}(A_0)$ and $u^{(m)} \in F_1^{\eta_1}(A_1)$, where the derivative is understood in the sense of generalized functions with values in $A_0 + A_1$. In this space we introduce a norm by the formula

$$\|u\|_{W_m} = \|u\|_{F_0^{\eta_0}(A_0)} + \|u^{(m)}\|_{F_1^{\eta_1}(A_1)}.$$

We leave it to the reader to prove the completeness of the normed space thus obtained.

For functions $u$ belonging to $W_m(\eta_0, F_0, A_0; \eta_1, F_1, A_1)$ we obviously have $u \in F_0^{\eta_0}(A_0 + A_1)$ and $u^{(m)} \in F_1^{\eta_1}(A_0 + A_1)$. Consequently Theorem 2.1 holds for them relative to the space $A = A_0 + A_1$. Therefore the following definition is meaningful.

DEFINITION 2.2. Let $j$ be a natural number with $j < m - \eta_1$. The *space of traces of order $j$* is the collection $T_j^m(\eta_0, F_0, A_0; \eta_1, F_1, A_1)$ of all elements $x$ belonging to $A_0 + A_1$ of the form

$$x = u^{(j)}(0), \quad \text{where } u \in W_m(\eta_0, F_0, A_0; \eta_1, F_1, A_1),$$

the norm being

$$\|x\|_{T_j^m} = \inf_{x = u^{(j)}(0)} \|u\|_{W_m(\eta_0, F_0, A_0; \eta_1, F_1, A_1)}. \tag{2.16}$$

In our case (2.7) implies the inequality

$$\|u^{(j)}(0)\|_{A_0+A_1} \leqslant c\left(\|u\|_{F_0^{\eta_0}(A_0)} + \|u^{(m)}\|_{F_1^{\eta_1}(A_1)}\right) = c\|u\|_{W_m}, \tag{2.17}$$

which in turn implies that the subspace of functions $u \in W_m$ for which $u^{(j)}(0) = 0$ is closed in $W_m$. The space $T_j^m$ of traces of order $j$ is isomorphic to the quotient space of $W_m$ modulo this subspace. Consequently $T_j^m$ is a Banach space.

REMARK 2.1. In calculating the quantity equivalent to the norm (2.16), the infimum may be taken over the set of all infinitely differentiable functions belonging to $W_m(\eta_0, F_0, A_0; \eta_1, F_1, A_1)$.

Indeed, if we choose the infinitely differentiable function $\varphi(t)$ with compact support so that $\int_0^\infty \tau^{-j}\varphi(\tau)\tau^{-1}\,d\tau = 1$, we can calculate the infimum in (2.16) only over functions of the form $\varphi * u(t)$, where $u \in W_m(\eta_0, F_0, A_0; \eta_1, F_1, A_1)$ and $u^{(j)}(0) = x$. By (2.11) we have $(\varphi * u)^{(j)}(0) = u^{(j)}(0) = x$, and by (2.10) and (8.38) of Chapter II we have $\|\varphi * u\|_{W_m} \leqslant c\|u\|_{W_m}$, where the constant $c$ depends only on the choice of $\varphi$. This implies the validity of Remark 2.1.

LEMMA 2.1. *If* $\eta_0 < j < m - \eta_1$, *then the space* $T_j^m(\eta_0, F_0, A_0; \eta_1, F_1, A_1)$ *is intermediate between* $A_0$ *and* $A_1$.

PROOF. The imbedding $T_j^m \subset A_0 + A_1$ follows from (2.17). Let $x \in A_0 \cap A_1$. We consider a smooth function $\varphi(t) = (1/j!)t^j\psi(t)$, where $\psi(0) = 1$, $\psi'(0) = \cdots = \psi^{(l-1)}(0) = 0$, $\psi(t) \equiv 0$ for $t \geqslant a > 0$, and $l$ is sufficiently large. We set $u(t) = \varphi(t)x$. Then the function

$$t^{\eta_0}\|u(t)\|_{A_0} = \frac{1}{j!}t^{\eta_0+j}|\psi(t)|\|x\|_{A_0}$$

belongs to $F_0$, since it belongs to $L_{1,*} \cap L_{\infty,*}$ by virtue of the condition $-\eta_0 < j$. The function

$$t^{\eta_1}\|u^{(m)}(t)\|_{A_1} = \frac{1}{j!}t^{\eta_1}\left[t^j\psi(t)\right]^{(m)}\|x\|_{A_1}$$

has order $t^{l-m+j+\eta_1}$ at zero, and so it belongs to $F_1$ for $l > m - j - \eta_1$, for the same reason. Finally, $u^{(j)}(0) = x$, i.e. $x \in T_j^m(\eta_0, F_0, A_0; \eta_1, F_1, A_1)$.

The lemma is proved.

The number of parameters on which the spaces $T_j^m$ depend can be decreased.

LEMMA 2.2. *If* $m \geqslant 2$ *and* $-\eta_0 < m - 1 < m - \eta_1$, *then*

$$T_{m-1}^m(\eta_0, F_0, A_0; \eta_1, F_1, A_1) = T_{m-2}^{m-1}(\eta_0 + 1, F_0, A_0; \eta_1, F_1, A_1)$$

*up to isomorphism.*

PROOF. Let $x = u^{(m-1)}(0)$ and $u \in W_m(\eta_0, F_0, A_0; \eta_1, F_1, A_1)$.

We apply the Hardy-Littlewood operator $H_l$ with sufficiently large $l$ to the function $u(t)$ and consider the function $w(t) = (l + m)(H_l u)'(t)$. According to (2.15) we have $w^{(m-2)}(0) = (l + m)(H_l u)^{(m-1)}(0) = u^{(m-1)}(0) = x$. Next, the function $tw(t)$ belongs to $F_0^{\eta_0}(A_0)$, i.e. $w \in F_0^{\eta_0+1}(A_0)$, and, finally, $w^{(m-1)}(t) = (l + m)(H_l u)^{(m)}(t) \in F_1^{\eta_1}(A_1)$. From these relations it follows that $x \in T_{m-2}^{m-1}(\eta_0 + 1, F_0, A_0; \eta_1, F_1, A_1)$.

Conversely, let $x$ have the last property. Then

$$x = w^{(m-2)}(0), \qquad w \in W_{m-1}(\eta_0 + 1, F_0, A_0; \eta_1, F_1, A_1).$$

We consider the function

$$u(t) = \frac{l+m-1}{m-1} tH_l w(t).$$

Then $H_l w \in F_0^{\eta_0+1}(A_0)$, and consequently $u \in F_0^{\eta_0}(A_0)$. Next, by (2.13)

$$u'(t) = \frac{l+m-1}{m-1}\left[w(t) - lH_l w(t)\right]$$

and by (2.14)

$$u^{(j)}(t) = \frac{l+m-1}{m-1}\left[w^{(j-1)}(t) - lH_{l+j-1}w^{(j-1)}(t)\right]. \tag{2.18}$$

In particular,

$$u^{(m-1)}(0) = \frac{l+m-1}{m-1}\left[w^{(m-2)}(0) - \frac{l}{l+m-1}w^{(m-2)}(0)\right] = w^{(m-2)}(0) = x$$

according to (2.15). Finally, from (2.18) for $j = m$ and properties of the Hardy-Littlewood operators it follows that $u^{(m)} \in F_1^{\eta_1}(A_1)$, and consequently $x \in T^m_{m-1}(\eta_0, F_0, A_0; \eta_1, F_1, A_1)$.

The lemma is proved.

LEMMA 2.3. *If* $-\eta_0 < j < m - \eta_1$ *and* $j \leqslant m-2$, *then*

$$T_j^m(\eta_0, F_0, A_0; \eta_1, F_1, A_1) = T_j^{m-1}(\eta_0, F_0, A_0; \eta_1 - 1, F_1, A_1)$$

*up to isomorphism.*

PROOF. Let $u \in W_m(\eta_0, F_0, A_0; \eta_1, F_1, A_1)$. Consider the operator

$$v(t) = \frac{l-m+j+1}{j+1-m}\left[u(t) - lH_{l-m}(t)\right].$$

For sufficiently large $l$ (for $l - m > \eta_0 - 1$) this operator acts in $F_0^{\eta_0}(A_0)$. Using (2.14), we obtain

$$v^{(m-1)}(t) = \frac{l-m+j+1}{j+1-m}\left[u^{(m-1)}(t) - lH_{l-1}u^{(m-1)}(t)\right].$$

Upon integration by parts we find that

$$v^{(m-1)}(t) = \frac{l-m+j+1}{j+1-m} tH_l u^{(m)}(t).$$

By a property of the Hardy-Littlewood operators we then have $v^{(m-1)} \in F_1^{\eta_1 - 1}(A_1)$. A calculation shows that $v^{(j)}(0) = u^{(j)}(0)$. Thus, if

$$x = u^{(j)}(0) \in T_j^m(\eta_0, F_0, A_0; \eta_1, F_1, A_1),$$

then $x = v^{(j)}(0) \in T_j^{m-1}(\eta_0, F_0, A_0; \eta_1, F_1, A_1)$.

The reverse inclusion can be obtained by means of the operator $u(t) = (l + j + 1)H_l v(t)$.

Indeed, if $v \in F_0^{\eta_0}(A_0)$, then $u \in F_0^{\eta_0}(A_0)$. Next, by (2.14)

$$u^{(m-1)}(t) = (l + j + 1)H_{l+m-1}v^{(m-1)}(t).$$

From this, by virtue of (2.13), we obtain

$$u^{(m)}(t) = (l + j + 1)\left[\frac{1}{t}v^{(m-1)}(t) - \frac{l+m}{t}H_{l+m-1}v^{(m-1)}(t)\right] \in F_1^{\eta_1}(A_1),$$

provided that $v^{(m-1)} \in F_1^{\eta_1-1}(A_1)$. Finally, $u^{(j)}(0) = v^{(j)}(0)$. Thus, if $x = v^{(j)}(0) \in T_j^{m-1}(\eta_0, F_0, A_0; \eta_1 - 1, F_1, A_1)$, then

$$x = u^{(j)}(0) \in T_j^m(\eta_0, F_0, A_0; \eta_1, F_1, A_1).$$

The lemma is proved.

REMARK. According to Remark 2.1, in the proof of the lemmas we assumed that all functions were as smooth as necessary.

From Lemmas 2.2 and 2.3 it follows that all spaces of traces which are intermediate between $A_0$ and $A_1$ can be obtained as spaces of type $T_0^1$. Indeed, the following corollary is true.

COROLLARY.

$$T_j^m(\eta_0, F_0, A_0; \eta_1, F_1, A_1) = T_0^1(\eta_0 + j, F_0, A_0; \eta_1 - m + j + 1, F_1, A_1)$$

*up to isomorphism.*

It turns out that the spaces of traces coincide with the spaces constructed by the $\mathcal{K}$- and $\mathcal{J}$-methods.

LEMMA 2.4. *If* $-\eta_0 < 0 < 1 - \eta_1$, *then*

$$T_0^1(\eta_0, F_0, A_0; \eta_1, F_1, A_1) = (A_0, A_1)^J_{F_0^{\eta_0}, F_1^{\eta_1-1}}$$

*up to isomorphism.*

PROOF. Let $x \in T_0^1(\eta_0, F_0, A_0; \eta_1, F_1, A_1)$. This means that $x = u(0)$, $u \in F_0^{\eta_0}(A_0)$, $u' \in F_1^{\eta_1}(A_1)$. Consider the convolution $\varphi * u = v$ of $u(t)$ with a smooth function $\varphi(t)$ with compact support normalized in such a way that $\int_0^\infty \varphi(s)s^{-1}\,ds = 1$. Then by (2.9) and (2.10) we have

$$v'(t) = t^{-1}\int_0^\infty \frac{t}{s}\varphi'\left(\frac{t}{s}\right)u(s)\frac{ds}{s} = \int_0^\infty \frac{s}{t}\varphi\left(\frac{t}{s}\right)u'(s)\frac{ds}{s}.$$

It follows from the first equality that $tv' \in F_0^{\eta_0}(A_0)$, and from the second that $v' \in F_1^{\eta_1}(A_1)$. Consequently, $tv' \in F_1^{\eta_1-1}(A_1)$. Finally, by (2.11) we have $v(0) = x$.

Next, by virtue of (8.44) of Chapter II, under the condition $-1 > -\eta_1$ the integral

$$\int_b^\infty v'(s)\,ds = \int_b^\infty sv'(s)\frac{ds}{s}$$

is absolutely convergent in $A_1$, and so $v(t)$ has a limit at infinity in the norm of $A_1$, and thus in the norm of $A_0 + A_1$ as well. Next, $v \in F_0^{\eta_0}(A_0)$, and hence $v \in F_0^{\eta_0}(A_0 + A_1)$. On the other hand, for the function $\|v(t)\|_{A_0+A_1}$ having a limit at infinity, the function $t^{\eta_0}\|v(t)\|_{A_0+A_1}$ can belong to $F_0 \subset L_{1,*} + L_{\infty,*}$ for $\eta_0 > 0$ only if the limit is equal to zero. Thus,

$$v(b) = -\int_b^\infty sv'(s)\frac{ds}{s}$$

in $A_0 + A_1$.

Since $v \in W_1(\eta_0, F_0, A_0; \eta_1, F_1, A_1)$, the limit $\lim_{b\to 0} v(b) = v(0) = x$ exists in $A_0 + A_1$ on the basis of Theorem 2.1, i.e.

$$v(0) = -\int_0^\infty sv'(s)\frac{ds}{s}\,.$$

In this representation $-sv'(s) \in F_0^{\eta_0}(A_0) \cap F_1^{\eta_1 - 1}(A_1)$, and consequently $x \in (A_0, A_1)^I_{F_0^{\eta_0}, F_1^{\eta_1-1}}$.

Conversely, let $x$ satisfy the last condition. Then $x = \int_0^\infty w(s)s^{-1}ds$, where $w \in F_0^{\eta_0}(A_0) \cap F_1^{\eta_1-1}(A_1)$, and the function $w(t)$ can be assumed smooth (see the proof of Theorem 2.3 in Chapter IV). Consider the function

$$u(t) = \int_t^\infty w(s)\frac{ds}{s}\,.$$

By virtue of (8.41) of Chapter II we have $u \in F_0^{\eta_0}(A_0)$. Next, $u'(t) = t^{-1}w(t) \in F_1^{\eta_1}(A_1)$. Finally, $u(0) = x$. Thus, $x \in T_0^1(\eta_0, F_0, A_0; \eta_1, F_1, A_1)$.

The lemma is proved.

Lemmas 2.1–2.4 imply the following theorem.

THEOREM 2.2. *If* $-\eta_0 < j < m - \eta_1$, *then*

$$T_j^m(\eta_0, F_0, A_0; \eta_1, F_1, A_1) = (A_0, A_1)^I_{F_0^{\eta_0+j}, F_1^{\eta_1+j-m}} \tag{2.19}$$

*up to isomorphism.*

COROLLARY 1. *If* $-\eta_0 < j < m - \eta_1$, *then*

$$T_j^m(\eta_0, F_0, A_0; \eta_1, F_1, A_1) = (A_0, A_1)_{\theta, F_0, F_1} \tag{2.20}$$

*up to isomorphism, where* $\theta = (j + \eta_0)/(m - \eta_1 + \eta_0)$.

This corollary immediately follows from Theorem 2.3 and the corollary to Lemma 2.4 of Chapter IV (taking account of Remark 2.3 of Chapter IV).

COROLLARY 2.

$$T_j^m(\eta_0, F_0, A_0; \eta_1, F_1, A_1) = T_j^m(\eta, F_0, A_0; \eta, F_1, A_1), \tag{2.21}$$

*up to isomorphism, where* $\eta = ((m-j)\eta_0 + j\eta_1)/(m - \eta_1 + \eta_0)$.

The space on the right side of (2.18) is intermediate between $A_0$ and $A_1$, since

$$j + \eta = m\frac{j + \eta_0}{m - \eta_1 + \eta_0} > 0 \quad \text{and} \quad m - \eta - j = m\frac{m - j - \eta_1}{m - \eta_1 + \eta_0} > 0.$$

According to (2.20) this space is also isomorphic to $(A_0, A_1)_{\theta, F_0, F_1}$ with $\theta = (j + \eta)/m = (j + \eta_0)/(m - \eta_1 + \eta_0)$.

COROLLARY 3. *For fixed* $\eta_0$, $\eta_1$, $F_0$, $F_1$, $m$ *and* $j$ *the functor* $T_j^m(\eta_0, F_0, A_0; \eta_1, F_1, A_1)$ *is an interpolation functor.*

This corollary follows easily from the definition of the spaces.

COROLLARY 4. *Under the conditions* $-\eta_0 < j < m - \eta_1$ *and* $0 \leqslant j \leqslant m - 1$, *the operation of taking the trace* $u^{(j)}(0)$ *is a continuous mapping of the space* $W_m(\eta_0, F_0, A_0; \eta_1, F_1, A_1)$ *onto the space* $(A_0, A_1)_{\theta, F_0, F_1}$, *where* $\theta = (j + \eta_0)/(m - \eta_1 + \eta_0)$.

**4. *Complete theorem on traces.*** If in the preceding constructions $\eta_0 > 0$, then the function $u \in W_m(\eta_0, F_0, A_0; \eta_1, F_1, A_1)$ has traces of orders $j = 0, \ldots, j_{\max}$, where $j_{\max}$ is the largest integer not exceeding $m - 1$ and smaller than $m - \eta_1$. According to the preceding, we have $u^{(j)}(0) \in (A_0, A_1)_{\theta_j, F_0, F_1}$, where $\theta_j = (j + \eta_0)/(m - \eta_1 + \eta_0)$. Thus, there arises the mapping $\gamma u = \{u(0), u'(0), \ldots, u^{(j_{\max})}(0)\}$ of the space $W_m(\eta_0, F_0, A_0; \eta_1, F_1, A_1)$ into the direct product $\prod_{j=0}^{j_{\max}} (A_0, A_1)_{\theta_j, F_0, F_1}$.

THEOREM 2.3. *The mapping* $\gamma u = \{u(0), u'(0), \ldots, u^{(j_{\max})}(0)\}$ *is a continuous mapping of* $W_m(\eta_0, F_0, A_0; \eta_1, F_1, A_1)$ *onto* $\prod_{j=0}^{j_{\max}} (A_0, A_1)_{\theta_j, F_0, F_1}$.

PROOF. For the proof of the theorem it suffices to construct, for every element $x \in (A_0, A_1)_{\theta_i, F_0, F_1}$ $(0 \leqslant i \leqslant j_{\max})$, a function $v_i \in W_m(\eta_0, F_0, A_0; \eta_1, F_1, A_1)$ such that $v_i^{(i)}(0) = x$ and $v_i^{(j)}(0) = 0$ for $j \neq i$ $(0 \leqslant j \leqslant j_{\max})$. According to Theorem 2.2 there exists a function $u \in W_m(\eta_0, F_0, A_0; \eta_1, F_1, A_1)$ such that $u^{(i)}(0) = x$. By means of the Hardy-Littlewood operators we construct the functions $w_{l_k}(t) = H_{l_k} u(t)$ $(k = 0, 1, \ldots, j_{\max})$ with sufficiently large $l_k$. These functions also belong to $W_m(\eta_0, F_0, A_0; \eta_1, F_1, A_1)$. By (2.15) we have

$$w_{l_k}^{(j)}(0) = \frac{1}{l_k + j + 1} u^{(j)}(0).$$

We now seek the function $v_i(t)$ in the form $v_i(t) = \Sigma\, \alpha_k w_{l_k}(t)$. For the determination of the numbers $\alpha_k$ we obtain the system

$$\sum_{k=0}^{j_{\max}} \frac{1}{l_k + j + 1}\, \alpha_k = \delta_{ij} \qquad (j = 0, \ldots, j_{\max}).$$

Using the arbitrariness in choosing the $l_k$, we can always make the determinant of this system different from zero. Then the numbers $\alpha_k$, and together with them the function $v_i(t)$, can be determined.

The theorem is proved.

We have proved that to every set of elements $x_i \in (A_0, A_1)_{\theta_i, F_\theta, F_1}$ ($i = 0, \ldots, j_{\max}$) there corresponds a function $u(t)$ belonging to $W_m(\eta_0, F_0, A_0; \eta_1, F_1, A_1)$ such that $x_i$ is the trace of order $i$ of $u(t)$. However an analysis of the proof shows that we have not guaranteed the linearity of the correspondence. In the following subsection we construct such a linear correspondence in a more special situation.

We also note that the assumption $\eta_0 > 0$ has been made for the simplicity of the formulation. If $\eta_0 < 0$, then Theorem 2.3 is true for the space

$$\prod_{j=j_{\min}}^{j_{\max}} (A_0, A_1)_{\theta_j, F_0, F_1},$$

where $j_{\min}$ is the smallest integer greater than $-\eta_0$ and not exceeding $m - 1$.

**5. *Functions smooth in t and smooth relative to an unbounded operator.*** Let $E$ be a Banach space in which a closed unbounded linear operator $B$ is given with domain $\mathfrak{D}(B)$ dense in $E$. For a function $u(t)$ $(0 < t < \infty)$ with values in $E$ two notions of smoothness may be considered: smoothness in $t$ and smoothness relative to the operator $B$ in the sense that the values of the function belong to spaces constructed from the powers of $B$ (smoothness in the space coordinates, so to say). If we assume that the operators $d/dt$ and $B$ are "of equal strength", then by total smoothness of order $m$ of the function $u(t)$ we must understand that $u(t)$ belongs to some space of functions of the form $B^l u^{(j)}(t)$, where $l + j \leqslant m$.

Suppose that $B$ satisfies the condition

$$\|(B + \lambda)^{-1}\|_E \leqslant a/(1 + \lambda) \qquad (\lambda \geqslant 0), \tag{2.29}$$

which is stronger than (2.15).

Clearly this condition implies (1.15): $\|\lambda(B + \lambda)^{-1}\|_E \leqslant a$, but, in addition, it also implies that $B$ has bounded inverse with $\|B^{-1}\|_E \leqslant a$. Under this assumption we have $\|B^l x\|_E \leqslant a^{r-l}\|B^r x\|_E$ for $r > l$, and therefore an equivalent norm can be introduced in the space $E_{B^r} = \mathfrak{D}(B^r)$ by the simpler formula

$$\|x\|_{E_{B^r}} = \|B^r x\|_E.$$

Now let $G$ be an interpolation space between $L_{1,*}$ and $L_{\infty,*}$, and let $\eta > 0$. We denote by $W_m(\eta, G, E, B)$ the space of functions $u(t)$ $(0 < t < \infty)$ with values in $E$ for which the functions $B^l u^{(j)}(t)$ belong to $G^\eta(E)$ for $l + j \leqslant m$. By virtue of what has been said above, it is sufficient to require that the functions $B^l u^{(j)}(t)$ belong to $G^\eta(E)$ for $l + j = m$. In the space $W_m(\eta, G, E, B)$ the norm

$$\|u\|_{W_m(\eta,G,E,B)} = \sum_{l=0}^{m} \|B^l u^{(m-l)}\|_{G^\eta(E)},$$

can be introduced, after which it becomes a Banach space. For functions $u(t)$ belonging to $W_m(\eta, G, E, B)$ we have $B^m u \in G^\eta(E)$ by definition, i.e. $u \in B^\eta(E_{B^m})$ and $u^{(m)} \in G^\eta(E)$, and so the imbedding

$$W_m(\eta, G, E, B) \subset W_m(\eta, G, E_{B^m}; \eta, G, E)$$

holds.

Then it follows from Theorem 2.1 that the functions $u \in W_m(\eta, G, E, B)$ have traces of orders $j = 0, \ldots, j_{\max}$, where $j_{\max}$ is the largest integer smaller than $m - \eta$. By Theorem 2.2 we have

$$T_j^m(\eta, G, E_{B^m}; \eta, G, E) = (E_{B^m}, E)_{\theta_j,G,G} = (E, E_{B^m})_{1-\theta_j,G,G}, \tag{2.23}$$

up to isomorphism, where $\theta_j = (j + \eta)/m$.

Let $x \in (E, E_{B^m})_{1-\theta_j,G,G}$. By Theorems 1.5 and 1.6 we have $x \in E_{B^r,G^{r-s}}$, where $s = (1 - \theta_j)m = m - j - \eta$ and $r$ is any integer greater than $s$. By the definition of the space $E_{B^r,G^{r-s}}$ this means that

$$t^{r-s}B^r(tB + I)^{-r}x \in G(E).$$

We choose $r = m - j > m - j - \eta = s$. Then the last membership relation can be written in the form

$$B^{m-j}(tB + I)^{-(m-j)}x \in G^\eta(E). \tag{2.24}$$

Consider the functions

$$u_{j,q}(t) = B^{-j}(tB + I)^{-q}x, \tag{2.25}$$

where the integer $q$ is sufficiently large. We calculate

$$B^l u_{j,q}^{(m-l)}(t)$$
$$= -q(-q - 1) \cdots (-q - m + l + 1)B^{m-j}(tB + I) - q - (m - l)x.$$

Since $(tB + I)^{-1}$ is uniformly bounded in $t$, it follows from (2.24) that, if $q \geqslant m - j$, then

$$B^l u_{j,q}^{(m-l)} \in G^\eta(E)$$

and consequently $u_{j,q} \in W_m(\eta, G, E, B)$. Thus, (2.25) enables us to constuct, for every $x \in (E, E_{B^m})_{1-\theta_j,G,G}$, a function belonging to $W_m(\eta, G, E, B)$, and then for $i = 0, \ldots, j_{\max}$ we have

$$u_{j,q}^{(i)}(0) = -q(-q-1)\cdots(-q-i+1)B^{i-j}x. \tag{2.26}$$

(In this calculation we have used the fact that on the basis of Theorem 1.6 it follows from the inequality $j_{\max} - j < m - \eta - j = s$ that the membership relation $x \in E_{B^r,G^{r-s}}$ implies that $x \in \mathfrak{D}(B^{j_{\max}-j})$.)

We are now ready to prove the following assertion.

THEOREM 2.4. *There exists a bounded linear operator $R$ mapping the space $\prod_{j=0}^{j_{\max}}(E_{B^m}, E)_{\theta_j,G,G}$ into the space $W_m(\eta, G, E, B)$ $(\theta_j = (j + \eta)/m)$ such that $\gamma R = I$, where $\gamma u = \{u(0), u'(0), \ldots, u^{(j_{\max})}(0)\}$.*

PROOF. It is sufficient to define the operator $R$ for elements of the form $\{x_0, x_1, \ldots, x_{j_{\max}}\}$, where $x_i = 0$ for $i \neq j$ and $x_j = x \in (E_{B^m}, E)_{\theta_j,G,G}$. Using the notation of (2.25), we set

$$R\{0, 0, \ldots, x_j, 0, \ldots, 0\} = \sum_{k=0}^{j_{\max}} \alpha_{k,j} u_{j,q_k}(t) = v_j(t)$$

where the $q_k$ are distinct sufficiently large integers. As we saw above, $v_j \in W_m(\eta, G, E, B)$. We choose the numbers $\alpha_{k,j}$ so that $v_j^{(i)}(0) = 0$ for $i \neq j$ and $v_j^{(j)}(0) = x$. By virtue of (2.26), for this it is sufficient that the $\alpha_{k,j}$ satisfy the system

$$\sum_{k=0}^{j_{\max}} - q_k(-q_k - 1)\cdots(-q_k - i + 1)\alpha_{k,j} = \delta_{ij} \qquad (i = 0, \ldots, j_{\max}) \tag{2.27}$$

It is easy to see that the determinant of this system coincides with the Vandermonde determinant constructed from the numbers $-q_k$, and so it is not zero. Thus, the $\alpha_{k,j}$ are determined uniquely. Finally, the operator $R$ is constructed according to the formula

$$R\{x_0, \ldots, x_{j_{\max}}\} = \sum_{j=0}^{j_{\max}} \sum_{k=0}^{j_{\max}} \alpha_{k,j} B^{-j}(tB + I)^{-q_k} x_j$$

where the $\alpha_{k,j}$ are solutions of (2.27).

The theorem is proved.

Our constructions simplify if only traces of order zero are considered. In this case we have

$$T_0^m(\eta, G, E_{B^m}; \eta, G, E) = (E, E_{B^m})_{1-\eta/m,G,G}.$$

For $x \in (E, E_{B^m})_{1-\eta/m,G,G}$ and $q \geqslant 2m$ the function $u_{0,q}(t) = (tB + I)^{-q}x$ belongs to $W_m(\eta, G, E, B)$, and $u_{0,q}(0) = x$. Since $q$ does not depend on $\eta$

and is bounded only from below, for $x \in T_0^{m_1}(\eta_1, G_1, E_{B^{m_1}}; \eta_1, G_1, E) = (E, E_B)_{1-\eta_1/m_2, G_1, G_1}$, where $1 \leqslant m_1 \leqslant m$ and $0 < \eta_1 < m_1$, we have $u_{0,q} \in W_{m_1}(\eta_1, G_1, E, B)$ and $u_{0,q}(0) = x$. We have arrived at the following assertion.

THEOREM 2.5. *The operator $\gamma_0$ of taking the trace of order zero ($\gamma_0 u = u(0)$) induces a right invertible mapping of the Banach couple*

$$W_{m_1}(\eta_1, G_1, E, B), \quad W_m(\eta, G, E, B) \quad (1 \leqslant m_1 < m,\ 0 < \eta_1 < m_1,\ 0 < \eta < m)$$

*onto the Banach couple*

$$(E, E_{B^{m_1}})_{1-\eta_1/m_1, G_1, G_1}, \qquad (E, E_{B^m})_{1-\eta/m, G, G}.$$

Theorem 4.10 of Chapter I implies the following corollary.

COROLLARY. *For any interpolation functor $\mathfrak{F}$ the operator $\gamma_0$ of taking the trace of order zero maps the space $\mathfrak{F}(W_{m_1}(\eta_1, G_1, E, B), W_m(\eta, G, E, B))$ onto the space*

$$\mathfrak{F}\big((E, E_{B^{m_1}})_{1-\eta_1/m_1, G_1, G_1}, (E, E_{B^m})_{1-\eta/m, G, G}\big).$$

*Moreover, the mapping $x \to (tB + I)^{-q}x$ ($q \geqslant 2m$) is a right inverse of $\gamma_0$.*

Similar assertions can be formulated for every trace of a fixed order.

## §3. Spaces of smooth functions of $n$ variables

**1. *Sobolev spaces.*** We shall consider scalar-valued functions defined on the space $\mathbf{R}^n$. Moreover, we shall use the following abbreviated notation: $s = (s_1, \ldots, s_n)$ is a point of $\mathbf{R}^n$, $|s| = (s_1^2 + \cdots + s_n^2)^{1/2}$, $(s, \xi) = s_1\xi_1 + \cdots + s_n\xi_n$, $x(s) = x(s_1, \ldots, s_n)$ is a function on $\mathbf{R}^n$, and

$$\int_{\mathbf{R}^n} \cdots \int_{\mathbf{R}^n} x(s_1, \ldots, s_n)\, ds_1 \ldots ds_n = \int x(s)\, ds.$$

In the space $L_p(\mathbf{R}^n)$ ($1 \leqslant p \leqslant \infty$) constructed for Lebesgue measure we consider the group

$$U_1(t)x(s) = x(s_1 + t, s_2, \ldots, s_n) \qquad (-\infty < t < \infty)$$

of isometric translation operators and determine the infinitesimal generator $D_1$ of this group. By assumption, a function $x \in L_p(\mathbf{R}^n)$ belongs to the domain of $D_1$ if the limit

$$\lim_{h \to 0} \frac{x(s_1 + h, s_2, \ldots, s_n) - x(s_1, \ldots, s_n)}{h} = D_1 u(s) \in L_p(\mathbf{R}^n)$$

exists in $L_p(\mathbf{R}^n)$.

We denote by $\mathfrak{D}(\mathbf{R}^n)$ the space of infinitely differentiable functions with compact support (the space of test functions). Let $\varphi \in \mathfrak{D}(\mathbf{R}^n)$; then

$$\int \frac{x(s_1 + h, s_2, \ldots, s_n) - x(s_1, \ldots, s_n)}{h} \varphi(s_1, \ldots, s_n)\, ds$$
$$= \int x(s_1, \ldots, s_n) \frac{\varphi(s_1 - h, s_2, \ldots, s_n) - \varphi(s_1, \ldots, s_n)}{h}\, ds.$$

Passage to the limit shows that

$$\int D_1 x(s) \varphi(s)\, ds = -\int x(s) \frac{\partial \varphi}{\partial s_1}\, ds$$

and consequently $D_1$ coincides with the differentiation $\partial/\partial s_1$ in the sense of generalized functions. Thus, $D_1 = \partial/\partial s_1$ on the domain of $D_1$, and $\partial x/\partial s_1 \in L_p(\mathbf{R}^n)$.

We introduce the operator $\hat{D}_1$ equal to $\partial/\partial s_1$ for all functions $x \in L_p(\mathbf{R}^n)$ for which $\partial x/\partial s_1 = L_p(\mathbf{R}^n)$. It is clear from the preceding that $\hat{D}_1$ is an extension of $D_1$. We show that $\hat{D}_1$ has resolvent for $\lambda > 0$. For this we consider the equation

$$\hat{D}_1 x - \lambda x = \partial x/\partial s_1 - \lambda x = y.$$

The function

$$x(s) = -\int_{s_1}^{\infty} e^{\lambda(s_1 - \tau)} y(\tau, s_2, \ldots, s_n)\, d\tau, \tag{3.1}$$

is a solution of this equation. This function can be written in convolution form:

$$x(s) = \int_{-\infty}^{\infty} \varphi(s_1 - \tau) y(\tau, s_2, \ldots, s_n)\, d\tau,$$

where $\varphi(u) = 0$ for $u > 0$ and $\varphi(u) = e^{\lambda u}$ for $u \leqslant 0$. By the Young inequality we have

$$\int_{-\infty}^{\infty} |x(s_1, \ldots, s_n)|^p\, ds_1 \leqslant \left( \int_{-\infty}^{\infty} |\varphi(u)|\, du \right)^p \int_{-\infty}^{\infty} |y(s_1, \ldots, s_n)|^p\, ds_1$$
$$= \frac{1}{\lambda^p} \int_{-\infty}^{\infty} |y(s_1, \ldots, s_n)|^p\, ds_1.$$

By integrating with respect to the variables $s_2, \ldots, s_n$, we obtain

$$\|x\|_{L_p(\mathbf{R}^n)} \leqslant \frac{1}{\lambda} \|y\|_{L_p(\mathbf{R}^n)}.$$

Thus, (3.1) determines the resolvent $R(\lambda, \hat{D}_1)$ of $\hat{D}_1$, and

$$\|R(\lambda, \hat{D}_1)\|_{L_p(\mathbf{R}^n) \to L_p(\mathbf{R}^n)} \leqslant 1/\lambda \qquad (\lambda > 0). \tag{3.2}$$

For $\lambda > 0$ the operator $D_1$ has a resolvent according to the general theory of semigroups (see §1.6), and for the operator $\hat{D}_1$ we have constructed the resolvent. Hence it follows that these operators coincide. Indeed, for any $x \in L_p(\mathbf{R}^n)$ we have

$$(\hat{D}_1 - \lambda I)R(\lambda, D_1)x = (D_1 - \lambda I)R(\lambda, D_1)x = x$$

because $\hat{D}_1$ is an extension of $D_1$. Applying $(\hat{D}_1 - \lambda I)^{-1}$ to both sides of the equality, we obtain that $R(\lambda, D_1)x = R(\lambda, \hat{D}_1)x$. Then the domains of $D_1$ and $\hat{D}_1$, and hence the operators themselves, coincide.

Thus, the infinitesimal generator of the group $U_1(t)$ is defined for all functions $x \in L_p(\mathbf{R}^n)$ having derivative $\partial x/\partial s_1$ (in the sense of generalized functions) belonging to $L_p(\mathbf{R}^n)$, and for these functions we have $D_1 x = \partial x/\partial s_1$.

Estimate (3.2) shows that the operator $B_1 = -D_1$ has the property (1.15), and for it $a = 1$. For each $j$ the group $U_j(t)$ of translations along the $j$th coordinate axis and its infinitesimal generator $D_j = \partial/\partial s_j$ can be considered similarly. It is obvious that these groups commute with each other, and we find ourselves under the conditions of §§1.7 and 1.8. Consider the space which was denoted by $\bigcap E_{B_j^r}$ in §1.7. In our case this is the space of functions for which $\partial^k x/\partial s_j^k \in L_p(\mathbf{R}^n)$ for all $j = 1, \dots, n$ and $k = 0, \dots, r$. It is denoted by $W^{r,p}(\mathbf{R}^n)$ and called a *Sobolev space*. The norm in it is defined by

$$\|x\|_{W^{r,p}} = \|x\|_{L_p} + \sum_{j=1}^{n} \left\| \frac{\partial^r x}{\partial s_j^r} \right\|_{L_p}.$$

Sometimes it is more convenient to introduce an equivalent norm by the formula

$$\|x\|_{W^{r,p}}^p = \|x\|_{L_p}^p + \sum_{j=1}^{n} \left\| \frac{\partial^r x}{\partial s_j^r} \right\|_{L_p}^p. \tag{3.3}$$

In accordance with what we said in §1.5 (see (1.21)), these norms are also equivalent to the norm

$$\|x\|_{W^{r,p}} = \left\{ \sum_{k=0}^{r} \sum_{j=1}^{n} \left\| \frac{\partial^k x}{\partial s_j^k} \right\|_{L_p} \right\}^{1/p}.$$

We introduce some notation. If $\alpha = (\alpha_1, \dots, \alpha_n)$ is a vector with nonnegative integral coordinates, then $|\alpha| = \alpha_1 + \cdots + \alpha_n$, $D^\alpha = D_1^{\alpha_1} \dots D_n^{\alpha_n}$ and $\alpha! = \alpha_1! \cdots \alpha_n!$.

The Sobolev spaces admit a description in terms of the Fourier transformation. We mention the necessary information from the theory of the Fourier transformation (see [42] and [48]).

The collection of infinitely differentiable functions on $\mathbf{R}^n$ for which all the norms

$$p_r(x) = \sup_{s \in \mathbf{R}^n} (1 + |s|^2)^r \sum_{|\alpha|=0}^{r} |D^\alpha x(s)| \qquad (r = 0, 1, \dots)$$

are finite is called the *Schwartz space* $S(\mathbf{R}^n)$.

The system of norms $p_r(x)$ defines a topology in $S(\mathbf{R}^n)$ in a natural way.

The *Fourier transformation*

$$\hat{u}(\xi) = \frac{1}{(2\pi)^{n/2}} \int e^{-i(\xi,s)} u(s)\, ds = \mathcal{F}(u),$$

is defined for functions belonging to $S(\mathbf{R}^n)$. It maps $S(\mathbf{R}^n)$ onto itself. The inverse transformation is given by

$$\check{v}(s) = \frac{1}{(2\pi)^{n/2}} \int e^{i(s,\xi)} v(\xi)\, d\xi = \mathcal{F}^{-1}(v).$$

We have the Parseval identity

$$\int u(s)v(s)\, ds = \int \hat{u}(\xi)\hat{v}(\xi)\, d\xi \quad \text{or} \quad \int u(s)\bar{v}(s)\, ds = \int \hat{u}(\xi)\,\overline{\hat{v}(\xi)}\, d\xi.$$

The space $\mathfrak{D}(\mathbf{R}^n)$ of infinitely differentiable functions with compact support is densely imbedded in $S(\mathbf{R}^n)$, and so the reverse imbedding $S'(\mathbf{R}^n) \subset \mathfrak{D}'(\mathbf{R}^n)$ holds for the dual spaces. The space $S'(\mathbf{R}^n)$ is called the *space of slowly increasing Schwartz distributions*. The spaces $L_p(\mathbf{R}^n)$ can be imbedded in $S'(\mathbf{R}^n)$ in a natural way. If $T$ is a distribution (generalized function) belonging to $S'(\mathbf{R}^n)$, then its Fourier transform $\mathcal{F}(T)$ is defined by

$$\mathcal{F}(T)(u) = \hat{T}(u) = T(\hat{u}) \qquad (u \in S(\mathbf{R}^n)).$$

In particular, the Fourier transformation $\mathcal{F}$ is defined for all functions belonging to $L_p(\mathbf{R}^n)$, $1 \leqslant p \leqslant \infty$. We have

$$\mathcal{F}(D_j x) = i\xi_j \mathcal{F}(x). \tag{3.4}$$

The *problem of multipliers* for the Fourier transformation is especially important in applications. It consists in the following: a function $\Phi(\xi)$ is given, from which the operator $\mathcal{F}^{-1}(\Phi(\xi)\mathcal{F}(x))$ is constructed. Under what conditions will this operator act boundedly in the spaces $L_p(\mathbf{R}^n)$?

S. G. Mihlin's well-known theorem (see [30] or [45]) contains sufficient conditions.

THEOREM 3.1. *If the function $\Phi(\xi)$ is continuous in $\mathbf{R}^n \setminus 0$ and has derivative $\partial^n\Phi/\partial\xi_1 \ldots \partial\xi_n$ at every point there, and all lower order derivatives are continuous and*

$$|\xi|^k \left| \frac{\partial^k \Phi}{\partial\xi_{i_1} \ldots \partial\xi_{i_k}} \right| \leqslant M,$$

*where* $k = 0, 1, \ldots, n$ and $1 \leqslant i_1 < i_2 < \cdots < i_k \leqslant n$, *then the operator* $\mathfrak{F}^{-1}\Phi\mathfrak{F}$ *is bounded in* $L_p(\mathbf{R}^n)$ *for* $1 < p < \infty$. *Moreover, the norm of this operator depends only on* $M$ *and* $p$.

Now we return to the Sobolev spaces. The fact that $\partial^r x/\partial s_j^r \in L_p(\mathbf{R}^n)$ for $x \in W^{r,p}(\mathbf{R}^n)$ may also be written, invoking (2.4), in the form

$$\mathfrak{F}^{-1}\big((i\xi_j)^r \mathfrak{F}(x)\big) \in L_p(\mathbf{R}^n). \tag{3.5}$$

From this we draw a number of conclusions. Since $x$ also belongs to $L_p(\mathbf{R}^n)$, we have

$$\mathfrak{F}^{-1}\big((1 + i\xi_j^r)\mathfrak{F}(x)\big) \in L_p(\mathbf{R}^n).$$

Now consider the function

$$\mathfrak{F}^{-1}\Big(\sqrt{1 + \xi_j^{2r}}\ \mathfrak{F}(x)\Big) = \mathfrak{F}^{-1}\left[\frac{\sqrt{1 + \xi_j^{2r}}}{1 + i\xi_j^r}\mathfrak{F}\right]\mathfrak{F}^{-1}\big((1 + i\xi_j^r)\mathfrak{F}(x)\big).$$

It is easy to verify that the function

$$\Phi(\xi) = \sqrt{1 + \xi_j^{2r}}\ /1 + i\xi_j^r$$

satisfies the assumptions of Theorem 3.1. Since $|\Phi(\xi)| = 1$, it is sufficient to show that $|\xi|\ |\partial\Phi/\partial\xi_j| \leqslant M$, and this can be verified by direct calculation. Then by Theorem 3.1 we have

$$\begin{aligned} \Big\|\mathfrak{F}^{-1}\Big(\sqrt{1 + \xi_j^{2r}}\ \mathfrak{F}(x)\Big)\Big\|_{L_p} &\leqslant C\big\|\mathfrak{F}^{-1}\big((1 + i\xi_j^r)\mathfrak{F}(x)\big)\big\|_{L_p} \\ &\leqslant C\|x\|_{W^{r,p}} \qquad (1 < p < \infty). \end{aligned} \tag{3.6}$$

Now we introduce the function

$$\mathfrak{F}^{-1}\big((1 + |\xi|^2)^{r/2}\mathfrak{F}(x)\big) = \sum_k \mathfrak{F}^{-1}\left[\frac{(1 + |\xi|^2)^{r/2}}{\Sigma_j\sqrt{1 + \xi_j^{2r}}}\sum_k\sqrt{1 + \xi_k^{2r}}\ \mathfrak{F}(x)\right]. \tag{3.7}$$

It can be shown that the function

$$\Phi(\xi) = (1 + |\xi|^2)^{r/2}/\sum\sqrt{1 + \xi_j^{2r}}$$

satisfies the hypotheses of Theorem 3.1, and therefore from (3.6) and (3.7) we obtain

$$\left\|\mathfrak{F}^{-1}\left((1+|\xi|^2)^{r/2}\mathfrak{F}(x)\right)\right\|_{L_p} \leqslant C\|x\|_{W^{r,p}}. \tag{3.8}$$

We show that the reverse inequality also holds. For this we write

$$\|x\|_{L_p} = \left\|\mathfrak{F}^{-1}\left(\frac{1}{(1+|\xi|^2)^{r/2}}\mathfrak{F}\right)\left(\mathfrak{F}^{-1}\left((1+|\xi|^2)^{r/2}\mathfrak{F}(x)\right)\right)\right\|_{L_p}$$

and

$$\left\|\frac{\partial^r x}{\partial \xi_j^r}\right\|_{L_p} = \left\|\mathfrak{F}^{-1}\left[\frac{(i\xi_j)^r}{(1+|\xi|^2)^{r/2}}\mathfrak{F}\right]\left(\mathfrak{F}^{-1}\left((1+|\xi|^2)^{r/2}\mathfrak{F}(x)\right)\right)\right\|_{L_p}.$$

Applying Theorem 3.1 again with the functions

$$(1+|\xi|^2)^{-r/2} \quad \text{and} \quad (i\xi_j)^r(1+|\xi|^2)^{-r/2},$$

we obtain

$$\|x\|_{W^{r,p}} \leqslant C\|\mathfrak{F}^{-1}\left((1+|\xi|^2)^{r/2}\mathfrak{F}(x)\right)\|_{L_p}.$$

Now we can also solve the problem of mixed derivatives of functions belonging to $W^{r,p}(\mathbf{R}^n)$. We have

$$\|D^\alpha x\|_{L_p} = \|\mathfrak{F}^{-1}(i^{|\alpha|}\xi^\alpha\mathfrak{F}(x))\|_{L_p}$$
$$= \left\|\mathfrak{F}^{-1}\left(\frac{i^{|\alpha|}\xi^\alpha}{(1+|\xi|^2)^{r/2}}\mathfrak{F}\right)\left(\mathfrak{F}^{-1}\left((1+|\xi|^2)^{r/2}\mathfrak{F}(x)\right)\right)\right\|_{L_p},$$

where $\xi^\alpha = \xi_1^{\alpha_1}\ldots\xi_n^{\alpha_n}$. If $|\alpha| \leqslant r$, then the function $\xi^\alpha(1+|\xi|^2)^{-r/2}$ satisfies the hypotheses of Theorem 3.1, and therefore, taking account of (3.8), we obtain $\|D^\alpha x\|_{L_p} \leqslant C\|x\|_{W^{r,p}}$ for $|\alpha| \leqslant r$.

We sum up our discussion in the following theorem.

THEOREM 3.2. *If $1 < p < \infty$, then in the space $W^{r,p}(\mathbf{R}^n)$ with norm* (2.3) *equivalent norms can be introduced by the formulas*

$$\|x\|_{W^{r,p}} = \left\|\mathfrak{F}^{-1}\left((1+|\xi|^2)^{r/2}\mathfrak{F}(x)\right)\right\|_{L_p} \tag{3.9}$$

*or*

$$\|x\|_{W^{r,p}} = \left\{\sum_{|\alpha|\leqslant r}\|D^\alpha x\|_{L_p}^p\right\}^{1/p}. \tag{3.10}$$

We note that the assertions of Theorem 3.2 are not true for $p = 1$ and $p = \infty$.

Theorem 3.2 has major significance for us, since in studying the spaces $W^{r,p}(\mathbf{R}^n)$ for $1 < p < \infty$ it enables us to get rid of the cumbersome scheme of several commuting operators (§§1.7 and 1.8) and to apply a scheme with a single operator (§1.5). Indeed, we consider the operator $B = \mathfrak{F}^{-1}(1 + |\xi|^2)^{1/2}\mathfrak{F}$, and show that it satisfies condition (2.19), which is stronger than (1.15).

For $\lambda > 0$ the operator $(B + \lambda I)^{-1}$ has the form $\mathfrak{F}^{-1}(((1 + |\xi|^2)^{1/2} + \lambda)^{-1})\mathfrak{F}$, and to estimate its norm we may use Theorem 3.1 again. For the function

$$\Phi(\xi) = \left(\sqrt{1 + |\xi|^2} + \lambda\right)^{-1} = (\sqrt{u} + \lambda)^{-1} \qquad (u = 1 + |\xi|^2)$$

we have $|\Phi(\xi)| \leqslant 1/(1 + \lambda)$. Next,

$$|\xi|^k \frac{\partial^k \Phi(\xi)}{\partial \xi_{i_1} \ldots \partial \xi_{i_k}} = 2^k |\xi|^k \xi_{i_1} \ldots \xi_{i_k} \left(\frac{1}{\sqrt{u} + \lambda}\right)$$

$$= |\xi|^k \xi_{i_1} \ldots \xi_{i_k} \frac{1}{(\sqrt{u}\,)^k(\sqrt{u} + \lambda)^2} \sum_{j=0}^{k-1} \frac{\alpha_j}{(\sqrt{u}\,)^j(\sqrt{u} + \lambda)^{k-1-j}}.$$

In denominators of the terms on the right the factors $(\sqrt{u}\,)^{k+j}(\sqrt{u} + \lambda)^{k-j}$ neutralize the growth of $\xi_{i_1} \ldots \xi_{i_k}|\xi|^k \leqslant |\xi|^{2k}$ at infinity, and $1/(\sqrt{u} + \lambda) \leqslant 1/(1 + \lambda)$ for the remaining factor. Therefore by Theorem 3.1 we have

$$\|(B + \lambda I)^{-1}\|_{L_p \to L_p} = \left\| \mathfrak{F}^{-1}\left(\frac{1}{\sqrt{1 + |\xi|^2} + \lambda}\right)\mathfrak{F} \right\|_{L_p \to L_p} \leqslant \frac{M_1}{1 + \lambda}. \quad (3.11)$$

Now it follows from Theorem 3.2 that $W^{r,p}(\mathbf{R}^n)$ is isomorphic to $E_{B^r}$ for $1 < p < \infty$, where $E = L_p(\mathbf{R}^n)$ and $B = \mathfrak{F}^{-1}((1 + |\xi|^2)^{1/2})\mathfrak{F}$.

We establish some more properties of Sobolev spaces.

LEMMA 3.1. *The collection $\mathfrak{D}(\mathbf{R}^n)$ of infinitely differentiable functions with compact support is dense in $W^{r,p}(\mathbf{R}^n)$ provided that $1 \leqslant p < \infty$.*

The lemma can be proved by a standard method which we describe briefly. A sequence of nonnegative functions $\rho_n(s) \in \mathfrak{D}(\mathbf{R}^n)$ with $\int \rho_n(s)\, ds = 1$ is considered, where the supports of the functions shrink to zero.

Then for $x \in W^{r,p}(\mathbf{R}^n)$ the functions

$$x_n(s) = \rho_n * x(s) = \int \rho_n(s - \tau)x(\tau)\, d\tau = \int \rho_n(\tau)x(s - \tau)\, d\tau$$

are infinitely differentiable, and

$$\mathfrak{D}^\alpha x_n = \rho_n \times D^\alpha x \to D^\alpha x \qquad (|\alpha| \leqslant r)$$

in $L_p(\mathbf{R}^n)$. Hence the infinitely differentiable functions $x_n(s)$ converge to $x(s)$ in $W^{r,p}(\mathbf{R}^n)$.

Now we choose a nonnegative function $\varphi(s) \in \mathfrak{D}(\mathbf{R}^n)$ equal to 1 for $|s| \leqslant 1$ and to 0 for $|s| \geqslant 2$. Then the functions $y_N(s) = \psi(s/N)x_n(s) \in \mathfrak{D}(\mathbf{R}^n)$ and converge to $x_n(s)$ in $L_p(\mathbf{R}^n)$, $1 \leqslant p < \infty$, as $N \to \infty$. Next,

$$\psi(s/N)D_i^r x_n(s) \to D_i^r x_n(s),$$

and $D_i^k[\psi(s/N)]D_i^{r-k}x_n(s) \to 0$ in $L_p(\mathbf{R}^n)$ for $1 \leqslant i \leqslant n$. Therefore $D_i^r y_N \to x_n$ in $L_p(\mathbf{R}^n)$ as $N \to \infty$.

The lemma is proved.

REMARK. In the first part of the proof we have established that the collection of infinitely differentiable functions is dense in $W^{r,p}(\mathbf{R}^n)$; moreover, the arguments remain valid for $p = \infty$. If a function $x$ belonging to $W^{r,\infty}(\mathbf{R}^n)$ is of compact support, then $x_n = \rho_n * x \in \mathfrak{D}(\mathbf{R}^n)$, and $x_n \to x$ in $W^{r,\infty}(\mathbf{R}^n)$.

LEMMA 3.2. *Let $\tau = \varphi(s)$ be a diffeomorphism of $\mathbf{R}^n$ onto itself such that its derivatives up to order $r$ are bounded in $\mathbf{R}^n$ and its Jacobian is bounded below and above by positive constants. Then for $1 < p < \infty$ the operator $\varphi^* x(s) = x(\varphi(s))$ determines an isomorphism of $W^{r,p}(\mathbf{R}^n)$ onto itself.*

If we introduce a norm in $W^{r,p}(\mathbf{R}^n)$ according to formula (3.10) and take into account that generalized derivatives change according to classical formulas under a coordinate transformation, then the assertion of the lemma becomes obvious.

REMARK. If for $p = 1$ and $p = \infty$ we define the spaces $W^{r,p}(\mathbf{R}^n)$ to be the spaces with norm (3.10), then Lemma 3.2 remains valid for them.

**2. *Spaces with intermediate smoothness.*** Let $G$ be an interpolation space between $L_{1,*}(0, \infty)$ and $L_{\infty,*}(0, \infty)$. We write

$$B_G^{\sigma,p}(\mathbf{R}^n) = \left(L_p(\mathbf{R}^n), W^{r,p}(\mathbf{R}^n)\right)_{\theta,G,G}, \quad \text{where } \sigma = \theta r, \quad 0 < \theta < 1.$$

According to Theorem 1.9 we have

$$B_G^{\sigma,p}(\mathbf{R}^n) = \bigcap_{j=1}^{n} \left(L_p(\mathbf{R}^n), L_p(\mathbf{R}^n)_{D_j^r}\right)_{\theta,G,G}$$

up to isomorphism.

REMARK 3.1. According to Theorem 1.9 we obtain spaces isomorphic to $B_G^{\sigma,p}(\mathbf{R}^n)$ if we introduce a norm in $W^{r,p}(\mathbf{R}^n)$ $(1 \leqslant p \leqslant \infty)$ by (3.10).

By Theorem 1.8 the norm in each of the spaces on the right can be written in explicit form. For this we write

$$\Delta_j^r(t, s; x) = \left(U_j(t) - I\right)^r x = \sum (-1)^k C_r^k x(s + kte_j)$$

($e_j$ is the $j$th unit basis vector) for the $r$th difference of $x(s)$ with respect to $s_j$. Then by Theorem 1.8 an equivalent norm can be introduced in $B_G^{\sigma,p}(\mathbf{R}^n)$ by the formula

$$\|x\|_{B_G^{\sigma,p}} = \|x\|_{L_p} + \sum_{j=1}^{n} \|t^{-\sigma}\|\Delta_j^r(t, s; x)\|_{L_p}\|_G.$$

As follows from the arguments of §1, the space $B_G^{\sigma,p}(\mathbf{R}^n)$ does not depend on $r > \sigma$ up to isomorphism. We may also use (1.41) and (1.42) for the equivalent norms. Then we obtain the following: if $\sigma$ is not an integer, $\sigma = m + \alpha$, $0 < \alpha < 1$, then

$$\|x\|_{B_G^{\sigma,p}} = \|x\|_{L_p} + \sum_{j=1}^{n} \|t^{-\alpha}\|\Delta_j^1(t, s; D_j^m x)\|_{L_p}\|_G;$$

if $\sigma$ is a positive integer, then

$$\|x\|_{B_G^{\sigma,p}} = \|x\|_{L_p} + \left\|t^{-1}\left\|\Delta_j^2\left(t, s; D_j^{\sigma-1}x\right)\right\|_{L_p}\right\|_G.$$

In particular, we may set $G = L_{q,*}(0, \infty)$ and denote the corresponding space by $B_q^{\sigma,p}(\mathbf{R}^n)$. Then the last two norms take the form

$$\|x\|_q^{\sigma,p} = \|x\|_{L_p} + \sum_{j=1}^{n} \left\{\int_0^\infty t^{-\alpha q-1}\left\|\Delta_j^1(t, s; D_j^m x)\right\|_{L_p}^q dt\right\}^{1/q}$$

$$(\sigma = m + \alpha, 0 < \alpha < 1) \tag{3.12}$$

and

$$\|x\|_{B_q^{\sigma,p}} = \|x\|_{L_p} + \sum_{j=1}^{n} \left\{\int_0^\infty t^{-q-1}\left\|\Delta_j^2(t, s; D^{\sigma-1}x)\right\|_{L_p}^q dt\right\}^{1/q}$$

$$(\sigma\text{-integral}). \tag{3.13}$$

According to Lemma 1.7, in all formulas for the equivalent norms, instead of norms of differences of $x(s)$ in $L_p(\mathbf{R}^n)$ we may write the corresponding moduli of continuity of $x(s)$ in $L_p(\mathbf{R}^n)$:

$$\omega_j^r(t; x; p) = \sup_{0<h<t} \|\Delta_j^r(h, s; x)\|_{L_p}.$$

Formulas (2.10) and (2.11) for the norms with just this alteration have been proposed by O. V. Besov, and the spaces $B_q^{\sigma,p}(\mathbf{R}^n)$ are called *Besov spaces*. For $q = \infty$ we obtain the *Nikol'skiĭ spaces* with norms

$$\|x\|_{B_\infty^{\sigma,p}} = \|x\|_{L_p} + \sum_{j=1}^{n} \sup_t \left\{t^{-\alpha}\omega_j^1(t; D_j^m x; p)\right\}$$

for $\sigma = m + \alpha$, $0 < \alpha < 1$, and

$$\|x\|_{B_\infty^{\sigma,p}} = \|x\|_{L_p} + \sum_{j=1}^{n} \sup_t \left\{t^{-1}\omega_j^2\left(t, D_j^{\sigma-1}x; p\right)\right\}$$

for $\sigma > 0$ integral.

We also note that if $p = q = \infty$ and $\sigma$ is not an integer, then we obtain the space $B_\infty^{\sigma,\infty}$ consisting of the functions whose derivatives of order $[\sigma]$ satisfy the Hölder condition of order $\alpha = \sigma - [\sigma]$, i.e. the space $C_{[\sigma]}^{\alpha}(\mathbf{R}^n)$.

We also use Theorem 1.11, which enables us to write the equivalent norm in $B_q^{\sigma,p}$ in invariant form not depending on the choice of coordinates in $\mathbf{R}^n$. For this we write

$$\Delta^r(\tau, s; x) = \sum (-1)^k C_r^k x(s + k\tau).$$

By Theorem 1.11 the norm

$$\|x\|_{B_G^{\sigma,p}} = \|x\|_{L_p} + \| |\tau|^{-\sigma} \|\Delta^r(\tau, s; x)\|_{L_p} \|_{G(\mathbf{R}^n_+)}$$

is equivalent to the preceding ones. In particular, for $G = L_{q,*}$ we obtain

$$\|x\|_{B_G^{\sigma,p}} = \|x\|_{L_p} + \left\{ \int |\tau|^{-\sigma q - n} \|\Delta^r(\tau, s; x)\|_{L_p}^q \, d\tau \right\}^{1/q}.$$

For $p = q$ the corresponding spaces were introduced by L. I. Slobodeckiĭ. They are often denoted by $W^{\sigma,p}(\mathbf{R}^n)$ $(= B_p^{\sigma,p}(\mathbf{R}^n))$ and are called *Sobolev-Slobodeckiĭ spaces*.

From Theorem 2.4 of Chapter IV we obtain the following theorem.

THEOREM 3.3 (interpolation theorem). *The triple*

$$\left(L_{p_1}(\mathbf{R}^{n_1}),\ W^{r_1,p_1}(\mathbf{R}^{n_1}),\ B_G^{\sigma_1,p_1}(\mathbf{R}^{n_1})\right)$$

*of spaces is an interpolation triple of type $\theta$ relative to the triple*

$$\left(L_{p_2}(\mathbf{R}^{n_2}),\ W^{r_2,p_2}(\mathbf{R}^{n_2}),\ B_G^{\sigma_2,p_2}(\mathbf{R}^{n_2})\right),$$

*provided that* $\sigma_1/r_1 = \sigma_2/r_2 = \theta$ $(0 < \theta < 1)$.

If we consider the two spaces $B_{G_0}^{\sigma_0,p}(\mathbf{R}^n)$ and $B_{G_1}^{\sigma_1,p}(\mathbf{R}^n)$, then for an integral $r > \max \sigma_i$ we have, by definition,

$$B_{G_0}^{\sigma_0,p} = (L_p, W^{r,p})_{\theta_0, G_0, G_0} \quad \text{and} \quad B_{G_1}^{\sigma_1,p} = (L_p, W^{r,p})_{\theta_1, G_1, G_1},$$

where $\theta_i = \sigma_i/r$. Then we can apply the reiteration Theorem 2.8 of Chapter IV, assuming that $E_0 = E_1 = G$. We then have $F_0 = F_1 = G$ and we obtain

$$\left(B_{G_0}^{\sigma_0,p}(\mathbf{R}^n),\ B_{G_1}^{\sigma_1,p}(\mathbf{R}^n)\right)_{\theta',G,G} = \left(L_p(\mathbf{R}^n),\ W^{r,p}(\mathbf{R}^n)\right)_{\theta,G,G} = B_G^{\sigma,p}(\mathbf{R}^n), \quad (3.14)$$

up to isomorphism, where $\sigma = (1 - \theta')\sigma_0 + \theta'\sigma_1$.

From this the reader may easily infer (Theorem 2.4 of Chapter IV) that a triple of spaces of the type $B_G^{\sigma,p}$ with the same $p$ and distinct indices $\sigma_0'$, $\sigma_1'$, and $\sigma'$ of smoothness is an interpolation triple of type $\theta'$ relative to another such triple with another $p$ in general and indices $\sigma_0''$, $\sigma_1''$, and $\sigma'''$ of smoothness provided that

$$\frac{\sigma' - \sigma_0'}{\sigma_1' - \sigma_0'} = \frac{\sigma'' - \sigma_0''}{\sigma_1'' - \sigma_0''}.$$

Moreover, the spaces $G$ occurring in the subscripts may be arbitrary.

Theorems 1.5 and 1.10 lead to the following isomorphism:

$$(W^{r_0,p}(\mathbf{R}^n), W^{r_1,p}(\mathbf{R}^n))_{\theta_1',G,G} = B_G^{\sigma,p}(\mathbf{R}^n) \qquad (r_0 < r_1), \tag{3.15}$$

where $\theta' = (\sigma - r_0)/(r_1 - r_0)$.

We apply Theorem 2.10 of Chapter IV to a couple of spaces with the same index of smoothness. We obtain

$$\left(B_{G_0}^{\sigma,p}, B_{G_1}^{\sigma,p}\right)_{\theta',F_0,F_1} = \left(L_p, W^{r,p}\right)_{\sigma/r,H,H} = B_H^{\sigma,p},$$

where $H = (G_0, G_1)_{\theta',F_0,F_1}$.

In particular, if $G_0 = L_{q_0,*}$, $G_1 = L_{q_1,*}$ and $F_0 = F_1 = L_{q,*}$, then

$$H = \left(L_{q_0,*}, L_{q_1,*}\right)_{\theta',L_{q,*},L_{q,*}} = L_{\bar{q},q}, \quad \text{where } \frac{1}{\bar{q}} = \frac{1-\theta'}{q_0} + \frac{\theta'}{q_1}.$$

From (2.12) it follows that

$$\left(B_{q_0}^{\sigma,p}, B_{q_1}^{\sigma,p}\right)_{\theta,L_{q,*},L_{q,*}} = B_{L_{\bar{q},q}}^{\sigma,p}$$

and, finally, if $q = \bar{q}$,

$$\left(B_{q_0}^{\sigma,p}, B_{q_1}^{\sigma,p}\right)_{\theta',L_{q,*},L_{q,*}} = B_{\bar{q}}^{\sigma,p}, \quad \text{where } \frac{1}{q} = \frac{1}{\bar{q}} = \frac{1-\theta'}{q_0} + \frac{\theta'}{q_1}.$$

As we mentioned in Remark 3.1, in studying the spaces $B_G^{\sigma,p}(\mathbf{R}^n)$ it may be assumed that the norm is introduced in $W^{r,p}(\mathbf{R}^n)$ ($1^G \leqslant p \leqslant \infty$) by (3.10). It is obvious that, with this norm, the operator $D^\alpha$ is a bounded operator from $W^{r,p}(\mathbf{R}^n)$ into $W^{r-|\alpha|,p}(\mathbf{R}^n)$ for $|\alpha| < r$. It follows from interpolation theorems and (3.13) that the operator $D^\alpha$ acts boundedly from $B_G^{\sigma,p}(\mathbf{R}^n)$ into $B_G^{\sigma-|\alpha|,p}(\mathbf{R}^n)$ for every $\sigma > |\alpha|$.

3. ***Imbedding theorem.*** We use results of §1.9 to obtain an imbedding theorem for the spaces $B_G^{\sigma,p}$. As the operator $B$ we again take $\mathfrak{F}^{-1}(\sqrt{1+|\xi|^2})\mathfrak{F}$. This operator is continuous in $S'(\mathbf{R}^n)$ and so is the operator $(B+\lambda I)^{-1} = \mathfrak{F}((\sqrt{1+|\xi|^2}+\lambda)^{-1})\mathfrak{F}^{-1}$ for $\lambda > 0$. Thus, this operator satisfies condition 1 of Theorem 1.13. If we assume that $E_0 = L_p$ and $E_1 = L_q$ ($0 < p, q < \infty$), then it follows from (3.11) that inequalities (1.56) (i.e. condition 2 of Theorem 1.11) are satisfied for $B$. Condition 4 is also satisfied, since the restrictions of $B$ to $L_p$ and $L_q$ are automatically defined on $S(\mathbf{R}^n)$, which is dense in $L_p$ and $L_q$. It remains to verify condition 3, i.e. to obtain (1.57). We

perform this verification relying on the following generalization of the multiplier Theorem 3.1 due to P. I. Lizorkin [217]:

THEOREM 3.4. *If the function* $\Phi(\xi)$ *is continuous together with the derivative* $\partial^n\Phi/\partial\xi_1\cdots\partial\xi_n$ *and all the preceding derivatives outside the coordinate planes* $\xi_j = 0$ $(j = 1, \ldots, n)$, *and if*

$$\left|\xi_{i_1}\cdots\xi_{i_k}\xi^\beta\frac{\partial^k\Phi}{\partial\xi_{i_1}\cdots\partial\xi_{i_k}}\right| \leqslant M,$$

*where* $k = 0, \ldots, n$, $1 \leqslant i_1 < i_2 < \cdots < i_k \leqslant n$, $\beta \in [0, 1)$ *and* $\xi^\beta = \xi_1^\beta\cdots\xi_n^\beta$, *then the operator* $\mathfrak{F}^{-1}\Phi\mathfrak{F}$ *acts boundedly from* $L_p(\mathbf{R}^n)$ *into* $L_q(\mathbf{R}^n)$, *where* $1/p - 1/q = \beta$ $(1 < p, q < \infty)$. *Moreover, the norm of the operator is proportional to* $M$.

We apply this theorem to the operator $\mathfrak{F}^{-1}((\sqrt{1+|\xi|^2}+\lambda)^{-1})\mathfrak{F}$. Repeating the arguments on p. 327 we arrive at the estimate

$$\left|\xi_{i_1}\cdots\xi_{i_k}\xi^\beta\frac{\partial^k\Phi}{\partial\xi_{i_1}\cdots\partial\xi_{i_k}}\right| \leqslant M_1\left|\frac{\xi^\beta}{\sqrt{1+|\xi|^2}+\lambda}\right| \leqslant M_1\lambda^{n\beta-1}\left|\frac{\eta^\beta}{1+|\eta|}\right|$$
$$(\xi = \lambda\eta).$$

If we now require that $n\beta < 1$, we have

$$\left|\xi_{i_1}\cdots\xi_{i_k}\xi^\beta\frac{\partial^k\Phi}{\partial\xi_{i_1}\cdots\partial\xi_{i_k}}\right| \leqslant M_1\lambda^{n\beta-1}.$$

It follows from Theorem 3.4 that for $p < q$ and $1/p - 1/q < 1/n$ the operator $B = \mathfrak{F}^{-1}(1+\xi^2)^{1/2}\mathfrak{F}$ satisfies condition 3 of Theorem 1.13 with $\nu = n/p - n/q$.

THEOREM 3.5. *If* $1 < p \leqslant q < \infty$, *then* $B_G^{\sigma,p}(\mathbf{R}^n)$ *is imbedded in* $B_G^{s,q}(\mathbf{R}^n)$ *for* $\sigma = s + n/p - n/q$.

PROOF. First we assume that $\nu = n/p - n/q < 1$. Then from the definition of $B_G^{\sigma,p}(\mathbf{R}^n)$, Theorem 3.2, and Theorem 1.13 it follows immediately that

$$B_G^{\sigma,p}(\mathbf{R}^n) = \left(L_p(\mathbf{R}^n), W^{r,p}(\mathbf{R}^n)\right)_{s/r,G,G} = \left(L_p(\mathbf{R}^n), W^{r,p}(\mathbf{R}^n)\right)_{s/r+\nu/r,G,G}$$
$$\subset \left(L_q(\mathbf{R}^n), W^{r,q}(\mathbf{R}^n)\right)_{s/r,G,G} = B_G^{s,q}(\mathbf{R}^n).$$

If now $\nu > 1$, then we partition the interval $[p, q]$ with points $p_0 = p < p_1 < \cdots < p_k = q$ so that $n/p_i - n/p_{i-1} < 1$ $(i = 1, \ldots, k)$. According to the preceding we have

$$B_G^{\sigma,p}(\mathbf{R}^n) \subset B_G^{\sigma-n(1/p-1/p_1),p_1}(\mathbf{R}^n) \subset \cdots$$
$$\subset B_G^{\sigma-n(1/p-1/p_k),p_k}(\mathbf{R}^n) = B_G^{s,q}(\mathbf{R}^n).$$

The theorem is proved.

It is natural to call the number $\kappa = n/p - \sigma$ the *dimension of the norm* in the space $B_G^{\sigma,p}$. If $p$ and $G$ are fixed, it is clear that a space with smaller dimension of the norm (with greater smoothness) is imbedded in any space with larger dimension of the norm (with less smoothness). Theorem 3.5 says that $p < q$ the space $B_G^{\sigma,p}$ is imbedded in $B_G^{s,q}$ if the dimensions of the norms are equal, and thus in the case where the dimension of the norm of the first space does not exceed the dimension of the norm of the second space.

**4. *Function spaces on regular domains.*** The Sobolev function spaces on an arbitrary domain $\Omega \subset \mathbf{R}^n$ can be defined as the collection $W^{r,p}(\Omega)$ of functions having derivatives $D^\alpha x$ (in the sense of the theory of generalized functions) for $|\alpha| \leqslant r$ and having finite norm

$$\|x\|_{W^{r,p}(\Omega)} = \left\{ \|x\|_{L_p(\Omega)}^p + \sum_{|\alpha| \leqslant r} \|D^\alpha x\|_{L_p(\Omega)}^p \right\}^{1/p}. \tag{3.16}$$

Since the operators of generalized differentiation are closed, this space is a Banach space.

If the function $x(s)$ belongs to $W^{r,p}(\mathbf{R}^n)$ and the norm in $W^{r,p}(\mathbf{R}^n)$ is introduced by (3.10), then the restriction $S_\Omega x = x|_\Omega$ obviously belongs to $W^{r,p}(\Omega)$, and $\|S_\Omega x\|_{W^{r,p}(\Omega)} \leqslant \|x\|_{W^{r,p}(\mathbf{R}^n)}$. Thus, the restriction operator $S_\Omega$ has norm 1 in all spaces $W^{r,p}$.

If $\tau = \varphi(s)$ is a diffeomorphism of the domain $\Omega$ onto a domain $\Omega_1$ having bounded derivatives up to order $r$ and Jacobian bounded above and below by positive constants, then the operator $\varphi^* x(s) = x(\varphi(s))$ is an isomorphism of $W^{r,p}(\Omega_1)$ onto $W^{r,p}(\Omega)$ (see Lemma 3.2).

In $W^{r,p}(\Omega)$ we may single out the important subspace $\overset{0}{W}{}^{r,p}(\Omega)$, which is the closure, in $W^{r,p}(\Omega)$, of the set of infinitely differentiable functions belonging to $W^{r,p}(\Omega)$ and having supports inside $\Omega$. It is obvious that the subspace $\overset{0}{W}{}^{r,p}(\Omega_1)$ is mapped onto the subspace $\overset{0}{W}{}^{r,p}(\Omega)$ by the diffeomorphism described above.

On the set of infinitely differentiable functions with support in $\Omega$ the norms of $W^{r,p}(\Omega)$ and $\overset{0}{W}{}^{r,p}(\Omega)$ coincide, and therefore every function belonging to $\overset{0}{W}{}^{r,p}(\Omega)$ extended to zero outside $\Omega$ belongs to $W^{r,p}(\mathbf{R}^n)$. It is clear from the proof of Lemma 3.1 that if a function from $W^{r,p}(\mathbf{R}^n)$ $(1 \leqslant p < \infty)$ has support in $\Omega$, then its restriction to $\Omega$ belongs to $\overset{0}{W}{}^{r,p}(\Omega)$. (If $\Omega$ is bounded, this is also true for $p = \infty$.) The following question arises: Does the last assertion hold for functions from $W^{r,p}(\mathbf{R}^n)$ with support in $\bar{\Omega}$? A positive answer to this question will be given for a special class of domains later (see Lemma 3.4).

We also note that $\overset{0}{W}{}^{r,p}(\Omega)$ (and thus $W^{r,p}(\Omega)$ as well) is densely imbedded in $L_p(\Omega)$.

DEFINITION. A domain $\Omega$ is said to be *r-regular* if there exists an operator $\Pi_r$ which extends functions from $\Omega$ to the whole of $\mathbf{R}^n$ and is a bounded right inverse of the restriction operator $S_\Omega$ in the spaces $W^{k,p}(\Omega)$:

$$S_\Omega \Pi_r x = x \quad (x \in W^{k,p}(\Omega)), \qquad \|\Pi_r x\|_{W^{k,p}(\mathbf{R}^n)} \leqslant C\|x\|_{W^{k,p}(\Omega)}.$$

If a domain $\Omega$ is $r$-regular for every $r = 0, 1, \ldots,$ we call it *regular*.

In an $r$-regular domain the space $W^{r,p}(\Omega)$ consists of restrictions to $\Omega$ of the functions belonging to $W^{r,p}(\mathbf{R}^n)$.

It follows from the definition and the remark to Lemma 3.1 that the restrictions of infinitely differentiable functions from $W^{r,p}(\mathbf{R}^n)$ to an $r$-regular domain $\Omega$ are dense in $W^{r,p}(\Omega)$.

From the fact that $S_\Omega$ is right invertible in the spaces $W^{k,p}$ provided that the domain $\Omega$ is $r$-regular, it follows that $W^{k,p}(\Omega)$ is isomorphic to the quotient space of $W^{k,p}(\mathbf{R}^n)$ modulo the subspace of functions equal to zero in $\Omega$ ($S_\Omega x = 0$). In $W^{k,p}(\Omega)$ an equivalent norm can be introduced by the formula

$$\|x\|_{W^{k,p}(\Omega)} = \inf_{x \in S_\Omega \tilde{x}} \|\tilde{x}\|_{W^{k,p}(\mathbf{R}^n)}.$$

The operator $\Pi_r S_\Omega$ projects the whole space $W^{k,p}(\mathbf{R}^n)$ onto a subspace isomorphic to $W^{k,p}(\Omega)$.

By means of Theorem 3.2 it is easy to see that for $1 < p < \infty$ and for an $r$-regular domain $\Omega$ the norm (3.16) is equivalent to the norm

$$\|x\|_{W^{r,p}(\Omega)} = \left\{ \|x\|^p_{L_p(\Omega)} + \sum_{j=1}^n \left\| \frac{\partial^r x}{\partial s_j^r} \right\|^p_{L_p(\Omega)} \right\}^{1/p}$$

in $W^{r,p}(\Omega)$. Now we introduce the spaces $B_G^{\sigma,p}(\Omega)$ by setting

$$B_G^{\sigma,p}(\Omega) = (L_p(\Omega), W^{r,p}(\Omega))_{\sigma/r,G,G}. \tag{3.17}$$

For an $r$-regular domain, $S_\Omega$ is a right invertible mapping of any couple $W^{k_0,p_0}(\mathbf{R}^n)$, $W^{k_1,p_1}(\mathbf{R}^n)$ onto the couple $W^{k_0,p_0}(\Omega)$, $W^{k_0,p_0}(\Omega)$ ($k_0$, $k_1 \leqslant r$), and so we can apply the results of §4.5 of Chapter I. In particular, applying Theorem 4.10 of Chapter I to the functor $(\ ,\ )_{\sigma/r,G,G}$, we conclude that $B_G^{\sigma,p}(\Omega)$

consists of restrictions to $\Omega$ of the functions belonging to $B_G^{\sigma,p}(\mathbf{R}^n)$. In $B_G^{\sigma,p}(\Omega)$ the norm is equivalent to

$$\|x\|_{B_G^{\sigma,p}(\Omega)} = \inf_{S_\Omega \tilde{x} = x} \|\tilde{x}\|_{B_G^{\sigma,p}(\mathbf{R}^n)}.$$

In other words, $B_G^{\sigma,p}(\Omega)$ is isomorphic to the quotient space of $B_G^{\sigma,p}(\mathbf{R}^n)$ modulo the space of functions from $B_G^{\sigma,p}(\mathbf{R}^n)$ equal to zero for $s \in \Omega$.

If we take account of the isomorphism (3.15), an application of Theorem 4.10 of Chapter I to $W^{r_0,p}(\mathbf{R}^n)$, $W^{r,p}(\mathbf{R}^n)$ and $W^{r_0,p}(\Omega)$, $W^{r,p}(\Omega)$ $(r_0 < r)$ leads to the isomorphism

$$(W^{r_0,p}(\Omega), W^{r,p}(\Omega))_{\theta',G,G} = B_G^{\sigma,p}(\Omega) \qquad (r_0 < r), \tag{3.18}$$

where $\theta' = (\sigma - r_0)/(r - r_0)$.

From what has been said it is clear that the interpolation Theorem 3.3 remains true if the $\mathbf{R}^{n_i}$ are replaced by $r_i$-regular domains $\Omega_i$ in $\mathbf{R}^{n_i}$ $(i = 1, 2)$.

The imbedding Theorem 3.5 with $\mathbf{R}^n$ replaced by an $r$-regular domain with $r > \sigma$ also remains valid.

A simple example of a regular domain is the half-space $\mathbf{R}^n_{n+} = \{s \colon s = (s', s_n), s_n > 0\}$.

THEOREM 3.6. *The half-space $\mathbf{R}^n_{n+}$ is a regular domain.*

PROOF. First we prove that the restrictions to $\mathbf{R}^n_{n+}$ of infinitely differentiable functions from $W^{r,p}(\mathbf{R}^n)$ form a dense set in $W^{r,p}(\mathbf{R}^n_{n+})$. Let $x \in W^{r,p}(\mathbf{R}^n_{n+})$. We consider the functions $x_h(s) = x(s_1, \ldots, s_{n-1}, s_n + h)$ $(h > 0)$. By virtue of the continuity of translation in the spaces $L_p$ and the fact that $D^\alpha x_h(s) = D^\alpha x(s_1, \ldots, s_{n-1}, s_n + h)$, the functions $x_h$ converge to $x$ in $W^{r,p}(\mathbf{R}^n_{n+})$ as $h \to 0$. Now let $\varphi(t)$ $(t \in \mathbf{R}^n)$ be a nonnegative infinitely differentiable function equal to 1 for $t \geqslant -h/2$ and 0 for $t \leqslant -h$. Then the function $y(s) = \varphi(s_n)x_h(s)$ for $s_n \geqslant -h$ and 0 for $s_n \leqslant -h$ belongs to $W^{r,p}(\mathbf{R}^n)$ and coincides with $x_h(s)$ for $s_n \geqslant 0$. By the remark to Lemma 3.1 the function $y(s)$ can be approximated with arbitrary accuracy by an infinitely differentiable function $z$ in $W^{r,p}(\mathbf{R}^n)$. Then the restriction of $z(s)$ to $\overline{\mathbf{R}}^n_{n+}$ approximates $x_h(s)$ (and consequently also $x(s)$ for sufficiently small $h$) with arbitrary accuracy in $W^{r,p}(\mathbf{R}^n_{n+})$.

We define the desired operator $\Pi_r$ by the Hestenes-Whitney method first for functions $x(s)$ infinitely differentiable on $\overline{\mathbf{R}}^n_{n+}$ by setting

$$\Pi_r x(s) = \begin{cases} x(s_1, \ldots, s_n) & \text{for } s_n \geqslant 0, \\ \sum_{k=1}^{r+1} \lambda_k x(s_1, \ldots, s_{n-1}, -ks_n) & \text{for } s_n \leqslant 0. \end{cases} \tag{3.19}$$

The numbers $\lambda_k$ are chosen so that

$$\frac{\partial^i}{\partial s_n^i}\Pi_r x(s_1, \ldots, s_{n-1}, 0) = \frac{\partial^i}{\partial s_n^i} x(s_1, \ldots, s_{n-1}, 0) \quad \text{for} \quad i = 0, 1, \ldots, r.$$

For this it is sufficient to determine them from the system

$$\sum_{k=1}^{r+1} (-k)^i \lambda_k = 1 \qquad (i = 0, 1, \ldots, r)$$

of linear equations with determinant different from zero (Vandermonde determinant).

The function $\Pi_r x(s)$ has continuous derivatives up to order $r$ inclusive, and therefore, if $x \in W^{r,p}(\mathbf{R}^n_{n+})$, then $\Pi_r x \in W^{r,p}(\mathbf{R}^n)$. An elementary calculation shows that

$$\|\Pi_r x\|_{W^{k,p}(\mathbf{R}^n)} \leqslant C\|x\|_{W^{k,p}(\mathbf{R}^n_{n+})} \qquad (k \leqslant r).$$

It follows from this and the preceding that $\Pi_r$ admits closure by continuity to the whole space $W^{r,p}(\mathbf{R}^n_{n+})$. From the continuity of $S_\Omega$ it follows that the relation $S_\Omega \Pi_r x = x$ remains valid for all $x \in W^{r,p}(\mathbf{R}^n_{n+})$.

The theorem is proved.

REMARK 1. It follows from Lemma 3.1 and the above proof that under the condition $1 \leqslant p < \infty$ the collection $\mathfrak{D}(\overline{\mathbf{R}}^n_{n+})$ of restrictions of infinitely differentiable functions on $\mathbf{R}^n$ is dense in $W^{r,p}(\mathbf{R}^n_{n+})$.

REMARK 2. It is clear from the definition of the extension operator (3.19) that if the function $x(s)$ has support in the half-ball $|s| \leqslant a$, $s_n \geqslant 0$, then the function $\Pi_r x(s)$ has support in the ball $|s| \leqslant a$.

We also mention another important property of $W^{r,p}(\mathbf{R}^n_{n+})$.

LEMMA 3.3. *The space* $\overset{0}{W}{}^{r,p}(\mathbf{R}^n_{n+})$ *consists of restrictions to* $\mathbf{R}^n_{n+}$ *of all functions belonging to* $W^{r,p}(\mathbf{R}^n)$ *with support in* $\overline{\mathbf{R}}^n_{n+}$.

PROOF. As we indicated in the definition of the spaces $\overset{0}{W}{}^{r,p}(\Omega)$, every element of $\overset{0}{W}{}^{r,p}(\mathbf{R}^n_{n+})$ falls in $W^{r,p}(\mathbf{R}^n_{n+})$ after it is defined to be zero outside $\mathbf{R}^n_{n+}$. Conversely, let $x \in W^{r,p}(\mathbf{R}^n)$ and let $x$ have support in $\overline{\mathbf{R}}^n_{n+}$. Then the functions $x_{-h}(s) = x(s_1, \ldots, s_{n-1}, s_n - h)$ $(h > 0)$ have supports in $\mathbf{R}^n_{n+}$, and $x_{-h} \to x$ in $W^{r,p}(\mathbf{R}^n)$ as $h \to 0$. By the remark to Lemma 3.1 the function $x_{-h}$ can be approximated by functions infinitely differentiable in $\mathbf{R}^n$ and the approximating functions can be chosen so that their supports lie in $\mathbf{R}^n_{n+}$. The restrictions of these functions to $\mathbf{R}^n_{n+}$ will approximate the restriction of $x(s)$ in $W^{r,p}(\mathbf{R}^n_{n+})$, i.e. $S_{\mathbf{R}^n_{n+}} x \in W^{r,p}(\mathbf{R}^n_{n+})$.

The lemma is proved.

Theorem 3.6 and Lemma 3.3 imply the following remarkable decomposition up to isomorphism:

$$W^{r,p}(\mathbf{R}^n) = W^{r,p}(\mathbf{R}^n_{n+}) \oplus \overset{0}{W}{}^{r,p}(\mathbf{R}^n_{n-}).$$

Indeed, since $\mathbf{R}^n_{n+}$ is regular, the space $W^{r,p}(\mathbf{R}^n_{n+})$ is isomorphic to the subspace of $W^{r,p}(\mathbf{R}^n)$ complementary to the subspace of functions equal to zero in $\mathbf{R}^n_{n+}$, i.e. with supports in $\overline{\mathbf{R}}^n_{n-}$. The last subspace is isometric to $\overset{0}{W}{}^{r,p}(\mathbf{R}^n_{n-})$ by virtue of Lemma 3.3.

Now we pass to a large class of domains $\Omega$ with boundary $\partial\Omega$ such that $\overline{\Omega} = \Omega \cup \partial\Omega$ is a $C^\infty$-submanifold of $\mathbf{R}^n$ with compact boundary.

We describe our assumptions more precisely.

We assume that there exists a finite covering of the boundary $\partial\Omega$ of $\Omega$ by bounded domains $U_1, \ldots, U_N$ in $\mathbf{R}^n$ such that the following conditions are satisfied:

a) The mappings $\varphi_i(s)$, $i = 1, \ldots, n$, are given, which are diffeomorphisms of the domains $U_i$ onto the open ball $V$ of $\mathbf{R}^n$ with coordinates $\tau = (\tau_1, \ldots, \tau_n)$, and the Jacobians of the diffeomorphisms are positive.

b) Under the mapping $\varphi_i$ the set $U_i \cap \overline{\Omega}$, $i = 1, \ldots, N$, is taken to $V_{n+} = V \cap \overline{\mathbf{R}}^n_{n+}$, where $\mathbf{R}^n_{n+} = \{\tau : \tau_n > 0\}$.

We shall consider the covering $\{U_i\}$ and the mappings $\{\varphi_i\}$ fixed. We add one more open set $U_0$ to the system $\{U_i\}$ so that $\overline{U}_0 \subset \Omega$ and the system $U_0, U_1, \ldots, U_N$ forms a covering of the whole domain $\Omega$.

We fix a partition of unity corresponding to the covering $\{U_i\}$ of $\overline{\Omega}$, i.e. we fix nonnegative infinitely differentiable functions $\{\eta_i(s)\}$ such that support of $\eta_i(s)$ is contained in $U_i$ and $\Sigma_0^N \eta_i(s) = 1$ for $s \in \overline{\Omega}$. By means of this partition of unity, to every function $x(s)$ defined on $\Omega$ there corresponds a collection of functions $\{\eta_i(s)x(s)\}_0^N$ defined in the domains $U_i \cap \Omega$. For $i \geqslant 1$ the mapping

$$\eta_i(s)x(s) \to \eta_i(\varphi_i^{-1}(\tau)x(\varphi_i^{-1}(\tau)))$$

converts these functions into functions defined in the half-ball $V_{n+}$. These functions are equal to zero in the neighborhood of the nonplanar boundary of $V_{n+}$. We extend them to zero in $\mathbf{R}^n_{n+}$ outside $V_{n+}$ and denote by $x_i^*(\tau)$ the functions thus obtained. It is easy to verify that the mapping $\Phi$ making the collection $\{x_i^*\}_1^N$ of functions correspond to the function $x \in W^{r,p}(\Omega)$ is a bounded operator from $W^{r,p}(\Omega)$ into $W^{r,p}(\mathbf{R}^n_{n+}) \times \cdots \times W^{r,p}(\mathbf{R}^n_{n+})$. Now we introduce the operator

$$\Pi_{r,\Omega}x(s) = \sum_{i=1}^{N} \Pi_r x_i^*(\varphi_i(s)) + \eta_0(s)x(s), \qquad s \in \bigcup_{i=0}^{N} U_i,$$

$$= 0, \qquad s \notin \bigcup_{i=0}^{N} U_i.$$

By Remark 2 to Theorem 3.6 the function $\Pi_r x_i^*(\tau)$ has support in $V$, and so $\Pi_r x_i^*(\varphi_i(s))$ is defined and has support in $U_i$. Its extension by zero outside $U_i$ belongs to $W^{r,p}(\mathbf{R}^n)$. Thus, $\Pi_{r,\Omega}x \in W^{r,p}(\mathbf{R}^n)$, provided that $x \in W^{k,p}(\Omega)$ $(k \leqslant r)$; moreover,

$$\|\Pi_{r,\Omega}x\|_{W^{k,p}(\mathbf{R}^n)} \leqslant C\|x\|_{W^{k,p}(\Omega)}.$$

It is clear from the construction that

$$\Pi_{r,\Omega}x(s) = \sum_{i=0}^{N} \eta_i(s)x(s) = x(s)$$

for $s \in \Omega$. We have arrived at the following assertion:

THEOREM 3.7. *Any domain $\Omega$ with properties* a) *and* b) *is regular.*

REMARK 1. The reader can easily find what minimum smoothness must be required of the mappings $\varphi_i$ in order that $\Omega$ be $r$-regular with given $r$.

The following lemma is easily obtained from Lemma 3.3.

LEMMA 3.4. *If a domain $\Omega$ has properties* a) *and* b), *then the space $\overset{0}{W}{}^{r,p}(\Omega)$ consists of restrictions to $\Omega$ of all functions from $W^{r,p}(\mathbf{R}^n)$ with support in $\overline{\Omega}$.*

PROOF. If $x \in W^{r,p}(\mathbf{R}^n)$ and $x$ has support in $\Omega$, then the functions $x_i^*$ $(i = 1, \ldots, N)$ constructed from the restrictions of $x$ to $U_i \cap \Omega$ as above and extended by zero outside $V_{n+}$ belong to $W^{r,p}(\mathbf{R}^n)$ and have supports in $V_{n+}$. By Lemma 3.3, they can be approximated in $W^{r,p}(\mathbf{R}^n_{n+})$ by functions $y_i(\tau)$ infinitely differentiable and having support in $V_{n+} \cap \mathbf{R}^n_{n+}$. Then the functions $y_i(\varphi_i(s))$, extended by zero to $\Omega$ outside $U_i \cap \Omega$ and infinitely differentiable with support in $U_i \cap \Omega$, belong to $W^{r,p}(\Omega)$, and approximate the functions $\eta_i(s)x(s)$. By virtue of Lemma 3.1 the function $\eta_0(x)x(s)$ can be approximated by infinitely differentiable functions $y_0(s)$ with support in $U_0$. Then the functions $y_0(s) + \sum_1^N y_i(\varphi_i(s))$ have support in $\Omega$ and approximate $x$ in $W^{r,p}(\Omega)$.

The lemma is proved.

Denote by $\Omega_e$ the domain complementary to $\overline{\Omega}$. If $\Omega$ has properties a) and b), so does $\Omega_e$. Theorem 3.7 and Lemma 3.3 imply the following decompositions up to isomorphism:

$$W^{r,p}(\mathbf{R}^n) = W^{r,p}(\Omega) \oplus \overset{0}{W}{}^{r,p}(\Omega_e)$$

and

$$W^{r,p}(\mathbf{R}^n) = \overset{0}{W}{}^{r,p}(\Omega) \oplus W^{r,p}(\Omega_e)$$

In the first decomposition, for example, the subspaces of $W^{r,p}(\mathbf{R}^n)$ isomorphic to the summands on the right are obtained by means of the projections $\Pi_{r\Omega}S_\Omega$ and $I - \Pi_{r\Omega}S_\Omega$, respectively.

We write

$$\overset{0}{B}{}^{\sigma,p}_G(\Omega) = \left(L_p(\Omega),\ \overset{0}{W}{}^{r,p}(\Omega)\right)_{\sigma/r,G,G}. \tag{3.20}$$

By virtue of the interpolation theorem applied to the operator of imbedding $\overset{0}{W}{}^{r,p}(\Omega)$ in $W^{r,p}(\Omega)$ we have

$$\overset{0}{B}{}^{\sigma,p}_G(\Omega) \subset B^{\sigma,p}_G(\Omega).$$

By Lemma 2.14 of Chapter IV the space $\overset{0}{W}{}^{r,p}(\Omega)$ is densely imbedded in $\overset{0}{B}{}^{\sigma,p}_G(\Omega)$ ($\sigma < r$), and since the infinitely differentiable functions from $W^{r,p}(\Omega)$ with support in $\Omega$ are dense in $\overset{0}{W}{}^{r,p}(\Omega)$, they are also dense in $\overset{0}{B}{}^{\sigma,p}_G(\Omega)$, i.e. $\overset{0}{B}{}^{\sigma,p}_G(\Omega)$ can be defined as the closure in $B^{\sigma,p}_G(\Omega)$ of the infinitely differentiable functions from $B^{\sigma,p}_G(\Omega)$ with support in $\Omega$.

By Theorem 4.10 of Chapter I, for a domain having properties a) and b) we have

$$B^{\sigma,p}_G(\mathbf{R}^n) = B^{\sigma,p}_G(\Omega) \oplus \overset{0}{B}{}^{\sigma,p}(\Omega_e)$$

and

$$B^{\sigma,p}_G(\mathbf{R}^n) = \overset{0}{B}{}^{\sigma,p}_G(\Omega) \oplus B^{\sigma,p}_G(\Omega_e)$$

up to isomorphism.

Next, the operator $I - \Pi_{r\Omega_e}S_{\Omega_e}$ projects $W^{r,p}(\mathbf{R}^n)$ onto a subspace isomorphic to $\overset{0}{W}{}^{r,p}_G(\Omega)$, and the isomorphism is established by the restriction operator $S_\Omega$. Therefore the same operator projects $B^{\sigma,p}(\mathbf{R}^n)$ onto $\overset{0}{B}{}^{\sigma,p}(\Omega)$, and the isomorphism is also established by the restriction operator $S_\Omega$. Thus, *for a domain with properties* a) *and* b) *the space* $\overset{0}{B}{}^{\sigma,p}_G(\Omega)$ *consists of the restrictions of all functions from* $B^{\sigma,p}_G(\mathbf{R}^n)$ *with support in* $\overline{\Omega}$.

**5. *Function spaces on the boundary and trace theorems.*** We need to view the space $W^{r,p}(\mathbf{R}^n_{n+})$ in a new way. If $x \in L_p(\mathbf{R}^n_{n+})$ for $p \geqslant 1$, then by Fubini's theorem we have

$$\int |x(s)|^p\, ds = \int_0^\infty \left\{ \int_{\mathbf{R}^{n-1}} |x(s', s_n)|^p\, ds' \right\} ds_n \qquad (s = (s', s_n))$$

where the inner integral on the right exists for almost all $s_n$. This means that the function $x(s) = x(s', s_n)$ generates the abstract function $\hat{x}(s_n)$ of $s_n$ with values in $L_p(\mathbf{R}^{n-1})$ defined for almost all $s_n$, and

$$\|x\|_{L_p(\mathbf{R}^n)} = \left\{ \int_0^\infty \|\hat{x}(s_n)\|^p_{L_p(\mathbf{R}^{n-1})} ds_n \right\}^{1/p}.$$

Conversely, every function $\hat{x} \in L_p(0, \infty; L_p(\mathbf{R}^{n-1}))$ induces some function $x(s', s_n)$, if for every $s_n$ we fix some function of $s'$ in the class of equivalent functions which is the value of $\hat{x}(s_n)$ in $L_p(\mathbf{R}^{n-1})$. There arise the (technically difficult) problems of strong measurability of the function $\hat{x}(s)$ constructed from $x(s, s')$ and of the measurability of $x(s)$ constructed from $\hat{x}(s)$. We shall not discuss these questions, but refer the reader to the book [11], where they are treated in a rather general situation. Thus, as far as the indicated problems are concerned, we have shown that the space $L_p(\mathbf{R}^n_{n+})$ can be considered isometric to $L_p(0, \infty, L_p(\mathbf{R}^{n-1}))$. Finally, the last space coincides with $L^{1/p}_{p,*}(0, \infty, L_p(\mathbf{R}^{n-1}))$.

We note that this isometric correspondence is lacking for $p = \infty$, although the formal arguments remain valid.

Now let the function $x \in L_p(\mathbf{R}^n_{n+})$ have generalized derivative $\partial x/\partial s_n \in L_p(\mathbf{R}^n_{n+})$ with respect to $s_n$. This means that for every function $\varphi$ belonging to $\mathfrak{D}(\mathbf{R}^n)$ with support in $\mathbf{R}^n_{n+}$ we have

$$\int \frac{\partial x}{\partial s_n}(s)\varphi(s)\, ds = -\int x(s)\frac{\partial \varphi}{\partial s_n}(s)\, ds.$$

We choose $\varphi(s) = \psi(s')\rho(s_n)$, where $\psi \in \mathfrak{D}(\mathbf{R}^{n-1})$ and $\rho \in \mathfrak{D}(\mathbf{R}^1)$, and $\rho$ has support in $(0, \infty)$. Then

$$\int_{\mathbf{R}^{n-1}} \psi(s') \int_0^\infty \frac{\partial x}{\partial s_n}(s', s_n)\rho(s_n) ds_n ds' = \left( -\int_{\mathbf{R}^{n-1}} \psi(s') \int_0^\infty x(s', s_n)\rho'(s_n)\, ds_n \right) ds.$$

Since $\psi \in \mathfrak{D}(\mathbf{R}^{n-1})$ is arbitrary, it now follows that

$$\int_0^\infty \frac{\partial x}{\partial s_n}(s', s_n)\rho(s_n)\, ds_n = -\int_0^\infty x(s', s_n)\rho'(s_n)\, ds_n$$

almost everywhere.

It can be verified by means of the Hölder inequality that the left and right sides of the last equality belong to $L_p(\mathbf{R}^{n-1})$, and so they coincide as elements of this space. This means that $x$ as an abstract function with values in $L_p(\mathbf{R}^{n-1})$ has generalized derivative $\partial x/\partial s_n$.

Conversely, let $y(s', s_n) \in L_p(\mathbf{R}^n_{n+})$ be the generalized derivative of $x(s', s_n) \in L_p(\mathbf{R}^n_{n+})$ as an abstract function with values in $L_p(\mathbf{R}^{n-1})$. This means that for every function $\varphi \in \mathfrak{D}(\mathbf{R}^1)$ with support in $(0, \infty)$ we have

$$\int_0^\infty y(s', s_n)\varphi(s_n)\, ds_n = -\int_0^\infty x(s', s_n)\varphi'(s_n)\, ds_n.$$

The equality is understood here as equality of two functions in $L_p(\mathbf{R}^{n-1})$. In particular, it is satisfied for almost all $s'$. Now let $\varphi_\alpha(s_n)$ be a family of functions belonging to $\mathfrak{D}(\mathbf{R}^n)$ with support in $(0, \infty)$ which is compact in $C^1(0, \infty)$. Then

$$\int_0^\infty y(s', s_n)\varphi_\alpha(s_n)\, ds_n = -\int_0^\infty x(s', s_n)\varphi'_\alpha(s_n)\, ds_n \tag{3.21}$$

almost everywhere for all $\alpha$. Indeed, choosing a countable net everywhere dense in the set $\varphi_\alpha$ (in the metric of $C^1(0, \infty)$), we can construct a set of measure zero in $\mathbf{R}^{n+1}$ such that (3.21) holds for all points $s'$ not in this set and for all functions $\varphi_\alpha$ of the net. Passage to the limit shows that the equality holds for $s'$ outside this set for all $\alpha$.

Now let $\varphi \in \mathfrak{D}(\mathbf{R}^n)$ and let $\varphi$ have support in $\mathbf{R}^n_{n+}$. The family of functions $\varphi_{s'}(s_n) = \varphi(s', s_n)$ is compact in $C^1(0, \infty)$, and so there exists a set $e$ of measure zero in $\mathbf{R}^{n-1}$ such that

$$\int_0^\infty y(s', s_n)\varphi_{\bar{s}'}(s_n)\, ds_n = -\int_0^\infty x(s', s_n)\frac{\partial \varphi_{\bar{s}'}}{\partial s_n}(s_n) ds_n,$$

for $s' \notin e$ and for all $\bar{s}' \in \mathbf{R}^{n-1}$. In particular, it also holds for $\bar{s}' = s'$, i.e.

$$\int_0^\infty y(s', s_n)\varphi(s', s_n)\, ds_n = -\int_0^\infty x(s', s_n)\frac{\partial \varphi}{\partial s_n}(s', s_n)\, ds_n$$

for $s' \notin e$. Integrating with respect to $s'$, we obtain

$$\int y(s', s_n)\varphi(s', s_n)\, ds = -\int x(s', s_n)\frac{\partial \varphi}{\partial s_n}(s', s_n)\, ds,$$

which means that $y = \partial x/\partial s_n$ in the sense of generalized functions.

Thus, the assumption that the generalized derivative $\partial x/\partial s_n$ belongs to $L_p(\mathbf{R}^n_{n+})$ is equivalent to the existence of the derivative of the abstract function

$$x(s', s_n) \in L_p\big(0, \infty, L_p(\mathbf{R}^{n-1})\big)$$

in the sense of generalized functions.

It is clear from what has been said that the assumption that $x$ belongs to $W^{r,p}(\mathbf{R}^n_{n+})$ $(1 \leqslant p < \infty)$ is equivalent to the assumption that the abstract function $x(s', s_n)$ of $s_n$ takes its values in $W^{r,p}(\mathbf{R}^{n-1})$ and its derivative $x^{(r)}(s', s_n)$ (with respect to $s_n$) takes its values in $L_p(\mathbf{R}^{n-1})$ and

$$\|x\|_{W^{r,p}(\mathbf{R}^n_{n+})} = \|x\|_{L^{1/p}_{p,*}(W^{r,p}(\mathbf{R}^{n-1}))} + \|x^{(r)}\|_{L^{1/p}_{p,*}(L_p(\mathbf{R}^{n-1}))}$$

Following the notation of §2, we may write

$$W^{r,p}(\mathbf{R}^n_{n+}) = W_r\left(1/p, L_{p,*}, W^{r,p}(\mathbf{R}^{n-1}); 1/p, L_{p,*}, L_p(\mathbf{R}^{n-1})\right). \quad (3.22)$$

If $1 < p < \infty$, then $W^{r,p}(\mathbf{R}^{n-1})$ coincides with the domain of the $r$th power of the operator $B' = \mathfrak{F}'^{-1}((1 + |\xi'|^2)^{1/2})\mathfrak{F}'$, where the prime indicates that the Fourier transform is taken in $\mathbf{R}^{n-1}$. According to (3.11), the operator $B'$ satisfies condition (2.22). Equality (3.10) shows that in this case

$$\begin{aligned} &W_r\left(1/p, L_{p,*}, W^{r,p}(\mathbf{R}^{n-1}); 1/p, L_{p,*}, L_p(\mathbf{R}^{n-1})\right) \\ &\quad = W_r\left(1/p, L_{p,*}, L_p(\mathbf{R}^{n-1}), B'\right) \end{aligned} \quad (3.23)$$

and we can apply Theorems 2.3 and 2.4.

THEOREM 3.8 (on traces). *If $1 < p < \infty$, then the operator defined on the infinitely differentiable functions from $W^{r,p}(\mathbf{R}^n_{n+})$ by the formula*

$$\mathrm{Tr}: x \to \left( x(s', 0), \frac{\partial x}{\partial s_n}(s', 0), \ldots, \frac{\partial^{r-1}x}{\partial s_n^{r-1}}(s', 0) \right)$$

*admits closure to a bounded operator from $W^{r,p}(\mathbf{R}^n_{n+})$ onto $\prod_{j=0}^{r-1} B^{r-j-1/p,p}(\mathbf{R}^{n-1})$. There exists a bounded linear right inverse of* Tr.

PROOF. It follows from (3.22) and (3.23) that in our case $\eta_0 = \eta_1 = \eta = 1/p$,

$$F_0 = F_1 = G = L_{p,*}, \qquad A_0 = E_{B'^r} = W^{r,p}(\mathbf{R}^{n-1}),$$

and

$$A_1 = E = L_p(\mathbf{R}^{n-1}).$$

Next, $j^{\max}$ is the largest integer not exceeding $r - 1$ and smaller than $r - 1/p$, i.e. $j^{\max} = r - 1$. The operator $\gamma$ defined in Theorem 2.3 coincides with the operator Tr on smooth functions, and according to the theorem it maps $W^{r,p}(\mathbf{R}^n_{n+})$ continuously into

$$\begin{aligned} &\prod_{j=0}^{r-1} \left( W^{r,p}(\mathbf{R}^{n-1}), L_p(\mathbf{R}^{n-1}) \right)_{\theta_j, L_{p,*}, l_{p,*}} \\ &= \prod_{j=0}^{r-1} \left( L_p(\mathbf{R}^{n-1}), W^{r,p}(\mathbf{R}_{n-1}) \right)_{1-\theta_j, L_{p,*}, L_{p,*}}, \end{aligned}$$

where $\theta_j = (j + 1/p)/r$. By definition, the last space is $\prod_{j=0}^{r-1} B^{r-j-1/p,p}(\mathbf{R}^{n-1})$. The first part of the theorem follows from this and Lemma 3.2. The second part follows from Theorem 2.4 immediately.

The theorem is proved.

COROLLARY 1. *Let $D^\alpha$ be differentiation of order $|\alpha| < r$. The operator $D^\alpha$ admits closure from the set of infinitely differentiable functions to a bounded operator from $W^{r,p}(\mathbf{R}^n_{n+})$ into $B_p^{r-|\alpha|-1/p,p}(\mathbf{R}^{n-1})$.*

Indeed, $D^\alpha = (\partial^{\alpha_n}/\partial s_n^{\alpha_n})D^{\alpha'}$, where $D^{\alpha'}$ contains derivatives only with respect to the variables $s_1, \dots, s_{n-1}$. The operator $D^{\alpha'}$ acts boundedly from $W^{r,p}(\Omega)$ into $W^{r-|\alpha'|,p}(\Omega)$, and according to the theorem so does the operator $\partial^{\alpha_n}/\partial s_n^{\alpha_n}$ from $W^{r-|\alpha|,p}(\Omega)$ into

$$B^{r-|\alpha'|-\alpha_n-1/p,p}(\mathbf{R}^{n-1}) = B_p^{r-|\alpha|-1/p}(\mathbf{R}^{n-1}).$$

COROLLARY 2. *If $B = \sum_{|\alpha|\leqslant 1} b_\alpha(s')D^\alpha$ is a boundary differential operator of order $k < r$ on $\mathbf{R}^{n-1}$ with bounded infinitely differentiable coefficients, then its closure is a bounded operator from $W^{r,p}(\mathbf{R}^n_{n+})$ into $B^{r-k-1/p,p}(\mathbf{R}^{n-1})$.*

We note that if the traces

$$\partial^j x/\partial s_n^j = \gamma_i(s') \in B^{r-j-1/p,p}(\mathbf{R}^{n-1}) \qquad (j = 0, \dots, r-1)$$

of $x \in W^{r,p}(\mathbf{R}^n_{n+})$ are known, then from them we can calculate the value of the boundary differential operator of order $k < r$ at this function. Indeed,

$$\begin{aligned} Bx_s &= \sum_{|\alpha|<k} b_\alpha(s')D^\alpha(x) = \sum_{|\alpha|<k} b_\alpha(s')D^{\alpha'}\frac{\partial^{\alpha_n}x}{\partial s_n^{\alpha_n}} \\ &= b_\alpha(s')D^{\alpha'}\gamma_{\alpha_n}(s'). \end{aligned}$$

Moreover, from what we said at the end of subsection 2 it follows that

$$Bx \in B_p^{r-k-1/p,p}(\mathbf{R}^{n-1}).$$

Soon we shall have to use this remark in the following situation. In the half-space $\mathbf{R}^n_{n+}$ let a new system $\sigma = (\sigma_1, \dots, \sigma_n)$ of coordinates be given in such a way that on the boundary plane $\mathbf{R}^{n-1}$ we have $\sigma_i = s_i$, $i = 1, \dots, n$, and let functions $\beta_j(s') \in B_p^{r-j-1/p,p}(\mathbf{R}^{n-1})$ be given. We would like to construct a function $x \in W^{r,p}(\mathbf{R}^n_{n+})$ such that

$$\left.\frac{\partial^i x}{\partial \sigma_n^i}\right|_{s_n=\sigma_n=0} = \beta_i(s^i).$$

The derivatives $\partial^j x/\partial s_n^j = \gamma_i(s')$ can be represented in the form of differential operators of order $j$ in the variables $\sigma_i$, and therefore by the above we may express them in terms of the functions $\beta_i(s')$. If we assume that the transition functions from coordinates $\sigma$ to coordinates $s$ and conversely have bounded derivatives up to order $r-1$, then the functions $\gamma_j(s')$ thus determined will also belong to the spaces $B_p^{r-j-1/p,p}(\mathbf{R}^{n-1})$. According to Theorem 3.8, we can construct from them a function belonging to $W^{r,p}(\mathbf{R}^n_{n+})$ such that $\partial^j x/\partial s_n^j = \gamma_j(s')$. It is obvious that we then have $\partial^i x/\partial\sigma_n^i = \beta_i(s')$.

By Theorem 2.5 the operator $\mathrm{Tr}_0\, x = x(s', 0)$ effects a right invertible mapping of the Banach couple $W^{1,p}(\mathbf{R}^n_{n+})$, $W^{r,p}(\mathbf{R}^n_{n+})$ onto the Banach couple $B_p^{1-1/p,p}(\mathbf{R}^{n-1})$, $B_p^{r-1/p}(\mathbf{R}^{n-1})$. We apply the corollary of Theorem 2.5 to the functor $(\ ,\ )_{\theta',G,G}$, where $\theta' = (\sigma-1)/(r-1)$ and $1<\sigma<r$. From (3.17) and (3.12) we find that

$$\left(W^{1,p}_{(\mathbf{R}^n_{n+})},\ W^{r,p}_{(\mathbf{R}^n_{n+})}\right)_{\theta',G,G} = B_G^{\sigma,p}(\mathbf{R}^n_{n+})$$

and

$$\left(B_p^{1-1/p,p}(\mathbf{R}^{n-1}),\ B_p^{r-1/p,p}(\mathbf{R}^{n-1})_{\theta',G,G}\right) = B_G^{\sigma-1/p,p}(\mathbf{R}^{n-1}).$$

From the indicated corollary we obtain the following theorem.

THEOREM 3.9. *If $1<p<\infty$, then for $\sigma>1$ the operator* $\mathrm{Tr}_0\, x = x(s', 0)$ *is a right invertible bounded operator from $B_G^{\sigma,p}(\mathbf{R}^n_{n+})$ onto $B_\sigma^{\sigma-1/p}(\mathbf{R}^{n-1}_{n+})$.*

REMARK 1. Theorem 3.9 holds for $\sigma > 1/p$. This fact can be proved by means of more complicated constructions.

REMARK 2. If we apply the preceding results to functions whose supports are concentrated in some half-ball $V_1 = \{s\colon |s| \leqslant a_i,\ s_n \geqslant 0\}$, then the right invertible operator occurring in the results can be constructed so that its range consists of functions with supports in an arbitrary half-ball $V \supset \overline{V}_1$.

Indeed, for this it is sufficient to multiply an arbitrary right inverse by an infinitely differentiable function equal to 1 on $\overline{V}_1$ and 0 outside $V$.

Now consider a domain $\Omega$ with properties a) and b) (see subsection 4). The partition of unity $\{\eta_i\}_0^N$ considered in subsection 4 induces a partition of unity $\{\overset{0}{\eta}_i\}_0^N$ relative to the covering $\{U_i\}_1^N$ of $\overline{\Omega}\setminus U_0$, and, in particular, relative to the covering of the boundary $\partial\Omega$ with the sets $U_i' = U_i \cap \partial\Omega$ $(i = 1, \ldots, N)$. We denote by $V'$ the intersection of $V$ with the space $\mathbf{R}^{n-1} = \{\tau\colon \tau_n = 0\}$. With every function $\psi(s)$ on $\partial\Omega$ functions $\psi_i^*$ can be associated, defined by $\psi_i^*(\tau') = \eta_i(\varphi_i^{-1}(\tau'))\psi(\varphi_i^{-1}(\tau'))$ for $\tau' \in V'$ and extended by zero to all of $\mathbf{R}^{n-1}$ outside $V'$. We denote by $W^{r,p}(\partial\Omega)$ the space of functions $\psi$ defined on $\partial\Omega$ for which all $\psi_i^*$ belong to $W^{r,p}(\mathbf{R}^{n-1})$. In this space we introduce the norm

$$\|\psi\|_{W^{r,p}(\partial\Omega)} = \left\{\sum_{i=1}^N \|\psi_i^*\|^p_{W^{r,p}(\mathbf{R}^{n-1})}\right\}^{1/p}.$$

Similarly we may define the spaces

$$B_G^{\sigma,p}(\partial\Omega) = \left(L_p(\partial\Omega), W^{r,p}(\partial\Omega)\right)_{\sigma/r,G,G}.$$

We need the following auxiliary assertion.

LEMMA 3.5. *Assume that the function* $\chi(\tau')$ *belonging to* $\overset{0}{W}{}^{r,p}(V')$ *has support in* $V'$. *The function* $\tilde{\chi}(s') = \chi(\varphi_i(s'))$ $(s' \in U_i')$, *extended by zero on* $\partial\Omega$ *outside* $U_i'$, *belongs to* $W^{r,p}(\partial\Omega)$.

PROOF. We consider the function $\tilde{\chi}_j(s') = \eta_j(s')\tilde{\chi}(s')$. Its support is contained in $U_i' \cap U_j'$, and so the support of the function

$$\tilde{\chi}_j^*(\tau') = \eta_j\left(\varphi_j^{-1}(\tau')\right)\chi\left(\varphi_i \circ \varphi_j^{-1}(\tau')\right)$$

lies in $\varphi_j(U_i' \cap U_j')$. The function $\tilde{\chi}_j^*(\tau')$ is obtained from the function $\eta_j(\varphi_i^{-1}(\tau'))\chi(\tau')$ by means of the diffeomorphism $\varphi_i \circ \varphi_j^{-1}$ of the domain $\varphi_j(U_i' \cap U_j')$ onto the domain $\varphi_i(U_i' \cap U_j')$. Since

$$\eta_j\left(\varphi_i^{-1}(\tau')\right)\chi(\tau') \in \overset{0}{W}{}^{r,p}\left(\varphi_i(U_i' \cap U_j')\right),$$

we have $\tilde{\chi}_j^* \in \overset{0}{W}{}^{r,p}(\varphi_j(U_i' \cap U_j'))$, and so $\tilde{\chi}_j^*$, extended by zero to $\mathbf{R}^{n-1}$ outside $V'$, belongs to $W^{r,p}(\mathbf{R}^{n-1})$.

The lemma is proved.

Denote by $V_1$ a ball containing all images of the supports of the functions $\eta_i(s)$ under the mappings $\varphi_i$ and such that $\overline{V}_1 \subset V$. Let $V_1'$ denote the intersection $V_1 \cap \mathbf{R}^{n-1}$. If $\psi \in W^{r,p}(\partial\Omega)$, then by construction the $\psi_i^*$ have supports in $V_1'$, and so they belong to $\overset{0}{W}{}^{r,p}(V_1')$. Now let $\{\chi_i(\tau')\}_1^N$ be a collection of functions belonging to $\overset{0}{W}{}^{r,p}(V_1')$. By Lemma 3.5 the functions $\chi_i(\varphi_i(s'))$, extended by zero to all of $\partial\Omega$ outside $\varphi_i(V_1')$, belong to $W^{r,p}(\partial\Omega)$, and so

$$\sum_{i=1}^{N} \chi_i(\varphi_i(s')) \in W^{r,p}(\partial\Omega).$$

It is clear that if $\psi \in W^{r,p}(\partial\Omega)$ and $\chi_i = \psi_i^*$, then $\sum_{i=1}^N \chi_i(\varphi_i(s')) = \psi(s')$. Thus we obtain a right invertible mapping of $\overset{0}{W}{}^{r,p}(V_1') \times \cdots \times \overset{0}{W}{}^{r,p}(V_1')$ ($N$ factors) onto $W^{r,p}(\partial\Omega)$. From (3.20), (3.24) and Theorem 4.10 of Chapter I it follows that the space $B_G^{\sigma,r}(\partial\Omega)$ consists of all functions $\psi$ defined on $\Gamma$ for which the restrictions of the functions $\psi_i^*$ to $V_1'$ belong to $\overset{0}{B}{}_G^{\sigma,p}(V_1')$ or, similarly, the functions $\psi_i^*$ themselves belong to $B^{r,p}(\mathbf{R}^{n-1})$. An equivalent norm is introduced in $B_G^{\sigma,p}(\partial\Omega)$ by the formula

$$\|\psi\|_{B_G^{\sigma,p}(\partial\Omega)} = \left\{\sum \|\psi_i^*\|^p_{B_G^{\sigma,p}(\mathbf{R}^{n-1})}\right\}^{1/p}.$$

We note that Lemma 3.5 remains valid if $W^{r,p}$ is replaced by $B_G^{\sigma,p}$ in its formulation.

We need a special coordinate system in the neighborhoods $U_i$ ($i = 1, \dots, N$). The mappings $\varphi_i$ induce a coordinate system $\tau = (\tau_1, \dots, \tau_n)$ in $U_i$. In particular, on the boundary $\partial\Omega$ we obtain the coordinate system $\tau' = (\tau_1, \dots, \tau_{n-1})$. Denote by $n(s)$ the distance of $s$ from $\partial\Omega$ and by $(\sigma_1, \dots, \sigma_{n-1})$ the coordinates of the point of $\partial\Omega$ closest to $s$. We shall assume, without loss of generality, that the covering $\{U_i\}$ is chosen so fine that the numbers $(\sigma_1, \dots, \sigma_{n-1}, n)$ yield a sufficiently smooth coordinate system in $U_i$.

Now let $x \in W^{r,p}(\Omega)$. We have $x_i^* \in W^{r,p}(\mathbf{R}^n_{n+})$, and the supports of the $x_i^*$ are contained in $V_1$. For the smooth function $x$ the traces on the boundary of the domain are defined in a natural way:

$$x(\tau', 0) = \psi^0(\tau'), \quad \frac{\partial x}{\partial n}(\tau', 0) = \psi^1(\tau'), \dots, \quad \frac{\partial^{r-1} x}{\partial n^{r-1}}(\tau', 0) = \psi^{r-1}(\tau').$$

By the Leibniz formula we have

$$\left.\frac{\partial^j(\eta_i x)}{\partial n^j}\right|_{\partial\Omega} = \sum_{k=0}^{n} C_j^k \left.\frac{\partial^k \eta_i}{\partial n^k}\right|_{\partial\Omega} \psi^{j-k}(\tau') \tag{3.25}$$

and, furthermore,

$$\psi^j(\tau') = \left.\frac{\partial^j x}{\partial n^j}\right|_{\partial\Omega} = \sum_{i=1}^{N} \left.\frac{\partial^j(\eta_i x)}{\partial n^j}\right|_{\partial\Omega}. \tag{3.26}$$

Remembering that $x_i^*(\tau) = \eta_i(\varphi_i^{-1}(\tau))x(\tau(\varphi_i^{-1}\tau))$, we obtain that the derivative $\partial^j(\eta_j x)/\partial n^j|_{\partial\Omega}$, upon application of the mapping $\varphi_i$, will be expressed in the form of a differential operator $B_j$ of order $j$ in the variables $\tau$ with infinitely differentiable coefficients, applied to $x_i^*(\tau)$. By Corollary 2 of Theorem 3.8 these operators act boundedly from $W^{r,p}(\mathbf{R}^n_{n+})$ into $B_p^{r-j-1/p,p}(\mathbf{R}_{n-1})$. Thus, if $x \in W^{r,p}(\Omega)$, then $x_i^* \in W^{r,p}(\mathbf{R}^n_{n+})$ and $B_j x_i^* \in B_p^{r-j-1/p}(\mathbf{R}^{n-1})$, and the support of the last function is contained in $V_1'$. Since Lemma 3.5 is valid for the spaces $B_p^{r-j-1/p,p}$, this implies that

$$\left.\frac{\partial^j(\eta_j x)}{\partial n^j}\right|_{\partial\Omega} \in B^{r-j-1/p,p}(\partial\Omega).$$

Then it follows from (3.26) that $\psi^j \in B^{r-j-1/p,p}(\partial\Omega)$. Thus the mapping $x \to (\psi^0, \psi^1, \dots, \psi^{r-1})$ admits extension by continuity to a bounded operator from $W^{r,p}(\Omega)$ into $\prod_{j=0}^{r-1} B_p^{r-j-1/p,p}(\partial\Omega)$.

Now we construct a right inverse of the mapping just obtained. If a system $\{\psi^j\}_0^{r-1}$ of functions is given and $\psi^j \subset B_p^{r-j-1/p,p}(\partial\Omega)$, then from it we can calculate the right side of (3.25), which will also belong to $B_p^{r-j-1/p,p}(\partial\Omega)$ and have support in $U_i'$. After the mapping $\varphi_i$ this function turns into a function $\beta_{ij} \in B_p^{r-j-1/p,p}(V_1')$. Now we are in the situation described on p. 344. In the half-space $\mathbf{R}_{n+}^n$ we can introduce a coordinate system $(\sigma_1, \ldots, \sigma_n = n)$ by "transferring" it from $U_i$ by means of $\varphi_i$. It will have the properties described on p. 344, and therefore we can construct a function $y_i \in W^{r,p}(\mathbf{R}_{n+}^n)$ such that $\partial y_i/\partial n^j = \beta_{ij}$ $(j = 0, \ldots, r-1)$. By Remark 2 this construction can be done so that the supports of the functions $y_i$ are contained in $V_{n+}$. The functions $y_i(\varphi_i(s)) = \tilde{y}_i(s)$ belong to $W^{r,p}(\Omega)$ and satisfy the relation

$$\left.\frac{\partial^j \tilde{y}_i}{\partial n^j}\right|_{\partial\Omega} = \sum_{k=0}^{n} C_j^k \left.\frac{\partial^k \eta_i}{\partial n^k}\right|_{\partial\Omega} \psi^{j-k}. \tag{3.27}$$

If we now consider the function $\tilde{y}(s) = \sum_{i=1}^N \tilde{y}_i(s) \in W^{r,0}(\Omega)$, then by (3.26) and (3.27) we have

$$\left.\frac{\partial^j \tilde{y}}{\partial n^j}\right|_{\partial\Omega} = \psi^j.$$

It is easy to check that the construction of $\tilde{y}$ from $\{\psi^k\}_0^{r-1}$ is effected by means of a bounded linear operator. We have arrived at the following assertion.

THEOREM 3.10. *The correspondence*

$$x \to \left\{ x|_{\partial\Omega}, \left.\frac{\partial x}{\partial n}\right|_{\partial\Omega}, \ldots, \left.\frac{\partial^{r-1} x}{\partial n^{r-1}}\right|_{\partial\Omega} \right\}$$

*is a bounded right invertible mapping of* $W^{r,p}(\Omega)$ *onto* $\prod_{j=0}^{r-1} B^{r-j-1/p,p}(\partial\Omega)$.

Similar arguments, using Theorem 3.9 and taking into account Remark 1 (which we did not prove), lead to the following assertion.

THEOREM 3.11. *If* $1 < p < \infty$ *and* $\sigma > 1/p$, *then the operator* $x \to x|_{\partial\Omega}$ *is a bounded right invertible operator from* $B_G^{\sigma,p}(\Omega)$ *onto* $B_G^{\sigma-1/p}(\partial\Omega)$.

# NOTES ON THE LITERATURE

## Chapter I

§1. As has been mentioned in the Foreword, the notion of imbedding of Banach spaces first appeared in a paper by Sobolev [355]. The important notion of relative completion has been used implicitly by S. G. Kreĭn and Petunin [184]; it has been introduced explicitly by Gagliardo [139] and studied in detail by Aronszajn and Gagliardo [55].

§2. Lemmas 2.1–2.3 and Theorem 2.3 are contained in [55], Theorem 2.1 in [184], and Theorem 2.2 in Berens' book [4].

§3. Intermediate spaces for a couple of Banach spaces were first considered by Lions [202]. Formula (3.6) and its consequences are due to Sedaev [41]. Lemma 3.4 is taken from Calderón [95]. The dual spaces of a sum or an intersection of spaces were studied by Aronszajn and Gagliardo [55] (see also [232]); they also introduced the sum and intersection of a family of Banach spaces.

§4. The notions of interpolation triples and interpolation spaces have been introduced in one form or another in all publications devoted to the abstract theory of interpolation of linear operators. The most important results of §4 are due to Aronszajn and Gagliardo [55]. In connection with subsections 4 and 5, see [37]. The important Theorems 4.9 and 4.10 were obtained in [55].

## Chapter II

*Introduction.* Ideal Banach lattices are also called Banach function spaces; their properties are described in [29] and [49] (see also [95], [188], [25] and [47]). Concerning Lebesgue space, see [305].

§1. Theorem 1.1 was obtained by Peetre [284] (concerning (1.7), see [122]). For Lemma 1.1, see [107]. For more detail on logarithmically convex functions, see the book [39]. Lemma 1.4 was obtained in [59].

§2. Theorem 2.1 was obtained by Kreĭn and Semenov [191]. The books [8], [17], and [50] have much information on rearrangements of measurable functions. Many properties of rearrangements have been established in [87], [88], [95], [172], [317], and [312]. Inequality (2.40) is published here for the first time.

§3. Formulas (3.4) and (3.5) were obtained by Peetre in [262] and [261] (see also Oklander [250]). The important inequality (3.14) was established by Lorentz and Shimogaki [227]. Orbits of the semigroup of contractive operators were studied by Ryff [311].

§4. Symmetric spaces with the additional assumption that the norm is semicontinuous were introduced by Lorentz under the name *spaces invariant under permutations* in the book [28] (see also Luxemburg [231]). Without this assumption they were studied by Semenov [331]; the main results of subsection 1 were obtained by him. The main theorem describing all interpolation spaces between $L_1$ and $L_\infty$ was proved by Calderón [96] by another method. This theorem has a long history: for integral operators in Orlicz spaces it was first proved by Orlicz [256], for integral operators in spaces invariant under permutations by Lorentz [28], and for arbitrary operators in separable spaces or symmetric spaces dual to separable spaces in [241] (see Theorems 4.9 and 4.10). The theorem has been generalized to the case of nonlinear operators satisfying a Lipschitz condition in [257], [227], and [349]. We note that Browder [85] has obtained a general theorem enabling us to obtain interpolation theorems for Lipschitz operators from interpolation theorems for linear operators.

The hypothesis of Theorem 4.3 can be formulated in the following way: the assumptions that $y \in E$, $x \in L_1 + L_\infty$, and $K(t, x) \leqslant K(t, y)$ imply that $x \in E$ and $\|x\|_E \leqslant \|y\|_E$. In connection with this there arose the following conjecture. Let $(A_0, A_1)$ and $(B_0, B_1)$ be two Banach couples. For the triple $(A_0, A_1, E)$ to be an interpolation triple relative to the triple $(B_0, B_1, F)$ it is necessary and sufficient that the assumptions $y \in E$, $x \in B_0 + B_1$, and $K(t, x, B_0, B_1) \leqslant K(t, y, A_0, A_1)$ imply that $x \in F$ and $\|x\|_F \leqslant C\|y\|_E$. This conjecture has been confirmed for the following couples: 1) $(L_q, L_\infty)$, $(L_q, L_\infty)$ (Lorentz and T. Shimogaki [228]; Sedaev [324]); 2) $(l_1^{a_0}, l_1^{a_1})$, $(l_1^{a_0}, l_1^{a_1})$ ($a_0$ and $a_1$ are weights) (Sedaev and Semenov [327]); 3) $(L_p^{w_0}, L_p^{w_1})$, $(L_p^{w_0}, L_p^{w_1})$ (Sadaev [324]); 4) $(A_0, A_1)$ $(L_\infty^{w_0}, L_\infty^{w_1})$ (Peetre [37] and Sadaev [41]); 5) $(\Lambda_\psi, L_\infty)$, $(L_1, L_\infty)$ (V. I. Dmitriev [117]); 6) $(L_1^{w_0} L_1^{w_1})$, $(L_1, L_\infty)$ (V. I. Dmitriev [119]); 7) $(L_{p_0}^{w_0}, L_{p_1}^{w_1})$, $(L_{p_0}^{w_0}, L_{p_1}^{w_1})$ (Sparr [361]). In the general case the conjecture could not be confirmed (see [37] and [117]). V. I. Dmitriev [119], [10] singled out a general class of couples of spaces (difference couples) for which the conjecture is true.

The action of dilation operators in symmetric spaces has been studied in [348], [350], and [83]. In terms of the norm of the dilation operator, the problem of interpolation of the property of complete continuity of linear operators in intermediate spaces for the couple $(L_1, L_\infty)$ has been solved in [348]. The upper and lower dilation exponents of symmetric spaces were introduced by Boyd [81] under the name *upper and lower indices*. Lemma 4.7 is due to Semenov and plays an important role in what follows. The notion of fundamental function was introduced in [331]. There are examples of spaces for which inequality (4.29) is strict, and, even more, the left and right sides have different asymptotics at infinity (see [350]).

§5. Lorentz spaces were introduced by Lorentz in [225]; he established that their dual spaces are the Marcinkiewicz spaces. Some properties of Lorentz spaces were obtained in [191]. The spaces $M_\psi^0$ were considered by Semenov [330]. The imbedding theorems were obtained in [331]. Theorem 5.9 is contained in [342] under different assumptions.

The first example of a noninterpolation symmetric space was constructed by Russu [307].

§6. Inequality (6.1′) is actually contained in [347]. The main interpolation Theorems 6.1 and 6.1′ are extensions of Marcinkiewicz' theorem [234], whose proof was published by Zygmund [407]. Many authors have dealt with the generalization of this theorem; see [64], [74], [94], [96], [106], [148], [161], [163], [190], [214], [250], [262], [342], [371], [406], etc. Here the versions of Kreĭn and Semenov are given from [192]. For the case of the $L_p$ spaces an analogue of Theorem 6.1 can be found in [334], where the conditions on the space $E$ are given in terms of the fundamental function $\varphi_E$. However, the proof contains an error and is true only in the case where the norm $\|\sigma_\tau\|_E$ of the dilation operator coincides with $M_{\varphi_E}$ (concerning (4.29), see above). A correction and strengthening of this result is expounded in §6 (see [192]). The first theorems on the optimality of interpolation triples of concrete symmetric spaces were obtained by Dikarev and Macaev [111] and Calderón [96]. Calderón's ideas lie at the base of the proofs of the optimality theorems of [192].

Hardy-Littlewood and Hilbert operators and majorant functions for symmetric spaces have been studied in [332], [347], [255], [144], [78], and [218]. A generalization of the Hardy inequality (6.41) has been obtained by E. A. Pavlov.

The spaces $L_{p,r}$ are special cases of the Lorentz spaces $\Lambda_{p,\psi}$ [225]. Interpolation theorems for them are contained in [96]. Theorems 6.12 and 6.13 are new. Applications of Theorem 6.1 to the convolution operator are indicated in [193]. Theorem 6.17 is a sharpening of a result of O'Neil [253].

§7. In discussing the properties of the Hilbert singular operator we have followed Zygmund's book [51]. The main formula (7.9) is due to Stein and Weiss [371]. Theorem 7.2 was proved by Boyd [78].

§8 has an auxiliary character. The operation of taking the dual in symmetric spaces has been studied in [133], [144], [254] and [333].

§9. For the classical Paley theorem, see [50]. The generalization of it in the present form is published here for the first time. The articles [313], [314], [316] and [333] are devoted to the generalization of the Hardy-Littlewood theorem on series with monotone coefficients by means of interpolation theorems. For properties of the Haar system, see [20]. Theorem 9.5 for the space $L_p$ was proved by F. Riesz (see [2] and [50]), for symmetric spaces it was proved by Semenov [333]. Theorem 9.6 for the $L_p$ spaces was obtained by Marcinkiewicz [235], for Orlicz spaces by V. F. Gapoškin, and for symmetric spaces by Semenov [336].(1) Concerning Theorem 9.8, see [304].

## Chapter III

§1. The properties of scales of Banach spaces are expounded in [186]. The notion of normal scale was introduced and studied by Kreĭn [181]. The problem of related Banach spaces was studied by Kreĭn and Petunin in [184]. The condensation of a normal scale by means of relative completion is considered here for the first time.

§2. Maximal scales have been studied in [181], and minimal and regular scales in [185]; their properties are discussed in detail in [186].

§3. The scale of Hölder spaces was considered in detail in [186]. V. Friedrich has kindly indicated to us that there are inaccuracies in [186], which we correct here. The new exposition of interpolation properties of the Hölder scale is based on the work of Petunin and Pličko [297]. We note that in a more general case a detailed exposition of the properties of the Hölder scale and its dual with application to the transportation problem is contained in Friedrichs' book [14].

## Chapter IV

*Introduction*. Theorem 1 was obtained by V. I. Dmitriev [116].

§1. The complex method of interpolation in the form expounded here was suggested independently by Calderón [95] and Lions [204]. The main results of subsections 3 and 4 are contained in Calderón [95]; the useful Remark 2 was made by Stafney [363]. Theorem 1.4 is also due to Calderón [95]; its proof has somewhat been simplified by I. Ja. Šneĭberg. In [95], besides the spaces $[A_0, A_1]_\alpha$, the interpolation family of spaces $[A_0, A_1]^{\alpha *}$ is also introduced, and the space $\mathfrak{F}(A_0, A_1)$ of functions in the strip $\Pi$ having the following properties is considered: 1) $\|f(z)\|_{A_0+A_1} < C(1 + |z|)$; 2) $f(z)$ is continuous in the norm of $A_0 + A_1$ in $\overline{\Pi}$; 3) $f(z)$ is analytic in $\Pi$; and 4) the difference $f(1 + it_2) - f(1 + it_1)$ belongs to $A_1$ and the difference

(1) Concerning [333] and [336], the remarks to §6 must be taken into account.

$f(it_2) - f(it_1)$ belongs to $A_0$; moreover,

$$\max\left\{\sup\left\|\frac{f(it_2)-f(it_1)}{t_2-t_1}\right\|_{A_0}, \sup\left\|\frac{f(1+it_2)-f(1+it_1)}{t_2-t_1}\right\|_{A_1}\right\} = \|f\|_{\mathfrak{F}} < \infty.$$

The space $[A_0, A_1]^\alpha$ consists of all $x \in A_0 + A_1$ for which $x = f'(\alpha), f \in \mathfrak{F}(A_0, A_1)$, and

$$\|x\|_{[A_0,A_1]^\alpha} = \inf_{f'(\alpha)=x} \|f\|_{\mathfrak{F}}.$$

It turns out [95] that $([A_0, A_1]_\alpha)'$ is isometrically isomorphic to the space $[A_0', A_1']^\alpha$ (if $A_0 \cap A_1$ is dense in $A_0$ and $A_1$). In subsection 6 the dual of $[A_0, A_1]_\alpha$ is described in terms of relative completion (Theorem 1.6, I. Ja. Šneĭberg). From what has been said we obtain the equality $[A_0', A_1']^\alpha = [A_0', A_1']_\alpha$. Changes have also been made in the proof of Calderón's reiteration Theorem 1.7. We note that in [320] it is indicated that the condition that $A_0 \cap A_1$ is dense in $A_\alpha \cap A_\beta$ is superfluous; however, the proof of this fact contains an error.

The notion of an analytic scale of spaces was introduced by Kreĭn [180]; he established the connection of this notion with the complex method of interpolation [186].

The theory of Hilbert scales of spaces was constructed independently (in different terms) by Lions [200] and Kreĭn [180]. An important role is played by families of topological linear spaces obtained from a Hilbert scale by means of projective or inductive limits. By means of them a number of delicate properties of spaces of analytic functions has been studied.

Theorems 1.11 and 1.13 were obtained from other considerations by E. Heinz, and were sharpened by T. Kato (see [23] and [24]). Heinz has proved an inequality more general than (1.49), in which the fractional power of the operators $j$ and $j_1$ is replaced by more general functions. Let us consider the class of functions positive on the semiaxis $[0, \infty)$, admitting analytic continuation to the complex plane with the negative semiaxis removed, which results in a function mapping the upper half-plane into itself. Let the function $\varphi(t)$ be such that $\varphi^2(t^{1/2})$ belongs to the indicated class. Under the conditions of Theorem 1.13 we have the inequality

$$(Tx, y) \leqslant \|\varphi(j)x\|_{H_0}\|\varphi_*(j_1)y\|_{H_0}, \quad \text{where } \varphi_*(t) = t/\varphi(t).$$

All interpolation spaces with interpolation constant 1 between a couple of Hilbert spaces have been described by means of functions of the indicated class (Foiaş and Lions [134], and Donoghue [120]; see also [160]).

The family of spaces $X_0^{1-\alpha}X_1^\alpha$ was introduced and studied by Calderón [95]. Here we do not discuss its connection with hyperscales (see [188]). Lozanovskiĭ [230] constructed an example in which the triples $(X_0, X_1, X_0^{1-\alpha}X_1^\alpha)$ and $(Y_0, Y_1, Y_0^{1-\alpha}Y_1^\alpha)$ of ideal lattices are not interpolation triples. Šestakov [340], [341] showed that in the general case $[X_0, X_1]_\alpha$ is the closure of $X_0 \cap X_1$ in $X_0^{1-\alpha}X_1^\alpha$, and consequently is a closed subspace of it. For further study of the family $X_0^{1-\alpha}X_1^\alpha$, see [401] and [229].

We note that Schechter [320], [319] constructed a generalization of the complex method of interpolation based on the idea that the intermediate space is constructed not from the values of functions analytic in a strip or their derivatives but rather from the values of some generalized function with compact support defined on these functions.

Interesting results concerning the unambiguous solvability of linear equations and the spectrum of linear operators in the family of spaces $[A_0, A_1]_\alpha$ have been obtained by Šneĭberg [353], [354] and Stafney [364].

§2. The methods of constants and averages originate in a paper by Lions and Peetre [214], where they are constructed for the case that $E_0$ and $E_1$ are $L_p$ spaces with power weights. The generalization of these methods to the case of arbitrary ideal lattices discussed here was proposed by Peetre in [35] and [264], and developed by Dmitriev in [114], [116] and [118]. The appearance of spaces of the Calderón scale in the reiteration Theorem 2.8 was unexpected (V. I. Dmitriev [59]).

The $\mathcal{K}$- and $\mathcal{J}$-methods were proposed by Peetre [262] in the case where $E$ is an $L_p$ space with power weight (see subsection 8) and have been the most widely disseminated of all real interpolation methods. These methods were generalized to the case of more general ideal lattices by Peetre [35], Bennett [67], and other authors. For a special case of Theorem 2.10, see [262] and [73].

Theorems 2.11 and 2.12 were proved by Lions and Peetre [214].

Theorem 2.13 on the number of parameters was obtained by Peetre [261]. It has been generalized to the case of arbitrary ideal lattices $E_0$ and $E_1$ by V. I. Dmitriev [118].

Extreme spaces (subsection 9) were studied by Hayakawa [156].

The duality of the methods of constants and averages (in the simplest case when they coincide) was studied by Lions and Peetre [214]. Here we have expounded the results of V. I. Dmitriev [116].

Applications of the methods of constants and averages to quasinormed spaces have been studied in [168] and [157].

The connection between the methods of constants and averages and the theory of scales and the almost interpolation properties of scales were studied by Petunin [186].

Lemma 2.20 and Theorem 2.23 were obtained by Lions and Peetre [214]. The problem of conditions on the commutativity of the functors corresponding to the method of averages and the complex method has been studied by Grisvard [153].

## Chapter V

§1. The approach to the construction, by means of an abstract approximation process, of intermediate spaces between a Banach space and the domain of an unbounded operator acting in it is expounded here for the first time. Concrete realizations of it have been studied by many authors. Closely related but different constructions are found in Berens' book [4]. It was apparently there that the connection between the behavior of an approximation process and relative completion (Lemma 1.1) was noticed for the first time. See [24] for more details on the subordinate operators of subsection 4. The approximation process constructed from a power of the resolvent was first studied in Grisvard's fundamental paper [153], the results of which are discussed here only partially. In particular, in it the relation of the intermediate spaces constructed there to the complex interpolation method is studied. Lemma 1.10 is due to Ljubič [220]. For operators satisfying condition (1.15) fractional powers are defined (see the books [23] and [24]), and therefore the theory expounded in subsections 5 and 7 can be carried over to spaces constructed from fractional powers of the operator. The connection between interpolation spaces and domains of fractional powers of operators was studied by Lions [209] for accretive operators in a Hilbert space, and in the general case in a long series of articles by Komatsu [177], and by Sobolevskiĭ in [356]–[359] (see also [72], [246], [392], and [398]).

The construction of interpolation spaces by means of bounded semigroups of operators was first done by Lions and Peetre [213], and since then it has been studied in many publications—see [3], [69]–[71], [91], [92], [151], [153], [222] and [392]. We note that a fairly complete exposition of properties of the spaces constructed from resolvents and semigroups of operators in the case $G = L_{p,*}$ is included in the book [7] by Butzer and Berens. Theorem 1.9 was obtained by Grisvard [153]; this theorem and Theorems 1.10 and 1.11 were generalized to the case of fractional powers of operators by Muramatu [246]. The proof of Theorem 1.12 given here is also due to Muramatu. Interesting but apparently not complete research has been carried out in the case where the operators do not commute and are infinitesimal operators of some representation of a Lie group (see [281]).

Grisvard in [153] began to consider unbounded operators in a couple of Banach spaces. The imbedding theorem given here, namely Theorem 1.13, is due to Yoshikawa [394]. It has been developed and applied in [245], [396] and [397].

§2. Spaces of traces were introduced and studied in a series of publications by Lions (see [201], [202], [205] and [207]). The study has been continued by Grisvard in [152] and [154]; the exposition given here is based on his articles.

§3. As we mentioned in the Foreword, Sobolev-Nikol'skiĭ-Slobodeckiĭ-Besov spaces and imbedding theorems for them served as a guideline for the construction of the corresponding abstract theory expounded in Chapter V in an incomplete form. In almost all publications mentioned in §§1 and 2 there are applications to the theory of the indicated spaces. In addition, we mention the series [379]–[386] of articles by Triebel. We have discussed only the facts which can be obtained from results of §§1 and 2; besides the abstract theory, we have here used Mihlin's [30] and P. I. Lizorkin's [217] theorems on multipliers for the Fourier transformation, the Hestenes-Whitney extension of smooth functions, and other techniques which have become standard in the theory of partial differential equations. Here we have used the monograph [27] in an essential way.

We entirely omitted other classes of smooth functions, in particular, Lebesgue spaces or spaces of Bessel potentials, which are connected with the complex method of interpolation. Their theory is expounded in Nikol'skiĭ's book [32] without this connection, and in connection with interpolation theory in Triebel's new book *Interpolation theory, function spaces, differential operators* mentioned in the Foreword (see also [93], [212] and [218]).

On other classes of functions, see [15], [86], [87], [90], [95], [99]–[101], [129], [171]–[174], [194], [198], [211], [212], [216], [218], [249], [279], [321], [365], [366], [373] and [374].

# BIBLIOGRAPHY


**Monographs, Dissertations, Technical Reports**

1. N. I. Ahiezer and I. M. Glazman, *The theory of linear operators in Hilbert space*, 2nd rev. ed., "Nauka", Moscow, 1966; English transl. of 1st ed., Vols. I, II, Ungar, New York, 1961.

2. N. K. Bari, *Trigonometric series*, Fizmatgiz, Moscow, 1961; English transl., Vols. I, II, Pergamon Press, Oxford, and Macmillan, New York, 1964.

3. Hubert Berens, *Approximationssätze für Halbgruppenoperatoren in intermediären Räumen*, Schr. Math. Inst. Univ. Münster No. 32 (1964).

4. ______, *Interpolationsmethoden zur Behandlung von Approximationsprozessen auf Banachräumen*, Lecture Notes in Math., Vol. 64, Springer-Verlag, 1968.

5. J. Bergh, *On the interpolation of normed linear spaces—a necessary condition*, Tech. Rep., Lund, 1971.

6. N. Bourbaki, *Espaces vectoriels topologiques*, Chaps. 1–5 and fascicule de résultats, Actualités Sci. Indust., Nos. 1189, 1229, 1230, Hermann, Paris, 1953, 1955.

7. Paul L. Butzer and Hubert Berens, *Semi-groups of operators and approximation*, Springer-Verlag, 1967.

8. K. M. Chong and N. M. Rice, *Equimeasurable rearrangements of functions*, Queen's Papers in Pure and Appl. Math., No. 28, Queen's Univ., Kingston, Ont., 1971.

9. Nimet Deutsch, *Interpolation dans les espaces vectoriels topologiques localement convexes*, Bull. Soc. Math. France Suppl. Mém. No. 13 (1968).

10. V. I. Dmitriev, *On the construction and description of interpolation spaces*, Author's Abstract of Candidate's Dissertation, Voronezh, 1975. (Russian)

11. Nelson Dunford and Jacob T. Schwartz, *Linear operators*. Vol. I, Interscience, 1958.

12. R. E. Edwards, *Functional analysis*, Holt, Reinhart and Winston, New York, 1965.

13. Dietrich Freitag, *Reelle Interpolation zwischen Banachideallen von beschränkten linearen operatoren*, Dissertaion, Jena, 1971.

14. V. Friedrich, *Stetige Transportoptimerung*, VEB Deutscher Verlag, Wiss., Berlin, 1972.

15. K. K. Golovkin, *Parametric-normed spaces and normed massives*, Trudy Mat. Inst. Steklov. **106** (1969); English transl., Proc. Steklov Inst. Math. **106** (1969) (1972).

16. Paul R. Halmos, *Measure theory*, Van Nostrand, Princeton, N. J., 1950.

17. G. H. Hardy, J. E. Littlewood and G. Pólya, *Inequalities*, Cambridge Univ. Press, 1934.

18. Einar Hille and Ralph S. Phillips, *Functional analysis and semi-groups*, rev. ed., Amer. Math. Soc., Providence, R. I., 1957.

19. Kenneth Hoffman, *Banach spaces of analytic functions*, Prentice-Hall, Englewood Cliffs, N. J., 1962.

20. Stefan Kaczmarz and Hugo Steinhaus, *Theorie der Orthogonalreichen*, PWN, Warsaw, 1935; reprint, Chelsea, New York, 1951.

21. Yitzhak Katznelson, *An introduction to harmonic analysis*, Wiley, New York, 1968.

22. M. A. Krasnosel'skiĭ and Ja. B. Rutickiĭ, *Convex functions and Orlicz spaces*, GITTL, Moscow, 1958; English transl., Noordhoff, 1961.

23. M. A. Krasnosel'skiĭ, P. P. Zabreĭko, E. I. Pustyl'nik and P. E. Sobolevskiĭ, *Integral operators in spaces of summable functions*, "Nauka", Moscow, 1966; English transl., Noordhoff, 1975.

24. S. G. Kreĭn, *Linear differential equations in Banach space*, "Nauka", Moscow, 1967; English transl., Amer. Math. Soc., Providence, R. I., 1971.

25. S. G. Kreĭn et al., *Functional analysis*, 2nd rev. ed., "Nauka", Moscow, 1972; English transl. of 1st ed., Noordhoff, 1972.

26. Joram Lindenstrauss and Lior Tzafriri, *Classical Banach spaces*, Lectures Notes in Math., Vol. 338, Springer-Verlag, 1973.

27. J. L. Lions and E. Magenes, *Prolèmes aux limites non homogènes et applications*. I, II, III, Dunod, Paris, 1968, 1970; English transl., Springer-Verlag, 1972, 1973.

28. G. G. Lorentz, *Berstein polynomials*, Univ. of Toronto Press, 1953.

29. W. A. J. Luxemburg, *Banach function spaces*, Van Gorcum, Assen, 1955.

30. S. G. Mihlin, *Multidimensional singular integrals and integral equations*, Fizmatgiz, Moscow, 1962; English transl., Pergamon Press, 1965.

31. M. A. Naĭmark, *Normed rings*, 2nd rev. ed., "Nauka", Moscow, 1968; English transl., *Normed algebras*, Noordhoff, 1972.

32. S. M. Nikol'skiĭ, *Approximation of functions of several variables and embedding theorems*, "Nauka", Moscow, 1969; English transl., Springer-Verlag, 1974.

33. Evelio T. Oklander, *Interpolación, espacios de Lorentz y teorema de Marcinkiewicz*, Cursos y Sem. Mat., Fasc. 20, Univ. Buenos Aires, Buenos Aires, 1965.

33a. A. M. Olevskiĭ, *Fourier series with respect to general orthogonal systems*, Springer-Verlag, 1975.

34. V. I. Ovčinnikov, *Spaces of measurable operators and interpolation*, Author's Abstract of Candidate's Dissertation, Voronezh, 1971. (Russian)

35. J. Peetre, *A theory of interpolation of normed spaces*, Lecture Notes, Univ. Brasília, Brasília, 1963; reprint, Notas de Mat., No. 39, Inst. Mat. Pura Apl., Conselho Nac. Pesquisas, Rio de Janeiro, 1968.

36. ______, *Interpolation i abstrakta rum*, Vorlesungs-ausarbeitung, Lund, 1966.

37. ______, *Banach couples*. I: *Elementary theory*, Tech. Rep., Lund, 1971.

38. Albrecht Pietsch, *Theorie der Operatorenideale* (*Zusammenfassung*), Wiss. Beiträge, Friedrich-Schiller-Univ., Jena, 1972.

39. I. I. Privalov, *Subharmonic functions*, ONTI, Moscow, 1937. (Russian)

40. Robert Schatten, *Norm ideals of completely continuous operators*, Springer-Verlag, 1960.

41. A. A. Sedaev, *Interpolation of linear operators and the geometries of Banach spaces*, Author's Abstract of Candidate's Dissertation, Voronezh, 1973. (Russian)

42. G. E. Šilov, *Mathematical analysis: Second special course*, "Nauka", Moscow, 1965; English transl., *Generalized functions and partial differential equations*, Gordon and Breach, New York, 1968.

43. S. L. Sobolev, *Applications of functional analysis in mathematical physics*, Izdat. Leningrad. Gos. Univ., Leningrad, 1950; English transl., Amer. Math. Soc., Providence, R. I., 1963.

44. Elias M. Stein, *Singular integrals and differentiability properties of functions*, Princeton Univ. Press, Princeton, N. J., 1970.

45. Elias M. Stein and Guido Weiss, *Introduction to Fourier analysis on Euclidean spaces*, Princeton Univ. Press, Princeton, N. J., 1971.

46. G. O. Thorin, *Convexity theorems generalizing those of M. Riesz and Hadamard with some applications*, Comm. Sem. Math. Univ. Lund. = Medd. Lunds Univ. Mat. Sem. **9** (1948).

47. B. Z. Vulih, *Introduction to the theory of partially ordered spaces*, Fizmatgiz, Moscow, 1961; English transl., Noordhoff, 1967.

48. Kôsaku Yosida, *Functional analysis*, Springer-Verlag, 1965.

49. P. P. Zabreĭko, *Nonlinear integral operators*, Voronež. Gos. Univ. Trudy Sem. Funkcional. Anal. Vyp. 8 (1966). (Russian)

50. A. Zygmund, *Trigonometric series*, 2nd rev. ed., Vols. I, II, Cambridge Univ. Press, 1959.

51. ______, *Intégrales singulières*, Lecture Notes in Math., Vol. 204, Springer-Verlag, 1971.

**Journal Papers**

52. Pedro Jorge Aranda and Enrique Paule Cattáneo, *Sur les espaces de Lorentz avec poids*, C. R. Acad. Sci. Paris Sér. A **264** (1967), A109–A112.

53. Pietro Arduini, *Sull'equivalenza di certi funzionali della teoria dell'interpolazione tra spazî di Banach*, Ricerche Mat. **11** (1962), 51–60.

54. N. Aronszajn, *Associated spaces, interpolation theorems and the regularity of solutions of differential problems*, Proc. Sympos. Pure Math., Vol. 4, Amer. Math. Soc., Providence, R. I., 1961, pp. 23–32.

55. N. Aronszajn and E. Gagliardo, *Interpolation spaces and interpolation methods*, Ann. Mat. Pura Appl. (4) **68** (1965), 51–117.

56. Michel Artola, *Sur un théorème d'interpolation*, J. Math. Anal. Appl. **41** (1973), 148–163.

57. Claudio Baiocchi, *Un teorema di interpolazione; applicazioni ai limiti per le equazioni differenziali a derivate parziali*, Ann. Mat. Pura Appl. (4) **73** (1966), 233–251.

58. Mohamed Salah Baouendi and Charles Goulaouic, *Commutation de l'intersection et des foncteurs d'interpolation*, C. R. Acad. Sci. Paris Sér. A-B **265** (1967), A313–A315.

59. N. K. Bari and S. B. Stečkin, *Best approximation and differential properties of two conjugate functions*, Trudy Moskov. Mat. Obšč. **5** (1956), 483–522. (Russian)

60. A. Benedek and R. Panzone, *The spaces $L^p$, with mixed norm*, Duke Math. J. **28** (1961), 301–324.

61. A. Benedek, A. P. Calderón and R. Panzone, *Convolution operators on Banach space valued functions*, Proc. Nat. Acad. Sci. U.S.A. **48** (1962), 356–365.

62. Colin Bennett, *A pair of indices for function spaces on the circle*, Trans. Amer. Math. Soc. **174** (1972), 289–304.

63. ______, *A Hausdorff-Young theorem for rearrangement-invariant spaces*, Pacific J. Math. **47** (1973), 311–328.

64. ______, *Estimates for weak-type operators*, Bull. Amer. Math. Soc. **79** (1973), 933–935.

65. ______, *Intermediate spaces and the class $L \log^+ L$*, Ark. Mat. **11** (1973), 215–228.

66. ______, *Banach function spaces and interpolation methods*. I, J. Functional Anal. **17** (1974), 409–440.

67. ______, *Banach function spaces and interpolation methods*. II, Linear Operators and Approximation. II (Proc. Conf., Oberwohlfach, 1974), Birkhäuser, 1974, pp. 129–130.

68. ______, *Banach function spaces and interpolation methods*. III, J. Approximation Theory **13** (1975), 267–275.

69. Hubert Berens, *Equivalent representations for the infinitesimal generator of higher orders in semi-group theory*, Nederl. Akad. Wetensch. Proc. Ser. A**68** = Indag. Math. **27** (1965), 497–512.

70. Hubert Berens and P. L. Butzer, *Approximation theorems for semi-group operators in intermediate spaces*, Bull. Amer. Math. Soc. **70** (1964), 689–692.

71. ______, *Über die Stetigkeit von Halbgruppen von Operatoren in intermediären Räumen*, Math. Ann. **163** (1966), 204–211.

72. H. Berens, P. L. Butzer and U. Westphal, *Representation of fractional powers of infinitesimal generators of semigroups*, Bull. Amer. Math. Soc. **74** (1968), 191–196.

73. Carlos Alberto Berenstein and Norberto Luis Kerzman, *Sur la réitération dans les espaces de moyenne*, C. R. Acad. Sci. Paris Sér. A **263** (1966), A609–A612.

74. C. A. Berenstein, M. Cotlar, N. Kerzman and P. Krée, *Some remarks on the Marcinkiewicz convexity theorem in the upper triangle*, Studia Math. **29** (1967/68), 79–95.

75. J. Bergh, *A generalization of Steffensen's inequality*, J. Math. Anal. Appl. **41** (1973), 187–191.

76. O. V. Besov, *Investigation of a family of function spaces in connection with theorems of imbedding and extension*, Trudy Mat. Inst. Steklov. **60** (1961), 42–81; English transl. in Amer. Math. Soc. Transl. (2) **40** (1964).

77. R. P. Boas, Jr., and S. Bochner, *On a theorem of M. Riesz for Fourier series*, J. London Math. Soc. **14** (1939), 62–73.

78. David W. Boyd, *The Hilbert transform on rearrangement-invariant spaces*, Canad. J. Math. **19** (1967), 599–616.

78a. ______, *A class of operators on the Lorentz spaces* $M(\varphi)$, Canad. J. Math. **19** (1967), 839–841.

79. ______, *Spaces between a pair of reflexive Lebesgue spaces*, Proc. Amer. Math. Soc. **18** (1967), 215–219.

80. ______, *The spectral radius of averaging operators*, Pacific J. Math. **24** (1968), 19–28.

81. ______, *Indices of function spaces and their relationship to interpolation*, Canad. J. Math. **21** (1969), 1245–1254.

82. ______, *Indices for the Orlicz spaces*, Pacific J. Math. **38** (1971), 315–323.

83. M. Š. Braverman, *On the geometric properties of symmetric spaces*, Sibirsk. Mat. Ž. **15** (1974), 675–679; English transl. in Siberian Math. J. **15** (1974).

84. M. Š. Braverman and E. M. Semenov, *Isometries of symmetric spaces*, Dokl. Akad. Nauk SSSR **217** (1974), 257–259; English transl. in Soviet Math. Dokl. **15** (1974).

85. Felix E. Browder, *Remarks on nonlinear interpolation in Banach spaces*, J. Functional Anal. **4** (1969), 390–403.

86. Ju. A. Brudnyĭ, *Spaces defined by means of local approximations*, Trudy Moskov. Mat. Obšč. **24** (1971), 69–132; English transl. in Trans. Moscow Math. Soc. **24** (1971).

87. ______, *On the rearrangement of a smooth function*, Uspehi Mat. Nauk **27** (1972), no. 2 (164), 165–166. (Russian)

88. ______, *Moduli of continuity and rearrangements*, Mat. Zametki **18** (1975), 63–66; English transl. in Math. Notes **18** (1975).

89. Ju. A. Brudnyĭ and V. K. Salašov, *Lipschitz spaces of functions*, Dokl. Akad. Nauk SSSR **197** (1971), 18–20; English transl. in Soviet Math. Dokl. **12** (1971).

90. ______, *Lipschitz spaces of functions*, Metric Questions of the Theory of Functions and Mappings, No. 4, "Naukova Dumka", Kiev, 1973, pp. 3–60. (Russian)

91. Paul L. Butzer, *Sur la théorie des demi-groupes et classes de saturation de certaines intégrales singulières*, C. R. Acad. Sci. Paris **243** (1956), 1473–1475.

92. ______, *Über den Grad der Approximation des Identitätsoperators durch Halbgruppen von linearen Operatoren und Anwendungen auf die Theorie der singulären Integrale*, Math. Ann. **133** (1957), 410–425.

93. A. P. Calderón, *Lebesgue spaces of differentiable functions and distributions*, Proc. Sympos. Pure Math., Vol. 4, Amer. Math. Soc., Providence, R. I., 1961, pp. 33–49.

94. ______, *Intermediate spaces and interpolation*, Studia Math.: Ser. Specjalna, No. 1 (1963), 31–34.

95. ______, *Intermediate spaces and interpolation, the complex method*, Studia Math. **24** (1964), 133–190.

96. ______, *Spaces between* $L^1$ *and* $L^\infty$ *and the theorems of Marcinkiewicz*, Studia Math. **26** (1966), 273–299.

97. A. P. Calderón and A. Zygmund, *A note on the interpolation of linear operations*, Studia Math. **12** (1951), 194–204.

98. ______, *A note on the interpolation of sublinear operations*, Amer. J. Math. **78** (1956), 282–288.

99. Sergio Campanato, *Teoremi di interpolazione per trasformazioni che applicano* $L^p$ *in* $C^{h,\alpha}$, Ann. Scuola Norm. Sup. Pisa Sci. Fis. Mat. (3) **18** (1964), 345–360.

100. ______, *Su un teorema di interpolazione di G. Stampacchia*, Ann. Scuola Norm. Sup. Pisa Sci. Fis. Mat. (3) **20** (1966), 649–652.

101. ______, *Maggiorazioni interpolatorie negli spazi* $H_\lambda^{m,p}(\Omega)$, Ann. Mat. Pura Appl. (4) **75** (1967), 261–276.

102. S. Campanato and M. K. V. Murthy, *Una generalizzazione del teorema di Riesz-Thorin*, Ann. Scuola Norm. Sup. Pisa Sci. Fis. Mat. (3) **19** (1965), 87–100.

103. Franco Chersi, *Ricerche per una teoria generale dell'interpolazione tra spazi lineari topologici*, Rend. Sem. Mat. Univ. Padova **34** (1964), 305–343.

104. M. Cotlar, *A general interpolation theorem for linear operations*, Rev. Mat. Cuyana **1** (1955), 57–84 (1956).

105. ______, *Some generalizations of the Hardy-Littlewood maximum theorem*, Rev. Mat. Cuyana **1** (1955), 85–104 (1956).

106. Mischa Cotlar and Maria Bruschi, *On the convexity theorems of Riesz-Thorin and Marcinkiewicz*, Univ. Nac. de La Plata Publ. Fac. Ci. Fisicomat. Ser. Segunda Rev. **5** (1954/57), 162–172.

107. Ju. P. Čuburin, *On a property of rearrangements*, Coll. of Papers by Graduate Students, Math. Fac., Izdat. Voronež. Gos. Univ., Voronezh, 1972, pp. 72–76. (Russian) RŽ Mat. **1973** #7B56.

108. Michael Cwikel, *On* $(L^{p_0}(A_0), L^{p_1}(A_1))_{\theta,q}$, Proc. Amer. Math. Soc. **44** (1974), 286–292.

109. Peter W. Day, *Decreasing rearrangements and doubly stochastic operators*, Trans. Amer. Math. Soc. **178** (1973), 383–392.

110. Ninet Deutsch, *Interpolation dans les espaces vectoriels topologiques localement convexes*, C. R. Acad. Sci. Paris **257** (1963), 3796–3799.

111. A. A. Dikarev and V. I. Macaev, *An exact interpolation theorem*, Dokl. Akad. Nauk SSSR **168** (1966), 986–988; English transl. in Soviet Math. Dokl. **7** (1966).

112. J. Dixmier, *Formes linéaires sur un anneau d'opérateurs*, Bull. Soc. Math. France **81** (1953), 9–39.

113. A. A. Dmitriev, *On the interpolation of one-dimensional operators*, Voronež. Gos. Univ. Trudy Naučno-Issled. Inst. Mat. VGU Vyp. 11 (1973), 31–43. (Russian)

114. V. I. Dmitriev, *On the Lions-Peetre method for the construction of interpolation spaces*, Dokl. Akad. Nauk SSSR **198** (1971), 747–750; English transl. in Soviet Math. Dokl. **12** (1971).

115. ______, *A generalization of the trace interpolation method*, Voronež. Gos. Univ. Trudy Naučno-Issled. Inst. Mat. VGU Vyp. 3 (1971), 19–26. (Russian)

116. ______, *Duality between the interpolation method of constants and that of means*, Dokl. Akad. Nauk SSSR **214** (1974), 22–24; English transl. in Soviet Math. Dokl. **15** (1974).

117. ______, *On the Calderón-Mitjagin interpolation theorem*, Dokl. Akad. Nauk SSSR **215** (1974), 518–521; English transl. in Soviet Math. Dokl. **15** (1974).

118. ______, *Theorems on parameters for the interpolation method of constants*, Dokl. Akad. Nauk SSSR **216** (1974), 257–258; English transl. in Soviet Math. Dokl. **15** (1974).

119. ______, *Interpolation spaces between* $(L_1^{w_0}, L_1^{w_1})$ *and* $(L_1, L_\infty)$, Mat. Zametki **17** (1975), 727–736; English transl. in Math. Notes **17** (1975).

120. William F. Donoghue, Jr., *The interpolation of quadratic norms*, Acta Math. **118** (1967), 251–270.

121. G. F. D. Duff, *Integral inequalities for equimeasurable rearrangements*, Canad. J. Math. **22** (1970), 408–430.

122. A. V. Efimov, *Linear methods of approximating continuous periodic functions*, Mat. Sb. **54(96)** (1961), 51–90. (Russian)

123. Angelo Favini, *Sulla teoria della interpolazione negli spazi vettoriali topologici*, Rend. Mat. (Rome) (6) **3** (1970), 351–380.

124. ______, *Alcuni risultati sulla interpolazione non-lineare*, Boll. Un. Mat. Ital. (4) **4** (1971), 918–936.

125. ______, *Su un metodo di interpolazione*, Boll. Un. Mat. Ital. (4) **4** (1971), 677–686.

126. ______, *Sulla interpolazione di operatori compatti*, Rend. Sem. Mat. Univ. Padova **45** (1971), 279–304.

127. ______, *Sulla interpolazione degli spazi $K\{M_p\}$*, Boll. Un. Mat. Ital. (4) **6** (1972), 440–449.

128. ______, *Sulla interpolazione multilineare: una precisazione*, Rend. Sem. Mat. Univ. Padova **49** (1973), 353–354.

129. ______, *Sulla interpolazione di certi spazi di Sobolev con peso*, Rend. Sem. Mat. Univ. Padova **50** (1973), 223–249.

130. C. Fefferman, N. M. Riviere and Y. Sagher, *Interpolation between $H^p$ spaces: The real method*, Trans. Amer. Math. Soc. **191** (1974), 75–81.

131. Franziska Fehér, *Approximationssätze auf rearrangement-invarianten Banach-Raümen über dem Torus*, Linear Operators and Approximation. II (Proc. Conf., Oberwohlfach, 1974), Birkhaüser, 1974, pp. 189–203.

132. Franziska Fehér, Dumitru Gaşpar and Hans Johnen, *Normkonvergenz von Fourierreichen in rearrangement invarianten Banachraümen*, J. Functional Anal. 13 (1973), 417–434.

133. ______, *Der Konjugiertenoperator auf rearrangement-invarianten Funktionenräumen*, Math. Z. **134** (1973), 129–141.

134. C. Foiaş and J. L. Lions, *Sur certains théorèmes d'interpolation*, Acta. Sci. Math. (Szeged) **22** (1961), 269–282.

135. D. H. Fremlin, *Stable subspaces of $L^1 + L^\infty$*, Proc. Cambridge Philos. Soc. **64** (1968), 625–643.

136. Emilio Gagliardo, *Interpolation d'espaces de Banach et applications*. I, II, III, C. R. Acad. Sci. Paris **248** (1959), 1912–1914, 3388–3390, 3517–3518.

137. ______, *Interpolazione di spazi di Banach e applicazioni*, Edizioni Sci., Genoa, 1959.*

138. ______, *Interpolazione di spazi di Banach e applicazioni*, Ricerche Mat. **9** (1960), 58–81; reprint, Pubbl. Inst. Mat. Univ. Genova No. 85 (1960).

139. ______, *Una struttura unitaria in diverse famiglie di spazi funzionali*. I, Ricerche Mat. **10** (1961), 244–281.

140. ______, *A common structure in various families of functional spaces*. II, Ricerche Mat. **12** (1963), 87–107.

141. ______, *Caratterizzazione construttiva di tutti gli spazi di interpolazione tra spazi di Banach*, Symposia Math., Vol. II (INDAM, Rome, 1968), Academic Press, 1969, pp. 95–106.

142. Christiane Gapaillard, *Caractérisation des espaces d'interpolation entre $\mathcal{C}^p$ et $\mathcal{C}^\infty$*, C. R. Acad. Sci. Paris Sér. A-B **279** (1974), A53–A55.

143. Christiane Gapaillard and Pham the Lai, *Remarques sur les propriétés de dualité et d'interpolation des idéaux de R. Schatten*, C. R. Acad. Sci. Paris Sér. A-B **274** (1972), 1794–1797.

144. Dumitru Gaspar and H. Johnen, *Über eine Klasse rearrangement invarianten Banachräume*, Linear Operators and Approximation. II (Proc. Conf., Oberwohlfach, 1974), Birkhäuser, 1974, pp. 181–188.

145. John E. Gilbert, *Interpolation between weighted $L^p$-spaces*, Ark. Mat. **10** (1972), 235–249.

146. ______, *Counter-examples in interpolation space theory from harmonic analysis*, Linear Operators and Approximation. II (Proc. Conf., Oberwohlfach, 1974), Birkhäuser, 1974, pp. 141–167.

147. J. P. Girardeau, *Sur l'interpolation entre un espace localement convexe et son dual*, Univ. Lisboa Rev. Fac. Ci. A (2) **11** (1964–66), 165–186.

148. K. K. Golovkin, *A generalization of Marcinkiewicz's interpolation theorem*, Trudy Mat. Inst Steklov. **102** (1967), 5–28; English transl. in Proc Steklov Inst. Math. **102** (1967) (1970).

149. Charles Goulaouic, *Interpolation entre les espaces $L^p$ avec poids; applications*, C. R. Acad. Sci. Paris Sér. A-B **262** (1966), 333–336.

---

* According to [138], this is "un'edizione provvisoria" of that paper.

150. ______, *Prolongements de foncteurs d'interpolation et applications*, Ann. Inst. Fourier (Grenoble) **18** (1968), fasc. 1, 1–98.

151. ______, *Interpolation entre des espaces localement convexes définis à l'aide de semi-groupes; cas des espaces de Gevrey*, Ann. Inst. Fourier (Grenoble) **19** (1969), fasc. 2, 269–278.

151a. N. E. Gretsky and J. J. Uhl, Jr., *Bounded linear operators on Banach function spaces of vector-valued functions*, Trans. Amer. Math. Soc. **167** (1972), 263–277.

152. Pierre Grisvard, *Identités entre espaces de traces*, Math. Scand. **13** (1963), 70–74.

153. ______, *Commutativité de deux foncteurs d'interpolation et applications*, J. Math. Pures Appl (9) **45** (1966), 143–290.

154. ______, *Spazi di tracce e applicazioni*, Rend. Mat. (Rome) (6) **5** (1972), 657–729.

155. Alexandre Grothendieck, *Réarrangements de fonctions et inégalités de convexité dans les algèbres de von Neumann munies d'une trace*, Sém. Bourbaki 1954/55; Exposé 113, Secrétariat Math., Paris, 1955; reprint, Benjamin, 1966.

156. Kantaro Hayakawa, *Interpolation by the real method preserves compactness of operators*, J. Math. Soc. Japan **21** (1969), 189–199.

157. Tord Holmstedt, *Interpolation of quasi-normed spaces*, Math. Scand. **26** (1970), 177–199.

158. Tord Holmstedt and Jaak Peetre, *On certain functionals arising in the theory of interpolation spaces*, J. Functional Anal. **4** (1969), 88–94.

159. Lars Hörmander, *Estimates for translation invariant operators in $L^p$ spaces*, Acta Math. **104** (1960), 93–140.

160. Edward Hughes, *A short proof of an interpolation theorem*, Canad. Math. Bull. **17** (1974), 127–128.

161. Richard A. Hunt, *An extension of the Marcinkiewicz interpolation theorem to Lorentz spaces*, Bull. Amer. Math. Soc. **70** (1964), 803–807.

162. ______, *On $L(p, q)$ spaces*, L'Enseignement Math. (2) **12** (1966), 249–276.

163. Richard A. Hunt and Guido Weiss, *The Marcinkiewicz interpolation theorem*, Proc. Amer. Math. Soc. **15** (1964), 996–998.

164. R. K. Juberg, *Measure of non-compactness and interpolation of compactness for a class of integral transformations*, Duke Math. J. **41** (1974), 511–525.

165. S. N. Kadampatta, *Scales of spaces of analytic functions*, Izv. Severo-Kavkaz. Naučn. Centra Vysš. Školy Ser. Estestv. Nauk **1975**, no. 4, 64–68 (Russian)

166. L. V. Kantorovič and G. Š. Rubinšteĭn, *On a function space and certain extremum problems*, Dokl. Akad. Nauk SSSR **115** (1957), 1058–1061. (Russian)

167. ______, *On a space of completely additive functions*, Vestnik Leningrad. Univ. **1958**, no. 7 (Ser. Mat. Meh. Astr. vyp. 2), 52–59. (Russian)

168. G. E. Karadžov, *An interpolation method of "means" for quasi-normed spaces*, Dokl. Akad. Nauk SSSR **209** (1973), 33–36; English transl. in Soviet Math. Dokl. **14** (1973).

169. ______, *On the commutativity of two interpolation functors*, Dokl. Akad. Nauk SSSR **223** (1975), 292–294; English transl. in Soviet Math. Dokl. **16** (1975).

170. ______, *The application of the theory of interpolation spaces for estimating the singular numbers of integral operators*, Problemy Mat. Anal., Vyp. 4, Izdat. Leningrad. Gos. Univ., Leningrad, 1973, pp. 37–45; English transl. in J. Soviet Math. **6** (1976), no. 1.

171. V. S. Klimov, *On imbedding theorems for symmetric spaces*, Mat. Sb. **82(124)** (1970), 371–386; English transl. in Math. USSR Sb. **11** (1970).

172. ______, *The rearrangements of differentiable functions*, Mat. Zametki **9** (1979), 629–638; English transl. in Math. Notes **9** (1970).

173. ______, *Imbedding theorems for symmetric spaces*, Mat. Sb. **79(121)** (1969), 171–178; English transl. in Math. USSR Sb. **8** (1969).

174. ______, *Imbedding theorems and intermediate Calderón spaces*, Funkcional. Anal. i Priložen. **8** (1974), no. 1, 79; English transl. in Functional Anal. Appl. **8** (1974).

175. Sumiyuki Koizumi, *Contributions to the theory of interpolation of operations*, Osaka J. Math. **8** (1971), 135–149.

176. ______, *A theorem about interpolation of operations on intermediate spaces*, Tôhoku Math. J. (2) **24** (1972), 245–249.

177. Hikosaburo Komatsu, *Fractional powers of operators*. In six parts:
I, II: Pacific J. Math. **19** (1966), 285–346; **21** (1967), 89–111.
II, IV: J. Math. Soc. Japan **21** (1969), 205–220, 221–228.
V. VI: J. Fac. Sci. Univ. Tokyo Sect IA Math. **17** (1970), 372–396; **19** (1972), 1–63.

178. M. A. Krasnosel'skiĭ, *On a theorem of M. Riesz*, Dokl. Akad. Nauk SSSR **131** (1960), 246–248; English transl. in Soviet Math. Dokl. **1** (1960).

179. Paul Krée, *Interpolation d'espaces vectoriels qui ne sont ni normés, ni complets. Applications*, Ann. Inst. Fourier (Grenoble) **17** (1967), fasc. 2, 137–174.

180. S. G. Kreĭn, *On an interpolation theorem in operator theory*, Dokl. Akad. Nauk SSSR **130** (1960), 491–494; English transl. in Soviet Math. Dokl. **1** (1960).

181. ______, *On the concept of a normal scale of spaces*, Dokl. Akad. Nauk SSSR **132** (1960), 510–513; English transl. in Soviet Math. Dokl. **1** (1960).

182. ______, *Interpolation of linear operators, and properties of the solutions of elliptic equations*, Elliptische Differentialgleichungen, Band II, Schriftenreihe Inst. Math. Deutsch. Akad. Wiss. Berlin, Reihe A, Heft 8, Akademie-Verlag, Berlin, 1971, pp. 155–166. (Russian)

183. S. G. Kreĭn and P. A. Kučment, *On an approach to the interpolation problem for linear operators*, Voronež. Gos. Univ. Trudy Naučno-Issled. Inst. Mat. VGU Vyp. 3 (1971), 54–60. (Russian)

184. S. G. Kreĭn and Ju. I. Petunin, *A criterion for relationship of two Banach spaces*, Dokl. Akad. Nauk SSSR **139** (1961), 1295–1298; English transl. in Soviet Math. Dokl. **2** (1961).

185. ______, *The concept of a minimal scale of spaces*, Dokl. Akad. Nauk SSSR **154** (1964), 30–33; English transl. in Soviet Math. Dokl. **5** (1964).

186. ______, *Scales of Banach spaces*, Uspehi Mat. Nauk **21** (1966), no. 2 (128), 89–168; English transl. in Russian Math. Surveys **21** (1966).

187. S. G. Kreĭn, Ju. I. Petunin and E. M. Semenov, *Hyperscales of Banach lattices*, Dokl. Akad. Nauk SSSR **170** (1966), 265–267; English transl. in Soviet Math. Dokl. **7** (1966).

188. ______, *Scales of Banach lattices of measurable functions*, Trudy Moskov. Mat. Obšč. **17** (1967), 293–322; English transl. in Trans. Moscow Math. Soc. **17** (1967).

189. ______, *Imbedding theorems and interpolation of linear operators*, Imbedding Theorems and Their Applications (Proc. All-Union Sympos., Baku, 1966), "Nauka", Moscow, 1970, pp. 98–104. (Russian)

190. S. G. Kreĭn and E. M. Semenov, *On a scale of spaces*, Dokl. Akad. Nauk SSSR **138** (1961), 763–766; English transl. in Soviet Math. Dokl. **2** (1961).

191. ______, *On a property of equimeasurable functions*, Voronež. Gos. Univ. Trudy Naučno-Issled. Inst. Mat. VGU Vyp. 5 (1972), 70–74. (Russian)

192. ______, *Interpolation of operators of weakened type*, Funkcional. Anal. i Priložen. **7** (1973), no. 2, 89–90; English transl. in Functional Anal. Appl. **7** (1973).

193. ______, *Some interpolation theorems of the theory of linear operators, and their applications*, Imbedding Theorems and Their Applications (Proc. All-Union Sympos., Alma-Ata, 1973), "Nauka", Alma-Ata, 1976, pp. 64–68. (Russian)

194. A. A. Kruzlov and M. Z. Solomjak, *Interpolation of operators in the spaces $V_p$*, Vestnik Leningrad. Univ. **1971**, no. 13 (Ser. Mat. Meh. Astr. vyp. 3), 54–60; English transl. in Vestnik Leningrad Univ. Math. **4** (1977).

195. R. A. Kunze, *$L_p$ Fourier transforms on locally compact unimodular groups*, Trans. Amer. Math. Soc. **89** (1958), 519–540.

196. Ju. G. Kuricyn, Ju. I. Petunin and E. M. Semenov, *Tchebycheff-type inequalities in symmetric spaces*, Mat. Zametki **10** (1971), 195–205; English transl. in Math. Notes **10** (1971).

197. A. I. Kuznecov, *On the structure of the space $L_1 + L_\infty$*, Collection of Papers by Graduate Students, Math. Fac., Izdat. Voronež. Gos. Univ., Voronezh, 1972, pp. 34–37. (Russian) RŽ Mat. **1973** #8Б657.

198. Marie-Thérèse Lacroix, *Espaces de traces des espaces de Sobolev-Orlicz*, J. Math. Pures Appl. (9) **53** (1974), 439–458.

199. K. de Leeuw, *Banach spaces of Lipschitz functions*, Studia Math. **21** (1961/62), 55–66.

200. Jacques-Louis Lions, *Espaces intermédiaires entre espaces hilbertiens et applications*, Bull. Math. Soc. Sci. Math. Phys. R. P. Roumaine (N.S) **2(50)** (1958), 419–432.

201. ______, *Un théorème de traces; applications*, C. R. Acad. Sci. Paris **249** (1959), 2259–2261.

202. ______, *Théorèmes de trace et d'interpolation*. In five parts:

I, II: Ann. Scuola Norm. Sup. Pisa Sci. Fis. Mat. (3) **13** (1959), 389–403; (3) **14** (1960), 317–331.

III: J. Math. Pures Appl. (9) **42** (1963), 195–203.

IV: Math. Ann. **151** (1963), 42–56.

V: An. Acad. Brasil. Ci. **35** (1963), 1–10.

203. ______, *Sur certains théorèmes d'interpolation*, C. R. Acad. Sci. Paris **250** (1960), 2104–2106.

204. ______, *Une construction d'espaces d'interpolation*, C. R. Acad. Sci. Paris **251** (1960), 1853–1855.

205. ______, *Sur les espaces d'interpolation: dualité*, Math. Scand. **9** (1961), 147–177.

206. ______, *On a trace problem*, Rend. Sem. Mat. Univ. Padova **31** (1961), 232–242.

207. ______, *Properties of some interpolation spaces*, J. Math. and Mech. **11** (1962), 969–977.

208. ______, *Remarques sur les espaces d'interpolation et les problèmes aux limites*, Les Équations aux Dérivées Partielles (Colloq. Internat. Centre Nat. Recherche Sci., No. 117, Paris, 1962), Éditions, C. N. R. S., Paris, 1963, pp. 75–86.

209. ______, *Espaces d'interpolation et domaines de puissances fractionnaires d'opérateurs*, J. Math. Soc. Japan **14** (1962), 233–241.

210. ______, *Une propriété de stabilité pour des espaces d'interpolation; applications*, C. R. Acad. Sci. Paris **256** (1963), 855–857.

211. ______, *Dérivées intermédiaires et espaces intermédiaires*, C. R. Acad. Sci. Paris **256** (1963), 4343–4345.

212. J. L. Lions and E. Magenes, *Problèmes aux limites non homogènes* (II, IV, VI, VII)/ *Problemi ai limiti non omogenei* (I, III, V). In seven parts:

I: Ann. Scuola Norm. Sup. Pisa Sci. Fis Mat (3) **14** (1960), 269–308.

II: Ann. Inst. Fourier (Grenoble) **11** (1961), 137–178.

III–V: Ann. Scuola Norm. Sup. Pisa Sci. Fis. Mat. (3) **15** (1961), 41–103, 311–326; (3) **16** (1962), 1–44.

VI: J. Analyse Math. **11** (1963), 165–188.

VII: Ann. Mat. Pura Appl. (4) **63** (1963), 201–224.

213. Jacques-Louis Lions and Jaak Peetre, *Propriétés d'espaces d'interpolation*, C. R. Acad. Sci. Paris **253** (1961), 747–749.

214. ______, *Sur une classe d'espaces d'interpolation*, Inst. Hautes Études Sci. Publ. Math. No. 19 (1964), 5–68.

215. Walter Littman, *Multipliers in $L^p$ and interpolation*, Bull. Amer. Math. Soc. **71** (1965), 764–766.

216. P. I. Lizorkin, *On the Fourier transform in Besov spaces. The zero scale $B^0_{p,\theta}$*, Dokl. Akad. Nauk SSSR **163** (1965), 1318–1321; English transl. in Soviet Math. Dokl. **6** (1965).

217. ______, *$(L_p, L_q)$-multipliers of Fourier integrals*, Dokl. Akad. Nauk SSSR **152** (1963), 808–811; English transl. in Soviet Math. Dokl. **4** (1963).

218. ______, *On a functional characterization of the interpolation spaces $(L_p(\Omega), W^1_p(\Omega))_{\theta,p}$*, Trudy Mat. Inst. Steklov. **134** (1975), 180–203; English transl. in Proc. Steklov Inst. Math. **134** (1975) (1977).

219. ______, *Interpolation of weighted $L_p$-spaces*, Dokl. Akad. Nauk SSSR **222** (1975), 32–35; English transl. in Soviet Math. Dokl. **16** (1975).

220. Ju. I. Ljubič, *On inequalities between the powers of a linear operator*, Izv. Akad. Nauk SSSR Ser. Mat. **24** (1960), 825–864; English transl. in Amer. Math. Soc. Transl. (2) **40** (1964).

221. Jörgen Löfström, *Some theorems on interpolation spaces with applications to approximation in $L_p$*, Math. Ann. **172** (1967), 176–196.

222. ______, *On certain interpolation spaces related to generalized semi-groups*, Math. Scand. **16** (1965), 41–54.

223. Lynn H. Loomis, *A note on the Hilbert transform*, Bull. Amer. Math. Soc. **52** (1946), 1082–1086.

224. G. G. Lorentz, *Some new functional spaces*, Ann. of Math. (2) **51** (1950), 37–55.

225. ______, *On the theory of spaces* $\Lambda$, Pacific J. Math. **1** (1951), 411–429.

226. ______, *Majorants in spaces of integrable functions*, Amer. J. Math. **77** (1955), 484–492.

227. George G. Lorentz and Tetsuya Shimogaki, *Interpolation theorems for operators in function spaces*, J. Functional Anal. **2** (1968), 31–51.

228. ______, *Interpolation theorems for the pairs of spaces* $(L^p, L^\infty)$ *and* $(L^1, L^q)$, Trans. Amer. Math. Soc. **159** (1971), 207–221.

229. G. Ja. Lozanovskiĭ, *Calderón's Banach lattices*, Dokl. Akad. Nauk SSSR **172** (1967), 1018–1020; English transl. in Soviet Math. Dokl. **8** (1967); erratum, Soviet Math. Dokl. **8** (1967), no. 2, p. v.

230. ______, *Remark on an interpolation theorem of Calderón*, Funkcional. Anal. i Priložen. **6** (1972), no. 4, 89–90; English transl. in Functional Anal. Appl. **6** (1972).

231. W. A. J. Luxemburg, *Rearrangement-invariant Banach function spaces*, Proc. Sympos. Anal., Queen's Papers in Pure and Appl. Math., No. 10, Queen's Univ., Kingston, Ont., 1967, pp. 83–144.

232. V. I. Luzin and Ja. B. Rutickiĭ, *Sums and intersections of Banach spaces*, Voronež. Gos. Univ. Trudy Naučno-Issled. Inst. Mat. VGU Vyp. 3 (1971), 84–87. (Russian)

233. Enrico Magenes, *Spazi d'interpolazione ed equazioni a derivate parziali*, Atti Settimo Congr. Un. Mat. Ital. (Genoa, 1963), Edizione Cremonese, Rome, 1965, pp. 134–197.

234. Jozeph [Józef] Marcinkiewicz, *Sur l'interpolation d'opérations*, C. R. Acad. Sci. Paris **208** (1939), 1272–1273; reprinted in his *Collected papers*, PWN, Warsaw, 1964, pp. 539–540.

235. ______, *Quelques théorèmes sur les series orthogonales*, Ann. Soc. Polon. Math. = Rocznik Polsk. Tow. Mat. **16** (1937), 84–96 (1938); reprinted in his *Collected papers*, PWN, Warsaw, 1964, pp. 307–318.

236. A. A. Mekler, *The Hardy-Littlewood property in Marcinkiewicz spaces*, Izv. Vysš. Učebn. Zaved. Matematika **1975**, no. 3 (154), 104–106; English transl. in Soviet Math. (Iz VUZ) **19** (1975).

237. Claude Mérucci and Pham the Lai, *Caractérisation, par interpolation, des idéaux de R. Schatten*, C. R. Acad. Sci. Paris Sér. A-B **274** (1972), A1912–A1914.

238. Stuart E. Mills, *Normed Köthe spaces as intermediate spaces of $L_1$ and $L_\infty$*, Pacific J. Math. **52** (1974), 157–173.

239. Carlo Miranda, *Sul teorema di Riesz-Thorin*, Ann. Mat. Pura Appl. (4) **84** (1970), 61–71.

240. N. V. Mirošin, *Properties of interpolation spaces of completely continuous operators*, Mat. Zametki **17** (1975), 293–300; English transl. in Math. Notes **17** (1975).

241. Boris S. Mitjagin, *An interpolation theorem for modular spaces*, Mat. Sb. **66(108)** (1965), 473–482. (Russian)

242. ______, *Normed ideals of intermediate type*, Izv. Akad. Nauk SSSR Ser. Mat. **28** (1964), 819–832; English transl. in Amer. Math. Soc. Transl. (2) **63** (1967).

243. ______, *Sur l'équivalence des bases inconditionnelles dans les échelles de Hilbert*, C. R. Acad. Sci. Paris Sér. A-B **269** (1969), A426–A428.

244. K. H. Moon, *On restricted weak type* (1, 1), Proc. Amer. Math. Soc. **42** (1974), 148–152.

245. Tosinobu Muramatu, *On imbedding theorems for Sobolev spaces and some of their generalizations*, Publ. Res. Inst. Math. Sci. Ser. A **3** (1967/68), 393–416.

246. ______, *Products of fractional powers of operators*, J. Fac. Sci. Univ. Tokyo Sect. 1A Math. **17** (1970), 581–590.

247. ______, *On Besov spaces of functions defined in general regions*, Publ. Res. Inst. Math. Sci. **6** (1970/71), 515–543.

248. M. A. Naĭmark and S. V. Fomin, *Continuous direct sums of Hilbert spaces and some of their applications*, Uspehi Mat. Nauk **10** (1955), no. 2(64), 111–143; English transl. in Amer. Math. Soc. Transl. (2) **5** (1957).

249. S. M. Nikolsky [Nikol'skiĭ], J. L. Lions and P. I. Lizorkin, *Integral representation and isomorphism properties of some classes of functions*, Ann. Scuola Norm. Sup. Pisa Sci. Fis. Mat. (3) **19** (1965), 127–178.

250. Evelio T. Oklander, *$L_{pq}$ interpolators and the theorem of Marcinkiewicz*, Bull. Amer. Math. Soc. **72** (1966), 49–53.

251. A. M. Olevskiĭ, *Fourier series and Lebesque functions*, Uspehi Mat. Nauk **22** (1967), no. 3 (135), 237–239. (Russian)

252. Rainer Oloff, *Interpolation zwischen den Klassen $\mathfrak{S}_p$ von Operatoren in Hilberträumen*, Math. Nachr. **46** (1970), 209–218.

253. Richard O'Neil, *Convolution operators and $L(p, q)$ spaces*, Duke Math. J. **30** (1963), 129–142.

254. ______, *Les fonctions conjuguées et les intégrales fractionnaires de la classe $L(\log^+ L)$*, C. R. Acad. Sci. Paris Sér. A-B **263** (1966), A463–A466.

255. R. O'Neil and G. Weiss, *The Hilbert transform and rearrangement of functions*, Studia Math. **23** (1963/64), 189–198.

256. W. Orlicz, *Ein Satz über die Erweiterung von linearen Operationen*, Studia Math. **5** (1934), 127–140.

257. ______, *On a class of operations over the space of integrable functions*, Studia Math. **14** (1953/54), 302–309.

258. V. I. Ovčinnikov, *Symmetric spaces of measurable operators*, Dokl. Akad. Nauk SSSR **191** (1970), 769–771; English transl. in Soviet Math. Dokl. **11** (1970).

259. ______, *s-numbers of measurable operators*, Funkcional. Anal. i Priložen. **4** (1970), no. 3, 78–85; English transl. in Functional Anal. Appl. **4** (1970).

260. ______, *Symmetric spaces of measurable operators*, Voronež. Gos. Univ. Trudy Naučno-Issled. Inst. Mat. VGU Vyp. 3 (1971), 88–107. (Russian)

261. Jaak Peetre, *Sur le nombre de paramètres dans la définition de certains espaces d'interpolation*, Ricerche Mat. **12** (1963), 248–261.

262. ______, *Nouvelles propriétés d'espaces d'interpolation*, C. R. Acad. Sci. Paris **256** (1963), 1424–1426.

263. ______, *On an interpolation theorem of Foiaş and Lions*, Acta Sci. Math. (Szeged) **25** (1964), 255–261.

264. ______, *Espaces d'interpolation, généralisations, applications*, Rend. Sem. Mat. e Fis. Milano **34** (1964), 133–164.

265. ______, *Applications de la théorie des espaces d'interpolation aux développements orthogonaux*, Rend. Sem. Mat. Univ. Padova **37** (1967), 133–145.

266. ______, *On interpolation functions*, Acta Sci. Math. (Szeged) **27** (1966), 167–171.

267. ______, *Relations entre deux méthodes d'interpolation*, Inst. Hautes Études Sci. Publ. Math. No. 29 (1966), 49–53.

268. ______, *Espaces d'interpolation et théorème de Soboleff*, Ann. Inst. Fourier (Grenoble) **16** (1966), fasc. 1, 279–317.

269. ______, *Applications de la théorie des espaces d'interpolation dans l'analyse harmonique*, Ricerche Mat. **15** (1966), 1–36.

270. ______, *Sur les espaces de Besov*, C. R. Acad. Sci. Paris Sér. A-B **264** (1967), A281–A283.

271. ______, *On interpolation of $L_p$ spaces with weight functions*, Acta Sci. Math. (Szeged) **28** (1967), 61–69.

272. ______, *On interpolation functions*. II, III, Acta Sci. Math. (Szeged) **29** (1968), 91–92; **30** (1969), 235–239.

273. ______, *ε-entropie, ε-capacité et espaces d'interpolation*, Ricerche Mat. **17** (1968), 216–220.

274. ______, *Sur la transformation de Fourier des fonctions à valeur vectorielles*, Rend. Sem. Mat. Univ. Padova **42** (1969), 15–26.

275. ______, *Exact interpolation theorems for Lipschitz continuous functions*, Ricerche Mat. **18** (1969), 239–259.

276. ______, *On the connection between the theory of interpolation spaces and approximation theory*, Proc. Conf. Constructive Theory of Functions (Budapest, 1969), Akad. Kiadó, Budapest, 1972, pp. 351–363.

277. ______, *Approximation of linear operators*, Constructive Theory of Functions (Proc. Internat. Conf., Varna, 1970), Izdat. Bolgar. Akad. Nauk, Sofia, 1972, pp. 245–263.

278. ______, *Interpolation functors and Banach couples*, Proc. Internat. Congr. Math. (Nice, 1970), Vol. 2, Gauthier-Villars, Paris, 1971, pp. 373–378.

279. ______, *Interpolationsräume*, Elliptische Differentialgleichungen, Band I, Schriftenreihe Inst. Math. Deutsch. Akad. Wiss. Berlin, Reihe A, Heft 7, Akademie-Verlag, Berlin, 1970, pp. 93–101.

280. ______, *Zur Interpolation von Operatorenräumen*, Arch Math. (Basel) **21** (1970), 601–608.

281. ______, *Non-commutative interpolation*, Le Matematiche (Catania) **25** (1970), 159–173.

282. ______, *Interpolation of Lipschitz operators and metric spaces*, Mathematica (Cluj) **12(35)** (1970), 325–334.

283. ______, *A new approach in interpolation spaces*, Studia Math. **34** (1970), 23–42.

284. ______, *Concave majorants of positive functions*, Acta Math. Acad. Sci. Hungar. **21** (1970), 327–333.

285. ______, *Sur l'utilisation des suites inconditionellement sommables dans la théorie des espaces d'interpolation*, Rend. Sem. Math. Univ. Padova **46** (1971), 173–190.

286. Jaak Peetre and Gunnar Sparr, *Interpolation of normed Abelian groups*, Ann. Mat. Pura Appl. (4) **92** (1972), 217–262.

287. ______, *Interpolation and non-commutative integration*, Ann. Mat. Pura Appl. (4) **104** (1975), 187–207.

288. Jaak Peetre and Vidar Thomée, *On the rate of convergence for discrete initial-value problems*, Math. Scand. **21** (1967), 159–176.

289. Arne Persson, *Compact linear mappings between interpolation spaces*, Ark. Mat. **5** (1963/65), 215–219.

290. Ju. I. Petunin, *The relationship of three Banach spaces*, Dokl. Akad. Nauk SSSR **170** (1966), 516–519; English transl. in Soviet Math. Dokl. 7 (1966).

291. ______, *On a nonlinear problem in the theory of scales of Banach spaces*, Appendix to [186], Russian pp. 161–166; English pp. 152–157.

292. ______, *Prenuclear transformations in scales of Banach and Hilbert spaces*, Dokl. Akad. Nauk SSSR **173** (1967), 40–43; English transl. in Soviet Math. Dokl. **8** (1967).

293. ______, *Two-dimensional scales of Banach spaces*, Voronež. Gos. Univ. Trudy Sem. Funkcional. Anal. Vyp. 9 (1967), 125–144. (Russian)

294. ______, *Transformation of the norm in a scale of Banach spaces*, Voronež. Gos. Univ. Trudy Sem. Funkcional. Anal. Vyp. 10 (1968), 92–100. (Russian)

295. ______, *Factorization of scales of Banach spaces*, Funkcional. Anal. i Priložen. **4** (1970), no. 4, 81–82; English transl. in Functional Anal. Appl. **4** (1970).

296. ______, *A problem of interpolation between quotient spaces*, Ukrain. Mat. Ž. **23** (1971), 157–167; English transl. in Ukrainian Math. J. **23** (1971).

297. Ju. I. Petunin and A. N. Pličko, *Interpolation property of the Hölder scale*, Proc. Sixth Winter School Math. Programming and Related Questions (Drogobych, 1973): Functional Anal. and Appl., Central. Èkonom.-Mat. Inst. Akad. Nauk SSSR, Moscow, 1975, pp. 229–235. (Russian)

298. Albrecht Pietsch, *Gegenbeispiele zur Interpolationstheorie der nuklearen und absolutsummierenden Operatoren*, Arch. Math. (Basel) **20** (1969), 65–71.

299. ______, *Interpolationsfunktoren, Folgenideale und Operatorenideale*, Czechoslovak Math. J. **21**(96) (1971), 644–652.

300. Albrecht Pietsch and Hans Triebel, *Interpolationstheorie für Banachideale von beschränkten linearen Operatoren*, Studia Math. **31** (1968), 95–109.

301. E. I. Pustyl'nik, *On the convergence of series of eigenfunctions of a completely continuous operator in Banach spaces*, Sibirsk. Mat. Ž. **4** (1963), 705–708. (Russian)

302. Marcel Riesz, *Sur les maxima des formes bilinéaires et sur les fonctionelles linéaires*, Acta Math. **49** (1926), 465–497.

303. N. M. Riviere and Y. Sagher, *Interpolation between $L^\infty$ and $H^1$, the real method*, J. Functional Anal. **14** (1973), 401–409.

304. V. A. Rodin and E. M. Semyonov [Semenov], *Rademacher series in symmetric spaces*, Anal. Math. **1** (1975), 207–222.

305. V. A. Rohlin, *On the fundamental ideas of measure theory*, Mat. Sb. **25**(67) (1949), 107–150; English transl. in Amer. Math. Soc. Transl. (1) **10** (1962).

306. S. Ju. Rotfel'd, *Analogs of the Mitjagin and Semenov interpolation theorems for operators in unnormed symmetric spaces*, Problemy Mat. Anal., Vyp. 4, Izdat. Leningrad. Gos. Univ., Leningrad, 1973, pp. 106–113; English transl. in J. Soviet Math. **6** (1976), no. 1.

307. G. I. Russu, *Symmetric spaces of functions that do not have the majorization property*, Mat. Issled. **4** (1969), vyp. 4 (14), 82–93. (Russian)

308. Ja. B. Rutickiĭ, *Scales of Orlicz spaces and interpolation theorems* Voronež. Gos. Univ. Trudy Sem. Funkcional. Anal. Vyp. 7 (1963), 112–129. (Russian)

309. ______, *Scales of Orlicz spaces and interpolation theorems*, Dokl. Akad. Nauk SSSR **149** (1963), 32–35; English transl. in Soviet Math. Dokl. **4** (1963).

310. ______, *Some remarks on the theory of scales of Orlicz spaces*, Voronež. Gos. Univ. Trudy Sem. Funkcional. Anal. Vyp. 10 (1968), 109–113. (Russian)

311. John V. Ryff, *Orbits of $L^1$-functions under doubly stochastic transformations*, Trans. Amer. Math. Soc. **117** (1965), 92–100.

312. ______, *Measure preserving transformations and rearrangements*, J. Math. Anal. Appl. **31** (1970), 449–458.

313. Yoram Sagher, *Some remarks on interpolation of operators and Fourier coefficients*, Studia Math. **44** (1972), 239–252.

314. ______, *An application of interpolation theory to Fourier series*, Studia Math. **41** (1972), 169–181.

315. ______, *On analytic families of operators*, Israel J. Math. **7** (1969), 350–356.

316. ______, *Norm inequalities on Fourier coefficients and interpolation theory*, Linear Operators and Approximation. II (Proc. Conf., Oberwohlfach, 1974), Birkhäuser, 1974, pp. 169–180.

317. V. K. Šalašov, *Lipschitz spaces and interpolation*, Uspehi Mat. Nauk **27** (1972), no. 6 (168), 255–256. (Russian)

318. Martin Schechter, *On $L^p$ estimates and regularity*. In three parts:
I: Amer. J. Math. **85** (1963), 1–13.
II: Math. Scand **13** (1963), 47–69.
III: Ricerche Mat. **13** (1964), 192–206.

319. ______, *Interpolation spaces by complex methods*, Bull. Amer. Math. Soc. **72** (1966), 526–533.

320. ______, *Complex interpolation*, Compositio Math. **18** (1967), 117–147.

321. Hans-Jürgen Schmeisser and Hans Triebel, *Anisotropic spaces*. I, Math. Nachr. **73** (1976), 107–123.

322. J. Schwartz, *A remark on inequalities of Calderón-Zygmund type for vector-valued functions*, Comm. Pure Appl. Math. **14** (1961), 785–799.

323. A. A. Sedaev, *The properties of operators in an interpolation pair of Banach spaces*, Voronež. Gos. Univ. Trudy Naučno-Issled. Inst. Mat. VGU Vyp. 3 (1971), 108–125. (Russian)

324. ______, *Description of interpolation spaces for the pair* $(L^p_{a_0}, L^p_{a_1})$ *and some related problems*, Dokl. Akad. Nauk SSSR **209** (1973), 798–800; English transl. in Soviet Math. Dokl. **14** (1973).

325. ______, *Weakened strong convergence in interpolation spaces*, Proc. Sixth Winter School Math. Programming and Related Questions (Drogobych, 1973): Functional Anal. and Appl., Central. Ekonom.-Mat. Inst. Akad. Nauk SSSR, Moscow, 1975, pp. 245–267. (Russian)

326. ______, *The Hardy operator in* $L_p$ *spaces with weight*, Coll. of Papers by Graduate Students Math. Fac., Izdat. Voronež. Gos. Univ., Voronezh, 1972, pp. 48–53. (Russian) RŽ Mat. **1973** #8Б681.

327. A. A. Sedaev and E. M. Semenov, *The possibility of describing interpolation spaces in terms of Peetre's K-method*, Optimizacija Vyp. 4 (21) (1971), 98–114. (Russian)

328. R. Seeley, *Interpolation in* $L^p$ *with boundary conditions*, Studia Math. **44** (1972), 47–60.

329. I. E. Segal, *A non-commutative extension of abstract integration*, Ann. of Math. (2) **57** (1953), 401–457; correction, (2) **58** (1953), 595–596.

330. E. M. Semenov, *On a scale of spaces with an interpolation property*, Dokl. Akad. Nauk SSSR **148** (1963), 1038–1041; English transl. in Soviet Math. Dokl. **4** (1963).

331. ______, *Imbedding theorems for Banach spaces of measurable functions*, Dokl. Akad. Nauk SSSR **156** (1964), 1292–1295; English transl. in Soviet Math. Dokl. **5** (1964).

332. ______, *Interpolation of linear operators in symmetric spaces*, Dokl. Akad. Nauk SSSR **164** (1965), 746–749; English transl. in Soviet Math. Dokl. **6** (1965).

333. ______, *Interpolation of linear operators and estimates of Fourier coefficients*, Dokl. Akad. Nauk SSSR **176** (1967), 1251–1254; English transl. in Soviet Math. Dokl. **8** (1967).

334. ______, *A new interpolation theorem*, Funkcional. Anal. i Priložen **2** (1968), no. 2, 68–80; English transl. in Functional Anal. Appl. **2** (1968).

335. ______, *Condensed scales of Banach spaces*, Voronež. Gos. Univ. Trudy Sem. Funkcional. Anal. Vyp. 10 (1968), 130–141. (Russian)

336. ______, *On a method for establishing interpolation theorems in symmetric spaces*, Dokl. Akad. Nauk SSSR **185** (1969), 1243–1246; English transl. in Soviet Math. Dokl. **10** (1969).

337. ______, *Some numerical characteristics of symmetric spaces*, Voronež. Gos. Univ. Trudy Naučno-Issled. Inst. Mat. VGU Vyp. 1 (1970), 137–144.

338. ______, *A survey of the interpolation theory of linear operators*, Proc. Sixth Winter School Math. Programming and Related Questions (Drogobych, 1973): Functional Anal. and Appl., Central Ekonom.-Mat. Inst. Akad. Nauk SSSR, Moscow, 1975, pp. 28–43. (Russian)

339. E. M. Semenov and O. A. Vasil'eva, $L_2$*-multipliers*, Voronež. Gos. Univ. Trudy Mat. Fak. Vyp. 11 (1973), 42–53. (Russian)

340. V. A. Sestakov, *Complex interpolation in Banach spaces of measurable functions*, Vestnik Leningrad. Univ. **1974**, no. 19 (Ser. Mat. Meh. Astr. vyp. 4), 64–68; English transl. in Vestnik Leningrad Univ. Math. 7 (1979).

341. ______, *Interpolation of linear operators in spaces of measurable functions*, Funkcional. Anal. i Priložen. **8** (1974), no. 3, 91–92; English transl. in Functional Anal. Appl. **8** (1974).

342. Robert Sharpley, *Spaces* $\Lambda_\alpha(X)$ *and interpolation*, J. Functional Anal. **11** (1972), 479–513.

343. ______, *Interpolation theorems for compact operators*, Indiana Univ. Math. J. **22** (1972/73), 965–984.

344. ______, *Interpolation of operators for* $\Lambda$ *spaces*, Bull. Amer. Math. Soc. **80** (1974), 259–261.

345. ______, *Characterization of intermediate spaces of* $M_\varphi$ *spaces*, Linear Operators and Approximation. II (Proc. Conf., Oberwohlfach, 1974), Birkhäuser, 1974, pp. 205–214.

346. ______, *Interpolation of n pairs and counterexamples employing indices*, J. Approximation Theory **13** (1975), 117–127.

347. Tetsuya Shimogaki, *Hardy-Littlewood majorants in function spaces*, J. Math. Soc. Japan **17** (1965), 365–373.

348. ______, *On the complete continuity of operators in an interpolation theorem*, J. Fac. Sci. Hokkaido Univ. Ser. I. Math. **20** (1968), 109–114.

349. ______, *An interpolation theorem on Banach function spaces*, Studia Math. **31** (1968), 233–240.

350. ______, *A note on norms of compression operators in function spaces*, Proc. Japan Acad. **46** (1970), 239–242.

351. I. B. Simonenko, *Interpolation and extrapolation of linear operators in Orlicz spaces*, Mat. Sb. **63 (105)** (1964), 536–553. (Russian)

352. G. Slenzak, *Elliptic problems in a refined scale of spaces*, Vestnik Moskov. Univ. Ser. I Mat. Meh. **1974**, no. 4, 48–58; English transl. in Moscow Univ. Math. Bull. **29** (1974).

353. I. Ja. Šneĭberg, *On the solvability of linear equations in interpolation familes of Banach spaces*, Dokl. Akad. Nauk SSSR **212** (1973), 57–59; English transl. in Soviet Math. Dokl. **14** (1973).

354. ______, *Spectral properties of linear operators in interpolation families of Banach spaces*, Mat. Issled. **9** (1974), no. 2 (32), 214–229. (Russian)

355. S. L. Sobolev, *On a theorem of functional analysis*, Mat. Sb. **4 (46)** (1938), 471–497; English transl. in Amer. Math. Soc. Transl. (2) **34** (1963).

356. P. E. Sobolevskiĭ, *On fractional norms in a Banach space generated by an unbounded operator*, Uspehi Mat. Nauk **19** (1964), no. 6 (120), 219–222. (Russian)

357. ______, *An interpolation theorem for fractional spaces that are generated by unbounded positive operators with incomparable domains of definition*, Voronež. Gos. Univ. Trudy Mat. Fak. Vyp. 9 (1973), 142–155. (Russian)

358. ______, *Interpolation theorems for fractional powers of operators acting in Banach spaces*, Voronež. Gos. Univ. Trudy Mat. Fak. Vyp. 13 (1974), 51–64. (Russian)

359. ______, *Some properties of solutions of differential equations in Banach spaces*, Voronež. Gos. Univ. Trudy Naučno-Issled. Inst. Mat. VGU Vyp. 14 (1974), 68–74. (Russian)

360. Sven Spanne, *Sur l'interpolation entre les espaces* $\mathfrak{L}_k^{p\Phi}$, Ann. Scoulo Norm. Sup. Pisa Sci. Fis. Mat. (3) **20** (1966), 625–648.

361. Gunnar Sparr, *Interpolation des espaces* $L_w^p$, C. R. Acad. Sci. Paris Sér. A-B **278** (1974), 491–492.

362. ______, *Interpolation of several Banach spaces*, Ann. Mat. Pura Appl. (4) **99** (1974), 247–316.

363. James D. Stafney, *Analytic interpolation of certain multiplier spaces*, Pacific J. Math. **32** (1970), 241–248.

364. ______, *The spectrum of an operator on an interpolation space*, Trans. Amer. Math. Soc. **144** (1969), 333–349.

365. Guido Stampacchia, $\mathfrak{L}^{(p,\lambda)}$*-spaces and interpolation*, Comm. Pure Appl. Math. **17** (1964), 293–306.

366. ______, *The spaces* $\mathfrak{L}^{(p,\lambda)}$, $N^{(p,\lambda)}$ *and interpolation*, Ann. Scuola Norm. Sup. Pisa Sci. Fis. Mat. (3) **19** (1965), 443–462.

367. Elias M. Stein, *Interpolation of linear operators*, Trans. Amer. Math. Soc. **83** (1956), 482–492.

368. ______, *Interpolation in polynomial classes and Markoff's inequality*, Duke Math. J. **24** (1957), 467–476.

369. ______, *On limits of sequences of operators*, Ann. of Math. (2) **74** (1961), 140–170.

370. E. M. Stein and Guido Weiss, *Interpolation of operators with change of measures*, Trans. Amer. Math. Soc. **87** (1958), 159–172.

371. ______, *An extension of a theorem of Marcinkiewicz and some of its applications*, J. Math. and Mech. **8** (1959), 263–284.

372. ______, *On the interpolation of analytic families of operators acting on* $H^p$*-spaces*, Tôhoku Math. J. (2) **9** (1957), 318–339.

373. Mitchell H. Taibleson, *Lipschitz classes of functions and distributions in $E_n$*, Bull. Amer. Math. Soc. **69** (1963), 487–493.

374. ______, *On the theory of Lipschitz spaces of distributions on Euclidean n-space*. I, II, III, J. Math. and Mech. **13** (1964), 407–479; **15** (1965), 821–839; **15** (1966), 973–981.

375. Luc Tartar, *Théorème d'interpolation non linéaire et applications*, C. R. Acad. Sci. Paris Sér. A-B **270** (1970), A1729–A1731.

376. G. O. Thorin, *An extension of a convexity theorem due to M. Riesz*, Kungl. Fysiograf. Sällskapets i Lund Förh. **8** (1938), 166–170; reprinted in Medd. Lunds Univ. Mat. Sem. **4** (1939).

377. N. G. Tomin, *Application of linear operator interpolation to questions of convergence of series of Fourier coefficients with respect to classical orthgonal polynomials*, Dokl. Akad. Nauk SSSR **212** (1973), 1074–1077; English transl. in Soviet Math. Dokl. **14** (1973).

378. Hans Triebel, *Zur Interpolation von Normidealen in Hilberträumen*, Wiss. Z. Friedrich-Schiller-Univ. Jena/Thüringen **18** (1969), Heft 2, 263–267.

379. ______, *Singuläre elliptische Differentialgleichungen und Interpolationssätze für Sobolev-Slobodeckij-Räume mit Gewichtsfunktionen*, Arch. Rational Mech. Anal. **32** (1969), 113–134.

380. ______, *Über die Verteilung der Approximationszahlen von Integraloperatoren in Sobolev-Besov-Räumen*, J. Math. and Mech. **19** (1969/70), 783–796.

381. ______, *Singuläre elliptische Differentialgleichungen und Interpolationssätze für Besov-Räume mit Gewichtsfunktionen*, Elliptische Differentialgleichungen, Band I, Schriftenreihe Inst. Math. Akad. Wiss. Berlin, Reihe A, Heft 7, Akademie-Verlag, Berlin, 1970, pp. 159–164.

382. ______, *Interpolationseigenschaften von Entropie—und Durchmesseridealen kompakter Operatoren*, Studia Math. **34** (1970), 89–107.

383. ______, *Eine Bemerkung zur Interpolation von Banachräumen*, Beiträge Anal. **2** (1971), 51–55.

384. ______, *Spaces of distributions of Besov type on Euclidean n-space. Duality, interpolation*, Ark. Mat. **11** (1973), 13–64.

385. ______, *Interpolation theory for function spaces of Besov type defined in domains*. I, II, Math. Nachr. **57** (1973), 51–85; **58** (1973), 63–86.

386. ______, *Über die Existenz von Schauderbasen in Sobolev-Besov-Räumen. Isomorphiebeziehungen*, Studia Math. **46** (1973), 83–100.

387. ______, *Interpolation properties of ε-entropy and diameters. Geometric characteristics of imbedding for function spaces of Sobolev-Besov type*, Mat. Sb. **98** (**140**) (1975), 27–41; English transl. in Math. USSR Sb. **27** (1975).

388. Hisaharu Umegaki, *Conditional expectation in an operator algebra*, Tôhoku Math. J. (2) **6** (1954), 177–181.

389. O. A. Vasil'eva and E. M. Semenov, *Method of interpolation of linear operators in a problem of multipliers*, Uspehi Mat. Nauk **26** (1971), no. 3 (159), 190. (Russian)

390. T. D. Ventcel', *Interpolation functions*, Vestnik Moskov. Univ. Ser. I Mat. Meh. **1969**, no. 5, 57–65; English transl. in Moscow Univ. Math. Bull. **24** (1969).

391. Guido Weiss, *An interpolation theorem for sublinear operators on $H_p$ spaces*, Proc. Amer. Math. Soc. **8** (1957), 92–99.

392. V. Westphal, *Ein Kalkül für gebrochene Potenzen infinitesimaler Erzeuger von Halbgruppen und Gruppen von Operatoren*. I, Compositio Math. **22** (1970), 67–103.

393. Vernon Williams, *Generalized interpolation spaces*, Trans. Amer. Math. Soc. **156** (1971), 309–334.

394. Atsushi Yoshikawa, *Remarks on the theory of interpolation spaces*, J. Fac. Sci. Univ. Tokyo Sect. I Math. Astr. Phys. Chem. **15** (1968), 209–251

395. ______, *Sur la théorie d'espaces de Banach—les espaces de moyenne de plusieurs espaces de Banach*, J. Fac. Sci. Univ. Tokyo Sect. I Math. Astr. Phys. Chem. **16** (1969/70), 407–468.

396. ______, *An abstract formulation of Sobolev type imbedding theorems and its applications to elliptic boundary value problems*, J. Fac. Sci. Univ. Tokyo Sect. I A Math. **17** (1970), 543–558.

397. ______, *An operator theoretical remark on the Hardy-Littlewood-Sobolev inequality*, J. Fac. Sci. Univ. Tokyo Sect. I A Math. **17** (1970), 559–566.

398. ______, *Fractional powers of operators, interpolation theory and imbedding theorems*, J. Fac. Sci. Univ. Tokyo Sect. I A Math. **18** (1971/72), 335–362.

399. Kyôichi Yoshinaga, *On a generalization of the interpolation method*, Bull. Kyushu Inst. Tech.: Math., Nat. Sci. No. 17 (1970), 1–23.

400. Robert M. Young, *Interpolation in a classical Hilbert space of entire functions*, Trans. Amer. Math. Soc. **192** (1974), 97–114.

401. P. P. Zabreĭko, *An interpolation theorem for linear operators*, Mat. Zametki **2** (1967), 593–598; English transl. in Math. Notes **2** (1967).

402. P. P. Zabreĭko and E. I. Pustyl'nik, *Interpolation of the property of complete continuity of an operator*, Kazan. Gos. Univ. Učen. Zap. **124** (1964), kn. 6, 114–118. (Russian)

403. V. P. Zaharjuta, *Bases and isomorphism of spaces of functions analytic in convex domains of several variables*, Teor. Funkciĭ, Funkcional. Anal. i Priložen. Vyp. 5 (1967), 5–12. (Russian)

404. ______, *Quasi-equivalence of bases at finite centers of Hilbert scales*, Dokl. Akad. Nauk SSSR **180** (1968), 786–788; English transl. in Soviet Math. Dokl. **9** (1968).

405. ______, *Extremal plurisubharmonic functions, Hilbert scales, and the isomorphism of spaces of analytic functions of several variables*. I, II, Teor. Funkciĭ, Funkcional. Anal. i Priložen. Vyp. 19 (1974), 133–157; Vyp. 21 (1974), 65–83. (Russian)

406. M. Zippin, *Interpolation of operators of weak type between rearrangement invariant function spaces*, J. Functional Anal. 7 (1971), 267–284.

407. A. Zygmund, *On a theorem of Marcinkiewicz concerning interpolation of operations*, J. Math. Pures Appl. (9) **35** (1956), 223–248.

# SUBJECT INDEX

# NOTATION INDEX